D1082088

Transforms and Applications Primer for Engineers with Examples and MATLAB®

ELECTRICAL ENGINEERING PRIMER SERIES

Series Editor

Alexander D. Poularikas

University of Alabama
Huntsville, Alabama

Transforms and Applications Primer for Engineers with Examples and MATLAB®,
Alexander D. Poularikas

Discrete Random Signal Processing and Filtering Primer with MATLAB®,
Alexander D. Poularikas

Signals and Systems Primer with MATLAB®, *Alexander D. Poularikas*

Adaptive Filtering Primer with MATLAB®, *Alexander D. Poularikas
and Zayed M. Ramadan*

Transforms and Applications Primer for Engineers with Examples and MATLAB®

Alexander D. Poularikas

CRC Press
Taylor & Francis Group
Boca Raton   London   New York

CRC Press is an imprint of the
Taylor & Francis Group, an **informa** business

MATLAB® and Simulink® are trademarks of The MathWorks, Inc. and are used with permission. The Math-Works does not warrant the accuracy of the text of exercises in this book. This book's use or discussion of MATLAB® and Simulink® software or related products does not constitute endorsement or sponsorship by The MathWorks of a particular pedagogical approach or particular use of the MATLAB® and Simulink® software.

CRC Press
Taylor & Francis Group
6000 Broken Sound Parkway NW, Suite 300
Boca Raton, FL 33487-2742

© 2010 by Taylor and Francis Group, LLC
CRC Press is an imprint of Taylor & Francis Group, an Informa business

No claim to original U.S. Government works

Printed in the United States of America on acid-free paper
10 9 8 7 6 5 4 3 2 1

International Standard Book Number: 978-1-4200-8931-8 (Paperback)

This book contains information obtained from authentic and highly regarded sources. Reasonable efforts have been made to publish reliable data and information, but the author and publisher cannot assume responsibility for the validity of all materials or the consequences of their use. The authors and publishers have attempted to trace the copyright holders of all material reproduced in this publication and apologize to copyright holders if permission to publish in this form has not been obtained. If any copyright material has not been acknowledged please write and let us know so we may rectify in any future reprint.

Except as permitted under U.S. Copyright Law, no part of this book may be reprinted, reproduced, transmit-ted, or utilized in any form by any electronic, mechanical, or other means, now known or hereafter invented, including photocopying, microfilming, and recording, or in any information storage or retrieval system, without written permission from the publishers.

For permission to photocopy or use material electronically from this work, please access www.copyright.com (http://www.copyright.com/) or contact the Copyright Clearance Center, Inc. (CCC), 222 Rosewood Drive, Danvers, MA 01923, 978-750-8400. CCC is a not-for-profit organization that provides licenses and registration for a variety of users. For organizations that have been granted a photocopy license by the CCC, a separate system of payment has been arranged.

Trademark Notice: Product or corporate names may be trademarks or registered trademarks, and are used only for identification and explanation without intent to infringe.

Library of Congress Cataloging-in-Publication Data

Poularikas, Alexander D., 1933-
    Transforms and applications primer for engineers with examples and MATLAB /
Alexander D. Poularikas.
        p. cm. -- (Electrical engineering primer series)
    Includes bibliographical references and index.
    ISBN 978-1-4200-8931-8
    1. Signal processing--Mathematics. 2. Transformations (Mathematics) 3. MATLAB.
I. Title. II. Series.

TK5102.9.P6835 2010
621.382'2--dc22   2009030139

Visit the Taylor & Francis Web site at
http://www.taylorandfrancis.com

and the CRC Press Web site at
http://www.crcpress.com

Contents

Preface

This book presents the most common and useful mathematical transforms for students and practicing engineers. It can be considered as a companion for students and a handy reference for practicing engineers who will need to use transforms in their work.

The Laplace transform, which undoubtedly is the most familiar example, is basic to the solution of initial value problems. The Fourier transform, being suited to solving boundary-value problems, is basic to the frequency spectrum analysis of time-varying signals. For discrete signals, we develop the z-transform and its uses. The purpose of this book is to develop the most important integral transforms and present numerous examples elucidating their use. Laplace and Fourier transforms are by far the most widely and most useful of all integral transforms. For this reason, they have been given a more extensive treatment in this book when compared to other books on the same subject.

This book is primarily written for seniors, first-year graduate students, and practicing engineers and scientists. To comprehend some of the topics, the reader should have a basic knowledge of complex variable theory. Advanced topics are indicated by a star (*).

The book contains several appendices to complement the main subjects. The extensive tables of the transforms are the most important contributions in this book. Another important contribution is the inclusion of an ample number of examples drawn from several disciplines. The included examples help the readers understand any of the transforms and give them the confidence to use it. Furthermore, it includes, wherever needed, MATLAB® functions and Book MATLAB functions developed by the author, which are included in the text.

MATLAB is a registered trademark of The MathWorks, Inc. For product information, please contact:

The MathWorks, Inc.
3 Apple Hill Drive Natick, MA 01760-2098 USA
Tel: 508 647 7000
Fax: 508-647-7001
E-mail: info@mathworks.com
Web: www.mathworks.com

Author

Alexander D. Poularikas received his PhD from the University of Arkansas, Fayetteville, and became a professor at the University of Rhode Island, Kingston. He became the chairman of the engineering department at the University of Denver, Colorado, and then became the chairman of the electrical and computer engineering department at the University of Alabama in Huntsville.

Dr. Poularikas has authored seven books and has edited two. He has served as the editor in chief of the Signal Processing series (1993–1997) with Artech House, and is now the editor in chief of the Electrical Engineering and Applied Signal Processing series as well as the Engineering and Science Primer series (1998 to present) with Taylor & Francis. He was a Fulbright scholar, is a lifelong senior member of the IEEE, and is a member of Tau Beta Pi, Sigma Nu, and Sigma Pi. In 1990 and in 1996, he received the Outstanding Educators Award of the IEEE, Huntsville Section. He is now a professor emeritus at the University of Alabama in Huntsville.

Dr. Poularikas has authored, coauthored, and edited the following books:

Electromagnetics, Marcel Dekker, New York, 1979.

Electrical Engineering: Introduction and Concepts, Matrix Publishers, Beaverton, OR, 1982.

Workbook, Matrix Publishers, Beaverton, OR, 1982.

Signals and Systems, Brooks/Cole, Boston, MA, 1985.

Elements of Signals and Systems, PWS-Kent, Boston, MA, 1988.

Signals and Systems, 2nd edn., PWS-Kent, Boston, MA, 1992.

The Transforms and Applications Handbook, CRC Press, Boca Raton, FL, 1995.

The Handbook for Formulas and Tables for Signal Processing, CRC Press, Boca Raton, FL, 1998, 2nd edn. (2000), 3rd edn. (2009).

Adaptive Filtering Primer with MATLAB, Taylor & Francis, Boca Raton, FL, 2006.

Signals and Systems Primer with MATLAB, Taylor & Francis, Boca Raton, FL, 2007.

Discrete Random Signal Processing and Filtering Primer with MATLAB, Taylor & Francis, Boca Raton, FL, 2009.

1

Signals and Systems

1.1 Introduction

The term **systems**, in general, has many meanings such as electronic systems, biological systems, communication systems, etc. The same is true for the term **signals**, since we talk about optical signals, intelligence signals, radio signals, bio-signals, etc.

The two terms mentioned above can have the following three interpretations: (1) An electric system is considered to be made of resistors, inductors, capacitors, and energy sources. Signals are the currents and voltages in the electric system. The signals are a function of time and they are related by a set of equations that are the product of physical laws (Kirchhoff's voltage and current laws). (2) We interpret the system based on the mathematical function it performs. For example, a resistor is a multiplier, an inductor is a differentiator, and a capacitor is an integrator. The signals are the result of the rules of the interconnected elements of the system. (3) If the operations can be performed digitally and in real time, then the analog system can be substituted by a computer. The system, under these circumstances, is a digital device (computer) whose input and output are sequences of numbers. Figure 1.1 illustrates three systems and their responses. The top part of the figure represents the ability of a filter to clear a signal from a superimposed noise. The middle part of the figure shows how a feedback configuration affects an input pulse. This is known as the step response of systems. The bottom part of the figure shows how a rectifier and a filter can produce a DC (direct current) source when the input is a sinusoidal signal as the one present in power transmission lines.

In addition to analog systems, we also have digital ones. These systems deal only with discrete signals and are presented later on in the book. A basic, but sophisticated, instrument is the **analog-to-digital** (A/D) converter, which most instruments nowadays contain.

1.2 Signals

A **signal** is a function representing a physical quantity. This can be a current, a voltage, heart signals (EKG), velocities of motion, music signals, economic time series, etc. In this chapter, we will concentrate only on one-dimensional signals, although images, for example, are two-dimensional signals.

A **continuous-time signal** is a function whose domain is every point in a specified interval.

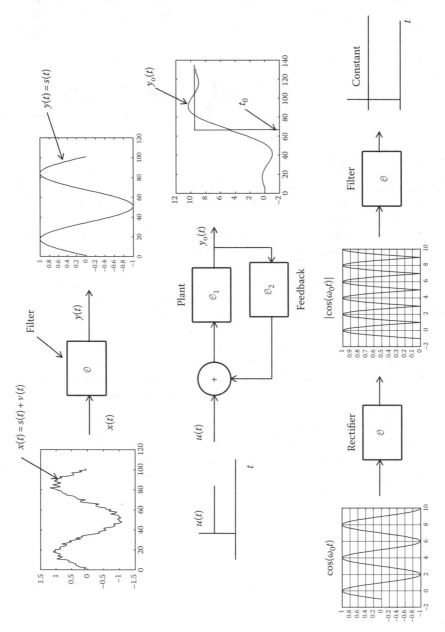

FIGURE 1.1

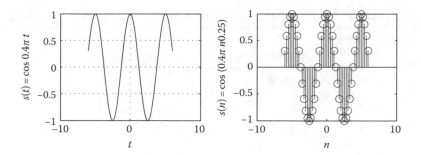

FIGURE 1.2

A **discrete-time signal** is a function whose domain is a set of integers. Therefore, this type of signal is a sequence of numbers denoted by $\{x(n)\}$. It is understood that the discrete-time signal is often formed by **sampling** a continuous-time signal $x(t)$. In this case and for equidistance samples, we write

$$x(n) = x(nT) \quad T = \text{sampling interval} \tag{1.1}$$

Figure 1.2 shows a transformation from a continuous-time signal to a discrete-time signal.

Some important and useful functions are given in Table 1.1.

If the above analog signals are sampled every T seconds, then we will obtain the corresponding discrete ones.

Approximation of a derivative

From Figure 1.3, we observe that we can approximate the samples $y(nT)$ of the derivative $y(t) = x'(t)$ of the signal $x(t)$ for a sufficiently small T as follows:

$$x'(t) \cong \frac{x(t) - x(t-T)}{T} \tag{1.2}$$

$$y(nT) = x'(nT) = \frac{x(nT) - x(nT-T)}{T} = \frac{1}{T}\Delta x(nT) \tag{1.3}$$

We observe that as $T \to 0$, the approximate derivative of $x(t)$, indicated by the inclination of line A, comes closer and closer to the exact one, indicated by the inclination of line B.

Approximation of an integral

The approximation of an integral with its discrete form is shown in Figure 1.4. Therefore, we write

$$y(nT) = \int_{0}^{nT-T} x(t)dt + \int_{nT-T}^{nT} x(t)dt \tag{1.4}$$

TABLE 1.1   Some Useful Mathematical Functions in Analog and Discrete Format

1. Signum function

$$\text{sgn}(t) = \begin{cases} 1 & t > 0 \\ 0 & t = 0; \\ -1 & t < 1 \end{cases} \quad \text{sgn}(nT) = \begin{cases} 1 & nT > 0 \\ 0 & nT = 0 \\ -1 & nT < 0 \end{cases}$$

2. Step function

$$u(t) = \frac{1}{2} + \frac{1}{2}\text{sgn}(t) = \begin{cases} 1 & t > 0 \\ 0 & t < 0 \end{cases}; \quad u(nT) = \begin{cases} 1 & nT > 0 \\ 0 & nT < 0 \end{cases}$$

3. Ramp function

$$r(t) = \int_{-\infty}^{t} u(x)dx = tu(t); \quad r(nT) = nTu(nT)$$

4. Pulse function

$$p_a(t) = u(t+a) - u(t-a) = \begin{cases} 1 & |t| < a \\ 0 & |t| > a \end{cases}; \quad p_a(nT) = u(nT + mT) - u(nT - mT)$$

5. Triangular pulse

$$\Lambda_a(t) = \begin{cases} 1 - \dfrac{|t|}{a} & |t| < a \\ 0 & |t| > a \end{cases}; \quad \Lambda_a(nT) = \begin{cases} 1 - \dfrac{|nT|}{mT} & |nT| < mT \\ 0 & |nT| > mT \end{cases}$$

6. Sinc function

$$\text{sinc}_a(t) = \frac{\sin at}{t} \quad -\infty < t < \infty; \quad \text{sinc}_a(nT) = \frac{\sin anT}{nT}$$

7. Gaussian function

$$g_a(t) = e^{-at^2} \quad -\infty < t < \infty$$

8. Error function

$$\text{erf}(t) = \frac{2}{\sqrt{\pi}} \int_0^t e^{-x^2}\,dx = \frac{2}{\sqrt{\pi}} \sum_{n=0}^{\infty} \frac{(-1)^n t^{2n+1}}{n!(2n+1)}$$

properties: $\text{erf}(\infty) = 1, \text{erf}(0) = 0, \text{erf}(-t) = -\text{erf}(t)$

$$\text{erfc}(t) = \text{complementary error function} = 1 - \text{erf}(t) = \frac{2}{\sqrt{\pi}} \int_t^{\infty} e^{-x^2}\,dx$$

9. Exponential and double exponential

$$f(t) = e^{-t}u(t) \quad t \geq 0; \quad f(t) = e^{-|t|} \quad -\infty < t < \infty$$
$$f(nT) = e^{-nT}u(nT) \quad nT \geq 0; \quad f(nT) = e^{-|nT|} \quad -\infty < nT < \infty$$

Note: T, sampling time; *n*, integer.

which becomes

$$y(nT) = y(nT - T) + \int_{nT-T}^{nT} x(t)dt \qquad (1.5)$$

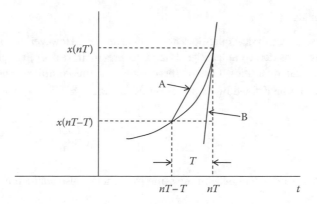

FIGURE 1.3

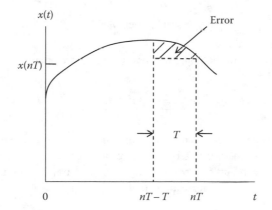

FIGURE 1.4

Approximating the integral in the above equation by the rectangle shown in Figure 1.4, we obtain its approximate discrete form:

$$y(nT) \cong y(nT - T) + Tx(nT) \quad n = 0, 1, 2, \ldots \qquad (1.6)$$

Trigonometric functions

Of special interest in the study of linear systems is the class of sine and cosine functions:

$$a \cos \omega t \quad b \sin \omega t \quad r \cos (\omega t + \varphi)$$

These functions are periodic with **a period** $2\pi/\omega$ **and a frequency** $f = \omega/2\pi$ cycles/s or Hz.

Complex signals

Signals representing physical quantities are, in general, real. However, in many cases it is convenient to consider complex signals and to use their real or imaginary parts to represent physical quantities. One of these signals is the **complex exponential** $e^{j\omega t}$. This function can be defined by its power series $\left(e^x = 1 + x + \frac{x^2}{2!} + \frac{x^3}{3!} + \cdots\right)$:

$$e^{j\omega t} = 1 + j\omega t + \frac{(j\omega t)^2}{2!} + \frac{(j\omega t)^3}{3!} + \cdots + \frac{(j\omega t)^n}{n!} + \cdots \tag{1.7}$$

The sum of two sine functions with the same frequency is also a sine function:

$$a\cos\omega t + b\sin\omega t = r\cos(\omega t + \varphi) \tag{1.8}$$

The discrete form of a sine function is

$$x(nT) = \cos\omega nT$$

By separating the real and the imaginary parts of (1.7), we obtain

$$\boxed{e^{j\omega t} = \cos\omega t + j\sin\omega t} \tag{1.9}$$

This fundamental identity can also be used to define the complex exponential $\exp(j\omega t)$ and to derive all its properties in terms of the properties of trigonometric functions.

We observe that $\exp(j\omega t)$ is a complex number with unity amplitude and phase ωt. The sample value of the complex exponential is

$$x(nT) = e^{j\omega nT}$$

This function is a geometric series whose ratio $e^{j\omega T}$ is a complex number of unit amplitude.

From (1.9), it follows that

$$e^{(a+j\omega)t} = e^{at}(\cos\omega t + j\sin\omega t) \tag{1.10}$$

Therefore, if $s = a + j\omega$ is a complex number, then e^{st} is a complex signal whose real part $e^{at}\cos\omega t$ and imaginary part $e^{at}\sin\omega t$ are exponentially decreasing $(a < 0)$ and increasing $(a > 0)$ sine functions.

From (1.9), we obtain

$$e^{-j\omega t} = \cos\omega t - j\sin\omega t$$

Adding and subtracting the last equation from (1.9), we find Euler's formula:

$$\cos \omega t = \frac{e^{j\omega t} + e^{-j\omega t}}{2} \qquad \sin \omega t \, \frac{e^{j\omega t} - e^{-j\omega t}}{2j} \tag{1.11}$$

A general complex signal $x(t)$ is a function of the form

$$x(t) = x_1(t) + jx_2(t)$$

where $x_1(t)$ and $x_2(t)$ are the real functions of the real variable t. The derivative of $x(t)$ is a complex signal given by

$$\frac{dx(t)}{dt} = \frac{dx_1(t)}{dt} + j\frac{dx_2(t)}{dt}$$

and, in general, for any s, real or complex, we have

$$\frac{de^{st}}{dt} = se^{st} \tag{1.12}$$

Impulse (delta) function

An important function in science and engineering is the **impulse** function also known as Dirac's **delta** function. The signal is represented graphically in Figure 1.5. The delta function is not an ordinary one. Therefore, some fundamental properties of these types of functions, and specifically those of the delta function are presented so that the reader uses it appropriately.

Property 1.1   The impulse function $\delta(t)$ is a signal with a unit area and is zero outside the point at the origin:

$$\begin{cases} \int\limits_{-\infty}^{\infty} \delta(t)dt = 1 \\ \delta(t) = 0 \qquad t \neq 0 \end{cases} \tag{1.13}$$

Property 1.2   The impulse function is the derivative of the step function $u(t)$:

$$\delta(t) = \frac{du(t)}{dt} \tag{1.14}$$

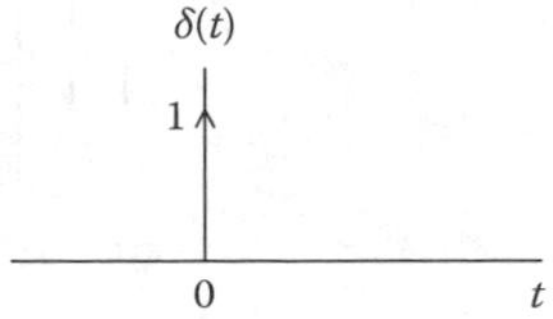

$\delta(t)$

1

0   t

FIGURE 1.5

Property 1.3   The area of the product $\varphi(t)\delta(t)$ equals $\varphi(0)$ for any regular function that is continuous at the origin:

$$\int_{-\infty}^{\infty} \varphi(t)\delta(t)dt = \varphi(0) \tag{1.15}$$

Property 1.4   The delta function can be written as a limit:

$$\delta(t) = \lim v_\varepsilon(t) \quad \varepsilon \to 0 \tag{1.16}$$

where $v_\varepsilon(t)$ is a family of functions with the unit area vanishing outside the interval $\left(-\dfrac{\varepsilon}{2}, \dfrac{\varepsilon}{2}\right)$:

$$\int_{-\varepsilon/2}^{\varepsilon/2} v_\varepsilon(t)dt = 1 \quad v_\varepsilon(t) = 0 \quad \text{for } t < -\frac{\varepsilon}{2} \quad \text{and} \quad t > \frac{\varepsilon}{2} \tag{1.17}$$

Figure 1.6 shows the approximation of the delta function by the pulse and the sufficiently small ε.

We can show (see Prob) that the impulse function is even. Hence,

$$\delta(t) = \delta(-t) \tag{1.18}$$

The impulse function $\delta(t - t_0)$ is centered at t_0 of area one. Therefore, from (1.18), we obtain

$$\delta(t - t_0) = \delta(t_0 - t) \tag{1.19}$$

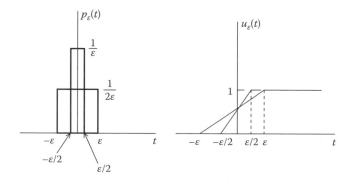

FIGURE 1.6

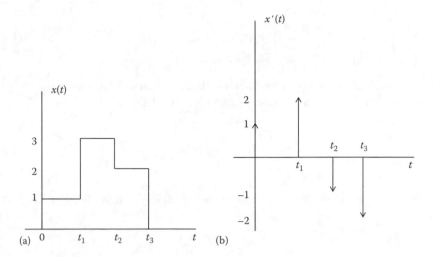

FIGURE 1.7

Using Property 1.2 above, we write

$$\delta(t - t_0) = \frac{du(t - t_0)}{dt} \tag{1.20}$$

Based on the above, the derivative of the function shown in Figure 1.7a is that shown in Figure 1.7b.

Considering Property 1.3, and taking into consideration the evenness of the delta function, we write

$$\boxed{\int_{-\infty}^{\infty} y(x)\delta(t - x)dx = y(t)} \tag{1.21}$$

The above integral is also known as the **convolution** integral. Therefore, we state that the convolution of a function with a delta function reproduces the function. Let us consider the function $y(t - t_0)$ to be convolved with the shifted delta function $\delta(t - a)$. From (1.21), we write

$$\int_{-\infty}^{\infty} y(x - t_0)\delta(t - x - a)dx = y(t - t_0 - a)$$

The identity (1.21) is basic. We can use it, for example, to define the derivative of the delta function. Because the two sides of the equation are functions of t, we can differentiate with respect to t to obtain

$$\int_{-\infty}^{\infty} y(x)\delta'(t-x)dx = y'(t) \tag{1.22}$$

Thus, the derivative of the delta function is such that the area of the product $y(x)\delta'(t-x)$, considered as a function of x, equals $y'(t)$. With $t=0$, (1.22) yields

$$\int_{-\infty}^{\infty} y(x)\delta'(-x)dx = y'(0) \tag{1.23}$$

From calculus we know that when a function is even its derivative is an odd function. Hence,

$$\delta'(-t) = -\delta'(t) \tag{1.24}$$

Inserting (1.24) in (1.23) and changing the dummy variable from x to t, we find

$$\boxed{\int_{-\infty}^{\infty} y(t)\delta'(t)dt = -y'(0)} \tag{1.25}$$

Additional delta functional properties are given in Table 1.2.

TABLE 1.2   Delta Functional Properties

1.	$\delta(at) = \dfrac{1}{|a|}\delta(t)$
2.	$\delta\left(\dfrac{t-t_0}{a}\right) = |a|\delta(t-t_0)$
3.	$\delta(at-t_0) = \dfrac{1}{|a|}\delta\left(t-\dfrac{t_0}{a}\right)$
4.	$\delta(-t+t_0) = \delta(t-t_0)$
5.	$\delta(-t) = \delta(t); \quad \delta(t) = $ even function
6.	$\int_{-\infty}^{\infty} \delta(t)f(t)dt = f(0)$
7.	$\int_{-\infty}^{\infty} \delta(t-t_0)f(t) = f(t_0)$
8.	$f(t)\delta(t) = f(0)\delta(t)$
9.	$f(t)\delta(t-t_0) = f(t_0)\delta(t-t_0)$
10.	$t\delta(t) = 0$
11.	$\int_{-\infty}^{\infty} A\delta(t)dt = \int_{-\infty}^{\infty} A\delta(t-t_0)dt = A$
12.	$f(t)*\delta(t) = $ convolution $= \int_{-\infty}^{\infty} f(t-\tau)\delta(\tau)d\tau = f(t)$
13.	$\delta(t-t_1)*\delta(t-t_2) = \int_{-\infty}^{\infty} \delta(\tau-t_1)\delta(t-\tau-t_2)d\tau = \delta[t-(t_1+t_2)]$
14.	$\sum_{n=-N}^{N} \delta(t-nT) * \sum_{n=-N}^{N} \delta(t-nT) = \sum_{n=-2N}^{2N} (2N+1-|n|)\delta(t-nT)$

TABLE 1.2 (continued) Delta Functional Properties

15. $\int_{-\infty}^{\infty} \dfrac{d\delta(t)}{dt} f(t) dt = -\dfrac{df(0)}{dt}$

16. $\int_{-\infty}^{\infty} \dfrac{d\delta(t - t_0)}{dt} f(t) dt = -\dfrac{df(t_0)}{dt}$

17. $\int_{-\infty}^{\infty} \dfrac{d^n\delta(t)}{dt^n} f(t) dt = (-1)^n \dfrac{d^n f(0)}{dt^n}$

18. $f(t) \dfrac{d\delta(t)}{dt} = -\dfrac{df(0)}{dt}\delta(t) + f(0)\dfrac{d\delta(t)}{dt}$

19. $t\dfrac{d\delta(t)}{dt} = -\delta(t)$

20. $t^n \dfrac{d^m\delta(t)}{dt^m} = \begin{cases} (-1)^n n! \delta(t), & m = n \\ (-1)^n \dfrac{m!}{m-n!} \dfrac{d^{m-n}\delta(t)}{dt^{m-n}}, & m > n \\ 0, & m < n \end{cases}$

21. $\int_{-\infty}^{\infty} \dfrac{d\delta(t)}{dt} = 0, \quad \dfrac{d\delta(t)}{dt} = \text{odd function}$

22. $f(t) * \dfrac{d\delta(t)}{dt} = \dfrac{df(t)}{dt}$

23. $f(t)\dfrac{d^n\delta(t)}{dt^n} - \sum_{k=0}^{n} (-1)^k \dfrac{n!}{k!(n-k)!} \dfrac{d^k f(0)}{dt^k} \dfrac{d^{n-k}\delta(t)}{dt^{n-k}}$

24. $\dfrac{\partial \delta(yt)}{\partial y} = -\dfrac{1}{y^2}\delta(t)$

25. $\delta(t) = \dfrac{du(t)}{dt}$

26. $\dfrac{d^n\delta(-t)}{dt^n} = (-1)^n \dfrac{d^n\delta(t)}{dt^n}, \left\{ \dfrac{d^n\delta(t)}{dt^n} \text{ is even if } n \text{ is even, and odd if } n \text{ is odd.} \right\}$

27. $(\sin \, at)\dfrac{d\delta(t)}{dt} = -a\delta(t)$

28. $\dfrac{d\delta(t)}{dt} = \dfrac{d^2 u(t)}{dt^2}$

29. $-\delta(t) = \dfrac{du(-t)}{dt}$

30. $\delta(t - t_0) = \dfrac{du(t - t_0)}{dt}$

31. $\dfrac{d\,\text{sgn}(t)}{dt} = 2\delta(t)$

32. $\delta[r(t)] = \sum_n \dfrac{\delta(t - t_n)}{\left|\dfrac{dr(t_n)}{dt}\right|}, \quad t_n = \text{zeros of } r(t), \dfrac{dr(t_n)}{dt} \neq 0$

33. $\dfrac{d\delta[r(t)]}{dt} = \sum_n \dfrac{\dfrac{d\delta(t - t_n)}{dt}}{\dfrac{dr(t)}{dt}\left|\dfrac{dr(t_n)}{dt}\right|}, \quad t_n = \text{zeros of } r(t), \dfrac{dr(t_n)}{dt} \neq 0, \dfrac{dr(t)}{dt} \neq 0$

34. $\delta(\sin t) = \sum_{n=-\infty}^{\infty} \delta(t - n\pi)$

35. $\delta(t^2 - 1) = \tfrac{1}{2}\delta(t - 1) + \tfrac{1}{2}\delta(t + 1)$

36. $\delta(t^2 - a^2) = \tfrac{1}{2a}[\delta(t + a) + \delta(t - a)]$

(continued)

TABLE 1.2 (continued)   Delta Functional Properties

37.
$$\delta(t) = \lim_{\varepsilon \to 0} \frac{e^{-t^2/\varepsilon}}{\sqrt{\varepsilon \pi}}$$

38.
$$\delta(t) = \lim_{\omega \to \infty} \frac{\sin \omega t}{\pi t}$$

39.
$$\delta(t) = \lim_{\varepsilon \to 0} \frac{1}{\pi} \frac{\varepsilon}{t^2 + \varepsilon^2}$$

40.
$$\delta(t) = \frac{1}{2\pi} \int_{-\infty}^{\infty} \cos \omega t \, d\omega$$

41.
$$\frac{df(t)}{dt} = \frac{d}{dt}[tu(t) - (t-1)u(t-1) - u(t-1)]$$
$$= t\delta(t) + u(t) - (t-1)\delta(t-1) - u(t-1) - \delta(t-1)$$

42.
$$\text{comb}_T(t) = \sum_{n=-\infty}^{\infty} \delta(t-nT), \quad f(t)\text{comb}_T(t) = \sum_{n=-\infty}^{\infty} f(nT)\delta(t-nT)$$
$$\text{COMB}_{\omega_0}(\omega) = \mathcal{F}\{\text{comb}_T(t)\} = \omega_0 \sum_{n=-\infty}^{\infty} \delta(\omega - n\omega_0), \quad \omega_0 = \frac{2\pi}{T}$$

$$\frac{d}{dt}([2 - u(t)]\cos t) = \frac{d}{dt}(2\cos t - u(t)\cos t)$$
$$= -2\sin t - \delta(t)\cos t + u(t)\sin t$$
$$= (u(t) - 2)\sin t - \delta(t)$$

$$\frac{d}{dt}\left(\left[u\left(t - \frac{\pi}{2}\right) - u(t - \pi)\right]\sin t\right) = \left[\delta\left(t - \frac{\pi}{2}\right) - \delta(t - \pi)\right]\sin t$$
$$+ \left[u\left(t - \frac{\pi}{2}\right) - u(t - \pi)\right]\cos t$$
$$= \delta\left(t - \frac{\pi}{2}\right) + \left[u\left(t - \frac{\pi}{2}\right) - u(t - \pi)\right]\cos t$$

Example

The values of the following integrals are

$$\int_{-\infty}^{\infty} e^{2t} \sin 4t \frac{d^2\delta(t)}{dt^2} dt = (-1)^2 \frac{d^2}{dt^2}[e^{2t} \sin 4t]|_{t=0} = 2 \times 2 \times 4 = 16$$

$$\int_{-\infty}^{\infty} (t^3 + 2t + 3)\left(\frac{d\delta(t-1)}{dt} + 2\frac{d^2\delta(t-2)}{dt^2}\right)dt = \int_{-\infty}^{\infty} (t^3 + 2t + 3)\frac{d\delta(t-1)}{dt} dt$$
$$+ 2\int_{-\infty}^{\infty} (t^3 + 2t + 3)\frac{d^2\delta(t-2)}{dt^2} dt$$
$$= (-1)(3t^2 + 2)|_{t=1} + (-1)^2 2(6t)|_{t=2}$$
$$= -5 + 24 = 19$$

Example

The values of the following integrals are

$$\int_0^4 e^{4t}\delta(2t - 3)dt = \int_0^4 e^{4t}\delta\left[2\left(t - \frac{3}{2}\right)\right]dt = \frac{1}{2}\int_0^4 e^{4t}\delta\left(t - \frac{3}{2}\right)dt = \frac{1}{2}e^{4\frac{3}{2}} = \frac{1}{2}e^6$$

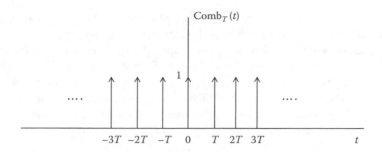

FIGURE 1.8

The comb function

The comb function is represented mathematically as follows:

$$\text{comb}_T(t) = \sum_{n=-\infty}^{\infty} \delta(t - nT) \tag{1.26}$$

This function is used extensively for studying the sampling of signals. Figure 1.8 shows the comb function pictorially.

1.3 Circuit Elements and Equation

In this text, we use the idealized model of physical devices, passive or active, which is specified in terms of its terminal properties. In Figure 1.9, we show the passive and active elements of electrical circuits.

A **circuit** or **network** is a combination of connected elements and external sources. The inputs are the sources (voltage or current) and the outputs are voltages or currents across elements and through elements, respectively. A network is a special form of an **analog** system since its inputs and outputs are continuous signals.

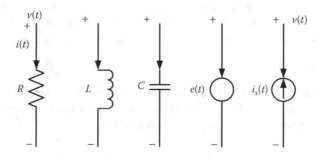

FIGURE 1.9

The **state** of a network at a certain time (taking the time $t = 0$ for simplicity) is the set of all the voltages across the capacitors and all the currents through the inductors. If we know the **initial state** of a network at $t = 0$ and all its inputs at $t = 0$, then we can determine all its responses for an all-time $t \geq 0$. If all the currents through the inductors and all the voltages across the capacitors are zero, the network is at a **zero initial state**. If the network is at a zero initial state, then its response is known as the **zero-state response**. If all the sources are zero, then its response is called the **zero-input response**. The zero-input response is due to the energy stored in the network.

The voltages and the currents of the passive elements are

Resistor

$$v(t) = Ri(t) \quad i(t) = Gv(t) \quad G = \frac{1}{R} \tag{1.27}$$

Inductor

$$v(t) = L\frac{di(t)}{dt} \quad i(t) = \frac{1}{L}\int_0^t v(x)dx + i(0) \tag{1.28}$$

Capacitor

$$i(t) = C\frac{dv(t)}{dt} \quad v(t) = \frac{1}{C}\int_0^t i(x)dx + v(0) \tag{1.29}$$

Voltage source

$$e(t) \equiv \text{known, independent of } i(t)$$

Current source

$$i_s(t) \equiv \text{known, independent of } v(t)$$

Initial conditions

Knowing the initial conditions of a network (voltages across the capacitors and the currents through the inductors) it is sufficient to find its response for $t \geq 0$. In this text we assume that there is a continuation of the initial conditions which means that the currents through the inductors or the voltages across the capacitors are the same at $t(0-)$ and $t(0+)$.

Impulse response

The following simple example will elucidate how a network responds to an impulse input source. Let the input voltage of a simple RL series circuit be a delta function as shown in Figure 1.10. Kirchhoff's voltage law of a network loop is

$$L\frac{di(t)}{dt} + Ri(t) = \delta(t) \tag{1.30}$$

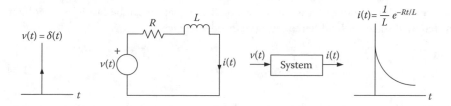

FIGURE 1.10

Integrating the above equation from (0−) to (0+), and taking into consideration that $i(t)$ is a continuous function, we obtain

$$L \int_{0-}^{0+} \frac{di(t)}{dt} dt + R \int_{0-}^{0+} i(t)dt = \int_{0-}^{0+} \delta(t)dt \quad \text{or} \quad L[i(0+) - i(0-)] + R0 = 1$$

$$\text{or} \quad L[i(0+) - i(0-)] = 1 \tag{1.31}$$

Since the input impulse function is a discontinuous one, the current is also a discontinuous function with a discontinuity such that $L[i(0+) - i(0-)]$ is equal to 1. If, in addition, the system (here the network) is causal, $i(0-) = 0$ and hence $i(0+) = 1/L$. Therefore, if the circuit is in the zero state and it is connected to a delta function source, the current $i(t)$ changes instantly from zero to $1/L$.

Derived initial conditions

Derived initial conditions are determined from the circuit equations, and, in general, depend also on the sources. Let us assume that there is an initial current, $i(0) = i_0$, in the RL circuit shown in Figure 1.10. In addition, let the voltage source be a constant, $v(t) = V$. In this case, the solution of (1.30) with a constant voltage is the function

$$i(t) = \underbrace{\frac{V}{R}(1 - e^{-Rt/L})}_{\text{zero-state response}} + \underbrace{i_0 e^{-Rt/L}}_{\text{zero-input response}} \tag{1.32}$$

This result will be derived in Chapter 7.

For an RC series circuit, Kirchhoff's mesh voltage law results in

$$Ri(t) + v_c(t) = v(t) \quad \text{or} \quad Ri(t) + \frac{1}{C} \int_{-\infty}^{0} i(x)dx + \frac{1}{C} \int_{0}^{t} i(x)dx = v(t) \quad \text{or}$$

$$Ri(t) + \frac{1}{C} \int_{0}^{t} i(x)dx + v_c(0) = v(t) \tag{1.33}$$

where $v_c(t)$ is the voltage across the capacitor. This equation can be cast into an ordinary differential equation by differentiating both sides. Hence,

$$R\frac{di(t)}{dt} + \frac{1}{C}i(t) = \frac{dv(t)}{dt} \tag{1.34}$$

To solve (1.34), we must find the initial value of the current, $i(0)$. Setting $t = 0$, we obtain

$$Ri(0) + v_c(0) = v(0) \quad \text{or} \quad i(0) = \frac{v(0) - v_c(0)}{R}$$

This is the **derived** initial condition, and it depends not only on the initial (state) voltage across the capacitor but also on the initial value of the voltage source.

As another example, Kirchhoff's mesh equation for a series RLC circuit with an initial current $i(0)$ through the inductor and the initial voltage $v_c(0)$ across the capacitor is

$$L\frac{di(t)}{dt} + Ri(t) + \frac{1}{C}\int_0^t i(x)dx + v_c(0) = v(t) \quad i(0) \tag{1.35}$$

Taking the derivative with respect to the independent variable t, we find

$$L\frac{d^2 i(t)}{dt^2} + R\frac{di(t)}{dt} + \frac{1}{C}i(t) = \frac{dv(t)}{dt} \tag{1.36}$$

Since the above equation is a second-order differential equation of the dependent variable i, we must find, in addition to its initial value $i(0)$, the initial value of its derivative $i'(0)$. Setting $t = 0$ in (1.35) and assuming that $v(t)$ does not have a discontinuity at $t = 0$, we obtain

$$Li'(0) + Ri(0) + v_c(0) = v(0) \quad \text{or} \quad i'(0) = \frac{1}{L}[v(0) - v_c(0) - Ri(0)]$$

which is a derived initial condition.

State equations of an RLC series circuit

The state variables are the current $i(t)$ through the inductor and the voltage $v_c(t)$ across the capacitor and they satisfy the following two first-order differential equations:

$$C\frac{dv_c(t)}{dt} = i(t) \qquad v_c(0)$$

$$L\frac{di(t)}{dt} + Ri(t) + v_c(t) = v(t) \qquad i(0) \tag{1.37}$$

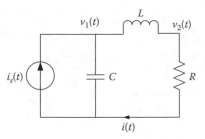

FIGURE 1.11

Node and state equations of the circuit in Figure 1.11

The circuit in Figure 1.11 has two variables: the voltage $v_1(t)$ across the capacitor and the current $i(t)$ through the inductor. The initial voltage across the capacitor is $v_1(0)$ and the initial current through the inductor is $i(0)$.

Node equations (the algebraic sum of currents at a node should be equal to zero)

For the node equation we use as primary unknowns, the node voltages $v_1(t)$ and $v_2(t)$. Hence,

$$C\frac{dv_1(t)}{dt} + \frac{v_2(t)}{R} = i_s(t) \quad v_1(0)$$
$$L\frac{di(t)}{dt} + Ri(t) - v_1(t) = 0 \quad \text{or} \quad \frac{L}{R}\frac{dv_2(t)}{dt} + v_2(t) - v_1(t) = 0 \quad v_2(0) = Ri(0) \tag{1.38}$$

State equations

$$C\frac{dv_1(t)}{dt} + i(t) = i_s(t) \qquad v_1(0)$$
$$L\frac{di(t)}{dt} + Ri(t) - v_1(t) = 0 \qquad i(0) \tag{1.39}$$

State equations for the circuit in Figure 1.12

State equations

$$L_1\frac{di_1(t)}{dt} + R_1 i_1(t) + v_c(t) = v(t)$$
$$L_2\frac{di_2(t)}{dt} + R_2 i_2(t) - v_c(t) = 0 \tag{1.40}$$
$$i_1(t) - i_2(t) - C\frac{dv_c(t)}{dt} = 0$$

Block diagrams of systems

Circuit diagrams describe the structure of a network. However, the block diagrams describe the terminal properties of the network (system). Inside the block we present

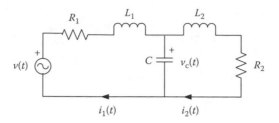

FIGURE 1.12

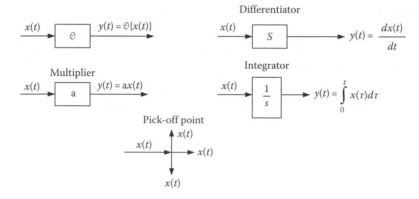

FIGURE 1.13

different identifiers that will characterize the system operation. In general, we introduce in the block a script $\odot$ to represent a general operator that operates on the input to the block to produce the output. In Figure 1.13, we show the block-diagram representation of, a **general** system, a **differentiator**, a **multiplier**, an **integrator**, and a **pick-off point**. The significance of s is given later in Chapter 7. Note that at the pick-off point the input quantity appears in all the branches without any variation of its magnitude.

Figure 1.14a depicts three basic ways that systems can be configured. It is assumed that the terminal properties of each system remain unchanged (no loading effect takes place). Figure 1.14b shows the equivalent input–output of the **cascade** and the **parallel** and the feedback configurations. In Figure 1.14a and for the first system, we obtain $y_1(t) = \odot_1 x(t)$ or $y(t) = \odot_2 y_1(t) = \odot_2\odot_1 x(t)$. The second expression characterizes the first system of part (b) of the figure. For the second system of part (a) of the figure, we find

$$y_1(t) = \odot_1 x(t) \quad \text{and} \quad y_2(t) = \odot_2 x(t), \quad \text{and, therefore, } y(t) = y_1(t) + y_2(t) = [\odot_1 + \odot_2]x(t)$$

which characterizes the second system of part (b) of the figure. For the feedback system of part (a), we obtain: $y_1(t) = x(t) \pm \odot_2 y(t)$ or $y(t) = \odot_1[x(t) \pm \odot_2 y(t)]$. Solving $y(t)$, and keeping in mind that we do not perform divisions with the operators but only use their inverse form, we find $y(t) = [1 \mp \odot_1\odot_2]^{-1}\odot_1 x(t)$. This expression characterizes the third

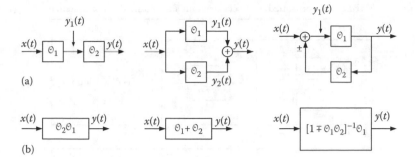

FIGURE 1.14

system of part (b) of the figure. If we substitute the two operators with constants a and b, the transfer functions of the three systems are

$$H_c = ab, \quad H_p = a + b, \quad H_f = \frac{a}{1 \mp ab} \tag{1.41}$$

Table 1.3 represents block-diagram transformations.

TABLE 1.3 Block-Diagram Transformations of Systems

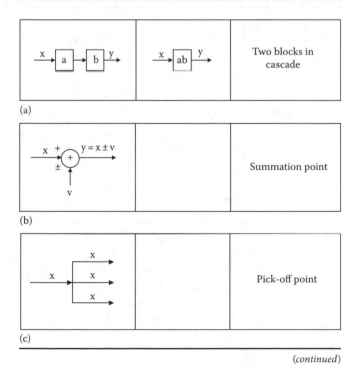

(continued)

TABLE 1.3 (continued)   Block-Diagram Transformations of Systems

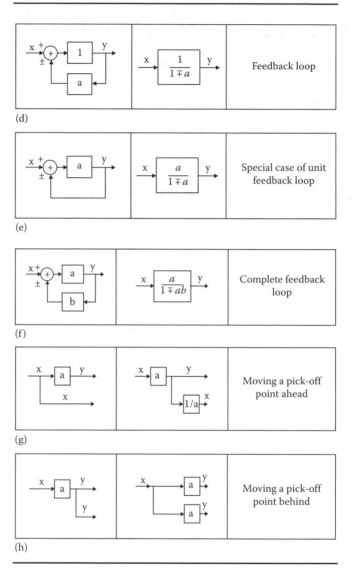

(d)

(e)

(f)

(g)

(h)

TABLE 1.3 (continued)   Block-Diagram Transformations of Systems

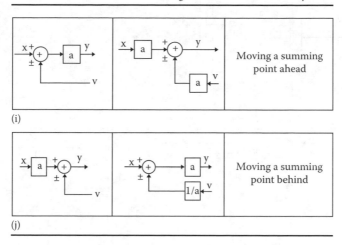

(i)

(j)

1.4 Linear Mechanical and Rotational Mechanical Elements

The linear mechanical systems with their equivalent circuit characterizations are shown in Figure 1.15. The rotating fundamental mechanical systems and their equivalent circuit characterizations are shown in Figure 1.16. The terminal properties of these signals are given below.

1.4.1 Linear Mechanical Systems

Damper

$$f(t) = Dv(t), \quad v(t) = \frac{1}{D}f(t), \quad D = \text{damping constant } (\text{N} \cdot \text{s/m})$$

$$f(t) = \text{force (N)}, \quad v(t) = \text{velocity (m/s)}$$

(1.42)

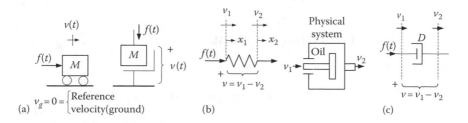

FIGURE 1.15

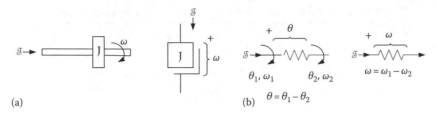

(a)

(b) $\theta = \theta_1 - \theta_2$

(c)   Viscous fluid

FIGURE 1.16

Spring

$$f(t) = Kx(t), \quad v(t) = \frac{1}{K}\frac{df(t)}{dt}, \quad x(t) = \text{displacement (m)}$$

$$K = \text{spring constant (N/m)}$$

(1.43)

Mass

$$f(t) = M\frac{dv(t)}{dt} = M\frac{d^2x(t)}{dt^2} \quad \text{Newton} = \text{kg} \cdot \text{m} \cdot \text{s}^{-2}(\text{N})$$

$$v(t) = \frac{1}{M}\int_{-\infty}^{t} f(x)dx \qquad M = \text{mass(kg)}$$

(1.44)

From the above equations, we observe the following analogies between the circuit elements and the linear mechanical elements: the mass and the capacitor, the spring and the inductor, and the damper and the resistor.

1.4.2 Rotational Mechanical Systems

Damper

$$\mathfrak{I}(t) = D\omega(t), \quad D = \text{damping costant (N} \cdot \text{s} \cdot \text{m/rad)}$$

$$\omega(t) = \frac{1}{D}\mathfrak{I}(t); \quad \mathfrak{I} = \text{torque (N} \cdot \text{m)}$$

(1.45)

Spring

$$\mathfrak{I}(t) = K\theta(t) = K \int_{-\infty}^{t} \omega(x)dx, \quad K = \text{spring constant (N} \cdot \text{m/rad)}$$

$$\omega(t) = \frac{1}{D}\frac{d\mathfrak{I}(t)}{dt}$$

(1.46)

Moment of inertia

$$\mathfrak{I}(t) = J\frac{d\omega(t)}{dt} = J\frac{d^2\theta(t)}{dt^2}, \quad J = \text{polar moment of inertia (kg} \cdot \text{m}^2)$$

$$\omega(t) = \frac{1}{J} \int_{-\infty}^{t} \mathfrak{I}(x)dx$$

(1.47)

For the rotation elements, we observe the following analogies between these elements and the circuit elements: the damper and the resistor, the inductor and the spring, and the mass and the moment of inertia. The current in circuits, the force in linear mechanical systems, and the torque in rotational mechanical elements are the **through variables**. The voltage in circuits, the velocity in linear mechanical systems, and the angular velocity in rotational mechanical systems are the **across variables**.

1.5 Discrete Equations and Systems

A **discrete** system is a process that relates the input discrete-time signal $x(n)$ (or $x(nT)$) with the discrete-time output signal $y(n)$ (or $y(nT)$). The elements with which we create discrete systems are shown in Figure 1.17. The special symbol z^{-1} indicates the delay which we discuss in Chapter 8. The delay element has a **memory.** This means that the output at any particular time depends on the value of the input one unit earlier. In the discrete systems, we also have pick-off points as we have defined them in the circuits case.

A simple first-order system is defined by the following discrete equation:

$$y(n) = -2y(n-1) + x(n)$$

(1.48)

Its block-type representation and its solution is shown in Figure 1.18. This is a recursive equation, and its solution is found by iteration, assuming (or defining), of course, its

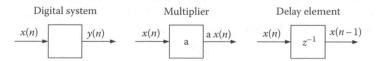

Digital system   Multiplier   Delay element

FIGURE 1.17

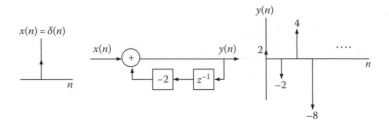

FIGURE 1.18

initial conditions. If we set $y(-1)=0$ and the input function to be a delta function, we obtain

$$y(0) = -2y(-1) + \delta(0) = 0 + 1 = 1$$
$$y(1) = -2y(0) + \delta(1) = -2 + 0 = -2$$
$$y(2) = -2y(1) + \delta(2) = -2(-2) + 0 = 4$$
$$y(3) = -2y(2) + \delta(3) = -2(4) + 0 = -8$$

$$\vdots$$

The **state** of a discrete system at a certain time n_0 is the set of values of the outputs $q_i(n-1)$ of all delay elements at $n = n_0$. Therefore, if we know the state of the system at $n = n_0$ and all its sources for $n > n_0$, then we can determine all its responses for any $n > n_0$.

The **initial state** of a system is its state at $n = 0$, where this time of origin is taken for convenience. Hence, the initial state of a system is the set of values $q_i(-1)$ of the inputs $q_i(n)$ to all delay elements at $n = -1$. If the system is at the zero state, then its responses for $n > 0$ are called **zero-state** responses. We therefore conclude that its responses are due only to its inputs (external sources). On the other hand, if all external sources are zero, its responses are only due to the energy sources of the system and they are called **zero-input** responses.

State equations

State variables are the inputs $q_i(n)$ to all the delay elements (or any linear transformation of these signals). The state variables are determined from the state equations resulting from the rules of the interconnected elementary systems. The **state** equations become a system of a specific number of equations equal to the number of the delay elements present in the system. To find the solution, besides the input sources, we need the **initial conditions** which are the values of the state variables at $n = -1$. It is apparent, for example, that the second-order discrete system

$$y(n) = 3.5y(n-1) + 5y(n-2) + x(n) \tag{1.49}$$

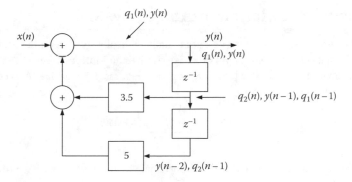

FIGURE 1.19

which is shown in the block-diagrammatic form in Figure 1.19, has the following state variables representation:

$$q_1(n) - 3.5q_2(n) - 5q_2(n-1) = x(n)$$
$$q_2(n) - q_1(n-1) = 0$$
$$\left.\begin{array}{l} q_1(-1) = y(-1) \\ q_2(-1) = y(-2) \end{array}\right\} \text{initial conditions}$$

(1.50)

Recursive and non-recursive systems

If a discrete (difference) equation, which represents the system, has one input and an output with additional delayed outputs, it is called a **recursive** one. We also call these systems **infinite impulse systems** (IIR). The difference equation of (1.49) represents a recursive system and it is shown in Figure 1.19.

If, however, we have the discrete system representation by the equation

$$y(n) = b_0 x(n) + b_1 x(n-1) + b_2 x(n-2) \tag{1.51}$$

we say that the system is not recursive. This type of system is also called the **finite impulse response** system (FIR). Figure 1.20 shows such a system. Note the feedback configuration of the IIR systems and the forward configuration of the FIR systems.

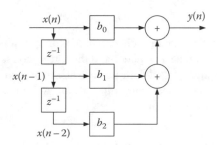

FIGURE 1.20

1.6 Digital Simulation of Analog Systems

Since the physical systems are represented mathematically by differential and integro-differential equations, we must approximate derivatives and integrals (see (1.3) and (1.6)). The approximations are derived by interrogating Figures 1.3 and 1.4. A second-order derivative is approximated in the form

$$\frac{d^2 y(t)}{dt^2} \cong \frac{y(nT) - 2y(nT - T) + y(nT - 2T)}{T^2} \tag{1.52}$$

To have the solution of a second-order differential equation, we must have the value of the derivative at $t = 0$. Therefore, we must substitute the analog derivative with an equivalent discrete one. Hence, we write

$$\left.\frac{dy(t)}{dt}\right|_{t=0} = \frac{dy(0)}{dt} \cong \frac{y(0T) - y(0T - T)}{T}$$

or

$$y(-T) = y(0) - T\frac{dy(0)}{dt} \tag{1.53}$$

1.7 Convolution of Analog Signals

The convolution operation on functions is one of the most useful operations encountered in the study of signals and systems. The importance of the convolution integral in systems studies stems from the fact that a knowledge of the output of the system to an impulse (delta) function excitation allows us to find its output to any input function (subject to some mild restrictions).

To help us develop the convolution integral, let us begin with the properties of the delta function. Based on the delta properties, we write

$$f(t) = \int_{-\infty}^{\infty} f(\tau)\delta(t - \tau)d\tau \tag{1.54}$$

Observe that, as far as the integral is concerned, the time t is a parameter (constant for the integral although it can take any value) and the integration is with respect to τ. Our next step is to represent the integral with its equivalent approximate form, the summation form, by dividing the τ axis into intervals of ΔT, then the above integral is represented approximately by the sum

$$f_a(t) = \lim_{\Delta T \to 0} \sum_{n=-\infty}^{\infty} f(n\Delta T)\delta(t - n\Delta T)\Delta T \tag{1.55}$$

As ΔT goes to zero and n increases to infinity, the product $n\Delta T$ takes the value of τ, ΔT becomes $d\tau$ and the summation becomes integral, thus recapturing (1.54).

Note: *The function $f(t)$ has been approximated with an infinite sum of shifted delta functions equal to $n\Delta T$ and their area is equal to $f(n\Delta T)\Delta T$.*

We define the response of a causal (system that reacts after being excited) and an LTI system to a delta function excitation by $h(t)$, known as the **impulse response** of the system. If the input to the system is $\delta(t)$ the output is $h(t)$, and when the input is $\delta(t-t_0)$ then the output is $h(t - t_0)$. Further, we define the output of a system by $g(t)$ if its input is $f(t)$. Based on the definitions discussed so far, it is obvious that if the input to the system is $f_a(t)$, the output is a sum of impulse functions shifted identically to the shifts of the input delta functions of the summation, and, therefore, the output is equal to

$$g(t) = \lim_{\Delta T \to 0} \sum_{n=-\infty}^{\infty} f(n\Delta T)h(t - n\Delta T)\Delta T$$

In the limit, as ΔT approaches zero, the summation becomes an integral of the form

$$g(t) = \int_{-\infty}^{\infty} f(\tau)h(t - \tau)d\tau \tag{1.56}$$

This is the **convolution integral** for any two functions $f(t)$ and $h(t)$.

Convolution is a general mathematical operation, and for any two real-valued functions, their convolution, indicated mathematically by the asterisk between the functions, is given by

$$\boxed{g(t) \stackrel{\Delta}{=} f(t) * h(t) = \int_{-\infty}^{\infty} f(\tau)h(t - \tau)d\tau = \int_{-\infty}^{\infty} f(t - \tau)h(\tau)d\tau} \tag{1.57}$$

Note: *Equation 1.57 tells us the following: given two functions in the time domain t, we find their convolution g(t) by doing the following steps: (1) rewrite one of the functions in the τ domain by just setting wherever there is t, the variable τ; the shape of the function is identical to that in the t domain; (2) to the second function substitute t-τ wherever there is t; this produces a function in the τ domain which is flipped (the minus sign in front of τ) and shifted by t (positive values of t shift the function to the right and negative values shift the function to the left); (3) multiply these two functions and find another function of τ, since t is a parameter and a constant as far as the integration is concerned; and (4) next find the area under the product function whose value is equal to the output of the convolution at t (in our case it is g(t)). By introducing the infinite values of t's, from minus infinity to infinity, we obtain the output function g(t).*

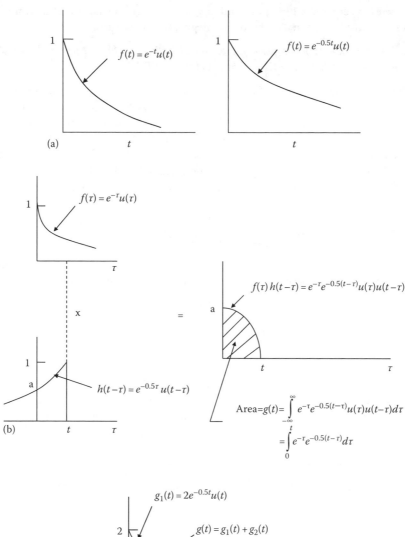

FIGURE 1.21

From the convolution integral, we observe that one of the functions does not change when it is mapped from the t to τ domain. The second function is reversed or folded over (mirrored with respect to the vertical axis) in the τ domain and it is shifted by an amount t, which is just a parameter in the integrand. Figure 1.21a and b shows two functions in the t and τ domains, respectively. We now write

$$g(t) = f(t) * h(t) = \int_{-\infty}^{\infty} e^{-\tau}u(\tau)e^{-0.5(t-\tau)}u(t-\tau)d\tau = \int_{0}^{t} e^{-\tau}e^{-0.5(t-\tau)}d\tau$$

$$= e^{-0.5t}\int_{0}^{t} e^{-0.5\tau}d\tau = 2(e^{-0.5t} - e^{-t})$$

Figure 1.21c shows the results of the convolution.

1.8 Convolution of Discrete Signals

As we have indicated in the above section, the convolution of continuous signals is defined as follows:

$$g(t) = \int_{-\infty}^{\infty} f(x)h(t-x)dx \tag{1.58}$$

The above equation is approximated as follows:

$$g(t) = \int_{-\infty}^{\infty} f(x)h(t-x)dx = \sum_{m=-\infty}^{\infty} \int_{mT-T}^{mT} f(x)h(t-x)dx \cong \sum_{m=-\infty}^{\infty} Tf(mT)h(t-mT)$$

or

$$\boxed{g(nT) = T\sum_{m=-\infty}^{\infty} f(mT)h(nT-mT) \quad n = 0, \pm 1, \pm 2, \ldots \quad m = 0, \pm 1, \pm 2, \ldots} \tag{1.59}$$

For $T = 1$, the above convolution equation becomes

$$g(n) = \sum_{m=-\infty}^{\infty} f(m)h(n-m) \quad n = 0, \pm 1, \pm 2, \ldots \quad m = 0, \pm 1, \pm 2, \ldots \tag{1.60}$$

If the input function to the system is the delta function

$$\delta(nT) = \begin{cases} 1 & n = 0 \\ 0 & n \neq 0 \end{cases} \quad \delta(nT - mT) = \begin{cases} 1 & n = m \\ 0 & n \neq m \end{cases} \tag{1.61}$$

then, (1.60) gives $g(n) = h(n)$.

Additional properties of the convolution process are shown in Table 1.4.

TABLE 1.4 Convolution Properties

1. Commutative

$$g(t) = \int_{-\infty}^{\infty} f(\tau)h(t-\tau)d\tau = \int_{-\infty}^{\infty} f(t-\tau)h(\tau)d\tau$$

2. Distributive

$$g(t) = f(t) * [h_1(t) + h_2(t)] = f(t) * h_1(t) + f(t) * h_2(t)$$

3. Associative

$$[f(t) * h_1(t)] * h_2(t) = f(t) * [h_1(t) * h_2(t)]$$

4. Shift invariance

$$g(t) = f(t) * h(t)$$

$$g(t-t_0) = f(t-t_0) * h(t) = \int_{-\infty}^{\infty} f(\tau-t_0)h(t-\tau)d\tau$$

5. Area property

$$A_f = \text{area of } f(t),$$

$$m_f = \int_{-\infty}^{\infty} tf(t)dt = \text{first moment}$$

$$K_f = \frac{m_f}{A_f} = \text{center of gravity}$$

$$A_g = A_f A_h, \quad K_g = K_f + K_h$$

6. Scaling

$$g(t) = f(t) * h(t)$$

$$f\left(\frac{t}{a}\right) * h\left(\frac{t}{a}\right) = |a|g\left(\frac{t}{a}\right)$$

7. Complex valued functions

$$g(t) = f(t) * h(t) = [f_r(t) * h_r(t) - f_i(t) * h_i(t)] + j[f_r(t) * h_i(t) + f_i(t) * h_r(t)]$$

8. Derivative

$$g(t) = f(t) * \frac{d\delta(t)}{dt} = \frac{df(t)}{dt}$$

9. Moment expansion

$$g(t) = m_{h0}f(t) - m_{h1}f^{(1)}(t) + \frac{m_{h2}}{2!}f^{(1)}(t) + \cdots + \frac{(-1)^{n-1}}{n-1!}m_{h(n-1)}f^{(n-1)}(t) + E_n$$

$$m_{hk} = \int_{-\infty}^{\infty} \tau^k h(\tau)d\tau$$

$$E_n = \frac{(-1)^n m_{hn}}{n!}f^{(n)}(t-\tau_0), \quad \tau_0 = \text{constant in the interval of integration}$$

10. Fourier transform

$$\mathcal{F}\{f(t) * h(t)\} = F(\omega) \, H(\omega)$$

(continued)

TABLE 1.4 (continued) Convolution Properties

11. Inverse Fourier transform

$$\frac{1}{2\pi}\int_{-\infty}^{\infty} F(\omega)H(\omega)e^{j\omega t}\,d\omega = \int_{-\infty}^{\infty} f(\tau)h(t-\tau)\,d\tau$$

12. Band-limited function

$$g(t) = \int_{-\infty}^{\infty} f(\tau)h(t-\tau)\,d\tau = \sum_{n=-\infty}^{\infty} Tf(nT)h_\sigma(t-nT)$$

$$h_\sigma(t) = \frac{1}{2\pi}\int_{-\sigma}^{\sigma} H(\omega)e^{j\omega t}\,d\omega, \quad f(t) = \sigma - \text{band limited} = 0, |t| > \sigma$$

13. Cyclical convolution

$$x(n) \otimes y(n) = \sum_{m=0}^{N-1} x((n-m) \bmod N)y(m)$$

14. Discrete-time

$$x(n) * y(n) = \sum_{m=-\infty}^{\infty} x(n-m)y(m)$$

15. Sampled

$$x(nT) * y(nT) = T\sum_{m=-\infty}^{\infty} x(nT-mT)y(mT)$$

where

$$H(e^{j\omega}) = \sum_{n=-\infty}^{\infty} h(n)e^{-j\omega n}.$$

Examples

Example 1.1

It is desired to plot the functions $x(t) = -2u(2-t)$, $x(t) = u(t-1) - 2u(t-3)$, and $x(t) = 2\delta(t-1) - \delta(t+1)$. These functions are plotted in Figure E.1.1. ∎

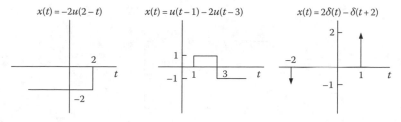

$x(t) = -2u(2-t)$ $x(t) = u(t-1) - 2u(t-3)$ $x(t) = 2\delta(t) - \delta(t+2)$

FIGURE E.1.1

Example 1.2

The evaluation of integrals, involving delta functions, is shown in the equations below:

$$\int_{-4}^{5} (t^2+2)[\delta(t)+3\delta(t-2)]dt = \int_{-4}^{5} (t^2+2)\delta(t)dt + \int_{-4}^{5} 3(t^2+2)\delta(t-2)dt = 2+18 = 20$$

$$\int_{-4}^{3} t^2[\delta(t+2)+\delta(t)+\delta(t-5)]dt = \int_{-4}^{3} t^2\delta(t+2)dt + \int_{-4}^{3} t^2\delta(t)dt + \int_{-4}^{3} t^2\delta(t-5)dt = 4+0+0 = 4$$

■

Example 1.3

A series RLC circuit is shown in Figure E.1.2 driven by the voltage source $v(t)$. The circuit has two state variables: the capacitor voltage $v_c(t)$ and the inductor current $i(t)$, with the initial conditions $v_c(0)$ and $i(0)$. The circuit has one mesh and one mesh current that satisfies Kirchhoff's voltage law:

$$L\frac{di(t)}{dt} + Ri(t) + \frac{1}{C}\int_0^t i(x)dx + v_c(0) = v(t) \quad i(0) \tag{1.62}$$

We next, reduce the above integrodifferential equation into a differential equation by differentiation:

$$L\frac{d^2i(t)}{dt^2} + R\frac{d(i)}{dt} + \frac{1}{C}i(t) = \frac{dv(t)}{dt} \tag{1.63}$$

To solve (1.63), we need the initial conditions $i(0)$ and its derivative at zero time $i'(0)$ since this is an equation of the second order. The $i(0)$ is the given initial state of the

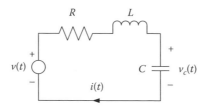

FIGURE E.1.2

system. To find the second initial condition, we must set $t = 0$ in (1.62). The substitution gives

$$Li'(0) + Ri(0) + \frac{1}{C}\int_0^0 i(x)dx + v_c(0) = v(0) \quad \text{or} \quad Li'(0) + Ri(0) + v_c(0) = v(0)$$

The above equation gives the desired initial condition:

$$i'(0) = \frac{1}{L}[v(0) - v_c(0) - Ri(0)]$$

State equations

The current through the capacitor based on Kirchhoff's law is equal to the current $i(t)$. Second, the algebraic sum of the voltages in the mesh should be equal to zero. Hence,

$$C\frac{dv_c(t)}{dt} = i(t) \qquad v_c(0)$$

$$L\frac{di(t)}{dt} + Ri(t) + v_c(t) = v(t) \qquad i(0)$$

(1.64)

and this is a system of two first-order differential equations. ∎

Example 1.4

Let the circuit (system) shown in Figure E.1.3 have the initial conditions $v_c(0)$, $i_1(0)$, and $i_2(0)$ of its state variables. To find the state equations, we sum algebraically the voltages in the two loops and the currents at the node. Hence,

State equations

$$L_1\frac{di_1(t)}{dt} + R_1 i_1(t) + v_c(t) = v(t)$$

$$L_2\frac{di_2(t)}{dt} + R_2 i_2(t) - v_c(t) = 0$$

$$i_1(t) - i_2(t) - C\frac{dv_c(t)}{dt} = 0$$

(1.65)

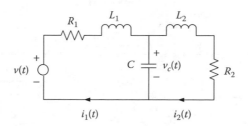

FIGURE E.1.3

Mesh equations

$$R_1 i_1(t) + L_1 \frac{di_1(t)}{dt} + \frac{1}{C} \int_0^t i_1(x)dx - \frac{1}{C} \int_0^t i_2(x)dx + v_c(0) = v(t)$$

(1.66)

$$-\frac{1}{C} \int_0^t i_1(x)dx + L_2 \frac{di_2(t)}{dt} + R_2 i_2(t) + \frac{1}{C} \int_0^t i_2(x)dx - v_c(0) = 0$$

∎

Example 1.5

It is desired to create the block-diagram representation of the following differential equation and its equivalent discrete representation. The equation is

$$3\frac{dy(t)}{dt} - y = v(t)$$

(1.67)

Its discrete representation is

$$y(nT) = \frac{1}{1 - \frac{T}{3}} y(nT - T) + \frac{T}{3\left(1 - \frac{T}{3}\right)} v(nT)$$

(1.68)

The block-diagram representation of the above two equations are given in Figure E.1.4.

∎

Example 1.6

It is required to find the differential equation of the linear mechanical system shown in Figure E.1.5a with respect to the distance traveled by the mass. This system is a rough representation, for example, of the spring-shock absorber system of a car. From the figure, the motion of the mass that is subjected to a spring and a damping force is described by the equation

$$-f(t) + f_M(t) + f_K(t) + f_D(t) = 0$$

(1.69)

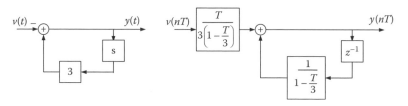

FIGURE E.1.4

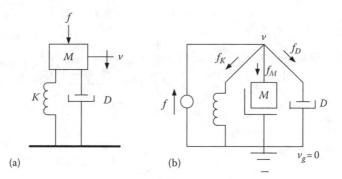

(a)   (b)

FIGURE E.1.5

or

$$M\frac{dv(t)}{dt} + K\int v(t)dt + Dv(t) = f(t) \qquad (1.70)$$

Since the velocity is related to the displacement x by the relation $v(t) = dx(t)/dt$, this equation takes the form

$$\frac{d^2x(t)}{dt^2} + \frac{D}{M}\frac{dx(t)}{dt} + \frac{K}{M}x(t) = \frac{1}{M}f(t) \qquad (1.71)$$

Because the velocity is an across variable, the velocity of the mass with respect to the ground, Figure E.1.5b represents the circuit representation of the system. We observe that the force is a through variable and the system is a node equivalent type circuit. ∎

Example 1.7

The system shown in Figure E.1.6 represents an idealized model of a stiff human limb as a step in assessing the passive control process of locomotive action. We try to find the movement of the system if the input torque is an exponential function. During the movement, we characterize the friction by the friction constant D. Furthermore, we assume that the initial conditions are zero, $\theta(0) = d\theta(0)/dt = 0$.

Applying D'Alembert's principle, which requires that the algebraic sum of the torques must be equal to zero at a node, we write

$$\boxed{\mathfrak{I}(t) = \mathfrak{I}_g(t) + \mathfrak{I}_D(t) + \mathfrak{I}_J(t)} \qquad (1.72)$$

where

$\mathfrak{I}(t)$ = input torque

$\mathfrak{I}_g(t)$ = gravity torque = $Mgl\sin\theta(t)$

$\mathfrak{I}_D(t)$ = frictional torque = $D\omega(t) = D\dfrac{d\theta(t)}{dt}$

$\mathfrak{I}_J(t)$ = inertial torque = $J\dfrac{d\omega(t)}{dt} = J\dfrac{d^2\theta(t)}{dt^2}$

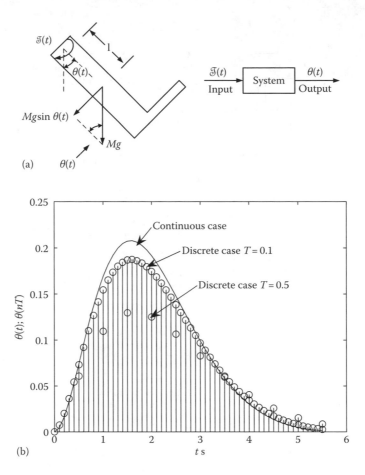

FIGURE E.1.6

Therefore, the equation that describes the system is

$$J\frac{d^2\theta(t)}{dt^2} + D\frac{d\theta(t)}{dt} + Mgl\sin\theta(t) = \Im(t) \qquad (1.73)$$

The above equation is nonlinear owing to the presence of the $\sin\theta(t)$ term in the expression of the gravity torque. To create a linear equation, we must assume that the system does not deflect much and the deflection angle stays below 30°. Under these conditions, (1.73) becomes

$$J\frac{d^2\theta(t)}{dt^2} + D\frac{d\theta(t)}{dt} + Mgl\theta = \Im(t) \qquad (1.74)$$

This is a second-order differential equation and, hence, its solution must contain two arbitrary constants, the values of which are determined from specified initial conditions. For the specific constants $J = 1$, $D = 2$, and $Mgl = 2$, the above equation becomes

$$\frac{d^2\theta(t)}{dt^2} + 2\frac{d\theta(t)}{dt} + 2\theta(t) = e^{-t}u(t) \tag{1.75}$$

We must first find the homogeneous solution from the homogeneous equation (the above equation equal to zero). If we assume a solution of the form $\theta_h(t) = Ce^{st}$, the solution requirements is

$$s^2 + 2s + 2 = 0$$

from which we find the roots $s_1 = -1 + j$ and $s_2 = -1 - j$. Therefore, the homogeneous solution is

$$\theta_h(t) = C_1 e^{s_1 t} + C_2 e^{s_2 t} \tag{1.76}$$

where C_i's are arbitrary unknown constants to be found by the initial conditions.

To find the particular solution, we assume a trial solution of the form $\theta_p(t) = Ae^{-t}$ for $t \geq 0$. Introducing the assumed solution in (1.75), we find

$$Ae^{-t} - 2Ae^{-t} + 2Ae^{-t} = e^{-t} \quad \text{or} \quad A = 1$$

The total solution is

$$\theta(t) = \theta_h(t) + \theta_p(t) = C_1 e^{s_1 t} + C_2 e^{s_2 t} + e^{-t} \qquad t \geq 0$$

Applying, next, the initial conditions in the above equation, we find the following system of equations.

$$\theta(0) = C_1 + C_2 + 1 = 0$$
$$\frac{d\theta(0)}{dt} = C_1 s_1 + C_2 s_2 - 1 = 0$$

Solving the unknown constants, we obtain $C_1 = (1 + s_2)/(s_1 - s_2)$, $C_2 = (1 + s_1)/(s_1 - s_2)$. Introducing, next, these constants into the total solution and the two roots, we find

$$\theta(t) = -\frac{1}{2}e^{-t}e^{jt} - \frac{1}{2}e^{-t}e^{-jt} + e^{-t} = (1 - \cos t)e^{-1} \qquad t \geq 0 \tag{1.77}$$

The digital simulation of (1.75) is deduced by employing (1.3), (1.52), and (1.53). Hence,

$$\frac{\theta(nT) - 2\theta(nT - T) + \theta(nT - 2T)}{T^2} + 2\frac{\theta(nT) - \theta(nT - T)}{T} + 2\theta(nT) = e^{-nT} \quad n = 0, 1, 2, \ldots$$

$$\tag{1.78}$$

After rearranging the above equation, we obtain

$$\theta(nT) = a(2 + 2T)\theta(nT - T) - a\theta(nT - 2T) + aT^2 e^{-nT}$$

$$a = \frac{1}{1 + 2T + 2T^2}, \quad n = 0, 1, 2, \dots \tag{1.79}$$

Using (1.53), we obtain that $\theta(-T) = 0$. Next, introducing this value and the initial condition $\theta(0T) = 0$ in (1.78), we find $\theta(-2T) = T^2$. The following m-file produces the desired output for the continuous case and for the two different sampling values, $T = 0.5$ and $T = 0.1$.

Book MATLAB® m-file for the Example 1.7: ex_1_5_1

```
%Book m-file for the Example 1.7: ex_1_5_1
t=0:0.1:5.5;
th=(1-cos(t)).*exp(-t);
T1=0.5;N1=5.5/T1;T2=0.1;N2=5.5/T2;
a1=1/(1+2*T1+2*T1^2);a2=1/(1+2*T2+2*T2^2);
thd1(2)=0;thd1(1)=T1^2;thd2(2)=0;thd2(1)=T2^2;
for n=0:N1
    thd1(n+3)=a1*(2+2*T1)*thd1(n+2)-a1*thd1(n+1)+T1^2*a1*exp(-n*T1);
end;
for n=0:N2
    thd2(n+3)=a2*(2+2*T2)*thd2(n+2)-a2*thd2(n+1)+T2^2*a2*exp(-n*T2);
end;
plot([0:55],th,'k');hold on;stem([0:5:N1*5],thd1(1,3:14),'k');
hold on;stem([0:N2],thd2(1,3:58),'k');                                   ∎
```

Example 1.8

It is desired to write the state equations for the system shown in Figure E.1.7 and express the output $y(n)$ in terms of the state variables. From the figure, we obtain

$$q_1(n) = 3q_2(n) + 2q_2(n - 1) + x(n)$$
$$q_2(n) = q_1(n - 1)$$
$$y(n) = 5q_1(n) + 4q_2(n) \qquad \blacksquare$$

Example 1.9

The convolution of the exponential function $f(t) = \exp(-t)u(t)$ and the pulse symmetric function $p_2(t)$ of width 4 is given by

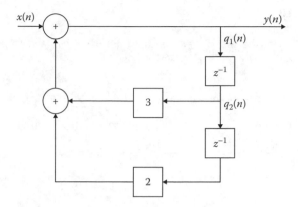

FIGURE E.1.7

$$g(t) = \int\limits_{-\infty}^{\infty} [u(x+2) - u(x-2)]e^{-(t-x)}u(t-x)dx$$

$$= e^{-t} \int\limits_{-\infty}^{\infty} u(x+2)e^{x}u(t-x)dx - e^{-t} \int\limits_{-\infty}^{\infty} u(x-2)e^{x}u(t-x)dx$$

$$= g_1(t) + g_2(t)$$

For $t < -2$, the exponential function and the step function $u(x+2)$ in the x-domain do not overlap and thus the integrand is zero in this range and the integral is also zero. Hence, $g_1(t) = 0$ for $t < -2$. For $t > -2$, there is an overlap from -2 to t for all ts from -2 to infinity. The integration gives

$$g_1(t) = e^{-t} \int\limits_{-2}^{t} e^{x}dx = 1 - e^{-2}e^{-t} \qquad -2 \le t < \infty$$

For the function $g_2(t)$, the exponential function overlaps the step function $-u(x-2)$ from 2 to infinity. Hence,

$$g_2(t) = -e^{-t} \int\limits_{2}^{t} e^{x}dx = -1 + e^{2}e^{-t} \qquad 2 \le t < \infty$$

Therefore, the function $g(t)$ is

$$g(t) = \begin{cases} 0 & t \le -2 \\ 1 - e^{-2}e^{-t} & -2 \le t \le 2 \\ (e^{2} - e^{-2})e^{-t} & 2 \le t < \infty \end{cases} \qquad \blacksquare$$

Example 1.10

The convolution of the two discrete functions $f(n) = 0.99^n u(n)$ and $h(n) = u(n-2)$ is given by

$$g(n) = \sum_{m=-\infty}^{\infty} 0.99^{n-m} u(n-m) u(m-2) = \sum_{m=2}^{n} 0.99^{n-m} = 0.99^n \sum_{m=2}^{n} 0.99^{-m}$$

$$= 0.99^n (0.99^{-2} + 0.99^{-3} + \cdots + 0.99^{-n}) = 0.99^n 0.99^{-2} (1 + 0.99^{-1} + 0.99^{-2} + \cdots + 0.99^{-n+2})$$

$$= 0.99^{n-2} \frac{1 - (0.99^{-1})^{n-1}}{1 - 0.99^{-1}} = 0.99^{n-2} \frac{1 - 0.99^{-n+1}}{1 - 0.99^{-1}} \qquad 2 \leq n < \infty \qquad \blacksquare$$

2

Fourier Series

2.1 Introduction

A **periodic** function is defined by the relation

$$f(t) = f(t + T) \quad T = \text{period}$$
$$f(t) = f(t + nT) \quad n = \pm 1, \pm 2, \ldots \tag{2.1}$$

The above relation is true for all t, and the smallest T which satisfies (2.1) is called the **period**.

Knowing the function within a period $f_p(t)$, the above equation can also be written in the form

$$f(t) = \sum_{n=-\infty}^{\infty} f_p(t - nT) \tag{2.2}$$

An important feature of a general periodic function is that it can be presented in terms of an infinite sum of sine and cosine functions. The functions that can be expressed by sine or cosine functions must at least obey the Dirichlet conditions, which are (a) only a finite number of maximums and minimums can be present, (b) the number of discontinuities must be finite, and (c) the discontinuities must be bounded, which implies that the function must be absolutely integrable with a value less than infinity.

2.2 Fourier Series in a Complex Exponential Form

Any periodic signal $f(t)$ that satisfies the Dirichlet conditions can be expressed as follows:

$$f(t) = \sum_{n=\infty}^{\infty} \alpha_n e^{jn\omega_0 t} = \sum_{n=\infty}^{\infty} |\alpha_n| e^{j(n\omega_0 t + \phi_n)} \quad -\infty < t < \infty$$

$$\alpha_n = \frac{1}{T} \int_a^{a+T} f(t) e^{-jn\omega_0 t} dt = \text{complex constant} = |\alpha_n| e^{j\phi_n}$$

$$= |\alpha_n| \cos \phi_n + j|\alpha_n| \sin \phi_n$$

$$\omega_0 = \frac{2\pi}{T}, \quad \phi_n = \tan^{-1}\left(\text{Im}\{\alpha_n\}/\text{Re}\{\alpha_n\}\right) \tag{2.3}$$

If the function is discontinuous at $t = a$, the function $f(t)$ will converge to $f(a) = [f(a+) + f(a-)]/2$, the mean value at the point of discontinuity (the arithmetic mean of the left-hand and right-hand limits). If $f(t)$ is real, then

$$\alpha_{-n} = \frac{1}{T} \int_a^{a+T} f(t) e^{jn\omega_0 t}\, dt = \left[\frac{1}{T} \int_a^{a+T} f(t) e^{-jn\omega_0 t}\, dt \right]^* = \alpha_n^* \tag{2.4}$$

This result, when combined with (2.3), yields

$$f(t) = \alpha_0 + \sum_{n=1}^{\infty} [(\alpha_n + \alpha_n^*) \cos n\omega_0 t + j(\alpha_n - \alpha_n^*) \sin n\omega_0 t] \tag{2.5}$$

2.3 Fourier Series in Trigonometric Form

The trigonometric form of the Fourier series is given by

$$f(t) = \frac{A_0}{2} + \sum_{n=1}^{\infty} (A_n \cos n\omega_0 t + B_n \sin n\omega_0 t) \quad -\infty < t < \infty$$

$$f(t) = \frac{A_0}{2} + \sum_{n=1}^{\infty} C_n \cos(n\omega_0 t + \phi_n)$$

$$A_0 = 2\alpha_0 = \frac{2}{T} \int_a^{a+T} f(t)\, dt$$

$$A_n = (\alpha_n + \alpha_n^*) = \frac{2}{T} \int_a^{a+T} f(t) \cos n\omega_0 t\, dt = C_n \cos \phi_n \tag{2.6}$$

$$B_n = j(\alpha_n - \alpha_n^*) = \frac{2}{T} \int_a^{a+T} f(t) \sin n\omega_0 t\, dt = -C_n \sin \phi_n$$

$$\phi_n = -\tan^{-1}(B_n/A_n)$$

$$C_n = (A_n^2 + B_n^2)^{1/2}$$

The coefficients C_n are known as the **amplitude spectrum** and the phase ϕ_n is the **phase spectrum**. Therefore, the frequency spectrum of a periodic function is discrete.

2.3.1 Differentiation of the Fourier Series

If $f(t)$ is continuous in $-T/2 \le t \le T/2$ with $f(-T/2) = f(T/2)$, and if its derivative $f'(t)$ is piecewise continuous and differentiable, then the trigonometric form of the Fourier series can be differentiated term by term to yield

$$f'(t) = \sum_{n=1}^{\infty} n\omega_0(-A_n \sin n\omega_0 t + B_n \cos n\omega_0 t) \tag{2.7}$$

2.3.2 Integration of the Fourier Series

If $f(t)$ is piecewise continuous in $-T/2 < t < T/2$, then the trigonometric form of the Fourier series can be integrated term by term to yield

$$\int_{t_1}^{t_2} f(t)dt = \frac{1}{2}A_0(t_2 - t_1) + \sum_{n=1}^{\infty} \frac{1}{n\omega_0}[-B_n(\cos n\omega_0 t_2 - \cos n\omega_0 t_1)$$

$$+ A_n(\sin n\omega_0 t_2 - \sin n\omega_0 t_1)] \tag{2.8}$$

2.4 Waveform Symmetries

Even function $[f(t) = f(-t)]$

If $f(t)$ is an even periodic function with a period T, then the trigonometric form of the Fourier series is

$$f(t) = \frac{A_0}{2} + \sum_{n=1}^{\infty} A_n \cos n\omega_0 t, \quad A_n = \frac{4}{T} \int_0^{T/2} f(t) \cos n\omega_0 t \, dt \tag{2.9}$$

Odd function $[f(t) = -f(-t)]$

If $f(t)$ is an odd function, then its trigonometric form is

$$f(t) = \sum_{n=1}^{\infty} B_n \sin n\omega_0 t, \quad B_n = \frac{4}{T} \int_0^{T/2} f(t) \sin n\omega_0 t \, dt \tag{2.10}$$

2.5 Some Additional Features of Periodic Continuous Functions

2.5.1 Power Content: Parseval's Theorem

The power content of a periodic function $f(t)$ in the period T is defined as the mean-square value:

$$\frac{1}{T} \int_{-T/2}^{T/2} [f(t)]^2 dt \tag{2.11}$$

If we assume the function as a voltage across an ohm resistor, then (2.11) represents the average power the source delivers to the resistor.

If $f(t)$ and $h(t)$ are two periodic functions with the same period T, then

$$\frac{1}{T} \int_{-T/2}^{T/2} f(t)h(t)dt = \sum_{n=-\infty}^{\infty} (\alpha_f)_n (\alpha_h)_{-n} \tag{2.12}$$

where

$$(\alpha_f)_n = \frac{1}{T} \int\limits_{-T/2}^{T/2} f(t)e^{-jn\omega_0 t}\,dt, \quad (\alpha_h)_n = \frac{1}{T} \int\limits_{-T/2}^{T/2} h(t)e^{-jn\omega_0 t}\,dt \tag{2.13}$$

If $f(t) = h(t)$, then the power content of the periodic function $f(t)$ is

$$\frac{1}{T} \int\limits_{-T/2}^{T/2} [f(t)]^2\,dt = \sum_{n=-\infty}^{\infty} |\alpha_n|^2 \quad \alpha_n = \frac{1}{T} \int\limits_{-T/2}^{T/2} f(t)e^{-jn\omega_0 t}\,dt \tag{2.14}$$

For a periodic function expanded in sine and cosine terms, the power content within a period is

$$\boxed{\frac{1}{T} \int\limits_{-T/2}^{T/2} [f(t)]^2\,dt = \frac{1}{4}A_0^2 + \frac{1}{2}\sum_{n=1}^{\infty} (A_n^2 + B_n^2)} \tag{2.15}$$

2.5.2 Output of an LTI System When the Input Is a Periodic Function

If the input periodic function is represented by the complex format of the Fourier series, then the output of an LTI system with a transfer function $H(\omega)$ is

$$f_o(t) = \sum_{n=-\infty}^{\infty} \alpha_n H(n\omega_0)e^{jn\omega_0 t} \tag{2.16}$$

If the input to an LTI system is a periodic signal in the form of sine and cosine series, then the output is

$$\boxed{\begin{aligned} f_o(t) &= \frac{A_0}{2}H(0) + \sum_{n=1}^{\infty} |H(n\omega_0)|[A_n \cos[n\omega_0 t + \phi(n\omega_0)] + B_n \sin[n\omega_0 t + \phi(n\omega_0)]] \\ \phi(n\omega_0) &= \tan^{-1}(\text{Im}\{H(n\omega_0)\}/\text{Re}\{H(n\omega_0)\}) \end{aligned}}$$

$$\tag{2.17}$$

2.5.3 Transmission without Distortion

If an LTI system has a transfer function of the form

$$H(\omega) = h_0 e^{jn\omega_0 t_0} \tag{2.18}$$

then the output signal will have a change in amplitude and will be shifted by t_0. Hence,

$$f_o(t) = h_0 f_{in}(t - t_0) \tag{2.19}$$

2.5.4 Band-Limited Periodic Signals

A band-limited periodic signal can be represented as follows:

$$f(t) = \sum_{n=-N}^{N} \alpha_n e^{jn\omega_0 t} = C_0 + \sum_{n=1}^{N} C_n \cos(n\omega_0 t + \phi_n) \tag{2.20}$$

The above signal contains only N harmonics in its expansion and, therefore, can be uniquely specified by its values at $2N+1$ instants of time in one period. This process produces a $2N+1$ system of equations with the $2N$ unknown C_n's and one C_0.

2.5.5 Sum and Difference of Functions

If the functions $f(t)$ and $h(t)$ are periodic with the same period, then we write

$$g(t) = C_1 f(t) \pm C_2 h(t) = \sum_{n=-\infty}^{\infty} [C_1 \beta_n \pm C_2 \gamma_n] e^{jn\omega_0 t}$$

$$= \sum_{n=-\infty}^{\infty} \alpha_n e^{jn\omega_0 t} \quad \alpha_n = C_1 \beta_n \pm C_2 \gamma_n \tag{2.21}$$

2.5.6 Product of Two Functions

If $f(t)$ and $h(t)$ are periodic functions with the same period, the product becomes

$$f(t)h(t) = \sum_{l=-\infty}^{\infty} \beta_l e^{jl\omega_0 t} \sum_{m=-\infty}^{\infty} \gamma_m e^{jm\omega_0 t} = \sum_{l=-\infty}^{\infty} \sum_{m=-\infty}^{\infty} \beta_l \gamma_m e^{j(l+m)\omega_0 t}$$

$$= \sum_{l=-\infty}^{\infty} \left(\sum_{m=-\infty}^{\infty} \beta_{n-m} \gamma_m \right) e^{jn\omega_0 t}, \quad l + m = n \tag{2.22}$$

Therefore, the Fourier coefficients of the complex exponential form of the product is

$$\frac{1}{T} \int_{-T/2}^{T/2} f(t)h(t) e^{-jn\omega_0 t} = \sum_{m=-\infty}^{\infty} \beta_{n-m} \gamma_m \tag{2.23}$$

The summation above indicates a convolution of two infinite sequences.

2.5.7 Convolution of Two Functions

A periodic convolution of two functions with the same period is defined by the relation

$$g(t) = \frac{1}{T} \int_{-T/2}^{T/2} f(x)h(t-x)dx \tag{2.24}$$

Therefore, we can expand the periodic function $g(t)$ in the Fourier series representation with coefficients:

$$\alpha_n = \frac{1}{T} \int_{-T/2}^{T/2} g(t)e^{-jn\omega_0 t}dt = \frac{1}{T^2} \int_{-T/2}^{T/2} \int_{-T/2}^{T/2} f(x)h(t-x)e^{-jn\omega_0 t}dt\,dx$$

$$= \frac{1}{T} \int_{-T/2}^{T/2} f(x)e^{-jn\omega_0 x}dx \frac{1}{T} \int_{-T/2}^{T/2} h(t-x)e^{-jn\omega_0(t-x)}dt \quad (\text{set } t-x = v)$$

$$= \frac{1}{T} \int_{-T/2}^{T/2} f(x)e^{-jn\omega_0 x}dx \frac{1}{T} \int_{-T/2-x}^{T/2-x} h(v)e^{-jn\omega_0 v}dv = \beta_n \gamma_n \tag{2.25}$$

The Fourier series expansion of $g(t)$ is

$$g(t) = \sum_{n=-\infty}^{\infty} \alpha_n e^{jn\omega_0 t} = \sum_{n=-\infty}^{\infty} \beta_n \gamma_n e^{jn\omega_0 t} \tag{2.26}$$

2.5.8 Gibbs' Phenomenon

A truncated form of a series expansion can be written in the form

$$f_N(t) = \sum_{n=-N}^{N} \beta_n e^{jn\omega_0 t} = \sum_{n=-\infty}^{\infty} \beta_n w_n e^{jn\omega_0 t} \quad w_n = \text{window} = \begin{cases} 1 & |n| \le N \\ 0 & |n| > N \end{cases} \tag{2.27}$$

Since w_n's can be considered as the coefficients' expansion of the function $h(t)$, we write

$$h(t) = \{e^{-jn\omega_0 t}(1 + e^{-j\omega_0 t} + (e^{-j\omega_0 t})^2 + \cdots + (e^{-j\omega_0 t})^{N-1})\} + \{1 + e^{j\omega_0 t} + \cdots + (e^{j\omega_0 t})^N\}$$

$$= \frac{e^{j(N+1/2)\omega_0 t} - e^{-j(N+1/2)\omega_0 t}}{e^{j\omega_0 t/2} - e^{-j\omega_0 t/2}} = \frac{\sin\left[\left(N + \dfrac{1}{2}\right)\omega_0 t\right]}{\sin\left(\omega_0 \dfrac{t}{2}\right)}$$

$$\text{Geometric series: } 1 + x + x^2 + \cdots + x^{N-1} = \frac{1 - x^N}{1 - x} \quad x < 1 \tag{2.28}$$

where the above relation is known as the **Fourier kernel**.

Comparing (2.27) with (2.26) we obtain

$$f_N(t) = \sum_{n=-N}^{N} \beta_n e^{jn\omega_0 t} = \frac{1}{T} \int_{-T/2}^{T/2} f(t-x) \frac{\sin\left[\left(N+\frac{1}{2}\right)\omega_0 x\right]}{\sin\left(\omega_0 \frac{x}{2}\right)} dx \qquad (2.29)$$

The convolution of (2.29) produces an approximation to $f_N(t)$, and at abrupt changes a ringing appears. This is known as the **Gibbs' phenomenon**.

In addition to using other types of windows, the following two smooth reconstructions of the finite expansion of the Fourier series are

Lanczos smooth expansion

$$f_N(t) = \frac{A_0}{2} + \sum_{n=1}^{N} \frac{\sin\left(\frac{n\pi}{N}\right)}{\frac{n\pi}{N}} [A_n \cos n\omega_0 t + B_n \sin n\omega_0 t] \qquad (2.30)$$

Fejer smooth expansion

$$f_N(t) = \frac{A_0}{2} + \sum_{n=1}^{N} \frac{N-n}{N} [A_n \cos n\omega_0 t + B_n \sin n\omega_0 t] \qquad (2.31)$$

2.5.9 Fourier Series of the Comb Function

Figure 1.8 shows the comb function. The figure indicates that the function is periodic with a period T. Therefore, its Fourier representation is (Table 2.1)

$$\text{comb}_T(t) = \sum_{n=-\infty}^{\infty} \delta(t-nT) \stackrel{\Delta}{=} \sum_{n=-\infty}^{\infty} \alpha_n e^{jn\omega_0 t} = \frac{1}{T} \sum_{n=-\infty}^{\infty} e^{jn\omega_0 t} = \frac{1}{T} + \frac{2}{T} \sum_{n=1}^{\infty} \cos n\omega_0 t$$

$$\alpha_n = \frac{1}{T} \int_{-T/2}^{T/2} \delta(t)dt = \frac{1}{T} \qquad (2.32)$$

Examples

Example 2.1

To expand the periodic function, shown in Figure E.2.1, in its complex and sinusoidal forms, we write

TABLE 2.1   Fourier Series Expansions of Some Periodic Functions
$(2L = T$, Period$)$

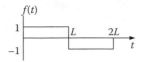

$$f(t) = \frac{4}{\pi} \sum_{n=1,3,5,\dots} \frac{1}{n} \sin \frac{n\pi t}{L}$$

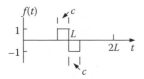

$$f(t) = \frac{2}{\pi} \sum_{n=1}^{\infty} \frac{(-1)^n}{n} \left(\cos \frac{n\pi c}{L} - 1 \right) \sin \frac{n\pi t}{L}$$

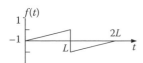

$$f(t) = \frac{c}{L} + \frac{2}{\pi} \sum_{n=1}^{\infty} \frac{(-1)^n}{n} \sin \frac{n\pi c}{L} \cos \frac{n\pi t}{L}$$

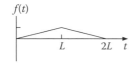

$$f(t) = \frac{2}{L} \sum_{n=1}^{\infty} \sin \frac{n\pi}{2} \frac{\sin(n\pi c / 2L)}{n\pi c / 2L} \sin \frac{n\pi t}{L}$$

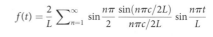

$$f(t) = \frac{2}{\pi} \sum_{n=1}^{\infty} \frac{(-1)^{n+1}}{n} \sin \frac{n\pi t}{L}$$

$$f(t) = \frac{1}{2} - \frac{4}{\pi^2} \sum_{n=1,3,5,\dots} \frac{1}{n^2} \cos \frac{n\pi t}{L}$$

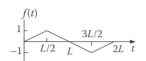

$$f(t) = \frac{8}{\pi^2} \sum_{n=1,3,5,\dots} \frac{(-1)^{(n-1)/2}}{n^2} \sin \frac{n\pi t}{L}$$

TABLE 2.1 (continued) Fourier Series Expansions of Some Periodic Functions
$(2L = T, \text{Period})$

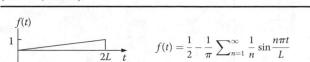

$$f(t) = \frac{1}{2} - \frac{1}{\pi} \sum_{n=1}^{\infty} \frac{1}{n} \sin \frac{n\pi t}{L}$$

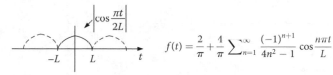

$$f(t) = \frac{1}{2}(1 + a) + \frac{2}{\pi^2 (1-a)} \sum_{n=1}^{\infty} \frac{1}{n^2} [(-1)^n \cos n\pi a - 1] \cos \frac{n\pi t}{L}$$

$$a = \frac{c}{2L}$$

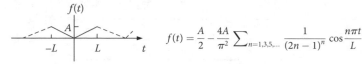

$$f(t) = \frac{1}{2} - \frac{4}{\pi^2 (1-2a)} \sum_{n=1,3,5,\ldots} \frac{1}{n^2} \cos n\pi a \cos \frac{n\pi t}{L} \qquad a = \frac{c}{2L}$$

$$f(t) = \frac{4A}{\pi} \sum_{n=1,3,5,\ldots} \frac{1}{n} \sin \frac{n\pi t}{L}$$

$$f(t) = \frac{2}{\pi} + \frac{4}{\pi} \sum_{n=1}^{\infty} \frac{(-1)^{n+1}}{4n^2 - 1} \cos \frac{n\pi t}{L}$$

$$f(t) = \frac{A}{2} - \frac{4A}{\pi^2} \sum_{n=1,3,5,\ldots} \frac{1}{(2n-1)^n} \cos \frac{n\pi t}{L}$$

$$f(t) = comb_T(t) = \sum_{n=-\infty}^{\infty} \delta(t - n2L) = \frac{1}{2L} \sum_{n=-\infty}^{\infty} e^{jn\frac{\pi}{L}t}$$

$$= \frac{1}{2L} + \frac{1}{L} \sum_{n=1}^{\infty} \cos n\frac{\pi}{L}t$$

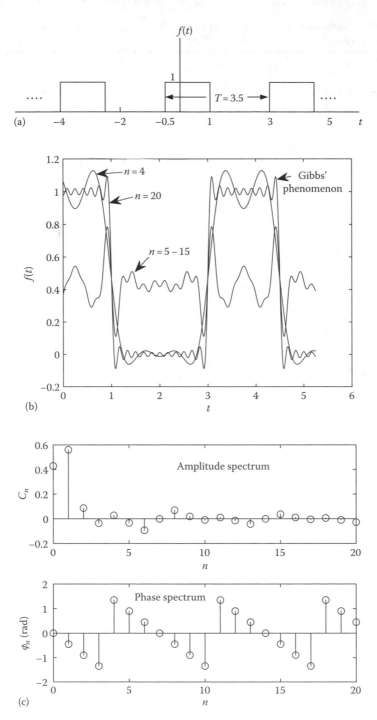

FIGURE E.2.1

$$\alpha_n = \frac{1}{3.5} \int\limits_{-0.5}^{3} f(t)e^{-jn\omega_0 t}dt = \frac{1}{3.5} \left(\int\limits_{-0.5}^{1} 1e^{-jn\omega_0 t}dt + \int\limits_{1}^{3} 0e^{-jn\omega_0 t}dt \right)$$

$$= \frac{1}{3.5(-jn\omega_0)}(e^{-jn\omega_0} - e^{j0.5n\omega_0}) \quad n \neq 0$$

$$\alpha_0 = \frac{1.5}{3.5} = \frac{3}{7}$$

$$\boxed{f(t) = \frac{3}{7} + \sum_{\substack{n=-\infty \\ n\neq 0}}^{\infty} \left[\frac{1}{3.5(-jn\omega_0)}(e^{-jn\omega_0} - e^{j0.5n\omega_0}) \right] e^{jn\omega_0 t}}$$

$$f(t) = \frac{3}{7} + \sum_{n=1}^{\infty} \left\{ \left[\frac{1}{3.5(-jn\omega_0)}(e^{-jn\omega_0} - e^{j0.5n\omega_0}) + \frac{1}{3.5(jn\omega_0)}(e^{jn\omega_0} - e^{-j0.5n\omega_0}) \right] \cos n\omega_0 t \right.$$

$$\left. + j\left[\frac{1}{3.5(-jn\omega_0)}(e^{-jn\omega_0} - e^{j0.5n\omega_0}) - \frac{1}{3.5(jn\omega_0)}(e^{jn\omega_0} - e^{-j0.5n\omega_0}) \right] \sin n\omega_0 t \right\}$$

$$\boxed{= \frac{3}{7} + \sum_{n=1}^{\infty} \left[\frac{2}{3.5n\omega_0}(\sin n\omega_0 + \sin 0.5n\omega_0)\cos n\omega_0 t - \frac{2}{3.5n\omega_0}(\cos n\omega_0 - \cos 0.5n\omega_0)\sin n\omega_0 t \right]}$$

$$\boxed{\begin{aligned} f(t) &= \frac{A_0}{2} + \sum_{n=1}^{\infty} C_n \cos(n\omega_0 t + \phi_n) \quad C_n = \left(A_n^2 + B_n^2\right)^{1/2} \quad \phi_n = -\tan^{-1}(B_n/A_n) \\ A_0 &= \frac{6}{7} A_n = \frac{4}{3.5n\omega_0}(\sin 0.75n\omega_0 \cos 0.25n\omega_0) \quad B_n = \frac{4}{3.5n\omega_0}(\sin 0.75n\omega_0 \sin 0.25n\omega_0) \end{aligned}}$$

$$(2.33)$$

The following Book MATLAB® m-file produces Figure E.2.1. The file produces the function f and the factors A_n, B_n, and $A_0/2$.

Book MATLAB m-File: ex_2_1_1

```
%Book MATLAB m-file: ex_2_1_1
%to compute any Fourier series we must
%supply a0,an,bn,n,T;
a0=3/7;T=3.5;
om=2*pi/T;
n=1:20;
t=0:0.005:1.5*T;
an=(4./(3.5*n*om)).*(sin(0.75*n*om).*cos(0.25*n*om));
bn=(4./(3.5*n*om)).*(sin(0.75*n*om).*sin(0.25*n*om));
cn=(an.^n+bn.^n).^(1/2);
f=a0+cos(t'*n*om)*an'+sin(t'*n*om)*bn';
     %f=desired function;to plot f we write plot(t,f);
     %cos(t'*n*om) is a t by n matrix and cos(t'*n*om)*an'
     %is a t by 1 vector;an and bn are raw vectors 1 by n;
```
■

Example 2.2

To find the Fourier series expansion of the function shown in Figure E.2.2a, we apply the differentiation approach of the function. The derivative of the function is shown in Figure E.2.2b. Let

$$f(t) = \frac{A_0}{2} + \sum_{n=1}^{\infty} A_n \cos n\omega_0 t + B_n \sin n\omega_0 t \quad \omega_0 = \frac{2\pi}{T} \tag{2.34}$$

and

$$f'(t) = \frac{a_0}{2} + \sum_{n=1}^{\infty} a_n \cos n\omega_0 t + b_n \sin n\omega_0 t \quad \omega_0 = \frac{2\pi}{T} \tag{2.35}$$

Differentiate (2.34) term by term and equate the results to (2.35) to obtain

$$A_n = -\frac{b_n}{n\omega_0}, \quad B_n = \frac{a_n}{n\omega_0} \tag{2.36}$$

Since the derivative of $f(t)$ is an odd function (see Figure E.2.2b),

$$a_n = 0, \quad n = 1, 2, \ldots$$

$$b_n = \frac{4}{T} \int_0^{T/2} f'(t) \sin n\omega_0 t \, dt = \frac{4}{T} \int_0^{T/2} \left(-A\delta\left(t - \frac{d}{2}\right) \right) \sin n\omega_0 t \, dt = -\frac{4A}{T} \sin n\omega_0 t |_{t=\frac{d}{2}}$$

$$= -\frac{4A}{T} \sin\left(\frac{n\omega_0 d}{2}\right) \tag{2.37}$$

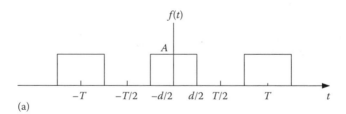

(a)

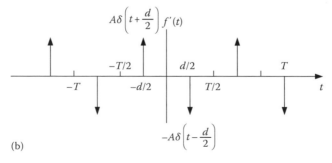

(b)

FIGURE E.2.2

Accordingly, from (2.36) we have

$$A_n = -\frac{b_n}{n\omega_0} = \frac{2Ad}{T}\frac{\sin(n\pi d/T)}{n\pi d/T} \quad B_n = 0 \qquad (2.38)$$

Since the constant term in the series expansion of $f(t)$ becomes zero during the differentiation, we must proceed to find it:

$$\frac{A_0}{2} = \frac{1}{T}\int_{-T/2}^{T/2} f(t)dt = \frac{Ad}{T}$$

$$\qquad (2.39)$$

$$f(t) = \frac{Ad}{T} + \frac{2Ad}{T}\sum_{n=1}^{\infty}\frac{\sin(n\pi d/T)}{n\pi d/T}\cos\left(n\frac{2\pi}{T}t\right)$$

Note that as we add more and more frequencies, we approximate better and better the function. In addition, we observe that the Gibbs' overshoot at the discontinuity does not decrease and remains the same, about 9%. When we filter the signal with a low-pass filter (retain only the low frequencies), the output is a smoother version of the input. If we retain only the high frequencies, we observe that the edges are primarily emphasized, and this is an indicator that the high frequencies are building up the sharp changes of the signal. ∎

Example 2.3

Let us assume that the periodic signal in Example 2.1 is the input current to the system shown in Figure E.2.3a. To find the output voltage, we must first find the transfer function of the system. Applying the Kirchhoff's node equation principle, we obtain

$$C\frac{dv(t)}{dt} + \frac{v(t)}{R} = i(t) \quad \text{or} \quad \frac{dv(t)}{dt} + 10v(t) = i(t) \qquad (2.40)$$

If the input is $i_0 e^{j\omega t}$, a complex sinusoidal signal input to a LTI system, the output is sinusoidal with the difference of amplitude and phase changes. Hence, we obtain for the output voltage $v(t) = v_0 e^{j\omega t}$, where v_0 is a complex number. Therefore, (2.40) becomes

$$v_0 j\omega e^{j\omega t} + 10v_0 e^{j\omega t} = i_0 e^{j\omega t} \quad \text{or} \quad H(j\omega) = \frac{v_0 e^{j\omega t}}{i_0 e^{j\omega t}} = \frac{1}{10 + j\omega} \qquad (2.41)$$

The following Book MATLAB m-file was used:

Book MATLAB m-File: ex_2_3_1

```
%Book MATLAB m-file: ex_2_3_1
%to compute any Fourier series we must
%supply a0,an,bn,n,T;
```

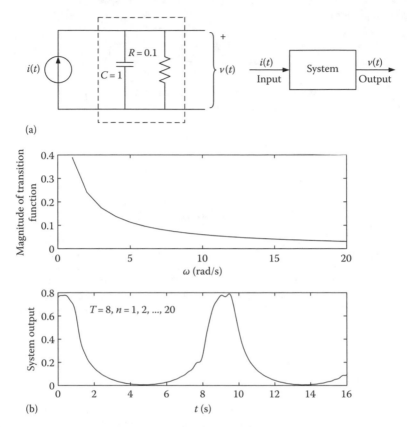

FIGURE E.2.3

```
a0 = 3/7; T = 8;
om = 2*pi/T;
n = 1:20;
t = 0:0.005:2*T;
an = (4./(3.5*n*om)).*(sin(0.75*n*om).*cos(0.25*n*om));
bn = (4./(3.5*n*om)).*(sin(0.75*n*om).*sin(0.25*n*om));
cn = (an.^n+bn.^n).^(1/2);
H = 1./(10+j*n*om);
phin = atan(imag(H)./real(H));
abH = abs(1./(1+2*n*om));
f = a0*(1/2)+cos(t'*(n*om+phin))*(abH.*an)'+sin(t'*(n*om+
phin))*(abH.*bn)';
        %f = desired function; to plot f we write plot(t,f);
        %cos(t'*n*om) is a t by n matrix and cos(t'*n*om)*an'
        %is a t by 1 vector; an and bn are raw vectors 1 by n;
```

Since the filter was a low pass one, the output is a smoother signal of its input, as was expected. ∎

Example 2.4

Consider the functions $x(t)$ and $h(t)$ that have the same period T. These are expressed in the form

$$x(t) = \sum_{n=-\infty}^{\infty} \alpha_n e^{jn\omega_0 t} \qquad h(t) = \sum_{m=-\infty}^{\infty} \beta_m e^{jm\omega_0 t} \qquad (2.42)$$

The mean value of the product of these two functions is

$$\frac{1}{T}\int_{-T/2}^{T/2} x(t)h^*(t)dt = \frac{1}{T}\sum_{n=-\infty}^{\infty}\sum_{m=-\infty}^{\infty}\alpha_n\beta_m^* \int_{-T/2}^{T/2} e^{j(n-m)\omega_0 t}dt = \begin{cases} 0 & n \neq m \\ \sum_{n=-\infty}^{\infty}\alpha_n\beta_m^* & n = m \end{cases}$$

$$(2.43)$$

If we set $x(t) = h(t)$, then the coefficients are the same and, hence,

$$\frac{1}{T}\int_{-T/2}^{T/2} |x(t)|^2 dt = \sum_{n=-\infty}^{\infty} |\alpha_n|^2 = \alpha_0^2 + \sum_{\substack{n=-\infty \\ n \neq 0}}^{\infty} \alpha_n\alpha_n^* = \frac{A_0^2}{4} + \sum_{n=1}^{\infty}(\alpha_n\alpha_n^* + \alpha_{-n}\alpha_{-n}^*)$$

$$= \frac{A_0^2}{4} + \sum_{n=1}^{\infty} 2\alpha_n\alpha_n^*$$

$$= \frac{A_0^2}{4} + \sum_{n=1}^{\infty} 2[(\text{Re}\{\alpha_n\})^2 + \text{Im}\{\alpha_n\}]^2$$

$$= \frac{A_0^2}{4} + \sum_{n=1}^{\infty} \left(\frac{A_n^2}{2} + \frac{B_n^2}{2}\right)$$

$$= \frac{A_0^2}{4} + \sum_{n=1}^{\infty} \frac{C_n^2}{2} \qquad (2.44)$$

■

Example 2.5

It is required to find the displacement $y(x, t)$ of a string of length L stretched between two points $(0, 0)$ and $(L, 0)$ if it is displaced initially into a position $y(x, 0) = f(x)$ and released from rest at this position without any external source acting.

The required displacement y is the solution of the following boundary value problem:

$$\frac{\partial^2 y}{\partial t^2} = a^2 \frac{\partial^2 y}{\partial x^2} \qquad (t > 0, 0 < x < L) \qquad \text{(a)}$$

$$y(0, t) = 0, \; y(L, t) = 0 \qquad (t \geq 0) \qquad \text{(b)}$$

$$y(x, 0) = f(x) \qquad (0 \leq x \leq L) \qquad \text{(c)} \qquad (2.45)$$

$$\frac{\partial y(x, 0)}{\partial t} = 0 \qquad (0 \leq x \leq L) \qquad \text{(d)}$$

Note that Equation 2.45a is the general form of the wave equation and is applicable to electromagnetic waves propagating in transmission lines, to electromagnetic waves propagating in space or enclosed spaces, to water waves, etc. Therefore, we must find the particular solutions of the partial differential equation (2.45a) which satisfy the homogeneous boundary conditions (2.45b) and (2.45d), and then determine a linear combination of those solutions which satisfy the nonhomogeneous boundary condition (2.45c).

We assume particular solutions of (2.45a) above of the form

$$y(x, t) = X(x)T(t) \tag{2.46}$$

Substituting (2.46) in (2.45a) above, we obtain

$$\frac{X''(x)}{X(x)} = \frac{T''(t)}{a^2 T(t)}$$

where the prime indicates differentiation. Since one side of the equation is a function of x only and the other is a function of t only, they both must be equal to a constant g. Hence, we obtain

$$X''(x) - gX(x) = 0 \tag{2.47}$$

$$T''(t) - ga^2 T(t) = 0 \tag{2.48}$$

If our particular solution is to satisfy condition (2.45b), $X(x)T(t)$ must vanish at $x = 0$ and $x = L$ for all values of t. Therefore, we must have

$$X(0) = 0 \quad X(L) = 0 \tag{2.49}$$

Similarly, if it is to satisfy condition (2.45d) then we have

$$T'(0) = 0 \tag{2.50}$$

Equations 2.47 and 2.48 are ordinary differential equations with constant coefficients. The characteristic equation of (2.47) is $D^2 - g = 0$ with the roots $D = \pm\sqrt{g}$. The general solution of (2.47) is

$$X(x) = C_1 e^{x\sqrt{g}} + C_2 e^{-x\sqrt{g}} \tag{2.51}$$

where the C_i's are constants. If g is positive, no value of C_1 and C_2 satisfies the boundary conditions (2.49).

If g is negative, we can write

$$g = -q^2$$

Introducing the above equation in (2.51), we obtain the general solution of (2.47):

$$X(x) = C_1 e^{jxq} + C_2 e^{-jxq} = (C_1 + C_2)\cos qx + j(C_1 - C_2)\sin qx = A\cos qx + B\sin qx \tag{2.52}$$

where A and B are constants. Furthermore, since $X(0) = 0$, the constant B must be zero. We must have A different than zero because we do not want to have the trivial solution $X(x) = 0$. Therefore, if $X(L) = 0$, then $\sin qL = 0$. The sine function is satisfied if $qL = n\pi$, or

$$q = \frac{n\pi}{L} \quad n = 1, 2, 3, \ldots \tag{2.53}$$

Hence, the solution of (2.47) is

$$X(x) = A\sin\frac{n\pi}{L} \quad n = 1, 2, 3, \ldots \tag{2.54}$$

The negative values of n do not provide new solutions.

Substituting $-n^2\pi^2/L^2$ for g in differential equation (2.48) and applying condition (2.50) we obtain

$$T(t) = C\cos\frac{n\pi a t}{L} \tag{2.55}$$

where C is a constant.

Therefore, all the functions

$$A_n \sin\frac{n\pi x}{L}\cos\frac{n\pi a t}{L} \tag{2.56}$$

are solutions of the partial differential equation (2.45a) and satisfy the linear homogeneous conditions (2.45b) and (2.45d) of the same equation. The constants $A_1, A_2, \ldots$ are arbitrary. Any finite linear combination of these solutions will also satisfy the same conditions. However, when $t = 0$, it will reduce to a finite linear combination of the functions $\sin(n\pi x/L)$. Thus, condition (2.45c) will not be satisfied unless the given function $f(x)$ has this particular character. Let us consider an infinite series of functions (2.56),

$$y(x, t) = \sum_{n=1}^{\infty} A_n \sin\frac{n\pi x}{L}\cos\frac{n\pi a t}{L} \tag{2.57}$$

This solution satisfies (2.45a), and conditions (2.45b) and (2.45d). It will satisfy the nonhomogeneous condition (2.45c) provided that the numbers A_n can be so determined that it satisfies the relation

$$f(x) = \sum_{n=1}^{\infty} A_n \sin\frac{n\pi x}{L} \tag{2.58}$$

This is a Fourier series expansion of $f(x)$ and, thus, the coefficients are determined by the relation

$$A_n = \frac{2}{L} \int_0^L f(x) \sin \frac{n\pi x}{L} dx \qquad (2.59)$$

Therefore, the solution is

$$y(x,t) = \sum_{n=1}^{\infty} \left(\frac{2}{L} \int_0^L f(x) \sin \frac{n\pi x}{L} dx \right) \sin \frac{n\pi x}{L} \cos \frac{n\pi at}{L} \qquad (2.60)$$

■

Example 2.6

Let the string be stressed between the points $(0, 0)$ and $(2, 0)$, and suppose we raise the middle point at height h above the x-axis. Then, the string is released from the rest, and hence, it is desired to find its subsequent position.

The function $f(x)$ is

$$f(x) = \begin{cases} hx & 0 \le x \le 1 \\ -hx + 2x & 1 \le x \le 2 \end{cases} \qquad (2.61)$$

The coefficients in solution (2.58) are given by

$$A_n = \int_0^2 f(x) \sin \frac{n\pi x}{2} dx = h \int_0^1 x \sin \frac{n\pi x}{2} dx + h \int_1^2 (-x+2) \sin \frac{n\pi x}{2} dx = \frac{8h}{\pi^2 n^2} \sin \frac{n\pi}{2} \qquad (2.62)$$

The solution is

$$y(x,t) = \sum_{n=1}^{\infty} \frac{8h}{\pi^2 n^2} \sin \frac{n\pi}{2} \sin \frac{n\pi x}{2} \sin \frac{n\pi at}{2} \qquad (2.63)$$

■

Example 2.7

The heat flow, as well as the electrical potential, obeys Laplace's equation:

$$\frac{\partial^2 u(x,y)}{\partial x^2} + \frac{\partial^2 u(x,y)}{\partial y^2} = 0 \qquad (2.64)$$

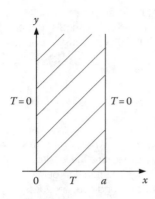

$T = 0$ $T = 0$

0 T a

FIGURE E.2.4

Figure E.2.4 shows a semi-infinite plate whose edges $x=0$ and $x=a$ are kept in zero temperature, and whose base $y=0$ is kept at a constant temperature T. It is desired to find the solution $u(x, y)$ that satisfies (2.64) and the boundary conditions:

$$u(0,y) = 0 \quad u(a,y) = 0 \quad u(x,0) = T \qquad (2.65)$$

From physical considerations we must have

$$\lim_{y \to \infty} u(x,y) = 0$$

Using the separation of variables approach, we assume a solution of the form

$$u(x,y) = X(x)Y(y) \qquad (2.66)$$

Introducing (2.66) into (2.64) and, dividing by $X(x)Y(y)$, we obtain

$$-\frac{X''(x)}{X(x)} = \frac{Y''(y)}{Y(y)} \qquad (2.67)$$

where primes indicate differentiation with respect to the independent variable. Because the first ratio is independent of y and the second is independent of x, their common ratio must be equal to a constant b^2. Therefore, we obtain the two ordinary differential equations:

$$\begin{aligned} X''(x) + b^2 X(x) &= 0 \\ Y''(y) - b^2 Y(y) &= 0 \end{aligned} \qquad (2.68)$$

whose solutions are respectively

$$\begin{aligned} X(x) &= A \cos bx + B \sin bx \\ Y(y) &= Ce^{by} + De^{-by} \end{aligned} \qquad (2.69)$$

Thus, for any value of b and any constants A, B, C, D,

$$(A \cos bx + B \sin bx)(Ce^{by} + De^{-by})$$

is a solution of (2.64). If the first boundary condition $u(0, y) = 0$ of (2.65) is to be satisfied,

$$u(0,y) = X(0)Y(y) = 0$$

which implies $X(0) = 0$ since $Y(y)$ is not identically zero. But $X(0) = A$, and hence, $A = 0$. To satisfy $u(a, y) = X(a)Y(y) = 0$, we must have $X(a) = 0$. But $X(a) = B \sin ba$, and since, $X(x)$ is not identically zero, $B \neq 0$. Hence, to make $X(a) = 0$ we must have $b = n\pi/a, n = 1, 2, 3, \ldots$. Therefore, the function

$$\sum_{n=1}^{\infty} \sin \frac{n\pi x}{a} \left(A_n e^{\frac{n\pi y}{a}} + B_n e^{-\frac{n\pi y}{a}} \right)$$

satisfies (2.64) and the first boundary conditions of (2.65) no matter what A_n and B_n may be. The physical condition $u(x, \infty) = 0$ implies that $A_n = 0$ because the positive exponential becomes infinity as y approaches the value of infinity. The function below satisfies the partial differential equation and all the boundary conditions besides $u(x, 0) = T$:

$$u(x, y) = \sum_{n=1}^{\infty} B_n \sin \frac{n\pi x}{a} e^{-\frac{n\pi y}{a}} \qquad (2.70)$$

Since the B_n are arbitrary, we choose them so that

$$T = u(x, 0) = \sum_{n=1}^{\infty} B_n \sin \frac{n\pi x}{a} e^{-\frac{n\pi y}{a}}$$

The above equation is a Fourier series expansion of the function $u(x, 0)$, and hence, its coefficients are given by

$$B_n = \frac{2}{a} \int_0^a T \sin \frac{n\pi x}{a} dx$$

$$B_n = \begin{cases} 0 & n = \text{even} \\ \dfrac{4T}{n\pi} & n = \text{odd} \end{cases}$$

The solution of the boundary value problem is

$$u(x, y) = \frac{4T}{\pi} \sum_{n=1}^{\infty} \frac{1}{n} \sin \frac{n\pi x}{a} e^{-\frac{n\pi y}{a}} \qquad (2.71)$$

■

Example 2.8

It is desired to find the temperature distribution, in space and in time, in a homogeneous slab bounded by two planes at $x = 0$ and $x = L$, which was kept at zero temperature. The initial temperature in the slab was $u(x, 0) = f(x)$. This problem is characterized by the diffusion-type partial differential (PDF) equation. The same equation characterizes the voltage in an RC transmission line. The one-dimensional equation characterizing the heat flow is

$$\frac{\partial u(x,t)}{\partial t} = k\frac{\partial^2 u(x,t)}{\partial t^2} \quad 0 < x < L \quad t > 0 \tag{2.72}$$

In addition, for this problem, the following boundary conditions must be satisfied:

$$u(0,t) = 0 \quad u(L,t) = 0 \quad t > 0 \tag{2.73}$$

$$u(x,0) = f(x) \quad 0 < x < L \tag{2.74}$$

To find the particular solutions of (2.72) that satisfy (2.73), we assume a solution of the form $u(x,t) = X(x)T(t)$. Substituting the assumed solution into differential equation and separating the factors with dependence only on x and the factors dependent only on t and setting the ratio equal to a constant a, we obtain the following ordinary differential equations:

$$X''(x) - aX(x) = 0 \quad T''(t) - akT(t) = 0 \tag{2.75}$$

If the function $X(x)T(t)$ is to satisfy conditions (2.73), then

$$X(0,0) = 0 \quad X(L,0) = 0 \tag{2.76}$$

The solution of the first differential equation is found from its characteristic equation $D^2 - a = 0$ to be $X(x,0) = Ae^{\sqrt{a}x} + Be^{-\sqrt{a}x}$. Applying the boundary condition $X(0,0) = 0$ we obtain $A = -B$, and hence, the solution is

$$X(x,0) = 2A\frac{e^{\sqrt{a}x} - e^{-\sqrt{a}x}}{2} = C\frac{e^{\sqrt{a}x} - e^{-\sqrt{a}x}}{2} = C \sinh x\sqrt{a}$$

This solution can satisfy the second condition of (2.76) if

$$a = -n^2 \quad n = 1, 2, 3, \ldots$$

Then $X(x,0) = C \sin x\sqrt{-n^2} = jC \sin nx = D \sin nx$. The solution of the second equation of (2.75) is $T(t) = Ee^{-n^2 kt}$. Therefore, the solutions of the PDF with the boundary conditions (2.73) are in the form $u(x,t) = X(x)T(t)$:

$$B_n e^{-n^2 kt} \sin nx \quad n = 1, 2, 3, \ldots \tag{2.77}$$

where the constants B_n are arbitrary. Since an infinite number of solutions will converge to $f(x)$ and since any solution satisfies the PDF and its boundary conditions, then the general solution is

$$u(x,t) = \sum_{n=1}^{\infty} B_n e^{-n^2 kt} \sin nx \tag{2.78}$$

For $t = 0$, the general solution becomes

$$u(x, 0) = \sum_{n=1}^{\infty} B_n \sin nx$$

The above equation is a sine Fourier series expansion of $u(x, 0)$ and, therefore, the coefficients are

$$B_n = \frac{2}{L} \int_0^L f(x) \sin nx \, dx \quad n = 1, 2, 3, \ldots \tag{2.79}$$

Therefore, the final form of the solution is

$$u(x, t) = \frac{2}{L} \sum_{n=1}^{\infty} e^{-n^2 kt} \sin nx \int_0^L f(x') \sin nx' \, dx' \quad n = 1, 2, 3, \ldots \tag{2.80}$$

■

3

Fourier Transforms

3.1 Introduction—Fourier Transform

Not all functions are Fourier transformable. However, the Dirichlet conditions provide sufficiency conditions for a function $f(t)$ to be Fourier transformable. These are

1. $\int_{-\infty}^{\infty} |f(t)| dt < \infty$
2. $f(t)$ has finite maxima and minima within any finite interval
3. $f(t)$ has a finite number of discontinuities within any finite interval

If these conditions are met, the function can be transformed uniquely. Although the delta function does not obey the above conditions, it belongs, however, to a specific set of functions known as **generalized functions** and it is transformable. The Fourier transform (FT) pair is defined as follows:

$$\mathcal{F}\{f(t)\} \stackrel{\Delta}{=} F(\omega) = \int_{-\infty}^{\infty} f(t)e^{-j\omega t} dt \qquad \text{direct Fourier transform (FT)}$$

$$\mathcal{F}^{-1}\{F(\omega)\} \stackrel{\Delta}{=} f(t) = \frac{1}{2\pi} \int_{-\infty}^{\infty} F(\omega)e^{j\omega t} d\omega \quad \text{inverse Fourier transform (IFT)}$$

(3.1)

3.2 Other Forms of Fourier Transform

3.2.1 $f(t)$ Is a Complex Function

If $f(t) = f_r(t) + jf_i(t)$, where the two functions are real, then we have

$$F(\omega) = \int_{-\infty}^{\infty} [f_r(t) \cos \omega t + f_i(t) \sin \omega t] dt - j \int_{-\infty}^{\infty} [f_r(t) \sin \omega t - f_i(t) \cos \omega t] dt \qquad (3.2)$$

Therefore,

$$F(\omega) = R(\omega) + jX(\omega)$$

$$R(\omega) = \int_{-\infty}^{\infty} [f_r(t)\cos\omega t + f_i(t)\sin\omega t]dt$$

$$X(\omega) = -\int_{-\infty}^{\infty} [f_r(t)\sin\omega t - f_i(t)\cos\omega t]dt$$

(3.3)

From (3.1) and Euler's expansion $e^{\pm j\omega t} = \cos\omega t \pm j\sin\omega t$, we obtain the inverse formulas:

$$f_r(t) = \frac{1}{2\pi}\int_{-\infty}^{\infty} [R(\omega)\cos\omega t - X(\omega)\sin\omega t]d\omega \qquad (3.4)$$

$$f_i(t) = \frac{1}{2\pi}\int_{-\infty}^{\infty} [R(\omega)\sin\omega t + X(\omega)\cos\omega t]d\omega \qquad (3.5)$$

3.2.2 Real Time Functions

If $f(t)$ is real, then

$$R(\omega) = \int_{-\infty}^{\infty} f(t)\cos\omega t\ dt \quad X(\omega) = -\int_{-\infty}^{\infty} f(t)\sin\omega t\ dt \qquad (3.6)$$

From (3.6), we obtain the relations

$$R(-\omega) = R(\omega) = \text{even} \quad X(-\omega) = -X(\omega) = \text{odd}, \quad F(-\omega) = F^*(\omega) \qquad (3.7)$$

Therefore, the inverse transform is

$$f_r(t) \overset{\Delta}{=} f(t) = \frac{1}{2\pi}\int_{-\infty}^{\infty} [R(\omega)\cos\omega t - X(\omega)\sin\omega t]d\omega \qquad (3.8)$$

3.2.3 Imaginary Time Functions

If $f(t) = jf_i(t)$, then

$$R(\omega) = \int_{-\infty}^{\infty} f_i(t)\sin\omega t\ dt \quad X(\omega) = \int_{-\infty}^{\infty} f_i(t)\cos\omega t\ dt \qquad (3.9)$$

From the above equation, we have

$$R(-\omega) = -R(\omega) \quad X(-\omega) = X(\omega) \quad F(-\omega) = -F^*(-\omega) \tag{3.10}$$

3.2.4 $f(t)$ Is Even

If $f(-t) = f(t)$, then $f(t) \cos \omega t$ is even and $f(t) \sin \omega t$ is odd with respect to t. Hence (see (3.6)),

$$R(\omega) = 2 \int_0^\infty f(t) \cos \omega t \ dt \quad X(\omega) = 0 \tag{3.11}$$

From (3.11) and (3.8), we obtain

$$f(t) = \frac{1}{\pi} \int_0^\infty R(\omega) \cos \omega t \ d\omega \tag{3.12}$$

3.2.5 $f(t)$ Odd

If $f(-t) = -f(-t)$, then

$$R(\omega) = 0 \quad X(\omega) = -2 \int_0^\infty f(t) \sin \omega t \ dt \tag{3.13}$$

The inversion of (3.8) becomes

$$f(t) = -\frac{1}{\pi} \int_0^\infty X(\omega) \sin \omega t \ d\omega \tag{3.14}$$

3.2.6 Odd and Even Representations

Any function can be decomposed into an odd and an even function as follows:

$$f_e(t) = \frac{f(t) + f(-t)}{2} \quad f_o(t) = \frac{f(t) - f(-t)}{2} \tag{3.15}$$

Hence,

$$f(t) = f_e(t) + f_o(t) \tag{3.16}$$

TABLE 3.1 Properties of $f(t)$ and $F(\omega)$

Property of $f(t)$	Characteristics of $F(\omega)$
$f(t) = f_e(t)$	Real transform and even
$f(t) = f_o(t)$	Imaginary transform and odd
$f(t) = $ real	Even real part and odd imaginary part
$f(t) = $ Complex and no symmetry	Complex transform and no symmetry
$f(t) = $ Complex even or odd	Complex and even transform or complex and odd

The following relations also hold (Table 3.1)

$$F(\omega) = R(\omega) + jX(\omega) = \int_{-\infty}^{\infty} f_e(t)\cos\omega t\, dt - j\int_{-\infty}^{\infty} f_o(t)\cos\omega t\, dt \overset{\Delta}{=} F_e(\omega) + F_o(\omega)$$

$$R(\omega) = 2\int_{0}^{\infty} f_e(t)\cos\omega t\, dt \quad X(\omega) = -2\int_{0}^{\infty} f_o(t)\sin\omega t\, dt \tag{3.17}$$

$$f_e(t) = \frac{1}{\pi}\int_{0}^{\infty} R(\omega)\cos\omega t\, d\omega \quad f_o(t) = -\frac{1}{\pi}\int_{0}^{\infty} X(\omega)\sin\omega t\, d\omega$$

3.2.7 Causal-Time Functions

A function is **causal** if it is zero for negative t:

$$f(t) = 0 \quad t < 0 \tag{3.18}$$

For $t > 0$, we have $f(-t) = 0$ and, therefore, from (3.15) we obtain

$$f(t) = 2f_e(t) = 2f_o(t) \quad t > 0 \tag{3.19}$$

From (3.19) and the bottom expressions of (3.17), we obtain

$$f(t) = \frac{2}{\pi}\int_{0}^{\infty} R(\omega)\cos\omega t\, d\omega = -\frac{2}{\pi}\int_{0}^{\infty} X(\omega)\sin\omega t\, d\omega \quad t > 0 \tag{3.20}$$

From (3.19), we obtain

$$F(\omega) = 2\int_{-\infty}^{\infty} f_e(t)\cos\omega t\, dt = 2F_e(\omega) \quad F(\omega) = -2j\int_{-\infty}^{\infty} f_o(t)\sin\omega t\, dt = 2F_o(\omega) \tag{3.21}$$

From the inverse transform and the relationships of even and odd real functions, we obtain

$$f(t) = \frac{1}{2\pi} \int\limits_{-\infty}^{\infty} 2F_e(\omega)e^{j\omega t}\,d\omega = \frac{1}{\pi} \int\limits_{-\infty}^{\infty} F_e(\omega)\cos\omega t\,d\omega = \frac{1}{\pi} \int\limits_{-\infty}^{\infty} R(\omega)\cos\omega t\,d\omega \quad t > 0$$

$$f(t) = \frac{1}{2\pi} \int\limits_{-\infty}^{\infty} 2F_o(\omega)e^{j\omega t}\,d\omega = \frac{j}{\pi} \int\limits_{-\infty}^{\infty} F_o(\omega)\sin\omega t\,d\omega = -\frac{1}{\pi} \int\limits_{-\infty}^{\infty} X(\omega)\sin\omega t\,d\omega \quad t > 0$$

$$(3.22)$$

Furthermore, the two functions $F(\omega)$ and $X(\omega)$ of a causal function are related to each other by the equations (Table 3.2)

$$R(\omega) = \frac{1}{\pi}\mathscr{P} \int\limits_{-\infty}^{\infty} \frac{X(\tau)}{\omega - \tau}\,d\tau$$

$$(3.23)$$

$$X(\omega) = -\frac{1}{\pi}\mathscr{P} \int\limits_{-\infty}^{\infty} \frac{R(\tau)}{\omega - \tau}\,d\tau$$

These are **Hilbert transform** relationships and the script $\mathscr{P}$, in front of the integrals, denotes the principal value of the integral (see Chapter 9).

3.3 Fourier Transform Examples

Example 3.1

The FT of the pulse function is given by

$$F(\omega) = \int\limits_{-\infty}^{\infty} p_a(t)e^{-j\omega t}\,dt = \int\limits_{-a}^{a} e^{-j\omega t}\,dt = \frac{e^{-j\omega a} - e^{j\omega a}}{-j\omega} = 2\frac{\sin\omega a}{\omega} \tag{3.24}$$

The signal and the amplitude spectrum are shown in Figure E.3.1. The amplitude is given by $|F(\omega)| = \left|\dfrac{2\sin\omega a}{\omega}\right|$. ∎

Example 3.2

The FT of $f(t) = e^{-at}\,u(t)\ a > 0$ is

$$F(\omega) = \int\limits_{-\infty}^{\infty} e^{-at}u(t)e^{-j\omega t}\,dt = \int\limits_{0}^{\infty} e^{-at}e^{-j\omega t}\,dt = \frac{1}{a+j\omega} = \frac{1}{\sqrt{a^2+\omega^2}\,e^{j\tan^{-1}(\omega/a)}} = \frac{1}{\sqrt{a^2+\omega^2}}e^{-j\tan^{-1}(\omega/a)}$$

$$(3.25)$$

TABLE 3.2 Fourier Transform Pairs

$F(\omega) = \mathscr{F}\{f(t)\},\, f(t) = \mathscr{F}^{-1}\{F(\omega)\}$

General time function

$\mathscr{F}\{f(t)\} \triangleq F(\omega) = \int_{-\infty}^{\infty} f(t)e^{-j\omega t}\,dt \quad \mathscr{F}^{-1}\{F(\omega)\} \triangleq f(t) = \frac{1}{2\pi}\int_{-\infty}^{\infty} F(\omega)e^{j\omega t}\,d\omega$

Complex function: $f(t) = f_r(t) + jf_i(t)$

$F(\omega) = \int_{-\infty}^{\infty} [f_r(t)\cos\omega t + f_i(t)\sin\omega t]\,dt - j\int_{-\infty}^{\infty} [f_r(t)\sin\omega t - f_i(t)\cos\omega t]\,dt$

$F(\omega) = R(\omega) + jX(\omega)$

$R(\omega) = \int_{-\infty}^{\infty} [f_r(t)\cos\omega t + f_i(t)\sin\omega t]\,dt$

$X(\omega) = -\int_{-\infty}^{\infty} [f_r(t)\sin\omega t - f_i(t)\cos\omega t]\,dt$

$f_r(t) = \frac{1}{2\pi}\int_{-\infty}^{\infty} [R(\omega)\cos\omega t - X(\omega)\sin\omega t]\,d\omega$

$f_i(t) = \frac{1}{2\pi}\int_{-\infty}^{\infty} [R(\omega)\sin\omega t + X(\omega)\cos\omega t]\,d\omega$

Real time function: $f(t) = f_r(t)$

$R(\omega) = \int_{-\infty}^{\infty} f(t)\cos\omega t\,dt$

$X(\omega) = -\int_{-\infty}^{\infty} f(t)\sin\omega t\,dt$

$f_r(t) \triangleq f(t) = \frac{1}{2\pi}\int_{-\infty}^{\infty} [R(\omega)\cos\omega t - X(\omega)\sin\omega t]\,d\omega$

Imaginary time function: $f(t) = jf_i(t)$

$R(\omega) = \int_{-\infty}^{\infty} f_i(t)\sin\omega t\,dt \quad X(\omega) = \int_{-\infty}^{\infty} f_i(t)\cos\omega t\,dt$

Real and even function

$R(\omega) = 2\int_0^{\infty} f(t)\cos\omega t\,dt \quad X(\omega) = 0$

$f(t) = \frac{1}{\pi}\int_0^{\infty} R(\omega)\cos\omega t\,d\omega$

Real and odd function

$R(\omega) = 0 \quad X(\omega) = -2\int_0^{\infty} f(t)\sin\omega t\,dt$

$f(t) = -\frac{1}{\pi}\int_0^{\infty} X(\omega)\sin\omega t\,d\omega$

Odd and even representation: $f(t) = f_e(t) + f_o(t) = \frac{f(t)+f(-t)}{2} + \frac{f(t)-f(-t)}{2}$

$F(\omega) = R(\omega) + jX(\omega) = \int_{-\infty}^{\infty} f_e(t)\cos\omega t\,dt - j\int_{-\infty}^{\infty} f_o(t)\cos\omega t\,dt \triangleq F_e(\omega) + F_o(\omega)$

$R(\omega) = 2\int_0^{\infty} f_e(t)\cos\omega t\,dt \quad X(\omega) = -2\int_0^{\infty} f_o(t)\sin\omega t\,dt$

$f_e(t) = \frac{1}{\pi}\int_0^{\infty} R(\omega)\cos\omega t\,d\omega \quad f_o(t) = -\frac{1}{\pi}\int_0^{\infty} X(\omega)\sin\omega t\,d\omega$

Causal-time functions: $f(t) = 0 \quad t < 0$

$F(\omega) = 2\int_{-\infty}^{\infty} f_e(t)\cos\omega t\,dt = 2F_e(\omega) \quad F(\omega) = -2j\int_{-\infty}^{\infty} f_o(t)\sin\omega t\,dt = 2F_o(\omega)$

$f(t) = \frac{1}{2\pi}\int_{-\infty}^{\infty} 2F_e(\omega)e^{j\omega t}\,d\omega = \frac{1}{\pi}\int_{-\infty}^{\infty} F_e(\omega)\cos\omega t\,d\omega = \frac{1}{\pi}\int_{-\infty}^{\infty} R(\omega)\cos\omega t\,d\omega \quad t > 0$

$f(t) = \frac{1}{2\pi}\int_{-\infty}^{\infty} 2F_o(\omega)e^{j\omega t}\,d\omega = \frac{j}{\pi}\int_{-\infty}^{\infty} F_o(\omega)\sin\omega t\,d\omega = -\frac{1}{\pi}\int_{-\infty}^{\infty} X(\omega)\sin\omega t\,d\omega \quad t > 0$

Hilbert transforms of causal-time functions

$R(\omega) = \frac{1}{\pi}\mathscr{P}\int_{-\infty}^{\infty} \frac{X(\tau)}{\omega-\tau}\,d\tau$

$X(\omega) = -\frac{1}{\pi}\mathscr{P}\int_{-\infty}^{\infty} \frac{R(\tau)}{\omega-\tau}\,d\tau$

Script $\mathscr{P}$ stands for the principal value of the integrals.

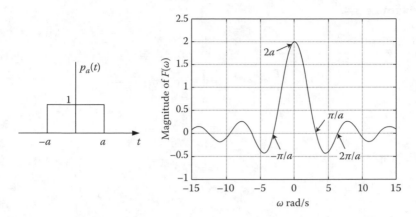

FIGURE E.3.1

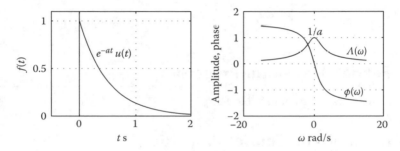

FIGURE E.3.2

The amplitude and the phase spectrums are given by

$$A(\omega) = \frac{1}{\sqrt{a^2 + \omega^2}} \quad \phi(\omega) = -\tan^{-1}(\omega/a)$$

The signal, the amplitude, and the phase spectra are shown in Figure E.3.2. ∎

Example 3.3

In this example, we obtain the FT of the function $f(t) = \text{sgn}(t)p_a(t)$. Therefore,

$$F(\omega) = \int_{-\infty}^{\infty} \text{sgn}(t)p_a(t)e^{-j\omega t}\,dt = \int_{-a}^{a} \text{sgn}(t)e^{-j\omega t}\,dt = -\int_{-a}^{0} e^{-j\omega t}\,dt + \int_{0}^{a} e^{-j\omega t}\,dt = -2j\frac{1 - \cos\omega a}{\omega}$$

(3.26)

∎

Example 3.4

The FT of $f(t) = e^{-a|t|}$ $-\infty < t < \infty$ is

$$F(\omega) = \int_{-\infty}^{0} e^{at} e^{-j\omega t} dt + \int_{0}^{\infty} e^{-at} e^{-j\omega t} dt = \frac{2a}{a^2 + \omega^2} \tag{3.27}$$

∎

Example 3.5

The FT of $f(t) = \text{sgn}(t) e^{-a|t|}$ $-\infty < t < \infty$ is

$$F(\omega) = -\int_{-\infty}^{0} e^{at} e^{-j\omega t} dt + \int_{0}^{\infty} e^{at} e^{-j\omega t} dt = \frac{-2j\omega}{a^2 + \omega^2} \tag{3.28}$$

∎

3.4 Fourier Transform Properties

The FT properties are given in Table 3.3.

3.5 Examples on Fourier Properties

Example 3.6

If in the inverse FT formula of a function $f(t)$ we interchange time to frequency and frequency to time and then we minus the omegas, Property #3 in Table 3.3 is obtained. Applying this property, for example, we find the relation

$$\frac{2a}{a^2 + t^2} \overset{\mathscr{F}}{\leftrightarrow} 2\pi e^{-a|\omega|}$$

This property is useful to find additional FTs of functions, most of the times difficult to obtain.

∎

Example 3.7

To prove the scaling Property #4 of Table 3.3, we proceed as follows:

$$\int_{-\infty}^{\infty} f(at) e^{-j\omega t} dt = \frac{1}{|a|} \int_{-\infty}^{\infty} f(x) e^{-j\omega x/a} dx = \frac{1}{|a|} F\left(\frac{\omega}{a}\right)$$

∎

TABLE 3.3   Commonly Used Fourier Properties

1. Linearity $af(t) \pm bh(t) \overset{\mathfrak{F}}{\leftrightarrow} aF(\omega) \pm bH(\omega)$

2. Time shifting $f(t \pm a) \overset{\mathfrak{F}}{\leftrightarrow} e^{\pm ja\omega} F(\omega)$

3. Symmetry $\begin{cases} F(t) \overset{\mathfrak{F}}{\leftrightarrow} 2\pi f(-\omega), & F(\omega) \overset{\mathfrak{F}}{\leftrightarrow} f(t) \\ 1 \overset{\mathfrak{F}}{\leftrightarrow} 2\pi\delta(-\omega) = 2\pi\delta(\omega) \end{cases}$

4. Time scaling $f(at) \overset{\mathfrak{F}}{\leftrightarrow} \dfrac{1}{|a|} F\left(\dfrac{\omega}{a}\right)$

5. Time reversal $f(-t) \overset{\mathfrak{F}}{\leftrightarrow} F(-\omega)$ (real time functions)

6. Frequency shifting $e^{\pm j\omega_0 t} \overset{\mathfrak{F}}{\leftrightarrow} F(\omega \mp \omega_0)$

7. Modulation $\begin{cases} f(t)\cos\omega_0 t \overset{\mathfrak{F}}{\leftrightarrow} \frac{1}{2}[F(\omega - \omega_0) + F(\omega + \omega_0)] \\ f(t)\sin\omega_0 t \overset{\mathfrak{F}}{\leftrightarrow} \frac{1}{2j}[F(\omega - \omega_0) - F(\omega + \omega_0)] \end{cases}$

8. Time differentiation $\dfrac{df(t)}{dt} \overset{\mathfrak{F}}{\leftrightarrow} j\omega F(\omega) \quad \dfrac{d^n t}{dt^n} \overset{\mathfrak{F}}{\leftrightarrow} (j\omega)^n F(\omega)$

9. Frequency differentiation $\begin{cases} (-jt)f(t) \overset{\mathfrak{F}}{\leftrightarrow} \dfrac{dF(\omega)}{d\omega} \\ (-jt)^n f(t) \overset{\mathfrak{F}}{\leftrightarrow} \dfrac{d^n F(\omega)}{d\omega^n} \end{cases}$

10. Time convolution $f(t)*h(t) \overset{\mathfrak{F}}{\leftrightarrow} F(\omega)H(\omega)$

11. Frequency convolution $f(t)h(t) \overset{\mathfrak{F}}{\leftrightarrow} \frac{1}{2\pi} F(\omega)*H(\omega) = \frac{1}{2\pi}\int_{-\infty}^{\infty} F(x)H(\omega - x)dx$

12. Correlation $f(t)^{\circledcirc}(t) = \int_{-\infty}^{\infty} f(x)h^*(x - t)dx \overset{\mathfrak{F}}{\leftrightarrow} |F(\omega)H^*(\omega)|$

13. Central ordinate $f(0) = \frac{1}{2\pi}\int_{-\infty}^{\infty} F(\omega)d\omega \quad F(0) = \int_{-\infty}^{\infty} f(t)dt$

14. Parseval's theorem $\begin{cases} \int_{-\infty}^{\infty} |f(t)|^2 dt = \frac{1}{2\pi}\int_{-\infty}^{\infty} |F(\omega)|^2 d\omega \\ \int_{-\infty}^{\infty} f(t)h^*(t)dt = \frac{1}{2\pi}\int_{-\infty}^{\infty} F(\omega)H^*(\omega)d\omega \end{cases}$

Example 3.8

Based on the properties of the delta function (see Section 1.2), we find

$$\mathfrak{F}\{\delta(t)\} \overset{\Delta}{=} \Delta(\omega) = \int_{-\infty}^{\infty} \delta(t)e^{-j\omega t}dt = e^{-j\omega 0} = 1 \tag{3.29}$$

Therefore, in connection with the symmetry Property #3 of Table 3.3 we obtain

$$\mathfrak{F}\{a\} = \int_{-\infty}^{\infty} ae^{-j\omega t}dt = a2\pi\delta(\omega) \quad a = \text{constant} \tag{3.30}$$

∎

Example 3.9

From the relation $\cos\omega_0 t = \frac{1}{2}(e^{j\omega t} + e^{-j\omega t})$ and the shifting Property #2 of Table 3.3, we obtain the FT of a modulated signal:

$$f(t)\cos\omega_0 t \overset{\mathfrak{F}}{\leftrightarrow} \frac{1}{2}[F(\omega + \omega_0) + F(\omega - \omega_0)] \tag{3.31}$$

Therefore, since $p_a(t) \overset{\mathcal{F}}{\leftrightarrow} 2 \sin a\omega/\omega$ then by (3.31) we obtain

$$p_a(t) \cos \omega_0 t \quad \overset{\mathcal{F}}{\leftrightarrow} \quad \frac{\sin a(\omega - \omega_0)}{\omega - \omega_0} + \frac{\sin a(\omega + \omega_0)}{\omega + \omega_0} \qquad (3.32)$$

∎

Example 3.10

The FT of the convolution of two functions is given by

$$f(t)*h(t) \quad \overset{\mathcal{F}}{\leftrightarrow} \quad \int_{-\infty}^{\infty} e^{-j\omega t} \left[\int_{-\infty}^{\infty} f(x)h(t-x)dx \right] dt = \int_{-\infty}^{\infty} f(x) \int_{-\infty}^{\infty} e^{-j\omega(y+x)} h(x) dx dy$$

$$= \int_{-\infty}^{\infty} f(x)e^{-j\omega x} dx \int_{-\infty}^{\infty} h(y)e^{-j\omega y} dy = F(\omega)H(\omega)$$

where we set $(t - x) = y$ (see Table 3.3, Property #10). ∎

Example 3.11

Applying the convolution property to an LTI system whose impulse response is $h(t)$ and its input is $f(t)$, we obtain its output spectrum to be equal to $F(\omega)H(\omega)$. ∎

Example 3.12

The FT of the pulse with height B and width a is $\mathcal{F}\{Bp_a(t)\} = 2B \sin lrba\omega)/\omega$. The convolution of the pulse to itself is equal to the triangle:

$$Bp_a(t)*Bp_a(t) = \begin{cases} B^2 2a(1+t) & -1 < t < 0 \\ B^2 2a(1-t) & 0 < t < 1 \end{cases} \qquad (3.33)$$

Therefore, the FT of the convolution (the triangle) is equal to

$$F(\omega) = \left(2B \frac{\sin a\omega}{\omega}\right)^2 = 4B^2 \frac{\sin^2 a\omega}{\omega^2} = B^2 L^2 \frac{\sin^2 \frac{L}{2}\omega}{\frac{L^2\omega^2}{4}} = B^2 \frac{2a}{2a} L^2 \frac{\sin^2 \frac{L}{2}\omega}{\frac{L^2\omega^2}{4}} = AL \frac{\sin^2 \frac{L}{2}\omega}{\frac{L^2\omega^2}{4}}$$

$$(3.34)$$

where first we substituted $L = 2a$ and next $B^2 2a = A$. For these substitutions and the spectrum see Figure E.3.3. ∎

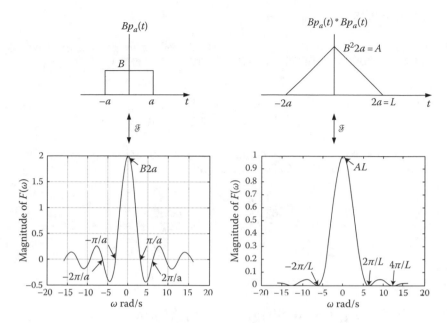

FIGURE E.3.3

Example 3.13

To find the FT of the function $f(t) = \dfrac{\sin at}{t}$, we use the symmetry property and, thus, we obtain

$$p_a(t) \overset{\mathcal{F}}{\longleftrightarrow} \frac{2\sin a\omega}{\omega}$$

$$\frac{2\sin at}{t} \overset{\mathcal{F}}{\longleftrightarrow} 2\pi p_a(\omega) \quad \text{or} \quad \frac{\sin at}{\pi t} \overset{\mathcal{F}}{\longleftrightarrow} p_a(\omega) \tag{3.35}$$

∎

Example 3.14

To obtain Parseval's identity, we proceed as follows:

$$\int\limits_{-\infty}^{\infty} |f(t)|^2 \, dt = \int\limits_{-\infty}^{\infty} f(t)f^*(t)dt = \int\limits_{-\infty}^{\infty} f(t)\left(\frac{1}{2\pi}\int\limits_{-\infty}^{\infty} F(\omega)e^{j\omega t}d\omega\right)^* dt$$

$$= \frac{1}{2\pi}\int\limits_{-\infty}^{\infty} F^*(\omega)\left(\int\limits_{-\infty}^{\infty} f(t)e^{-j\omega t}dt\right)d\omega$$

$$= \frac{1}{2\pi}\int\limits_{-\infty}^{\infty} F^*(\omega)F(\omega)d\omega = \frac{1}{2\pi}\int\limits_{-\infty}^{\infty} |F(\omega)|^2 \, d\omega$$

∎

Example 3.15

Suppose the FT $F(\omega)$ of a function $f(t)$ is truncated above to $|\omega| = a$. This would imply that we have obtained the band-limited function:

$$F_a(\omega) = \left\{ \begin{array}{cc} F(\omega) & |\omega| < a \\ 0 & |\omega| > a \end{array} \right\} = F(\omega)p_a(\omega) \tag{3.36}$$

Therefore, the IFT of both sides of (3.36), taking into consideration (3.35) and the convolution Property #11 of Table 3.3, we obtain

$$f_a(t) = \int_{-\infty}^{\infty} f(x) \frac{\sin a(t - x)}{\pi(t - x)} dx = f(t) \overset{*}{} \frac{\sin at}{\pi t} \overset{\mathcal{F}}{\leftrightarrow} F(\omega)p_a(\omega) \tag{3.37}$$

∎

3.6 FT Examples of Singular Functions

The basic properties of the delta function are

$$\int_{-\infty}^{\infty} \delta(t - t_0)\phi(t)dt = \phi(t_0) \qquad \int_{-\infty}^{\infty} \frac{d^n\delta(t - t_0)}{dt^n}\phi(t)dt = (-1)^n \frac{d^n\phi(t_0)}{dt^n} \tag{3.38}$$

where $\phi(t)$ is an arbitrary function and continuous at a given point t_0 (Table 3.4).

Example 3.16

Based on the above properties of the delta function, its FT is

$$\Delta(\omega) = \int_{-\infty}^{\infty} \delta(t)e^{-j\omega t} dt = e^{-j\omega 0} = 1 \qquad \delta(t) \overset{\mathcal{F}}{\leftrightarrow} 1 \tag{3.39}$$

Using the symmetry property we obtain the relations

$$\delta(t) \overset{\mathcal{F}}{\leftrightarrow} 1$$

$$1 \overset{\mathcal{F}}{\leftrightarrow} 2\pi\delta(\omega) \tag{3.40}$$

From the second relation of (3.39), we also obtain

$$\delta(t) = \frac{1}{2\pi} \int\limits_{-\infty}^{\infty} e^{j\omega t} d\omega = \frac{1}{2\pi} \int\limits_{-\infty}^{\infty} \cos \omega t \, d\omega \qquad (3.41)$$

Furthermore,

$$\delta(t - t_0) \overset{\mathcal{F}}{\leftrightarrow} e^{-j\omega t_0} \Rightarrow e^{j\omega_0 t} \overset{\mathcal{F}}{\leftrightarrow} \delta(\omega - \omega_0)$$

$$\cos \omega_0 t = \frac{1}{2}(e^{j\omega_0 t} + e^{-j\omega_0 t}) \overset{\mathcal{F}}{\leftrightarrow} \pi[\delta(\omega - \omega_0) + \delta(\omega + \omega_0)] \qquad (3.42)$$

■

Example 3.17

The FT of the sgn(t) is

$$\text{sgn}(t) = \lim_{a \to 0} \begin{cases} e^{-at} & t > 0 \\ 0 & t = 0 \\ e^{at} & t < 0 \end{cases} \Rightarrow \mathcal{F}\{\text{sgn}(t)\} = \lim_{a \to 0} \left[\int\limits_{\infty-}^{0} e^{at} e^{-j\omega t} dt + \int\limits_{0}^{\infty} e^{-at} e^{-j\omega t} dt \right] = \frac{2}{j\omega}$$

$$(3.43)$$

■

Example 3.18

From $\sin at / t \overset{\mathcal{F}}{\leftrightarrow} \pi p_a(\omega)$ Parseval's relation gives

$$\int\limits_{-\infty}^{\infty} \frac{\sin^2 at}{t^2} dt = \frac{1}{2\pi} \int\limits_{-a}^{a} \pi^2 d\omega = a\pi \qquad (3.44)$$

■

Example 3.19

The FT of the unit step function is

$$\mathcal{F}\{u(t)\} = \mathcal{F}\left\{ \frac{1}{2} + \frac{1}{2}\text{sgn}(t) \right\} = \pi\delta(\omega) + \frac{1}{j\omega} \Rightarrow \mathcal{F}\{e^{j\omega_0 t} u(t)\}$$

$$= \pi\delta(\omega - \omega_0) + \frac{1}{j(\omega - \omega_0)} \qquad (3.45)$$

■

Example 3.20

To find the FT of the $comb_T(t)$ function (see Figure 1.8), we first expand the periodic comb function into its FS and then we take its FT. Hence,

$$comb_T(t) = \sum_{n=-\infty}^{\infty} a_n e^{jn\omega_0 t} \quad a_n = \frac{1}{T} \int_{-T/2}^{T/2} \delta(t) e^{-jn\omega_0 t} dt = \frac{1}{T} e^{-jn\omega_0 0} = \frac{1}{T};$$

$$\mathfrak{F}\{comb_T(t)\} = COMB_{\omega_0}(\omega) = \frac{1}{T} \sum_{n=-\infty}^{\infty} \int_{-\infty}^{\infty} e^{jn\omega_0 t} e^{-j\omega t} dt = \frac{1}{T} \sum_{n=-\infty}^{\infty} \int_{-\infty}^{\infty} e^{-j(\omega-n\omega_0)t} dt$$

$$= \frac{2\pi}{T} \sum_{n=-\infty}^{\infty} \delta(\omega - n\omega_0) \quad \omega_0 = \frac{2\pi}{T} \tag{3.46}$$

Therefore, the FT of a comb function is another comb function in the frequency domain whose amplitudes are $2\pi/T$ and whose location is at points $n\omega_0 = n2\pi/T$ for $n = \dots, -1, -2, 0, 1, 2, \dots$ ∎

Example 3.21

The FT of a periodic pulse function with period T is given by (see Figure E.3.4)

$$\mathfrak{F}\{f(t)\} = \mathfrak{F}\{f_0(t) * comb_T(t)\} = \frac{2\sin a\omega}{\omega} \frac{2\pi}{T} \sum_{n=-\infty}^{\infty} \delta(\omega - n\omega_0)$$

$$= \frac{4\pi}{T} \sum_{n=-\infty}^{\infty} \frac{\sin an\omega_0}{n\omega_0} \delta(\omega - n\omega_0) \quad T > 2a \tag{3.47}$$

The FT of the pulse is $\mathfrak{F}\{f_0(t)\} \overset{\Delta}{=} F_0(\omega) = 2\sin\omega a/\omega$ and the FT of the periodic function $f(t)$ is shown to be discrete as was anticipated since the function is periodic. ∎

Example 3.22 (Gibbs' phenomenon)

Let us truncate the spectrum $U(\omega)$ of the unit step:

$$U_{\omega_0}(\omega) = \begin{cases} U(\omega) & |\omega| \leq \omega_0 \\ 0 & \text{elsewher} \end{cases} \quad \text{or} \quad U_{\omega_0}(\omega) = U(\omega) p_{\omega_0}(\omega) \tag{3.48}$$

The approximate reconstruction of the unit step function, $u_a(t)$, is

$$u_a(t) = \mathfrak{F}^{-1}\{U(\omega) p_{\omega_0}(\omega)\} = u(t) * \mathfrak{F}^{-1}\{p_{\omega_0}(\omega)\} = u(t) * \frac{\sin\omega_0 t}{\pi t}$$

$$= \frac{\omega_0}{\pi} \int_{-\infty}^{t} \frac{\sin\omega_0 x}{\omega_0 x} dx = \frac{1}{\pi} \int_{-\infty}^{0} \frac{\sin y}{y} dy + \frac{1}{\pi} \int_{0}^{\omega_0 t} \frac{\sin y}{y} dy = \frac{1}{2} + \frac{1}{\pi} Si(\omega_0 t) \tag{3.49}$$

where $Si(\ )$ is the sine integral. Properties of $Si(x)$ include: $Si(-x) = Si(x)$ and $Si(\infty) = Si(-\infty) = \pi/2$. The following MATLAB® Book program produces the Figure E.3.5.

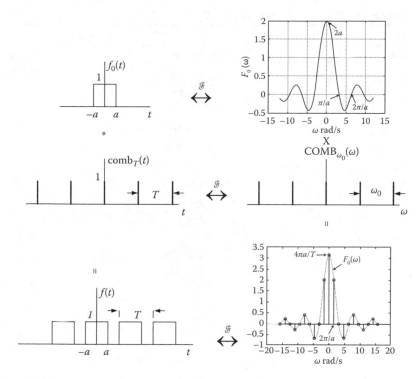

FIGURE E.3.4

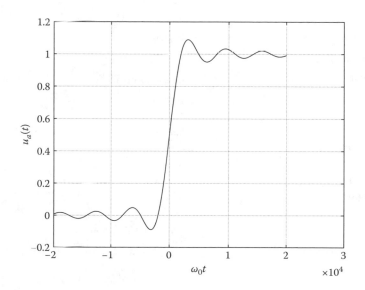

FIGURE E.3.5

TABLE 3.4 Fourier Transforms of Some Common Functions

$f(t)$	$F(w) = \mathfrak{F}[f(t)]|_\omega|$
$p_\alpha(t)$	$\frac{2}{\omega}\sin(\alpha\omega)$
$(\alpha - |t|)p_\alpha(t)$	$\left(\frac{2\sin\left(\frac{\alpha}{2}\omega\right)}{\omega}\right)^2$
$\sqrt{1-t^2}\,p_1(t)$	$\frac{\pi}{\omega}J_1(\omega)$
$(1-t^2)^{\alpha-1}p_1(t)$	$\Gamma(\alpha)\sqrt{\pi}\left(\frac{2}{|\omega|}\right)^{\alpha-\frac{1}{2}}J_{\alpha-\frac{1}{2}}(|\omega|)$
$\operatorname{sgn}(t)p_\alpha(t)$	$-2j\frac{1-\cos(\alpha\omega)}{\omega}$
$\cos\left(\frac{\pi}{2\alpha}t\right)p_\alpha(t)$	$\frac{4\pi\alpha}{\pi^2-4\alpha^2\omega^2}\cos(\alpha\omega)$
$\operatorname{Rect}_{(\beta,\gamma)}(t)$	$\frac{j}{\omega}[e^{-j\gamma\omega} - e^{-j\beta\omega}]$
$e^{-(\alpha+j\beta)t}u(t)$	$\frac{1}{\alpha+j\beta+j\omega}$
$t^{\nu-1}e^{-(\alpha+j\beta)t}u(t)$	$\frac{\Gamma(\nu)}{(\alpha+j\beta+j\omega)^\nu}$
$e^{(\alpha+j\beta)t}u(-t)$	$\frac{1}{(\alpha+j\beta-j\omega)}$
$(-1)^{\nu-1}e^{(\alpha+j\beta)t}u(-t)$	$\frac{\Gamma(\nu)}{(\alpha+j\beta-j\omega)^\nu}$
$e^{-\alpha|t|}$	$\frac{2\alpha}{\alpha^2+\omega^2}$
$\operatorname{sgn}(t)\,e^{-\alpha|t|}$	$\frac{-2j\omega}{\alpha^2+\omega^2}$
$e^{-\alpha t^2}$	$\sqrt{\frac{\pi}{\alpha}}\exp\left[-\frac{1}{4\alpha}\omega^2\right]$
$e^{-\alpha t^2+\beta t}$	$\sqrt{\frac{\pi}{\alpha}}\exp\left[-\frac{1}{4\alpha}\omega^2 - j\frac{\beta}{2\alpha}\omega + \frac{\beta^2}{4\alpha}\right]$
$\frac{e^{-\lambda t}}{\alpha+j\beta+e^{-t}}$	$\pi(\alpha+j\beta)^{\lambda-1+j\omega}\csc(\pi\lambda+j\pi\omega)$
$\operatorname{sech}(\alpha t)$	$\frac{\pi}{\alpha}\operatorname{sech}\left(\frac{\pi}{2\alpha}\omega\right)$
$e^{\pm j\alpha t2}$	$\sqrt{\frac{\pi}{\alpha}}\exp\left[\mp j\frac{1}{4\alpha}(\omega^2-\alpha\pi)\right]$
$e^{-\theta t^2}$	$\left[\sqrt{|\theta|+\alpha} - j\operatorname{sgn}(\beta)\sqrt{|\theta|-\alpha}\right]$
$(w/\theta = \alpha + j\beta)$	$\times\frac{1}{|\theta|}\sqrt{\frac{\pi}{2}}\exp\left(-\frac{1}{4\theta}\omega^2\right)$
$\frac{1}{t}\sin(\alpha t)$	$\pi p_\alpha(\omega)$
$\left(\frac{1}{t}\sin(\alpha t)\right)^2$	$\frac{\pi}{2}(2\alpha - |\omega|)p_{2\alpha}(\omega)$
$\frac{1}{|t|}\sin(\alpha t)$	$-j\operatorname{sgn}(\omega)\ln\left|\frac{|\omega|+\alpha}{|\omega|-\alpha}\right|$
1	$2\pi\delta(\omega)$
t^n	$j^n 2\pi\delta^{(n)}(\omega)$
$e^{j\beta t}$	$2\pi\delta(\omega-\beta)$
$\delta(t-\beta)$	$e^{-j\beta\omega}$
$\delta^{(n)}(t)$	$(j\omega)^n$
$\sin(\alpha t)$	$-j\pi[\delta(\omega-\alpha) - \delta(\omega+\alpha)]$
$\cos(\alpha t)$	$\pi[\delta(\omega-\alpha) + \delta(\omega+\alpha)]$
$\sin(\alpha t^2)$	$-\sqrt{\frac{\pi}{\alpha}}\sin\left[\frac{1}{4\alpha}(\omega^2-\alpha\pi)\right]$
$\cos(\alpha t^2)$	$\sqrt{\frac{\pi}{\alpha}}\cos\left[\frac{1}{4\alpha}(\omega^2-\alpha\pi)\right]$
$e^{-\alpha t^2}\cos(\nu t^2)$	$\frac{1}{|\theta|}\sqrt{\frac{\pi}{2}}\exp\left(-\frac{\alpha}{4|\theta|^2}\omega^2\right)\times$
$(w/\theta = \alpha + j\nu)$	$\left[\sqrt{|\theta|+\alpha}\cos\left(\frac{\nu\omega^2}{4|\theta|^2}\right) + \sqrt{|\theta|-\alpha}\sin\left(\frac{\nu\omega^2}{4|\theta|^2}\right)\right]$

TABLE 3.4 (continued)   Fourier Transforms of Some Common Functions

$f(t)$	$F(w) = f[f(t)]\|_{\omega}\|$
$e^{-\alpha t^2}\sin(\nu t^2)$	$\frac{1}{\|\theta\|}\sqrt{\frac{\pi}{2}}\exp\left(-\frac{\alpha}{4\|\theta\|}\omega^2\right)\times$
$(w/\theta = \alpha + j\nu)$	$\left[\sqrt{\|\theta\| + \alpha}\cos\left(\frac{\nu\omega^2}{4\|\theta\|^2}\right) - \sqrt{\|\theta\| - \alpha}\sin\left(\frac{\nu\omega^2}{4\|\theta\|^2}\right)\right]$
$\mathrm{comb}_\alpha(t)$	$\frac{2\pi}{\alpha}\mathrm{comb}_{\frac{2\pi}{\alpha}}(\omega)$
$\|\sin(\alpha t)\|$	$\sum_{k=-\infty}^{\infty}\frac{4}{1-4k^2}\delta(\omega - 2\alpha k)$
$\|\cos(\alpha t)\|$	$\sum_{k=-\infty}^{\infty}(-1)^k\frac{4}{1-4k^2}\delta(\omega - 2\alpha k)$
$\mathrm{saw}(t)$	$j\sum_{\substack{n=-\infty\\n\neq 0}}^{\infty}(-1)^n\frac{2}{n}\delta(\omega - n\pi)$
$\sum_{m=-\infty}^{\infty}p_\alpha(t - m\nu)$	$\frac{4\pi\alpha}{\nu}\delta(\omega) + \sum_{\substack{k=-\infty\\k\neq 0}}^{\infty}\frac{2}{k}\sin\left(\frac{2\pi k\alpha}{\nu}\right)\delta\left(\omega - \frac{2\pi k}{\nu}\right)$
$(w/2\alpha \leq \nu)$	
$\mathrm{sgn}(t)$	$-j\frac{2}{\omega}$
$u(t)$	$\pi\delta(\omega) - j\frac{1}{\omega}$
$\frac{1}{t}$	$-j\pi\,\mathrm{sgn}(\omega)$
t^{-n}	$-j\pi\frac{(-j\omega)^{n-1}}{(n-1)!}\,\mathrm{sgn}(\omega)$
$\|t\|$	$-\frac{2}{\omega^2}$
$t^n\mathrm{sgn}(t)$	$(-j)^{n+1}\frac{2(n!)}{\omega^{n+1}}$
$\mathrm{ramp}(t)$	$j\pi\delta'(\omega) - \frac{1}{\omega^2}$
$t^n\,u(t)$	$j^n\pi\delta^{(n)}(\omega) + n!\left(\frac{-j}{\omega}\right)^{n+1}$
$\|t\|^{-1/2}$	$\sqrt{2\pi}\|\omega\|^{-1/2}$
$\|t\|^{\lambda-1}$	$2\Gamma(\lambda)\cos\left(\frac{\lambda\pi}{2}\right)\|\omega\|^{-\lambda}$
$J_0(t)$	$\frac{2}{\sqrt{1-\omega^2}}p_1(\omega)$
$Y_0(\|t\|)$	$\frac{2}{\sqrt{\omega^2-1}}[1 - p_1(\omega)]$
$J_{2n}(t)$	$\frac{2\cos[2n\arcsin(\omega)]}{\sqrt{1-\omega^2}}p_1(\omega)$
$J_{2n+1}(t)$	$\frac{-2j\sin[(2n+1)\arcsin(\omega)]}{\sqrt{1-\omega^2}}p_1(\omega)$
$J_n(t)$	$\frac{2(-j)^n T_n(\omega)}{\sqrt{1-\omega^2}}p_1(\omega)$
$\frac{1}{t^n}J_n(t)$	$\frac{2(1-\omega^2)^{n-\frac{1}{2}}}{1\cdot 3\cdot 5\cdots(2n-1)}p_1(\omega)$
$\|t\|^{-\alpha+\frac{1}{2}}J_{\alpha-\frac{1}{2}}(\|t\|)$	$\frac{\sqrt{2\pi}}{\Gamma(\alpha)}\left(\frac{1-\omega^2}{2}\right)^{\alpha-1}p_1(\omega)$
$\frac{1}{t}J_n(t)$	$(-j)^n\frac{2j}{n}\sqrt{1-\omega^2}U_{n-1}(\omega)p_1(\omega)$
$t^{-1/2}J_{n+\frac{1}{2}}(t)$	$(-j)^n\sqrt{2\pi}P_n(\omega)p_1(\omega)$
$\mathrm{sgn}(t)J_0(t)$	$j\frac{2}{\sqrt{\omega^2-1}}\mathrm{sgn}(\omega)[1 - p_1(\omega)]$
$J_0(t)u(t)$	$\frac{p_1(\omega)+j\mathrm{sgn}(\omega)[1-p_1(\omega)]}{\sqrt{\|1-\omega^2\|}}$

Notes: $\alpha, \beta, \gamma, \lambda, \nu$, and n denote real numbers with $\alpha > 0$, $0 < \lambda < 1$, $\nu > 0$, and $n = 1$, 2, 3,... $\Gamma(x)$, the Gamma function; $P_n(x)$, the nth Legendre polynomial; J_ν, the Bessel function of the first kind of order ν; Y_ν, the Bessel function of the second kind of order ν; $T_n(x)$, the nth Chebyshev polynomial of the first kind; $U_n(x)$, the nth Chebyshev polynomial of the second kind; $\mathrm{saw}(t)$ is the saw function.

TABLE 3.5   Graphical Representation of Some Fourier Transforms

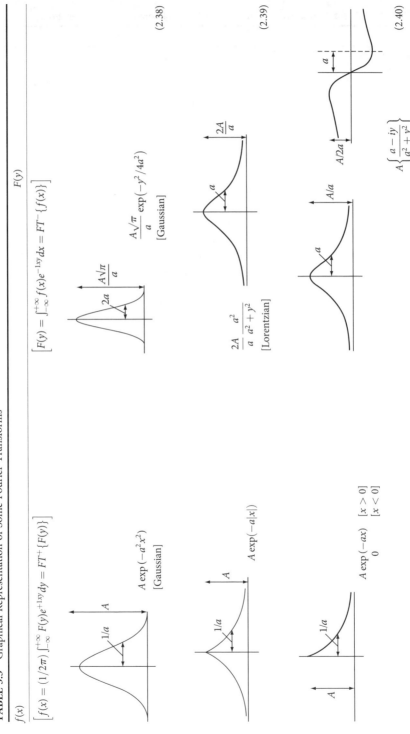

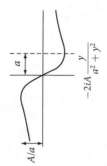

$$A \exp(-ax) \quad [x>0]$$
$$-A \exp(-a|x|) \quad [x<0]$$

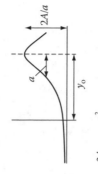

$$-2iA\,\frac{y}{a^2+y^2}$$

(2.41)

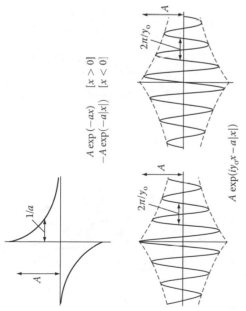

$$A \exp(iy_o x - a|x|)$$

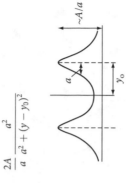

$$\frac{2A}{a}\,\frac{a^2}{a^2+(y-y_0)^2}$$

(2.42)

$$A\cos y_0 x \exp(-a|x|)$$

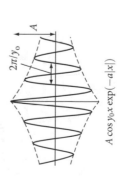

$$\frac{A}{a}\left\{\frac{a^2}{a^2+(y-y_0)^2}+\frac{a^2}{a^2+(y+y_0)^2}\right\}$$
$$=\frac{A}{a}\left\{\frac{2a^2(a^2+y_0^2+y^2)}{(a^2+y_0^2-y^2)^2+4a^2y^2}\right\}$$

(2.43)

(continued)

TABLE 3.5 (continued)   Graphical Representation of Some Fourier Transforms

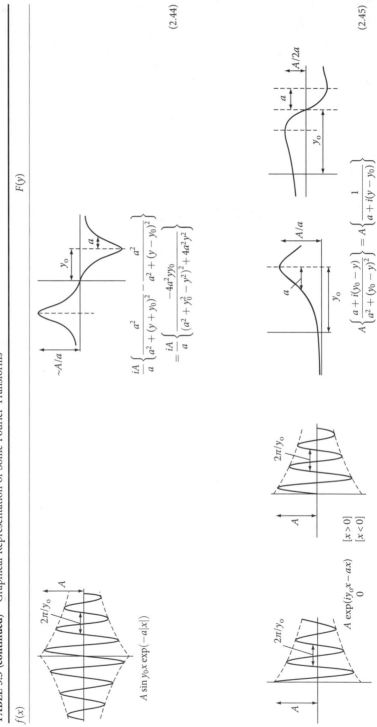

$f(x)$

$F(y)$

$A \sin y_0 x \exp(-a|x|)$

$$\frac{iA}{a}\left\{\frac{a^2}{a^2+(y+y_0)^2}-\frac{a^2}{a^2+(y-y_0)^2}\right\}$$
$$=\frac{iA}{a}\left\{\frac{-4a^2 y y_0}{(a^2+y_0^2-y^2)^2+4a^2 y^2}\right\}$$

(2.44)

$A\exp(iy_0 x - ax)$ $[x>0]$
0 $[x<0]$

$$A\left\{\frac{a+i(y_0-y)}{a^2+(y_0-y)^2}\right\} = A\left\{\frac{1}{a+i(y-y_0)}\right\}$$

(2.45)

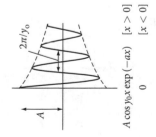

$$A\cos y_0 x \exp(-ax) \quad [x > 0]$$
$$0 \quad [x < 0]$$

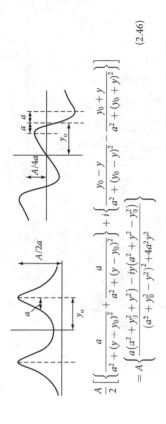

$$\frac{A}{2}\left[\left\{\frac{a}{a^2+(y+y_0)^2}+\frac{a}{a^2+(y-y_0)^2}\right\}+i\left\{\frac{y_0-y}{a^2+(y_0-y)^2}-\frac{y_0+y}{a^2+(y_0+y)^2}\right\}\right]$$
$$=A\left\{\frac{a(a^2+y_0^2+y^2)-iy(a^2+y^2-y_0^2)}{(a^2+y_0^2-y^2)^2+4a^2y^2}\right\}$$

(2.46)

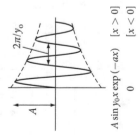

$$A\sin y_0 x \exp(-ax) \quad [x > 0]$$
$$0 \quad [x < 0]$$

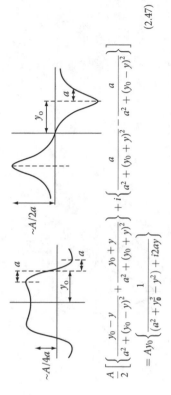

$$\frac{A}{2}\left[\left\{\frac{y_0-y}{a^2+(y_0-y)^2}+\frac{y_0+y}{a^2+(y_0+y)^2}\right\}+i\left\{\frac{a}{a^2+(y_0+y)^2}-\frac{a}{a^2+(y_0-y)^2}\right\}\right]$$
$$=Ay_0\left\{\frac{1}{(a^2+y_0^2-y^2)^2+i2ay}\right\}$$

(2.47)

(*continued*)

TABLE 3.5 (continued) Graphical Representation of Some Fourier Transforms

$f(x)$	$F(y)$

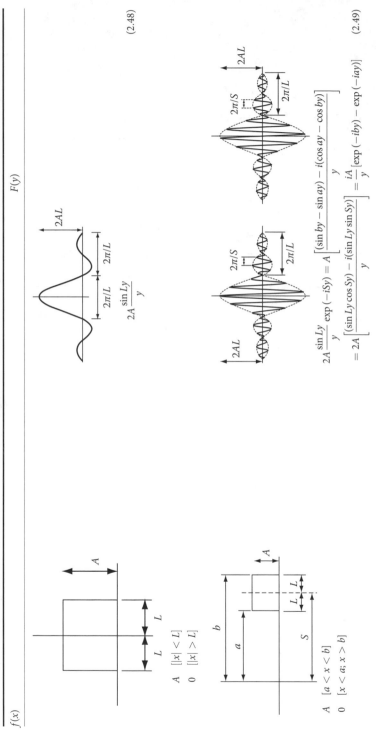

$$A \quad [|x| < L]$$
$$0 \quad [|x| > L]$$

$$2A \frac{\sin Ly}{y} \tag{2.48}$$

$$A \quad [a < x < b]$$
$$0 \quad [x < a; x > b]$$

$$2A \frac{\sin Ly}{y} \exp(-iSy) = A\left[\frac{(\sin Ly \cos Sy) - i(\sin Ly \sin Sy)}{y}\right] = 2A\left[\frac{(\sin by - \sin ay) - i(\cos ay - \cos by)}{y}\right] = \frac{iA}{y}\left[\exp(-iby) - \exp(-iay)\right] \tag{2.49}$$

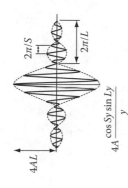

$$A \quad [(S-L) < |x| < (S+L)]$$
$$0 \quad [\text{otherwise}]$$

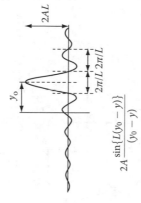

$$4A \frac{\cos Sy \sin Ly}{y} \qquad (2.50)$$

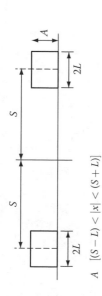

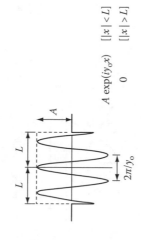

$$A \exp(iy_0 x) \quad [|x| < L]$$
$$0 \quad [|x| > L]$$

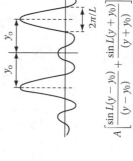

$$2A \frac{\sin\{L(y_0 - y)\}}{(y_0 - y)} \qquad (2.51)$$

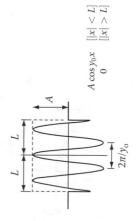

$$A \cos y_0 x \quad [|x| < L]$$
$$0 \quad [|x| > L]$$

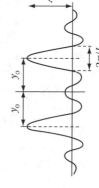

$$A\left[\frac{\sin L(y - y_0)}{(y - y_0)} + \frac{\sin L(y + y_0)}{(y + y_0)}\right] \qquad (2.52)$$

(continued)

TABLE 3.5 (continued) Graphical Representation of Some Fourier Transforms

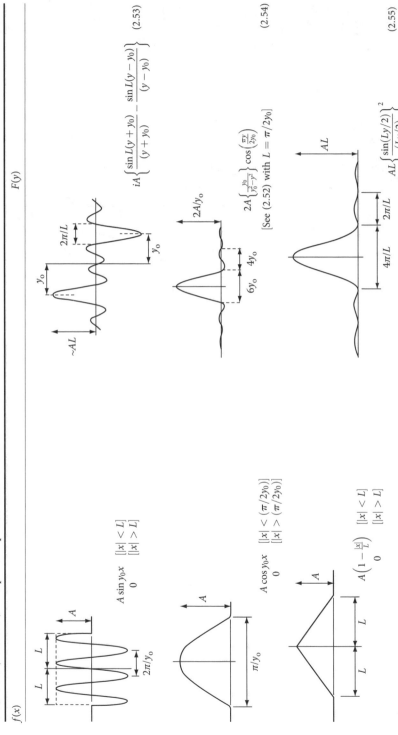

$f(x)$	$F(y)$

$A\sin y_0 x \quad \begin{bmatrix} |x| < L \\ |x| > L \end{bmatrix}$

$$iA\left\{ \frac{\sin L(y+y_0)}{(y+y_0)} - \frac{\sin L(y-y_0)}{(y-y_0)} \right\} \quad (2.53)$$

$A\cos y_0 x \quad \begin{bmatrix} |x| < (\pi/2y_0) \\ 0 \quad |x| > (\pi/2y_0) \end{bmatrix}$

$$2A\left\{ \frac{y_0}{y_0^2 - y^2} \right\} \cos\left(\frac{\pi y}{2y_0} \right) \quad (2.54)$$

[See (2.52) with $L = \pi/2y_0$]

$A\left(1 - \frac{|x|}{L}\right) \quad \begin{bmatrix} |x| < L \\ 0 \quad |x| > L \end{bmatrix}$

$$AL\left\{ \frac{\sin(Ly/2)}{(Ly/2)} \right\}^2 \quad (2.55)$$

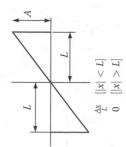

$$\frac{Ax}{L} \quad [|x| < L] \qquad 0 \quad [|x| > L]$$

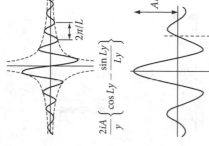

$$\frac{2iA}{y}\left\{\cos Ly - \frac{\sin Ly}{Ly}\right\} \tag{2.56}$$

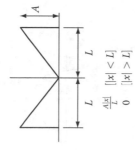

$$\frac{A|x|}{L} \quad [|x| < L] \qquad 0 \quad [|x| > L]$$

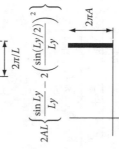

$$2AL\left\{\frac{\sin Ly}{Ly} - 2\left(\frac{\sin(Ly/2)}{Ly}\right)^2\right\} \tag{2.57}$$

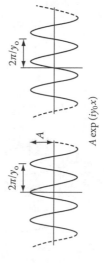

$$A\exp(iy_0 x)$$

$$2\pi A\delta(y - y_0). \tag{2.58}$$

(continued)

TABLE 3.5 (continued)   Graphical Representation of Some Fourier Transforms

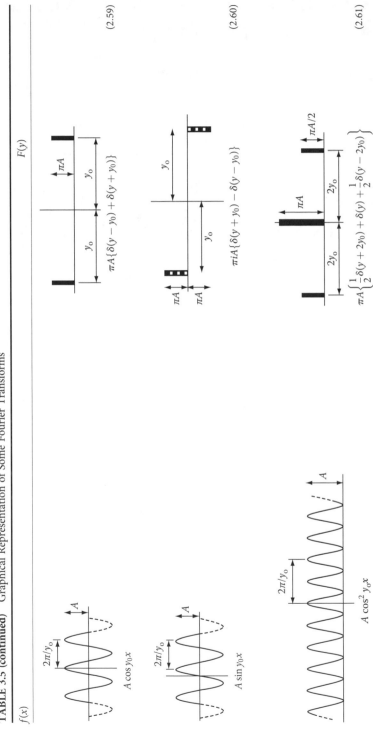

(2.62)

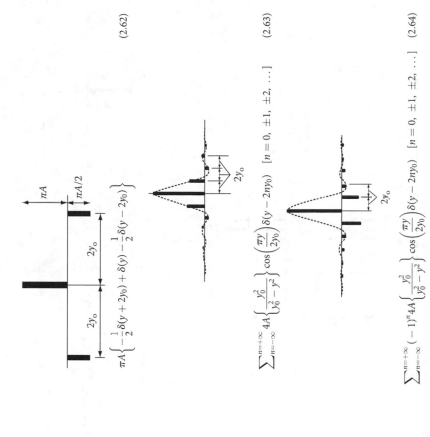

$$\pi A \left\{ -\frac{1}{2}\delta(y + 2y_0) + \delta(y) - \frac{1}{2}\delta(y - 2y_0) \right\}$$

$$\sum_{n=-\infty}^{n=+\infty} 4A \left\{ \frac{y_0^2}{y_0^2 - y^2} \right\} \cos\left(\frac{\pi y}{2y_0}\right) \delta(y - 2ny_0) \quad [n = 0, \pm 1, \pm 2, \ldots] \tag{2.63}$$

$$\sum_{n=-\infty}^{n=+\infty} (-1)^n 4A \left\{ \frac{y_0^2}{y_0^2 - y^2} \right\} \cos\left(\frac{\pi y}{2y_0}\right) \delta(y - 2ny_0) \quad [n = 0, \pm 1, \pm 2, \ldots] \tag{2.64}$$

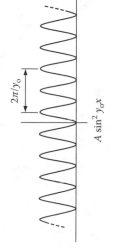

$A \sin^2 y_0 x$

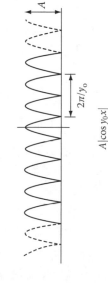

$A|\cos y_0 x|$

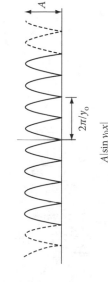

$A|\sin y_0 x|$

(continued)

TABLE 3.5 (continued)   Graphical Representation of Some Fourier Transforms

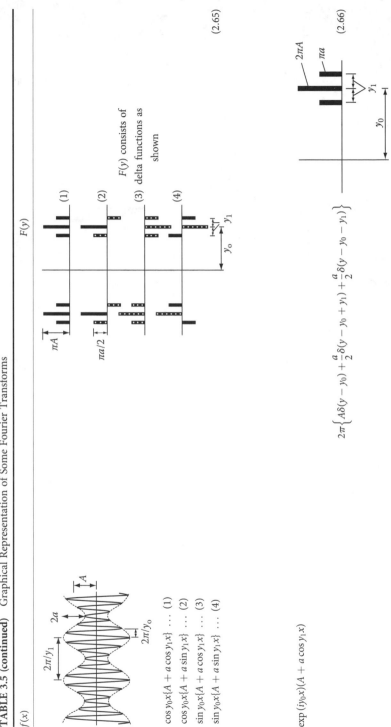

$f(x)$

$F(y)$

$\cos y_0 x\{A + a \cos y_1 x\} \ \cdots \ (1)$

$\cos y_0 x\{A + a \sin y_1 x\} \ \cdots \ (2)$

$\sin y_0 x\{A + a \cos y_1 x\} \ \cdots \ (3)$

$\sin y_0 x\{A + a \sin y_1 x\} \ \cdots \ (4)$

$F(y)$ consists of delta functions as shown

$$2\pi\left\{A\delta(y - y_0) + \frac{a}{2}\delta(y - y_0 + y_1) + \frac{a}{2}\delta(y - y_0 - y_1)\right\}$$ (2.65)

$\exp(iy_0 x)(A + a \cos y_1 x)$

(2.66)

$$\exp(iy_0 x)(A + a \sin y_1 x)$$

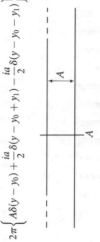

$$2\pi \left\{ A\delta(y - y_0) + \frac{ia}{2}\delta(y - y_0 + y_1) - \frac{ia}{2}\delta(y - y_0 - y_1) \right\}$$ (2.67)

$$A\delta(x)$$

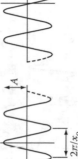

$$A$$ (2.68)

$$A\delta(x - x_0)$$

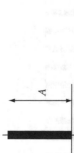

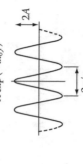

$$A \exp(-ix_0 y)$$ (2.69)

$$A\{\delta(x - x_0) + \delta(x + x_0)\}$$

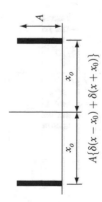

$$2A \cos x_0 y$$ (2.70)

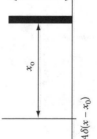

(continued)

TABLE 3.5 (continued)   Graphical Representation of Some Fourier Transforms

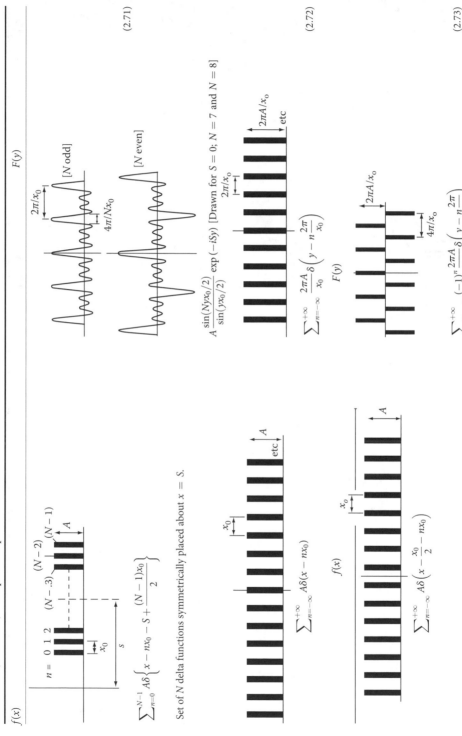

$f(x)$

$F(y)$

$$\sum_{n=0}^{N-1} A\delta\left\{x - nx_0 - S + \frac{(N-1)x_0}{2}\right\}$$

(2.71)

Set of N delta functions symmetrically placed about $x = S$.

$$A\frac{\sin(Nyx_0/2)}{\sin(yx_0/2)}\exp(-iSy) \quad \text{[Drawn for } S = 0; N = 7 \text{ and } N = 8]$$

$$\sum_{n=-\infty}^{+\infty} A\delta(x - nx_0)$$

$$\sum_{n=-\infty}^{+\infty}\frac{2\pi A}{x_0}\delta\left(y - n\frac{2\pi}{x_0}\right)$$

(2.72)

$$\sum_{n=-\infty}^{+\infty} A\delta\left(x - \frac{x_0}{2} - nx_0\right)$$

$$\sum_{n=-\infty}^{+\infty}(-1)^n\frac{2\pi A}{x_0}\delta\left(y - n\frac{2\pi}{x_0}\right)$$

(2.73)

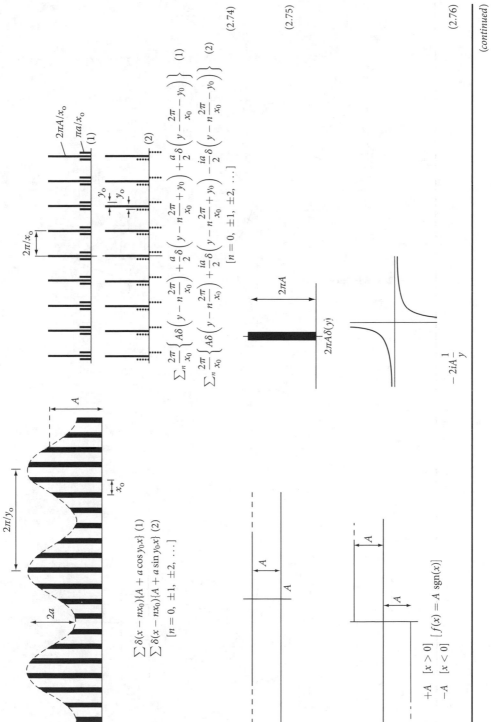

$$\sum_n \frac{2\pi}{x_0}\left\{ A\delta\left(y - n\frac{2\pi}{x_0}\right) + \frac{a}{2}\delta\left(y - n\frac{2\pi}{x_0} + y_0\right) + \frac{a}{2}\delta\left(y - \frac{2\pi}{x_0} - y_0\right)\right\} \quad (1)$$

$$\sum_n \frac{2\pi}{x_0}\left\{ A\delta\left(y - n\frac{2\pi}{x_0}\right) + \frac{ia}{2}\delta\left(y - n\frac{2\pi}{x_0} + y_0\right) - \frac{ia}{2}\delta\left(y - n\frac{2\pi}{x_0} - y_0\right)\right\} \quad (2)$$

$$[n = 0, \pm 1, \pm 2, \ldots]$$

(2.74)

(2.75)

$-2iA\dfrac{1}{y}$

(2.76)

$$\sum \delta(x - nx_0)\{A + a\cos y_0 x\} \quad (1)$$
$$\sum \delta(x - nx_0)\{A + a\sin y_0 x\} \quad (2)$$

$$[n = 0, \pm 1, \pm 2, \ldots]$$

$+A \quad [x > 0]$
$-A \quad [x < 0]$ $\quad [f(x) = A\,\mathrm{sgn}(x)]$

(continued)

TABLE 3.5 (continued) Graphical Representation of Some Fourier Transforms

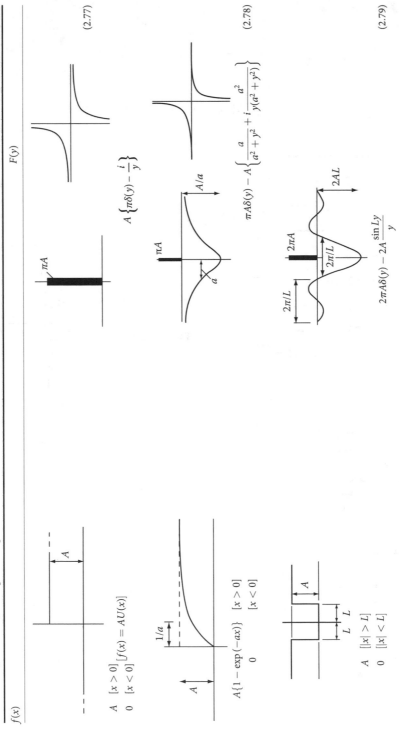

$f(x)$	$F(y)$	
$A \begin{cases} x > 0 \\ 0 \quad [x < 0] \end{cases}$ $[f(x) = AU(x)]$	$A\left\{ \pi\delta(y) - \dfrac{i}{y} \right\}$	(2.77)
$A\{1 - \exp(-ax)\} \quad [x > 0]$ $\qquad\qquad\qquad 0 \quad [x < 0]$	$\pi A\delta(y) - A\left\{ \dfrac{a}{a^2 + y^2} + i\,\dfrac{a^2}{y(a^2 + y^2)} \right\}$	(2.78)
$A \quad [|x| > L]$ $0 \quad [|x| < L]$	$2\pi A\delta(y) - 2A\,\dfrac{\sin Ly}{y}$	(2.79)

$A \exp\{i(a\cos y_0 x + bx)\}$

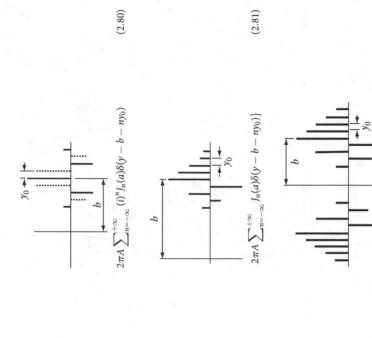

$$2\pi A \sum_{n=-\infty}^{+\infty} (i)^n J_n(a)\delta(y - b - ny_0) \qquad (2.80)$$

$A \exp\{i(a\sin y_0 x + bx)\}$

$$2\pi A \sum_{n=-\infty}^{+\infty} J_n(a)\delta(y - b - ny_0)\} \qquad (2.81)$$

$A \cos(a\sin y_0 x + bx)$

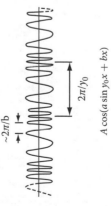

$$\pi A \sum_{n=-\infty}^{+\infty} \{J_n(a)\delta(y - b - ny_0) + J_n(a)\delta(y + b + ny_0)\} \qquad (2.82)$$

(continued)

TABLE 3.5 (continued)   Graphical Representation of Some Fourier Transforms

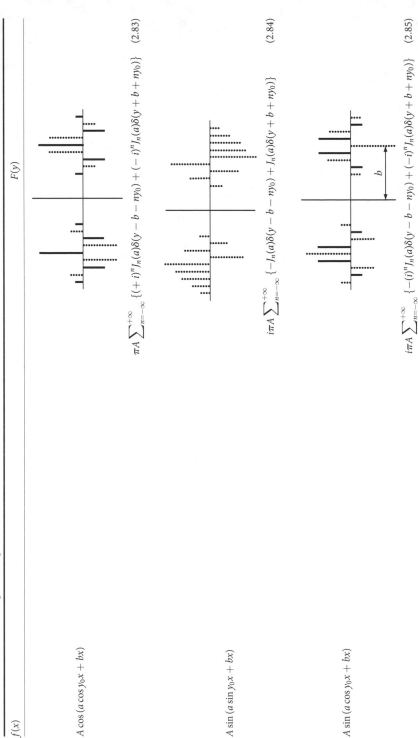

$f(x)$	$F(y)$
$A\cos(a\cos y_0 x + bx)$	$\pi A \sum_{n=-\infty}^{+\infty} \{(+i)^n J_n(a)\delta(y - b - ny_0) + (-i)^n J_n(a)\delta(y + b + ny_0)\}$ (2.83)
$A\sin(a\sin y_0 x + bx)$	$i\pi A \sum_{n=-\infty}^{+\infty} \{-J_n(a)\delta(y - b - ny_0) + J_n(a)\delta(y + b + ny_0)\}$ (2.84)
$A\sin(a\cos y_0 x + bx)$	$i\pi A \sum_{n=-\infty}^{+\infty} \{-(i)^n J_n(a)\delta(y - b - ny_0) + (-i)^n J_n(a)\delta(y + b + ny_0)\}$ (2.85)

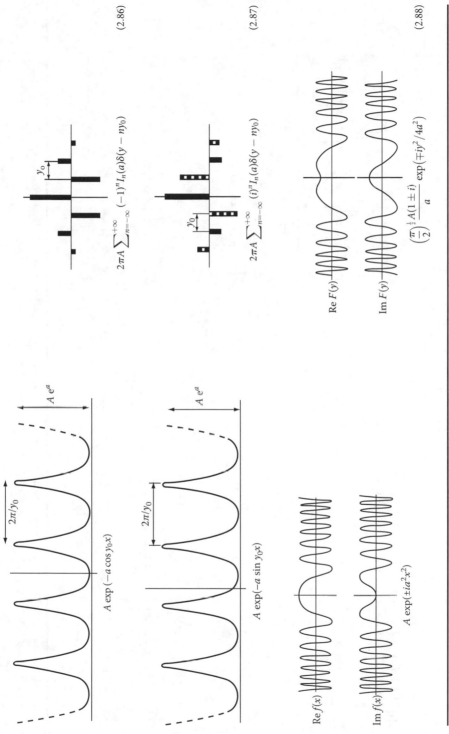

$$A \exp(-a \cos y_0 x)$$

$$2\pi A \sum_{n=-\infty}^{+\infty} (-1)^n I_n(a)\delta(y - ny_0) \tag{2.86}$$

$$A \exp(-a \sin y_0 x)$$

$$2\pi A \sum_{n=-\infty}^{+\infty} (i)^n I_n(a)\delta(y - ny_0) \tag{2.87}$$

$$A \exp(\pm i a^2 x^2)$$

$$\left(\frac{\pi}{2}\right)^{\frac{1}{2}} \frac{A(1 \pm i)}{a} \exp(\mp i y^2 / 4a^2) \tag{2.88}$$

(continued)

TABLE 3.5 (continued) Graphical Representation of Some Fourier Transforms

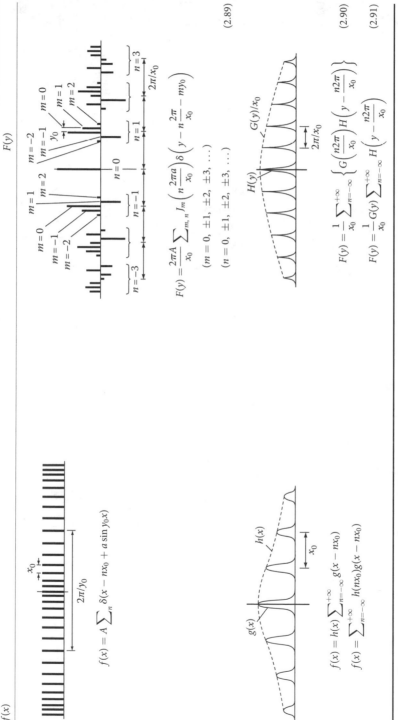

$$f(x) = A \sum_n \delta(x - nx_0 + a \sin y_0 x)$$

$$F(y) = \frac{2\pi A}{x_0} \sum_{m,n} J_m\left(n\frac{2\pi a}{x_0}\right) \delta\left(y - n\frac{2\pi}{x_0} - m y_0\right)$$

$$(m = 0, \pm1, \pm2, \pm3, \dots)$$

$$(n = 0, \pm1, \pm2, \pm3, \dots)$$

(2.89)

$$f(x) = h(x) \sum_{n=-\infty}^{+\infty} g(x - nx_0)$$

$$f(x) = \sum_{n=-\infty}^{+\infty} h(nx_0) g(x - nx_0)$$

$$F(y) = \frac{1}{x_0} \sum_{n=-\infty}^{+\infty} \left\{ G\left(\frac{n2\pi}{x_0}\right) H\left(y - \frac{n2\pi}{x_0}\right) \right\}$$

(2.90)

$$F(y) = \frac{1}{x_0} G(y) \sum_{n=-\infty}^{+\infty} H\left(y - \frac{n2\pi}{x_0}\right)$$

(2.91)

Source: Champeney, D.C., *Fourier Transforms and Their Physical Applications*, Academic Press, New York, 1973. With permission.

Note: $J_n(-a) = J_{-n}(a) = (-1)^n J_n(a)$. See Appendix H for some properties of Bessel functions.

Book MATLAB m-File: ex_3_6_7

```
%Book m-file: ex_3_6_7
dt = 0.001;
n = 0:20000;
f = (sin(dt*n)./(dt*n+eps));
for m = 0:20000
    ua(m+1) = .5+0.001*sum(f(1,1:1+m))/pi;
end;
k = -20000:20001;
plot(k,[1-fliplr(ua) ua],'k')
grid on;
xlabel('\omega_0t');ylabel('u_a(t)')
```

∎

Example 3.23

The FT of an integral $g(t) = \int_{-\infty}^{t} f(x)dx$ is found as follows:

$$f(t)*u(t) = \int_{-\infty}^{\infty} f(x)u(t-x)dx = \int_{-\infty}^{t} f(x)dx = g(t) \quad u(t-x) = 0 \quad \text{for } x > t \quad (3.50)$$

Using the convolution property, we find

$$\mathcal{F}\{g(t)\} = \mathcal{F}\left\{ \int_{-\infty}^{t} f(x)dx \right\} = \mathcal{F}\{f(t)\}\mathcal{F}\{u(t)\} = F(\omega)\left(\pi\delta(\omega) + \frac{1}{j\omega} \right)$$

$$= \frac{1}{j\omega}F(\omega) + \pi F(\omega)\delta(\omega) = \frac{1}{j\omega}F(\omega) + \pi F(0)\delta(\omega) \quad (3.51)$$

∎

3.7 Duration of a Signal and the Uncertainty Principle

It is appropriate to take the moments of a signal with respect to origin since any shift affects only the phase. From the frequency differentiation Property #9 Table 3.3, we obtain

$$(-jt)f(t) \overset{\mathcal{F}}{\leftrightarrow} \frac{dF(\omega)}{d\omega} = \frac{d}{d\omega}\{A(\omega)e^{j\phi(\omega)}\} = \left(\frac{dA(\omega)}{d\omega} + jA(\omega)\frac{d\phi(\omega)}{d\omega} \right)e^{j\phi(\omega)} \quad (3.52)$$

From Parseval's formula, we find the RMS duration of a signal to be

$$\int_{-\infty}^{\infty} t^2|f(t)|^2 dt = \frac{1}{2\pi}\int_{-\infty}^{\infty} \left|\frac{dF(\omega)}{d\omega}\right|^2 d\omega = \frac{1}{2\pi}\int_{-\infty}^{\infty} \left[\left(\frac{dA}{d\omega}\right)^2 + A^2\left(\frac{d\phi}{d\omega}\right)^2 \right] d\omega \quad (3.53)$$

Without loss of generality, we assume that the energy of the signal is one. Hence,

$$\int_{-\infty}^{\infty} |f(t)|^2 dt = \frac{1}{2\pi} \int_{-\infty}^{\infty} A^2(\omega)d\omega = 1 \tag{3.54}$$

Define the duration of $f(t)$ and $F(\omega)$ by

$$D_t^2 = \int_{-\infty}^{\infty} t^2 |f(t)|^2 dt \quad D_\omega^2 = \int_{-\infty}^{\infty} \omega^2 |F(\omega)|^2 d\omega \tag{3.55}$$

If $\lim_{t \to \pm\infty} \sqrt{t}\, f(t) = 0$ then the **uncertainty principle** states that

$$D_t D_\omega \geq \sqrt{\frac{\pi}{2}} \tag{3.56}$$

and the equality holds only for the Gaussian signals

$$f(t) = \sqrt{\frac{a}{\pi}} e^{-at^2} \tag{3.57}$$

3.8 Applications to Linear-Time Invariant Systems

Example 3.24

To find the transfer function of a relaxed (zero initial condition) RLC series circuit (see Figure E.1.3), we first write Kirchhoff's voltage law equation and then take the FT of both sides of the equation, applying, of course, the linearity property. Hence,

$$\mathscr{F}\left\{L\frac{di(t)}{dt}\right\} + \mathscr{F}\{Ri\} + \mathscr{F}\left\{\frac{1}{C}\int_0^t i(t)dt\right\} = \mathscr{F}\{v(t)\}$$

or

$$j\omega L I(\omega) + R I(\omega) + \frac{I(\omega)}{j\omega C} = V(\omega)$$

or

$$H(\omega) = \frac{I(\omega)}{V(\omega)} = \frac{j\omega C}{1 + j\omega RC - \omega^2 LC} \tag{3.58}$$

■

Example 3.25 (Distortional filter)

The output of a distortionless system (filter) is given by

$$g(t) = H_0 f(t - t_0) \tag{3.59}$$

which indicates that the output is a shifted version of the input multiplied by a constant, always less than one for passive systems. Taking the FT of the above equation, we find

$$G(\omega) = H_0 e^{-j\omega t_0} F(\omega) \tag{3.60}$$

But, since any output of an LTI system is $G(\omega) = H(\omega) F(\omega)$, then from the above equation we obtain the desired transfer function:

$$H(\omega) \triangleq |H(\omega)| e^{j\theta_h(\omega)} = H_0 e^{-j\omega t_0} \tag{3.61}$$

If $|H(\omega)|$ is not constant, we say that the filter is **amplitude distorted**, and if $\theta_h(\omega)$ is a nonlinear function, the filter is known as **phase distorted**. ∎

Example 3.26 (Ideal low-pass filter)

An ideal low-pass filter is one that has a transfer function of the form

$$H(\omega) = H_0 p_{\omega_0}(\omega) e^{-j\omega t_0} \triangleq \text{Ideal low-pass filter} \tag{3.62}$$

Therefore, the impulse response of this filter is found by taking the inverse FT of (3.62). Hence,

$$h(t) = \frac{1}{2\pi} \int_{-\infty}^{\infty} H_0 p_{\omega_0}(\omega) e^{-j\omega t_0} e^{j\omega t} d\omega = \frac{1}{2\pi} H_0 \int_{-\omega_0}^{\omega_0} e^{j\omega(t-t_0)} d\omega = \frac{H_0}{\pi} \frac{\sin \omega_0(t - t_0)}{t - t_0} \tag{3.63}$$

Since the impulse response is defined for negative and positive times, it indicates that the ideal filter is noncausal and, therefore, is not physically realizable; that is, it is not possible to build an ideal filter using any combination of resistors, capacitors, and inductors. ∎

Example 3.27 (Step response of an ideal low-pass filter)

The FT of a unit step function is found to be

$$U(\omega) \triangleq \mathcal{F}\{u(t)\} = \pi\delta(\omega) + \frac{1}{j\omega} \tag{3.64}$$

But we have found above that the output of a system in the frequency domain to a unit step function is given by $G(\omega) = H(\omega)U(\omega)$. Therefore, the output of an ideal low-pass filter due to a unit step function is

$$g(t) \overset{\Delta}{=} \mathcal{F}^{-1}\{G(\omega)\} = \mathcal{F}^{-1}\{H(\omega)U(\omega)\} = \frac{1}{2\pi} \int\limits_{-\infty}^{\infty} H_0 p_{\omega_0}(\omega)e^{-j\omega t_0}\left[\pi\delta(\omega) + \frac{1}{j\omega}\right]e^{j\omega t}\,d\omega$$

$$= \frac{H_0}{2} \int\limits_{-\omega_0}^{\omega_0} \delta(\omega)e^{j\omega(t-t_0)}\,d\omega + \frac{H_0}{2j\pi} \int\limits_{-\omega_0}^{\omega_0} \frac{e^{j\omega(t-t_0)}}{\omega}\,d\omega \tag{3.65}$$

The first integral is equal to $H_0/2$, by the properties of the delta function. The integrand of the second integral can be expanded into Euler's form of $\cos\omega(t - t_0)/\omega + j\sin\omega$ $(t - t_0)/\omega$. The first factor of this expansion is an odd function; since the integration is symmetric around the origin, the integral vanishes. Thus, (3.65) becomes

$$g(t) = \frac{H_0}{2} + \frac{H_0}{\pi} \int\limits_{0}^{\omega_0} \frac{\sin\omega(t - t_0)}{\omega}\,d\omega = \frac{H_0}{2} + \frac{H_0}{\pi} \int\limits_{0}^{\omega_0(t-t_0)} \frac{\sin x}{x}\,dx$$

$$= \frac{H_0}{2} + \frac{H_0}{\pi} Si(\omega_0(t - t_0)) \tag{3.66}$$

where $x = \omega(t - t_0)$ and $Si( )$ is the **sine integral.** Its graphical representation of the output function $g(t)$ is shown in Figure E.3.6. We observe the following for this response function:

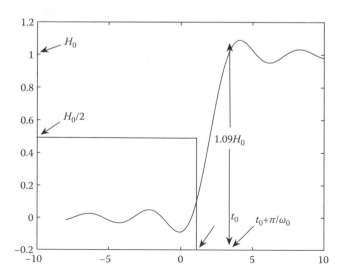

FIGURE E.3.6

1. The output signal is shifted from 0 by t_0 in the time domain.
2. The response is not identically zero for $t < 0$, as expected for non-realizable filters.
3. The response has a gradual rise, in contrast to the abrupt input rise.
4. The **rise time** in the time interval $t_0 - \pi/\omega < t < t_0 + \pi/\omega$ is $t_r = 2\pi/\omega_0 = 1/f_0$ is equal to the reciprocal of the cutoff frequency of the filter. (This is not the only definition of rise time used in engineering literature. For example, the rise time defined in electronic circuits is from $0.1H_0$ to $0.9H_0$ of its height.)
5. The response is oscillatory with a frequency equal to the cutoff frequency.

Book MATLAB Function for Figure E.3.6:
f = taa_fig_ex_3_8_4(tmin,tmax,dt,H0,w0,dw,t0)

```
function[g] = taa_fig_ex_3_4(tmin,tmax,dt,H0,w0,dw,t0)
for t = tmin:tmax
    g(t-tmin+1) = (H0/2)+(H0/pi)*dw*sum((sin([0:dw:w0]*...
        (t*dt-t0)+eps)./([0:dw:w0]+eps)));
end;
```

Figure E.3.6 was plotted for the following values: $H_0 = 1$, $t_0 = 2$, $w_0 = 1.5$, $dt = 0.01$, $dw = 0.0001$, $t_{min} = -800$, $t_{max} = 1000$. With these values we created 1801 values of the output function, and therefore, to plot it out we created first the vector $t = [-800:1000]$ *0.01 and then we wrote in the command window: plot(t,g,'k'). ∎

Example 3.28

It is desired to find the impulse response and the frequency characteristics of a seismic instrument (system, filter) shown in Figure E.3.7a. The input is a delta force function $f(t) = \delta(t)$ and the output is the displacement x. The forces act on a straight line and are shown in Figure E.3.7b. Since the algebraic sum of the forces must be equal to zero, the equation governing the system is

$$M\frac{d^2h(t)}{dt^2} + D\frac{dh(t)}{dt} + Kh(t) = \delta(t) \tag{3.67}$$

Since the input function is a delta one, we follow the convention to substitute $x(t)$ with $h(t)$ (the impulse response). If the system is initially relaxed, we have the initial conditions

$$\boxed{h(0-) = \frac{dh(0-)}{dt} = 0} \tag{3.68}$$

Next, we integrate (3.67) between the limits $t = 0-$ and $t = 0+$. This is written as

$$M\int_{0-}^{0+} \frac{d^2h}{dt^2}dt + D\int_{0-}^{0+} \frac{dh}{dt}dt + K\int_{0-}^{0+} hdt = \int_{0-}^{0+} \delta(t)dt \tag{3.69}$$

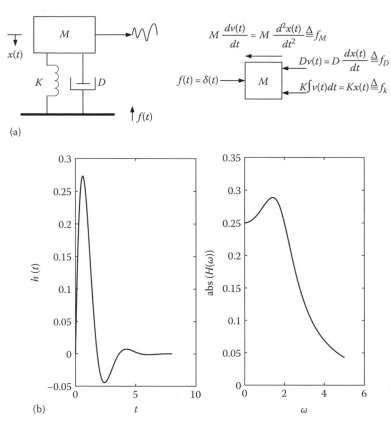

FIGURE E.3.7

We assume that both h and dh/dt do not include infinite values (are regular functions), and (3.69) gives

$$M\left[\frac{dh(0+)}{dt} - \frac{dh(0-)}{dt}\right] + D[h(0+) - h(0-)] = 1 \quad \text{or} \quad \boxed{M\frac{dh(0+)}{dt} + Dh(0+) = 1} \quad (3.70)$$

Integrate (3.67) twice to find

$$M\int_{0-}^{0+}\frac{dh}{dt}dt + D\int_{0-}^{0+}hdt = \int_{0-}^{0+}\int\delta(t)dtdt \quad \text{or} \quad M[h(0+) - h(0-)] = 0 \quad \text{or} \quad \boxed{h(0+) = h(0-)} \quad (3.71)$$

Combining (3.71), (3.70), and (3.68), we obtain

$$\frac{dh(0+)}{dt} = \frac{1}{M} \quad (3.72)$$

The above equation indicates that the impulse response of the velocity $dx/dt \overset{\Delta}{=} dh/dt$ jumps from 0 to $1/M$ instantaneously while the distance $x(t)$ remains at zero.

We are now ready to solve our problem, which becomes

$$\frac{d^2h(t)}{dt^2} + \frac{D}{M}\frac{dh(t)}{dt} + \frac{K}{M}h(t) = 0 \quad t > 0 \quad \text{(a)}$$

$$h(0+) = 0 \quad \frac{dh(0+)}{dt} = \frac{1}{M} \quad \text{(b)}$$

(3.73)

Set $D/M = 2a$ and $b^2 = K/M$, and with $D > 0$, $M > 0$, $K > 0$, the solution is

$$h(t) = e^{-at}\left(C_1 \cos \sqrt{b^2 - a^2}\, t + C_2 \sin \sqrt{b^2 - a^2}\, t\right) \quad t > 0 \qquad (3.74)$$

It is here assumed that $b > a$ is the **underdamped** case for the system. The constants are easily found using the initial conditions given by (3.73b). These lead to $C_1 = 0$ and $C_2 = 1/\left(M\sqrt{b^2 - a^2}\right)$ and thus, the solution is

$$x(t) = \frac{1}{M\sqrt{b^2 - a^2}} e^{-at} \sin \sqrt{b^2 - a^2}\, t \quad t > 0$$

$$2a = \frac{D}{M} \quad b^2 = \frac{K}{M} \quad b > a \quad K > 0 \quad M > 0 \quad D > 0$$

(3.75)

If we take the FT of (3.67), then the transfer function is

$$H(\omega) = \frac{1}{M}\frac{1}{b^2 - \omega^2 + j2a\omega} = \frac{1}{M}\frac{1}{\left[\left(1 - \left(\frac{\omega}{b}\right)^2\right)^2 + 4\left(\frac{a}{b}\right)^2\left(\frac{\omega}{b}\right)^2\right]^{1/2}}$$

$$\times \exp\left[-j\tan^{-1}\left(2\left(\frac{a}{b}\right)\frac{\frac{\omega}{b}}{1 - \left(\frac{\omega}{b}\right)^2}\right)\right] \qquad (3.76)$$

Figure E.3.7b shows the impulse response and the amplitude of its transfer function. The constants used for the graphs were $M = 1$, $b = 2$, $a/b = 0.5$, and $a = 1$. ∎

Example 3.29

To find the transfer function $V_o(\omega)/V_i(\omega) = H(\omega)$, for the relaxed system shown in Figure E.3.8a, we first write the equation of the system

$$L\frac{di(t)}{dt} + \frac{1}{C}\int i(t)dt + Ri(t) = v_i(t) \qquad (3.77)$$

Taking the FT of the above equation, we obtain

$$j\omega LI(\omega) + \frac{1}{j\omega C}I(\omega) + RI(\omega) = V_i(\omega) \quad \text{or} \quad \frac{I(\omega)}{V_i(\omega)} = \frac{j\omega C}{LC(j\omega)^2 + j\omega RC + 1} \qquad (3.78)$$

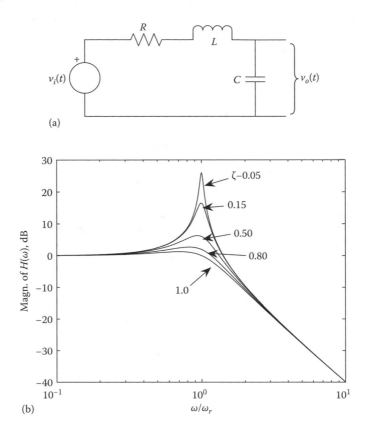

FIGURE E.3.8

The voltage across the capacitor, which is the output voltage, is given by

$$v_o(t) = \frac{1}{C}\int i(t)dt \quad \text{or in FT domain } V_o(\omega) = \frac{1}{j\omega C}I(\omega) \tag{3.79}$$

Combining this result with (3.78), we have

$$\frac{\left(\dfrac{I(\omega)}{j\omega C}\right)}{V_i(\omega)} = \frac{V_o(\omega)}{V_i(\omega)} \triangleq H(\omega) = \frac{1}{LC(j\omega)^2 + RC(j\omega) + 1} = \frac{\omega_r^2}{(j\omega)^2 + j2\varsigma\omega_r\omega + \omega_r^2}$$

$$= \frac{1}{1 + j2\varsigma\left(\dfrac{\omega}{\omega_r}\right) - \left(\dfrac{\omega}{\omega_r}\right)^2} \tag{3.80}$$

$$\omega_r = 1/\sqrt{LC} \quad \varsigma = (R/2L)/\omega_r$$

Figure E.3.9b shows the amplitude of the transfer function versus ω/ω_r for the different values of ς. ∎

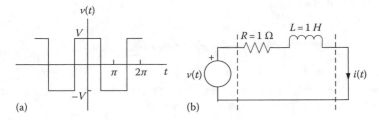

FIGURE E.3.9

Example 3.30 (Response to $e^{j\omega t}$)

Since the LTI system is characterized by a differential equation, we conclude that its output due to an exponential input is also an exponential multiplied by a constant (most of the times, complex). Therefore, we have

$$a_p \frac{d^p y(t)}{dt^p} + a_{p-1} \frac{d^{p-1} y(t)}{dt^{p-1}} + \cdots + a_0 y(t) = b_q \frac{d^q x(t)}{dt^q} + b_{q-1} \frac{d^{q-1} x(t)}{dt^{q-1}} + \cdots + b_0 x(t)$$

or

$$a_p (j\omega)^p H e^{j\omega t} + a_{p-1} (j\omega)^{p-1} H e^{j\omega t} + \cdots + a_0 H e^{j\omega t}$$
$$= b_q (j\omega)^q e^{j\omega t} + b_{q-1} (j\omega)^{q-1} e^{j\omega t} + \cdots + b_0 e^{j\omega t}$$

or

$$A(\omega) y(t) = B(\omega) e^{j\omega t}$$

or

$$y(t) = \text{output} = \frac{B(\omega)}{A(\omega)} e^{j\omega t} = H(\omega) e^{j\omega t} \tag{3.81}$$

Therefore, in operational form we write

$$\Theta\{e^{j\omega t}\} = H(\omega) e^{j\omega t} \tag{3.82}$$

If the input is a constant K, from (3.82) we conclude that

$$\Theta\{K\} = K H(0) \tag{3.83}$$

∎

Example 3.31 (Response to a periodic function)

Since the input function to the system is periodic with period T, it can be expressed in FS as follows:

$$f_i(t) = \sum_{n=-\infty}^{\infty} \alpha_n e^{jn\omega_0 t} \qquad \omega_0 = \frac{2\pi}{T} \qquad \alpha_n = \frac{1}{T} \int_{-T/2}^{T/2} f_i(t) e^{-jn\omega_0 t} dt \tag{3.84}$$

It follows from (3.82) that

$$f_{on}(t) = H(n\omega_0)\alpha_n e^{jn\omega_0 t} \quad \text{and} \quad f_o(t) = \sum_{n=-\infty}^{\infty} H(n\omega_0)\alpha_n e^{jn\omega_0 t} \tag{3.85}$$

Therefore, if the input to the LTI system is periodic, then the output is also periodic. It must be noted that the response is the steady-state response. ∎

Example 3.32

For the circuit (system) shown in Figure E.3.9b, it is desired to find its output (current) if its input is the signal shown in Figure E.3.9a. The FS expansion of the input signal is

$$v(t) = \frac{4V}{\pi}\left[\cos t - \frac{1}{3}\cos 3t + \frac{1}{5}\cos 5t - \cdots\right] \quad \omega_0 = \frac{2\pi}{T} = 1 \tag{3.86}$$

The transfer function of the system is

$$H(n\omega_0) = \frac{I(n\omega_0)}{V(n\omega_0)} = \frac{1}{1+jn} = \frac{1}{\sqrt{1+n^2}} e^{-j\tan^{-1}(n)} \tag{3.87}$$

From the superposition principle, we obtain

$$i(t) = \frac{4V}{\pi}\left[\frac{1}{\sqrt{2}}\cos(t - \tan^{-1}1) - \frac{1}{3\sqrt{10}}\cos(3t - \tan^{-1}3)\right.$$
$$\left. + \frac{1}{5\sqrt{26}}\cos(5t - \tan^{-1}5) + \cdots\right] \tag{3.88}$$

∎

Example 3.33 (Ideal high-pass filter)

The frequency characteristic of an ideal high-pass filter is given by

$$H(\omega) = [H_0 - H_0 p_{\omega_0}(\omega)]e^{-j\omega t_0} \tag{3.89}$$

The corresponding impulse response function, which is obtained by taking the inverse FT of the above equation, is (see Table 3.5)

$$h(t) = H_0\delta(t - t_0) - \frac{H_0}{\pi}\frac{\sin\omega_0(t - t_0)}{t - t_0} \quad -\infty < t < \infty \tag{3.90}$$

Note that the filter is noncausal. ∎

3.9 Applications to Communication Signals

The method of processing a signal for more efficient transmission is called **modulation**. To accomplish this, we multiply the signal with a sinusoidal signal.

Example 3.34 (Modulation)

From tables, we know that

$$\mathcal{F}\{\cos \omega_c t\} = \pi\delta(\omega - \omega_c) + \pi\delta(\omega + \omega_c)$$

and, therefore, by the frequency convolution property, we obtain

$$\mathcal{F}\{f(t) \cos \omega_c t\} = \frac{1}{2\pi} F(\omega) * [\pi\delta(\omega - \omega_c) + \pi\delta(\omega + \omega_c)]$$

$$= \frac{1}{2}[F(\omega) * \delta(\omega - \omega_c) + F(\omega) * \delta(\omega + \omega_c)]$$

$$= \frac{1}{2}F(\omega - \omega_c) + \frac{1}{2}F(\omega + \omega_c) \tag{3.91}$$

The above equation indicates that the spectrum of a modulated signal is translated by $\pm\omega_c$. ∎

Example 3.35 (Ordinary AM signal)

The frequency spectrum of an amplitude modulated (AM) signal is given by

$$F(\omega) = \mathcal{F}\{K[1 + m(t)] \cos \omega_c t] = \mathcal{F}\{K \cos \omega_c t\} + \mathcal{F}\{Km(t) \cos \omega_c t\}$$

$$= K\pi\delta(\omega - \omega_c) + K\pi\delta(\omega + \omega_c) + \frac{1}{2}KM(\omega - \omega_c) + \frac{1}{2}KM(\omega + \omega_c) \quad \mathcal{F}\{m(t)\} = M(\omega)$$

$$\tag{3.92}$$

The spectrum for a band-limited signal is shown in Figure E.3.10a. We observe that the spectrum of the ordinary modulated signal consists of the carrier spectrum and the

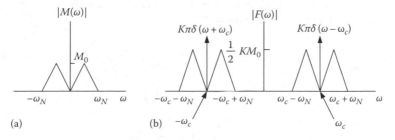

FIGURE E.3.10

spectrum of the modulated signal has a bandwidth twice of the original signal (see Figure E.3.10b). The portion of the spectrum above ω_c is called the **upper sideband** spectrum, and the symmetrical portion below ω_c is called the **lower sideband**. Note that the sidebands are the information-bearing components of the modulated signal.

A **double-sideband-suppressed carrier** (DSBSC) AM signal is given by

$$f(t) = m(t) \cos \omega_c t \tag{3.93}$$

and its spectrum is

$$F(\omega) = \mathcal{F}\{m(t) \cos \omega_c t\} = \frac{1}{2} M(\omega - \omega_c) + \frac{1}{2} M(\omega + \omega_c) \tag{3.94}$$

∎

Example 3.36 (Demodulation)

The modulated signal $f(t) = m(t) \cos \omega_c t$ received by an antenna is multiplied by the carrier to obtain the signal

$$f(t) \cos \omega_c t = m(t) \cos^2 \omega_c t = m(t) \frac{1}{2}(1 + \cos 2\omega_c t) = \frac{1}{2} m(t) + \frac{1}{2} m(t) \cos 2\omega_c t \tag{3.95}$$

The spectrum of the above signal is

$$\mathcal{F}\{f(t) \cos \omega_c t\} = \mathcal{F}\{m(t) \cos^2 \omega_c t\} = \mathcal{F}\left\{\frac{1}{2} m(t)\right\} + \mathcal{F}\left\{\frac{1}{2} m(t) \cos 2\omega_c t\right\}$$

$$= \frac{1}{2} M(\omega) + \frac{1}{4} M(\omega - 2\omega_c) + \frac{1}{4} M(\omega + 2\omega_c) \tag{3.96}$$

If the spectrum of $m(t)$ is narrow (band-limited), the first term indicates a spectrum concentrated around the origin (zero frequency) and the second term indicates a spectrum shifted by $2\omega_c$. Therefore, a low-pass filter will eliminate the shifted part of the spectrum and retain only the spectrum around the origin, which is the desired signal $m(t)$. ∎

Example 3.37 (Pulse amplitude modulation [PAM])

If $g(t)$ is a train of periodic rectangular pulses of width $2a$ seconds and repeated every T seconds, then the spectrum of a PAM is

$$F(\omega) = \mathcal{F}\{f(t)\} = \mathcal{F}\{m(t)g(t)\} = \frac{1}{2\pi} M(\omega) * G(\omega) \tag{3.97}$$

From Example 3.21 and by setting

$$T = \frac{1}{2f_N} = \frac{\pi}{\omega_N} \qquad \omega_0 = \frac{2\pi}{T} = 2\omega_N$$

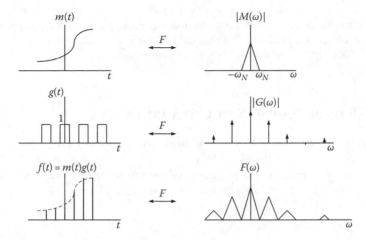

FIGURE E.3.11

we obtain

$$G(\omega) = 2 \sum_{n=-\infty}^{\infty} \frac{\sin a2n\omega_N}{n} \delta(\omega - 2n\omega_N)$$

$$F(\omega) = \frac{1}{\pi} M(\omega)^* \sum_{n=-\infty}^{\infty} \frac{\sin a2n\omega_N}{n} \delta(\omega - 2n\omega_N) = \frac{1}{\pi} \sum_{n=-\infty}^{\infty} \frac{\sin a2n\omega_N}{n} M(\omega)^* \delta(\omega - 2n\omega_N)$$

$$= \frac{1}{\pi} \sum_{n=-\infty}^{\infty} \frac{\sin a2n\omega_N}{n} M(\omega - 2n\omega_N) \tag{3.98}$$

For a band-limited signal $m(t)$, Figure E.3.11 graphically shows the above development.

∎

Example 3.38

Let a PAM signal be the product of the band-limited modulated signal $m(t)$ and a periodic signal $g(t)$ of period $T\,(<\pi/\omega_N)$ seconds. Expanding the periodic function $g(t)$ in FS, we obtain

$$f(t) = m(t)g(t) = m(t) \sum_{n=-\infty}^{\infty} \alpha_n e^{jn\omega_0 t} = \sum_{n=-\infty}^{\infty} \alpha_n m(t) e^{jn\omega_0 t} \qquad \omega_0 = \frac{2\pi}{T}$$

$$F(\omega) = \mathcal{F}\{f(t)\} = \mathcal{F}\left\{ \sum_{n=-\infty}^{\infty} \alpha_n m(t) e^{jn\omega_0 t} \right\} = \sum_{n=-\infty}^{\infty} \alpha_n \mathcal{F}\{m(t) e^{jn\omega_0 t}\} = \sum_{n=-\infty}^{\infty} \alpha_n M(\omega - n\omega_0)$$

$$\tag{3.99}$$

where the convolution property of the FT was used in the last step. From (3.99), we infer that the amplitude spectrum of the PAM signal is periodic, repetitive of the spectrum of $m(t)$, but each one's amplitude modified by the coefficients α_n's of the Fourier coefficients of $g(t)$.   ∎

3.10 Signals, Noise, and Correlation

Often signals are contaminated with noise. By noise, we normally mean any spurious or undesired disturbances that tend to obscure or mask the signal transmitted. The noise signal encounter in practice, has random amplitude variations. In the present development we assume that the noise signal has zero mean value, for example,

$$\lim_{T \to \infty} \frac{1}{T} \int_{-1/T}^{1/T} v(t)dt = 0 \tag{3.100}$$

In general, the two signals are said to be **uncorrelated** if

$$\bar{r}_{x,h}(\tau) = \lim_{T=\infty} \frac{1}{T} \int_{-T/2}^{T/2} x(t)h(t-\tau)dt$$

$$= \left[\lim_{T=\infty} \frac{1}{T} \int_{-T/2}^{T/2} x(t)dt \right] \left[\lim_{T=\infty} \frac{1}{T} \int_{-T/2}^{T/2} h(t)dt \right] \tag{3.101}$$

From the last two equations we conclude that the cross-correlation of any signal with noise that has zero mean value is zero for all τ.

Example 3.39

The average autocorrelation function of $f(t) = s(t) + v(t)$ ($v(t) = $ noise uncorrelated with $s(t)$) is

$$\bar{r}_{ff}(\tau) = \lim_{T=\infty} \frac{1}{T} \int_{-T/2}^{T/2} s(t)v(t-\tau)dt = \lim_{T=\infty} \frac{1}{T} \int_{-T/2}^{T/2} [s(t) + v(t)][s(t-\tau) + v(t-\tau)]dt$$

$$= \bar{r}_{ss}(\tau) + \bar{r}_{vv}(\tau) + \bar{r}_{sv}(\tau) + \bar{r}_{vs}(\tau) \quad \text{or} \quad \bar{r}_{ff}(\tau) = \bar{r}_{ss}(\tau) + \bar{r}_{vv}(\tau) \tag{3.102}$$

                                                                                                                       ∎

Example 3.40

Let $g(t)$ and $f(t)$ be the transmitted and the received signals, respectively. Then, $f(t) = s(t) + v(t)$, where $s(t)$ is the desired signal and $v(t)$ is the noise signal. If we now cross-correlate the received signal with the transmitted one, we obtain

$$\bar{r}_{fg}(\tau) = \lim_{T=\infty} \frac{1}{T} \int_{-T/2}^{T/2} [s(t) + v(t)]g(t - \tau)dt = \bar{r}_{sg}(\tau) + \bar{r}_{vg}(\tau) \quad \text{or} \quad \bar{r}_{fg}(\tau) = \bar{r}_{sg}(\tau) \quad (3.103)$$

This is because the noise and the signal are not correlated, and therefore, their cross-correlation is zero. ∎

3.11 Average Power Spectra, Random Signals, Input–Output Relations

The **power spectral density** (PSD) of a function is defined (Wiener–Khinchin theorem) as the FT of the average autocorrelation function of the function. Thus, we have

$$S(\omega) = \mathcal{F}\{\bar{r}_{ff}(\tau)\} = \int_{-\infty}^{\infty} \bar{r}_{ff}(\tau)e^{-j\omega\tau}d\tau$$

$$(3.104)$$

$$\bar{r}_{ff}(\tau) = \mathcal{F}^{-1}\{S(\omega)\} = \frac{1}{2\pi} \int_{\infty}^{\infty} S(\omega)e^{j\omega\tau}d\omega$$

It is assumed that the energy of the signal is finite, which is equivalent to saying that the left-hand side of Equation 3.107 is finite.

Example 3.41

To find the total average power (mean-square value) of a function $f(t)$ we write (see (3.104))

$$\bar{r}_{ff}(0) = \frac{1}{2\pi} \int_{-\infty}^{\infty} S(\omega)d\omega \qquad (3.105)$$

From the definition of the autocorrelation function, we also have

$$\bar{r}_{ff}(0) = \lim_{T\to\infty} \frac{1}{T} \int_{-T/2}^{T/2} [f(t)]^2 dt \qquad (3.106)$$

Equating the right-hand side of the last two equations, we obtain

$$\lim_{T \to \infty} \frac{1}{T} \int_{-T/2}^{T/2} [f(t)]^2 dt = \frac{1}{2\pi} \int_{-\infty}^{\infty} S(\omega) d\omega \tag{3.107}$$

■

Example 3.42 (White noise)

White noise is defined as any signal whose power spectral density is a constant for all frequencies. Hence, $S(\omega) = C$ and, therefore, its autocorrelation function is

$$\bar{r}(\tau) = \mathscr{F}^{-1}\{S(\omega)\} = \frac{1}{2\pi} \int_{-\infty}^{\infty} S(\omega) e^{j\omega\tau} d\omega = C \frac{1}{2\pi} \int_{-\infty}^{\infty} e^{j\omega\tau} d\omega = C\delta(\tau) \tag{3.108}$$

■

Example 3.43 (Input–output relations of LTI)

If an input signal $x(t)$ is a random one, the output of an LTI system $y(t)$ is also a random signal. We, therefore, have the relationships (see Chapter 1)

$$y(t) = \int_{-\infty}^{\infty} h(\tau)x(t - \tau)d\tau = \int_{-\infty}^{\infty} h(\lambda)x(t - \lambda)d\lambda \quad \text{and}$$

$$y(t - \tau) = \int_{-\infty}^{\infty} h(\sigma)x(t - \tau - \sigma)d\sigma \tag{3.109}$$

From the definition of the average autocorrelation function (3.101), we write

$$\bar{r}_{yy}(\tau) = \lim_{T \to \infty} \frac{1}{T} \int_{-T/2}^{T/2} y(t)y(t - \tau)dt \tag{3.110}$$

Substituting the expressions of (3.109) in the equation above, we have

$$\bar{r}_{yy}(\tau) = \lim_{T \to \infty} \frac{1}{T} \int_{-T/2}^{T/2} \left[\int_{-\infty}^{\infty} h(\lambda)x(t - \lambda)d\lambda \int_{-\infty}^{\infty} h(\sigma)x(t - \tau - \sigma)d\sigma \right] dt$$

$$= \int_{-\infty}^{\infty} h(\lambda) \int_{-\infty}^{\infty} h(\sigma) \left[\lim_{T \to \infty} \frac{1}{T} \int_{-T/2}^{T/2} x(t - \lambda)x(t - \tau - \sigma)dt \right] d\sigma d\lambda$$

$$= \int_{-\infty}^{\infty} h(\lambda) \int_{-\infty}^{\infty} h(\sigma)\bar{r}_{xx}(\tau + \sigma - \lambda)d\sigma d\lambda \tag{3.111}$$

■

Example 3.44 (Input–output power spectra)

If the input random function is $x(t)$ and the output function is $y(t)$, then the output power spectrum is given by (see (3.104))

$$S_y(\omega) = \mathcal{F}\{\bar{r}_{yy}(\tau)\} = \int_{-\infty}^{\infty} \bar{r}_{yy}(\tau)e^{-j\omega\tau}d\tau$$

$$= \int_{-\infty}^{\infty}\left[\int_{-\infty}^{\infty} h(\lambda)\int_{-\infty}^{\infty} h(\sigma)\bar{r}_{xx}(\tau+\sigma-\lambda)d\sigma d\lambda\right]e^{-j\omega\tau}d\tau \quad (\text{set } \mu = \tau+\sigma-\lambda)$$

$$= \int_{-\infty}^{\infty} h(\lambda)e^{-j\omega\lambda}d\lambda\int_{-\infty}^{\infty} h(\sigma)e^{j\omega\sigma}d\sigma\int_{-\infty}^{\infty} \bar{r}_{xx}(\mu)e^{-j\omega\mu}d\mu = H(\omega)H^{*}(\omega)S_{xx}(\omega)$$

$$= |H(\omega)|^2 S_{xx}(\omega) \tag{3.112}$$

∎

Example 3.45

The transfer function of an RC circuit and the power spectrum of the input white noise are

$$H(\omega) = \frac{1/RC}{j\omega + (1/RC)} \quad S_x(\omega) = K = \text{constant} \tag{3.113}$$

Therefore, the output power density is given by (see (3.112))

$$S_y(\omega) = |H(\omega)|^2 S_x(\omega) = \frac{(1/RC)^2}{\omega^2 + (1/RC)^2}K \tag{3.114}$$

The mean-square output voltage may be evaluated from $S_y(\omega)$ as follows:

$$\lim_{T\to\infty}\frac{1}{T}\int_{-T/2}^{T/2}[y(t)]^2dt = \frac{1}{2\pi}\int_{-\infty}^{\infty} S_y(\omega)d\omega = \frac{K}{2\pi(RC)^2}\int_{-\infty}^{\infty}\frac{d\omega}{\omega^2+(1/RC)^2} = \frac{K}{2RC} \tag{3.115}$$

where the integral was found using mathematical tables of integrals. ∎

3.12 FT in Probability Theory

A random variable (rv) X (in this section we present the rv's with capital letters and their values with lower case letters) taking real values is characterized by its **cumulative probability distribution function** (cdf) $F(x)$ and it is defined by

$$F(x) = \Pr\{X < x\} = \text{probability that the rv } X \text{ assumes a value less than } x \tag{3.116}$$

If $F(x)$ is differentiable, then we define the **probability density function (pdf)** $f(x)$ by

$$f(x) = \frac{dF(x)}{dx} \tag{3.117}$$

The main properties of these two functions are

$$
\begin{align}
&F(-\infty) = 0 \quad F(\infty) = 1 &&(1)\\
&F(x_1) < F(x_2) \quad \text{if } x_1 < x_2 &&(2)\\
&\Pr\{x_1 < X < x_2\} = F(x_2) - F(x_1) &&(3)\\
&F(x) = \int_{-\infty}^{x} f(x)dx &&(4)\\
&\int_{-\infty}^{\infty} f(x)dx = 1 &&(5)\\
&\Pr\{x_1 X < x_2\} = \int_{x_1}^{x_2} f(x)dx &&(6)\\
&f(x) > 0 &&(7)
\end{align}
\tag{3.118}
$$

The expectations and moments of random variables are

$$E\{X\} = \int_{-\infty}^{\infty} x f(x)dx = \text{expectation or mean value of } X \tag{1}$$

$$E\{g(X)\} = \int_{-\infty}^{\infty} g(x)f(x)dx \tag{2}$$

$$E\{X^2\} = \int_{-\infty}^{\infty} x^2 f(x)dx = \text{mean-squared value of } X \tag{3}$$

$$E\{X^n\} = \int_{-\infty}^{\infty} x^n f(x)dx = m_n = n\text{th moment of } X \tag{4}$$

$$E\{(X - (E\{X\})^2\} = E\{X^2\} - (E\{X\})^2 = \text{variance of } X \tag{5}$$

$$\sigma = \sqrt{E\{(X - (E\{X\})^2\}} = \text{standard deviation} \tag{6}$$

$$(3.119)$$

3.12.1 Characteristic Function

The characteristic function of a random variable X is defined as the FT of its pdf $f(x)$. Hence, the FT pair is

$$\Phi(\omega) = E\{e^{j\omega X}\} = \int_{-\infty}^{\infty} f(x)e^{j\omega x} dx$$

$$(3.120)$$

$$f(x) = \frac{1}{2\pi} \int_{-\infty}^{\infty} \Phi(\omega)e^{-j\omega x} d\omega$$

Example 3.46

Starting from the expansion of the exponential function:

$$e^{j\omega x} = 1 + \frac{j\omega x}{1!} + \frac{(j\omega x)^2}{2!} + \cdots + \frac{(j\omega x)^n}{n!} + \cdots \quad (3.121)$$

we write

$$\Phi(\omega) = \int_{-\infty}^{\infty} f(x)e^{j\omega x} dx = \int_{-\infty}^{\infty} f(x)\left[1 + \frac{j\omega x}{1!} + \frac{(j\omega x)^2}{2!} + \cdots + \frac{(j\omega x)^n}{n!} + \cdots\right] dx$$

$$= \int_{-\infty}^{\infty} f(x)dx + j\omega \int_{-\infty}^{\infty} xf(x)dx + \cdots + \frac{(j\omega)^n}{n!} \int_{-\infty}^{\infty} x^n f(x)dx + \cdots$$

$$= 1 + j\omega m_1 + \cdots + \frac{(j\omega)^n}{n!} m_n + \cdots$$

$$\left.\frac{d^n \Phi(\omega)}{d\omega^n}\right|_{\omega=0} = j^n m_n \quad (3.122)$$

■

3.12.2 Joint Cumulative Distribution Function

The cdf of two rv's X and Y is defined as follows:

$$F(x, y) = \Pr\{X < x, Y < y\} \quad (3.123)$$

Assuming the cdf is differentiable, the pdf is

$$f(x, y) = \frac{\partial^2 F(x, y)}{\partial x \partial y} \quad (3.124)$$

If the two rv's are independent, then

$$F(x, y) = F(x)F(y) \quad f(x, y) = f(x)f(y) \tag{3.125}$$

Additional definitions are

$$E\{XY\} = E\{X\}E\{Y\} \quad X, Y \text{ are uncorrelated}$$
$$E\{XY\} = 0 \quad X, Y \text{ are orthogonal} \tag{3.126}$$

It is easy to show that if two variables are independent, they are also uncorrelated. The reverse is always true.

3.12.3 Characteristic Function of Two Variables

The characteristic function of two rv's is a double FT and, therefore, the pair is given by

$$\Phi(\omega_1, \omega_2) = E\{e^{j(\omega_1 X + \omega_2 Y)}\} = \int\int f(x, y)e^{j(\omega_1 x + \omega_2 y)} dx dy$$
$$f(x, y) = \frac{1}{(2\pi)^2} \int\int \Phi(\omega_1, \omega_2)e^{-j(\omega_1 x + \omega_2 y)} d\omega_1 d\omega_2 \tag{3.127}$$

If the rv's are independent, the characteristic function is the product of the one-dimensional characteristic function:

$$\Phi(\omega_1, \omega_2) = \Phi(\omega_1)\Phi(\omega_2) \tag{3.128}$$

Example 3.47

Let the rv Z be the sum of the two independent rv's X and Y: $Z = X + Y$. Therefore, we have

$$\Phi_z(\omega) = E\{e^{j\omega Z}\} = E\{e^{j\omega(X+Y)}\} = E\{e^{j\omega X}\}E\{e^{j\omega Y}\} = \Phi_x(\omega)\Phi_y(\omega) \tag{3.129}$$

where the independent property was used (see (3.128)).
Applying the convolution property of the FT, we obtain

$$f_z(z) = \mathcal{F}^{-1}\{\Phi_z(\omega)\} = \mathcal{F}^{-1}\{\Phi_x(\omega)\Phi_y(\omega)\} = f_x(x) * f_y(y) = \int_{-\infty}^{\infty} f_x(x)f_y(z-x)dx \tag{3.130}$$

which indicates that the pdf of the rv Z is equal to the convolution of the pdf's of the two rv's. ∎

Relatives to the Fourier Transform

4.1 Infinite Fourier Sine Transform

The infinite Fourier sine Transform (FST) of $f(t), 0 < t < \infty$, and its inverse are defined by

$$F_s(\omega) = \mathcal{F}_s\{f(t)\} = \int_0^\infty f(t) \sin \omega t \, dt$$

$$f(t) = \mathcal{F}_s^{-1}\{F_s(\omega)\} = \frac{2}{\pi} \int_0^\infty F_s(\omega) \sin \omega t \, d\omega$$

(4.1)

assuming, of course, that the integrals exist (see Table 4.2).

4.2 Infinite Fourier Cosine Transform

The infinite Fourier cosine transform (FCT) of $f(t), 0 < t < \infty$, and its inverse are defined by

$$F_c(\omega) = \mathcal{F}_c\{f(t)\} = \int_0^\infty f(t) \cos \omega t \, dt$$

$$f(t) = \mathcal{F}_c^{-1}\{F_c(\omega)\} = \frac{2}{\pi} \int_0^\infty F_c(\omega) \cos \omega t \, d\omega$$

(4.2)

assuming, of course, that the integrals exist (see Table 4.1).

Example 4.1

The FST of the function

$$f(t) = \begin{cases} t & 0 < t < 1 \\ 2 - t & 1 < t < 2 \\ 0 & t > 2 \end{cases}$$

(4.3)

TABLE 4.1 Fourier Cosine Transforms

$f(t)$		$F_c(\omega) = \int_0^\infty f(t)\cos\omega t\, dt$	$\omega > 0$
General properties			
1	$F_c(t)$		$(\pi/2)f(\omega)$	
2	$f(at)$	$a > 0$	$(1/a)F_c(\omega/a)$	
3	$f(at)\cos bt$	$a, b > 0$	$(1/2a)\left[F_c\left(\dfrac{\omega + b}{a}\right) + F_c\left(\dfrac{\omega - b}{a}\right)\right]$	
4	$f(at)\sin bt$	$a, b > 0$	$(1/2a)\left[F_s\left(\dfrac{\omega + b}{a}\right) - F_s\left(\dfrac{\omega - b}{a}\right)\right]$	
5	$t^{2n}f(t)$		$(-1)^n\dfrac{d^{2n}}{d\omega^{2n}}F_c(\omega)$	
6	$t^{2n+1}f(t)$		$(-1)^n\dfrac{d^{2n+1}}{d\omega^{2n+1}}F_s(\omega)$	
7	$\int_0^\infty f(r)[g(t+r) \\ +g(|t-r|)]dr$		$2F_c(\omega)G_c(\omega)$	
8	$\int_t^\infty f(r)dr$		$(1/\omega)F_s(\omega)$	
9	$f(t+a)-f_o(t-a)$		$2F_s(\omega)\sin a\omega$	$a > 0$
10	$\int_0^\infty f(r)[g(t+r) \\ +g_o(t-r)]dr$		$2F_s(\omega)\, G_s(\omega)$	
Algebraic functions			
1	$(1/\sqrt{t})$		$\sqrt{(\pi/2)}(1/\omega)^{1/2}$	
2	$(1/\sqrt{t})[1-U(t-1)]$		$(2\pi/\omega)^{1/2}C(\omega)$	
3	$(1/\sqrt{t})U(t-1)$		$(2\pi/\omega)^{1/2}[1/2 - C(\omega)]$	
4	$(t+a)^{-1/2}$	$\|\arg a\| < \pi$	$(\pi/2\omega)^{1/2}\{\cos a\omega[1 - 2C(a\omega)] + \\ \sin a\omega[1 - 2S(a\omega)]\}$	
5	$(t-a)^{-1/2}U(t-a)$		$(\pi/2\omega)^{1/2}[\cos a\omega - \sin a\omega]$	
6	$a(t^2+a^2)^{-1}$	$a > 0$	$(\pi/2)\exp(-a\omega)$	
7	$t(t^2+a^2)^{-1}$	$a > 0$	$-1/2[e^{-a\omega}\overline{\mathrm{Ei}}(a\omega) + e^{a\omega}\mathrm{Ei}(a\omega)]$	
8	$(1-t^2)(1+t^2)^{-2}$		$(\pi/2)\omega\exp(-\omega)$	
9	$-t(t^2-a^2)^{-1}$	$a > 0$	$\cos a\omega\,\mathrm{Ci}(a\omega) + \sin a\omega\,\mathrm{Si}(a\omega)$	
Exponential and logarithmic functions			
1	e^{-at}	$\mathrm{Re}\,a > 0$	$a(a^2+\omega^2)^{-1}$	
2	$(1+t)e^{-t}$		$2(1+\omega^2)^{-2}$	
3	$\sqrt{t}\,e^{-at}$	$\mathrm{Re}\,a > 0$	$\dfrac{\sqrt{\pi}}{2}(a^2+\omega^2)^{-3/4}\cos[3/2\tan^{-1}(\omega/a)]$	
4	$e^{-at}/\sqrt{t}$	$\mathrm{Re}\,a > 0$	$\sqrt{(\pi/2)}(a^2+\omega^2)^{-1/2} \\ \bullet[(a^2+\omega^2)^{1/2}+a]^{1/2}$	
5	$t^n\,e^{-at}$	$\mathrm{Re}\,a > 0$	$n![a/(a^2+\omega^2)]^{n+1} \\ \bullet\sum_{2m=0}^{n+1}(-1)^m\left(\dfrac{n+1}{2m}\right)\left(\dfrac{\omega}{a}\right)^{2m}$	

TABLE 4.1 (continued) Fourier Cosine Transforms

	$f(t)$		$F_c(\omega) = \int_0^\infty f(t)\cos\omega t\,dt$	$\omega > 0$
6	$\exp(-at^2)/\sqrt{t}$	Re $a > 0$	$\pi(\omega/8a)^{1/2}\exp(-\omega^2/8a)$	
			$\bullet I_{-1/4}(-\omega^2/8a)$	
7	$t^{2n}\exp(-a^2t^2)$	$\lvert\arg a\rvert < \pi/4$	$(-1)^n\sqrt\pi 2^{-n-1}a^{-2n-1}$	
			$\bullet\exp[-(\omega/2a)^2]\,\mathrm{He}_{2n}(2^{-1/2}\omega/a)$	
8	$t^{-3/2}\exp(-a/t)$	Re $a > 0$	$(\pi/a)^{1/2}\exp[-(2a\omega)^{1/2}]\cos(2a\omega)^{1/2}$	
9	$t^{-1/2}\exp(-a/\sqrt t)$	Re $a > 0$	$(\pi/2\omega)^{1/2}[\cos(2a\sqrt\omega) - \sin(2a\sqrt\omega)]$	
10	$t^{-1/2}\ln t$		$-(\pi/2\omega)^{1/2}[\ln(4\omega) + C + \pi/2]$	
11	$(t^2 - a^2)^{-1}\ln t$	$a > 0$	$(\pi/2\omega)\{\sin(a\omega)[\mathrm{ci}(a\omega) - \ln a]$	
			$\quad -\cos(a\omega)[\mathrm{si}(a\omega) - \pi/2]\}$	
12	$t^{-1}\ln(1+t)$		$(1/2)\{[\mathrm{ci}(\omega)]^2 + [\mathrm{si}(\omega)]^2\}$	
13	$\exp(-t/\sqrt 2)$		$(1+\omega^4)^{-1}$	
	$\sin(\pi/4 + t/\sqrt 2)$			
14	$\exp(-t/\sqrt 2)$		$\omega^2(1+\omega^4)^{-1}$	
	$\cos(\pi/4 + t/\sqrt 2)$			
15	$\ln\dfrac{a^2 + t^2}{1 + t^2}$	$a > 0$	$(\pi/\omega)[\exp(-\omega) - \exp(-a\omega)]$	
16	$\ln[1 + (a/t)^2]$	$a > 0$	$(\pi/\omega)[1 - \exp(-a\omega)]$	
Trigonometric functions				
1	$t^{-1}e^{-t}\sin t$		$(1/2)\tan^{-1}(2\omega^{-2})$	
2	$t^{-2}\sin^2(at)$	$a > 0$	$(\pi/2)(a - \omega/2)$	$\omega < 2a$
			0	$\omega > 2a$
3	$\left(\dfrac{\sin t}{t}\right)^n$	$n = 2, 3, \dots$	$\dfrac{n\pi}{2^n}\displaystyle\sum_{r>0}^{r<(\omega+n)/2}\dfrac{(-1)^r(\omega + n - 2r)^{n-1}}{r!(n-r)!},$	
			$0 < \omega < n$	
4	$\exp(-\beta t^2)\cos at$	Re $\beta > 0$	$(1/2)(\pi/\beta)^{1/2}\exp\left(-\dfrac{a^2 + \omega^2}{4\beta}\right)\cosh\left(\dfrac{a\omega}{2\beta}\right)$	
5	$(a^2 + t^2)^{-1}(1 - 2\beta$		$(1/2)(\pi/a)(1 - \beta^2)^{-1}(e^a - \beta)^{-1}$	
	$\cos t + \beta^2)^{-1}$			
	Re $a > 0,\ \lvert\beta\rvert < 1$		$\bullet(e^{a-a\omega} + \beta e^{a\omega})$	$0 \le \omega < 1$
6	$\sin(at^2)$	$a > 0$	$\left(\dfrac{1}{4}\right)\left(\dfrac{2\pi}{a}\right)^{1/2}\left[\cos\left(\dfrac{\omega^2}{4a}\right) - \sin\left(\dfrac{\omega^2}{4a}\right)\right]$	
7	$\sin[a(1 - t^2)]$	$a > 0$	$-(1/2)(\pi/a)^{1/2}\cos[a + \pi/4 + \omega^2/(4a)]$	
8	$\cos(at^2)$	$a > 0$	$\left(\dfrac{1}{4}\right)\left(\dfrac{2\pi}{a}\right)^{1/2}\left[\cos\left(\dfrac{\omega^2}{4a}\right) + \sin\left(\dfrac{\omega^2}{4a}\right)\right]$	
9	$\cos[a(1 - t^2)]$	$a > 0$	$(1/2)(\pi/a)^{1/2}\sin[a + \pi/4 + \omega^2/(4a)]$	
10	$\tan^{-1}(a/t)$	$a > 0$	$(2\omega)^{-1}[e^{-a\omega}\mathrm{Ei}(a\omega) - e^{a\omega}\mathrm{Ei}(-a\omega)]$	

TABLE 4.2   Fourier Sine Transforms

$f(t)$		$F_s(\omega) = \int_0^\infty f(t) \sin \omega t \, dt$	$\omega > 0$
General properties			
1	$F_s(t)$	$(\pi/2)f(\omega)$	
2	$f(at) \; a > 0$	$(1/a)F_s(\omega/a)$	
3	$f(at) \cos bt$	$a, b > 0$	$\left(\dfrac{1}{2a}\right)\left[F_s\left(\dfrac{\omega+b}{a}\right) + F_s\left(\dfrac{\omega-b}{a}\right)\right]$
4	$f(at) \sin bt$	$a, b > 0$	$-\left(\dfrac{1}{2a}\right)\left[F_c\left(\dfrac{\omega+b}{a}\right) - F_c\left(\dfrac{\omega-b}{a}\right)\right]$
5	$t^{2n} f(t)$	$(-1)^n \dfrac{d^{2n}}{d\omega^{2n}} F_s(\omega)$	
6	$t^{2n+1} f(t)$	$(-1)^{n+1} \dfrac{d^{2n+1}}{d\omega^{2n+1}} F_c(\omega)$	
7	$\int_0^\infty f(r) \int_{|t-r|}^{t+r} g(s) \, ds \, dr$	$(2/\omega)F_s(\omega)G_s(\omega)$	
8	$f_o(t+a) + f_o(t-a)$	$2F_s(\omega) \cos a\omega$	
9	$f_e(t-a) - f_e(t+a)$	$2F_c(\omega) \sin a\omega$	
10	$\int_0^\infty f(r)[g(|t-r|)$ $-g(t+r)]dr$	$2F_s(\omega) \, G_c(\omega)$	
Algebraic functions			
1	$1/t$	$\pi/2$	
2	$1/\sqrt{t}$	$(\pi/2\omega)^{1/2}$	
3	$1/\sqrt{t}[1- U(t-1)]$	$(2\pi/\omega)^{1/2} \, S(\omega)$	
4	$(1/\sqrt{t})U(t-1)$	$(2\pi/\omega)^{1/2}[1/2 - S(\omega)]$	
5	$(t+a)^{-1/2}$	$|\arg a| < \pi$	$(\pi/2\omega)^{1/2} \{\cos a\omega[1 - 2S(a\omega)]$ $- \sin a\omega[1 - 2C(a\omega)]\}$
6	$(t-a)^{-1/2}U(t-a)$	$(\pi/2\omega)^{1/2}(\sin a\omega + \cos a\omega)$	
7	$t(t^2+a^2)^{-1}$	$a > 0$	$(\pi/2) \exp(-a\omega)$
8	$t(a^2-t^2)^{-1}$	$a > 0$	$-(\pi/2) \cos a\omega$
9	$t(a^2+t^2)^{-2}$	$a > 0$	$(\pi\omega/4a) \exp(-a\omega)$
10	$a^2[t(a^2+t^2)]^{-1}$	$a > 0$	$(\pi/2)[1 - \exp(-a\omega)]$
11	$t(4+t^4)^{-1}$		$(\pi/4) \exp(-\omega) \sin \omega$
Exponential and logarithmic functions			
1	e^{-at}	Re $a > 0$	$\omega(a^2 + \omega^2)^{-1}$
2	te^{-at}	Re $a > 0$	$(2a\omega)(a^2 + \omega^2)^{-2}$
3	$t(1+at)e^{-at}$	Re $a > 0$	$(8a^3\omega)(a^2 + \omega^2)^{-3}$
4	$e^{-at}\sqrt{t}$	Re $a > 0$	$\sqrt{(\pi/2)}(a^2 + \omega^2)^{-1/2} \bullet [(a^2 + \omega^2)^{1/2} - a]^{1/2}$
5	$t^{-3/2}e^{-at}$	Re $a > 0$	$(2\pi)^{1/2}[(a^2 + \omega^2)^{1/2} - a]^{1/2}$
6	$\exp(-at^2)$	Re $a > 0$	$-j(1/2)(\pi/a)^{1/2} \exp(-\omega^2/4a) \; \text{Erf}\left(\dfrac{j\omega}{2\sqrt{a}}\right)$
7	$t \exp(-t^2/4a)$	Re $a > 0$	$2a\omega\sqrt{(\pi a)} \exp(-a\omega^2)$
8	$t^{-3/2} \exp(-a/t)$	$|\arg a| < \pi/2$	$(\pi/a)^{1/2} \exp[-(2a\omega)^{1/2}] \sin(2a\omega)^{1/2}$

TABLE 4.2 (continued) Fourier Sine Transforms

$f(t)$		$F_s(\omega) = \int_0^\infty f(t)\,\sin \omega t\,dt$	$\omega > 0$
9	$t^{-3/4}\exp(-a/\sqrt{t})$	$\lvert \arg a\rvert < \pi/2$	$-(\pi/2)(a/\omega)^{1/2}[J_{1/4}(a^2/8\omega)\cdot\cos(\pi/8+a^2/8\omega)$ $+Y_{1/4}(a^2/8\omega)\cdot\sin(\pi/8+a^2/8\omega)]$
10	$t^{-1}\ln t$		$-(\pi/2)[C+\ln\omega]$
11	$t(t^2-a^2)^{-1}\ln t$	$a>0$	$-(\pi/2)\{\cos a\omega[\text{Ci}(a\omega)-\ln a]$ $+\sin a\omega[\text{Si}(a\omega)-\pi/2]\}$
12	$t^{-1}\ln(1+a^2t^2)$	$a>0$	$-\pi\,\text{Ei}(-\omega/a)$
13	$\ln\dfrac{t+a}{\lvert t-a\rvert}$	$a>0$	$(\pi/\omega)\sin a\omega$

is

$$\mathscr{F}\{f(t)\} = \int_0^\infty f(t)\sin\omega t\,dt = \int_0^1 f(t)\sin\omega t\,dt + \int_1^2 f(t)\sin\omega t\,dt + \int_2^\infty f(t)\sin\omega t\,dt$$

$$= \int_0^1 t\sin\omega t\,dt + \int_1^2 (2-t)\sin\omega t\,dt + \int_2^\infty 0\sin\omega t\,dt$$

$$= \left[-\frac{t\cos\omega t}{\omega}+\frac{\sin\omega t}{\omega^2}\right]_0^1 + \left[\frac{2\cos\omega t}{\omega}\right]_1^2 - \left[-\frac{t\cos\omega t}{\omega}+\frac{\sin\omega t}{\omega}\right]_1^2 = \frac{2(1-\cos\omega)\sin\omega}{\omega^2}$$

$$\tag{4.4}$$

■

Example 4.2

The FCT of the function $f(t) = e^{-at}u(t)$ is

$$I = \int_0^\infty e^{-at}\cos\omega t\,dt = \frac{1}{-a}\int_0^\infty d(e^{-at})\cos\omega t$$

$$= -\frac{1}{a}\left(e^{-at}\cos\omega t\Big|_0^\infty - \int_0^\infty e^{-at}(-\omega)\sin\omega t\,dt\right)$$

$$= -\frac{1}{a}\left(-1+\omega\int_0^\infty e^{-at}\sin\omega t\,dt\right) = \frac{1}{a}-\frac{\omega}{a}\left(\frac{1}{-a}\int_0^\infty d(e^{-at})\sin\omega t\right)$$

$$= \frac{1}{a}-\frac{\omega}{a}\left(-\frac{1}{a}\left(e^{-at}\sin\omega t\Big|_0^\infty - \omega\int_0^\infty e^{-at}\cos\omega t\,dt\right)\right) = \frac{1}{a}+\frac{\omega}{a^2}(0-\omega I)$$

$$= \frac{1}{a}-\frac{\omega^2}{a^2}I \quad\text{or}$$

$$\left(\frac{1}{a}+\frac{\omega^2}{a^2}\right)I = \frac{1}{a} \quad\text{or}\quad I = \frac{a}{a^2+\omega^2} \tag{4.5}$$

Following similar steps, its FST is

$$I = \int_0^\infty e^{-at} \sin \omega t \, dt = \frac{\omega}{a^2 + \omega^2} \tag{4.6}$$

∎

Example 4.3

To find the FCT of the function $f(t) = e^{-t^2} u(t)$, we write

$$I = \int_0^\infty e^{-t^2} \cos \omega t \, dt \quad \text{or} \quad \frac{dI}{d\omega} = \frac{1}{2} \int_0^\infty (-2te^{-t^2}) \sin \omega t \, dt = \frac{1}{2} \int_0^\infty \frac{de^{-t^2}}{dt} \sin \omega t \, dt$$

$$= \frac{1}{2} \left(\left[e^{-t^2} \sin \omega t \right]_0^\infty - \omega \int_0^\infty e^{-t^2} \cos \omega t \, dt \right) = -\frac{\omega}{2} \int_0^\infty e^{-t^2} \cos \omega t \, dt = -\frac{\omega}{2} I \Rightarrow$$

$$\frac{dI}{d\omega} = -\frac{\omega}{2} I \quad \text{or} \quad \frac{dI}{I} = -\frac{\omega}{2} d\omega \quad \text{or} \quad \int \frac{dI}{I} = -\int \frac{\omega}{2} d\omega \quad \text{or} \quad \ln I = -\frac{\omega^2}{4} + \ln A \quad \text{or}$$

$$I = Ae^{-\omega^2/4}; \quad \text{when } \omega = 0 \; I = \int_0^\infty e^{-t^2} \cos 0 \, dt = \int_0^\infty e^{-t^2} dt = \frac{\sqrt{\pi}}{2};$$

$$\text{therefore } A = \frac{\sqrt{\pi}}{2} \quad \text{and} \quad I = \frac{\sqrt{\pi}}{2} e^{-\omega^2/4} \tag{4.7}$$

∎

Example 4.4

To find the FST of $f(t) = \dfrac{t}{1 + t^2}$, we first find the FCT of $g(t) = \dfrac{1}{1 + t^2}$. Hence, we write

$$F_c \left\{ \frac{1}{1 + t^2} \right\} = \int_0^\infty \frac{\cos \omega t}{1 + t^2} dt = I(i) \quad \text{or} \quad \frac{dI}{d\omega} = \int_0^\infty -\frac{t \sin \omega t}{1 + t^2} dt = F_s \left\{ \frac{-t}{1 + t^2} \right\} \text{ (ii)} \quad \text{or}$$

$$\frac{dI}{d\omega} = \int_0^\infty -\frac{t^2 \sin \omega t}{t(1 + t^2)} dt = -\int_0^\infty \frac{(1 + t^2 - 1)\sin \omega t}{t(1 + t^2)} dt = -\int_0^\infty \frac{\sin \omega t}{t} dt + \int_0^\infty \frac{\sin \omega t}{t(1 + t^2)} dt$$

$$= -\frac{\pi}{2} + \int_0^\infty \frac{\sin \omega t}{t(1 + t^2)} dt \text{ (iii)} \quad \frac{d^2 I}{d\omega^2} = \int_0^\infty \frac{\cos \omega t}{(1 + t^2)} dt = I \quad \text{or} \quad \frac{d^2 I}{d\omega^2} - I = 0 \quad \text{or}$$

$$I = Ae^{-\omega} + Be^{\omega} \text{ (iv) putting } \omega = 0 \text{ in (i) we get } I = \int_0^\infty \frac{1}{(1 - t^2)} dt = \left[\tan^{-1}(t) \right]_0^\infty$$

$$= \frac{\pi}{2}; \quad \text{from (iv) } \frac{\pi}{2} = A + B \text{ (v) putting } \omega = 0 \text{ in (iii) we get } \frac{dI}{d\omega} = -\frac{\pi}{2}; \quad \text{from (iv)}$$

$$\frac{dI}{d\omega} = -Ae^{-\omega} + Be^{\omega}; \quad \text{hence} -\frac{\pi}{2} = -A + B; \quad \text{from (iv) and (v) we find } B = 0 \text{ and}$$

$$A = \frac{\pi}{2}; \quad \text{hence } I = Ae^{-\omega} + Be^{\omega} = \frac{\pi}{2} e^{-\omega} + 0 \text{ (vi); thus} \quad \boxed{F_c \left\{ \frac{1}{(1 + t^2)} \right\} = \frac{\pi}{2} e^{-\omega}}$$

(vii) from (vi)

$$\frac{dI}{d\omega} = -\frac{\pi}{2} e^{-\omega}; \quad \text{using this in (ii)} -\frac{\pi}{2} e^{-\omega} = F_s \left\{ \frac{-t}{(1 + t^2)} \right\} \Rightarrow \boxed{F_s \left\{ \frac{t}{(1 + t^2)} \right\} = \frac{\pi}{2} e^{-\omega}} \tag{4.8}$$

∎

Example 4.5

The IFST of $F(\omega) = e^{-\pi\omega}$ is

$$f(t) = \frac{2}{\pi} \int_0^\infty e^{\pi\omega} \sin \omega t \, d\omega = \frac{2}{\pi} \frac{t}{\pi^2 + t^2} \tag{4.9}$$

where the integral was taken from the mathematical tables. ∎

Example 4.6

If the FCT is $1/(1 + \omega^2)$, then its IFCT is

$$f(t) = \frac{2}{\pi} \int_0^\infty \frac{1}{1 + \omega^2} \cos \omega t \, d\omega \text{ (i)} \Rightarrow \frac{df(t)}{dt} = \frac{2}{\pi} \int_0^\infty \frac{-\omega \sin \omega t}{1 + \omega^2} \, d\omega$$

$$= -\frac{2}{\pi} \int_0^\infty \frac{(1 + \omega^2 - 1)}{(1 + \omega^2)\omega} \sin \omega t \, d\omega \text{ (ii)}$$

$$= -\frac{2}{\pi} \int_0^\infty \frac{\sin \omega t}{\omega} \, d\omega + \frac{2}{\pi} \int_0^\infty \frac{\sin \omega t}{(1 + \omega^2)\omega} \, d\omega = -\frac{2}{\pi} \frac{\pi}{2} + \frac{2}{\pi} \int_0^\infty \frac{\sin \omega t}{(1 + \omega^2)\omega} \, d\omega \text{ (iii)} \Rightarrow$$

$$\frac{d^2 f(t)}{dt^2} = \frac{2}{\pi} \int_0^\infty \frac{\cos \omega t}{1 + \omega^2} \, d\omega = f(t) \Rightarrow \frac{d^2 f(t)}{dt^2} - f(t) = 0 \Rightarrow f(t) = Ae^{-t} + Be^t \text{ (iv)};$$

$$\text{for } t = 0 \text{ in (i)} \rightarrow f(0) = \frac{2}{\pi} \int_0^\infty \frac{1}{1 + \omega^2} \, d\omega - \frac{2}{\pi} [\tan^{-1} \omega]_0^\infty = 1 \text{ (v)};$$

$$\text{for } t = 0 \text{ in (iii)} \frac{df(0)}{dt} = -1$$

$$\Rightarrow A + B = 1 \quad -A + B = -1 \Rightarrow B = 0 \quad A = 1 \Rightarrow \boxed{f(t) = e^{-t}u(t)} \tag{4.10}$$

∎

Example 4.7

The FST of $f(t) = [e^{-at}/t]u(t)$ is

$$I = \int_0^\infty \frac{e^{-at}}{t} \sin \omega t \, dt \text{ (i)} \Rightarrow \frac{dI}{dt} = \int_0^\infty e^{-at} \cos \omega t \, dt = \frac{a}{a^2 + \omega^2}, \text{ integrating}$$

$$I = \tan^{-1} \left(\frac{\omega}{a} \right) + A$$

For $\omega = 0, I = A$, and for $\omega = 0$ (i) indicates that $I = 0 \Rightarrow A = 0 \Rightarrow I = \tan^{-1} \left(\frac{\omega}{a} \right)$ (4.11)

∎

Example 4.8

The FST of the function

$$f(t) = \begin{cases} \sin t & 0 < t < a \\ 0 & t > a \end{cases}$$

is

$$I = \int_0^a \sin t \sin \omega t \, dt = \frac{1}{2} \int_0^a [\cos (1 - \omega)t - \cos (1 + \omega)t] dt$$

$$= \frac{1}{2} \left[\frac{\sin (1 - \omega)a}{1 - \omega} - \frac{\sin (1 + \omega)a}{1 + \omega} \right] \qquad (4.12)$$

∎

Example 4.9

The IFST of the function $F(\omega) = e^{-a\omega}/\omega$ is

$$f(t) = \frac{2}{\pi} \int_0^\infty \frac{e^{-a\omega}}{\omega} \sin \omega t \, d\omega \text{ (i)} \Rightarrow \frac{df(t)}{dt} = \frac{2}{\pi} \int_0^\infty \cos \omega t \, d\omega = \frac{2}{\pi} \frac{a}{t^2 + a^2}$$

$$\Rightarrow f(t) = \frac{2}{\pi} \int \frac{a}{t^2 + a^2} dt$$

$$= \frac{2}{\pi} \tan^{-1} \left(\frac{t}{a} \right) + A, \text{ when } t = 0 \Rightarrow f(0) = 0 \text{ from (i)} \qquad (4.13)$$

$$\Rightarrow \boxed{f(t) = \frac{2}{\pi} \tan^{-1} \left(\frac{t}{a} \right)} \text{ (ii); Since}$$

$$\mathscr{F}_s^{-1} \left(\frac{1}{\omega} \right) = \mathscr{F}_s^{-1} \left(\frac{e^{-a\omega}}{\omega} \right)_{a=0} \Rightarrow \text{ from (ii) } \boxed{\mathscr{F}_s^{-1} \left(\frac{1}{\omega} \right) = \frac{2}{\pi} \tan^{-1} (\infty) = 1}$$

∎

Example 4.10

The FCT of the function

$$f(t) = \begin{cases} 1 & 0 < t < a \\ 0 & t \geq a \end{cases}$$

is

$$F(\omega) = \int_0^a 1 \cos \omega t \, dt = \frac{\sin a\omega}{\omega} \qquad (4.14)$$

∎

Example 4.11

The FST and FCT of the signal $f(t) = tu(t)$ are

$$G(\omega) = \int_0^\infty te^{-j\omega t}\,dt, \text{ let } j\omega t = y \Rightarrow j\omega\,dt = dy \Rightarrow G(\omega) = \int_0^\infty \frac{y}{j\omega}e^{-y}\frac{dy}{j\omega} = -\frac{1}{\omega^2}\int_0^\infty y^{2-1}e^{-y}\,dy$$

$$= -\frac{\Gamma(2)}{\omega^2} = -\frac{1}{\omega^2} \Rightarrow \int_0^\infty te^{-j\omega t}\,dt = -\frac{1}{\omega^2} \Rightarrow \int_0^\infty t(\cos\omega t - j\sin\omega t)\,dt = -\frac{1}{\omega^2}$$

$\Rightarrow$ equating real and imaginary parts

$$\Rightarrow F_c(\omega) = \int_0^\infty t\cos\omega t\,dt = -\frac{1}{\omega^2}, \quad F_s(\omega) = \int_0^\infty t\sin\omega t\,dt = 0 \tag{4.15}$$

■

4.3 Applications to Boundary-Value Problems

Example 4.12 (Waves in LC transmission lines and vibrating strings)

Let the initial displacement of an infinite string be $f(x)$, for $-\infty < x < \infty$, and its initial velocity be zero. The displacement of the string $u(x, t)$ satisfies the one-dimensional wave equation:

$$\frac{\partial^2 u(x,t)}{\partial x^2} - \frac{1}{c^2}\frac{\partial^2 u(x,t)}{\partial t^2} = 0 \tag{4.16}$$

and the initial conditions:

$$u(x,0) = f(x) \quad -\infty < x < \infty, \quad \left.\frac{\partial u(x,t)}{\partial t}\right|_{t=0} = 0 \tag{4.17}$$

Let the FT of the solution $u(x, t)$ with respect to x and its inverse be given by

$$U(\omega, t) = \int_{-\infty}^\infty u(x,t)e^{-j\omega t}\,dt \quad u(x,t) = \frac{1}{2\pi}\int_{-\infty}^\infty U(\omega,t)e^{j\omega t}\,d\omega \tag{4.18}$$

We assume that the solution and its first derivative with respect to x are approaching zero as x approaches minus or plus infinity. Let

$$u_{xx}(x,t) = \frac{\partial^2 u(x,t)}{\partial x^2} \quad u_x(x,t) = \frac{\partial u(x,t)}{\partial x} \quad u_{tt}(x,t) = \frac{\partial^2 u(x,t)}{\partial t^2} \quad u_t(x,t) = \frac{\partial u(x,t)}{\partial t}$$

Therefore, we have

$$\mathcal{F}\{u_{xx}(x,t)\} = \int_{-\infty}^{\infty} u_{xx}(x,t)e^{-j\omega x}dx = \int_{-\infty}^{\infty} du_x(x,t)e^{-j\omega x} = du_x(x,t)e^{-j\omega x}\Big|_{-\infty}^{\infty}$$

$$- \int_{-\infty}^{\infty} u_x(x,t)e^{-j\omega x}(-j\omega)dx$$

$$= j\omega \int_{x=-\infty}^{x=\infty} e^{-j\omega x}du(x,t) = j\omega e^{-j\omega x}u(x,t)\Big|_{-\infty}^{\infty} - j\omega(-j\omega)\int_{-\infty}^{\infty} u(x,t)e^{-j\omega x}dx$$

$$= -\omega^2 U(\omega,t) \tag{4.19}$$

since $u_x(\pm\infty, t) = u(\pm\infty, t) = 0$. Similarly, we find

$$\mathcal{F}\{u_{tt}(x,t)\} = \int_{-\infty}^{\infty} u_{tt}(x,t)e^{-j\omega x}dx = \frac{\partial^2}{\partial t^2} \int_{-\infty}^{\infty} u(x,t)e^{-j\omega x}dx = U_{tt}(\omega,t) \tag{4.20}$$

Applying the FT to the wave equation (4.16) and introducing the results of (4.19) and (4.20), we obtain

$$-\omega^2 U(\omega,t) - \frac{1}{c^2}U_{tt}(\omega,t) = 0 \quad \text{or} \quad \frac{\partial^2 U(\omega,t)}{\partial t^2} + \omega^2 c^2 U(\omega,t) = 0 \tag{4.21}$$

which is an ordinary differential equation. The general solution of (4.21) is

$$U(\omega,t) = A(\omega)e^{j\omega ct} + B(\omega)e^{-j\omega ct} \tag{4.22}$$

Applying the FT to the initial conditions (4.17), we obtain

$$U(\omega,0) = \mathcal{F}\{u(x,0)\} = \int_{-\infty}^{\infty} u(x,0)e^{-j\omega x}dx = \int_{-\infty}^{\infty} f(x)e^{-j\omega x}dx = F(\omega)$$

$$U_t(\omega,0) = \mathcal{F}\{u_t(x,t)|_{t=0}\} = 0 \tag{4.23}$$

Therefore, using (4.23) we can evaluate the constants A and B of (4.22):

$$F(\omega) = U(\omega,0) = A(\omega) + B(\omega)$$
$$0 = U_t(\omega,0) = j\omega c[A(\omega) - B(\omega)]$$

or

$$A(\omega) = B(\omega) = \frac{1}{2}F(\omega)$$

Hence, the solution (4.22) becomes

$$U(\omega, t) = \frac{1}{2}F(\omega)e^{j\omega ct} + \frac{1}{2}F(\omega)e^{-j\omega ct} \tag{4.24}$$

The desired solution $u(x, t)$ is the IFT of $U(\omega, t)$, namely,

$$u(x, t) = \mathfrak{F}^{-1}\{U(\omega, t)\} = \frac{1}{2}\mathfrak{F}^{-1}\{F(\omega)e^{j\omega ct}\} + \frac{1}{2}\mathfrak{F}^{-1}\{F(\omega)e^{-j\omega ct}\}$$

$$= \frac{1}{2}f(x + ct) + \frac{1}{2}f(x - ct) \tag{4.25}$$

The solution indicates that there exists two traveling disturbances in opposite directions.   ∎

Example 4.13 (Voltage propagating in RC transmission lines and heat flow)

The flow of heat in a body is governed by the heat PDF:

$$\nabla^2 u(x, y, z, t) - \frac{1}{c^2}\frac{\partial u(x, y, z, t)}{\partial t} = 0 \quad \text{or} \quad \frac{\partial^2 u}{\partial x^2} + \frac{\partial^2 u}{\partial y^2} + \frac{\partial^2 u}{\partial z^2} - \frac{1}{c^2}\frac{\partial u(x, y, z, t)}{\partial t} = 0 \tag{4.26}$$

Let the initial temperature distribution in an infinite bar be $f(x)$, for $\infty < x < \infty$. Therefore, the temperature $u(x, t)$ satisfies the one-dimensional PDF equation:

$$\frac{\partial^2 u}{\partial x^2} - \frac{1}{c^2}\frac{\partial u(x, y, z, t)}{\partial t} = 0 \tag{4.27}$$

and the initial condition:

$$u(x, 0) = f(x) \quad -\infty < x < \infty \tag{4.28}$$

It is assumed that the temperature and its derivative are infinitesimally small at the limit as x approaches minus plus infinity. The FT of the solution $u(x, t)$ with respect to x (see (4.19)) is

$$\mathfrak{F}\{u_{xx}(x, t)\} = \int_{-\infty}^{\infty} u_{xx}(x, t)e^{-j\omega x}dx = -\omega^2 U(\omega, t) \tag{4.29}$$

The FT of $u_t(x, t)$ is

$$\mathcal{F}\{u_t(x, t)\} = \int_{-\infty}^{\infty} u_t(x, t)e^{-j\omega x}\,dx = \frac{\partial}{\partial t}\int_{-\infty}^{\infty} u(x, t)e^{-j\omega x}\,dx = \frac{\partial}{\partial t}U(\omega, t) = U_t(\omega, t) \quad (4.30)$$

Now, applying the FT to the heat equation (4.27), we obtain

$$-\omega^2 U(\omega, t) - \frac{1}{c^2}U_t(\omega, t) = 0 \quad \text{or} \quad U_t(\omega, t) + c^2\omega^2 U(\omega, t) = 0 \quad (4.31)$$

The solution of (4.31) is

$$U(\omega, t) = U(\omega, 0)e^{-c^2\omega^2 t} \quad U(\omega, 0) \equiv \text{initial condition} \quad (4.32)$$

Next, applying the FT to the initial condition (4.28), we obtain

$$U(\omega, 0) = \int_{-\infty}^{\infty} u(x, 0)e^{-j\omega x}\,dx = \int_{-\infty}^{\infty} f(x)e^{-j\omega x}\,dx = \int_{-\infty}^{\infty} f(y)e^{-j\omega y}\,dy \quad (4.33)$$

Substituting (4.33) into (4.32), we have

$$U(\omega, t) = e^{-c^2\omega^2 t}\int_{-\infty}^{\infty} f(y)e^{-j\omega y}\,dy \quad (4.34)$$

The IFT of (4.34) gives us the desired solution. Hence, we obtain

$$u(x, t) = \frac{1}{2\pi}\int_{-\infty}^{\infty} U(\omega, t)e^{j\omega x}\,d\omega = \frac{1}{2\pi}\int_{-\infty}^{\infty} e^{(j\omega x - c^2\omega^2 t)}\left[\int_{-\infty}^{\infty} f(y)e^{-j\omega y}\,dy\right]d\omega$$

$$= \frac{1}{2\pi}\int_{-\infty}^{\infty} f(y)\left[\int_{-\infty}^{\infty} f(y)e^{[j\omega(x-y)-c^2\omega^2 t]}\,d\omega\right]dy \quad (4.35)$$

From mathematical tables $\int_{-\infty}^{\infty} e^{-w^2}\,dw = \sqrt{\pi}$ and, therefore, we have

$$\int_{-\infty}^{\infty} e^{[j\omega(x-y)-c^2\omega^2 t]}\,d\omega = \int_{-\infty}^{\infty} \exp\left[\left(\frac{x-y}{2c\sqrt{t}} + jc\omega\sqrt{t}\right)^2 - \left(\frac{x-y}{2c\sqrt{t}}\right)^2\right]d\omega$$

$$= e^{-(x-y)^2/(4c^2 t)}\int_{-\infty}^{\infty} \exp\left(\frac{x-y}{2c\sqrt{t}} + jc\omega\sqrt{t}\right)^2 d\omega$$

Introducing a new variable of integration w by setting $(x-y)/(2c\sqrt{t}) + jc\omega\sqrt{t} = jw$, we obtain

$$\int_{-\infty}^{\infty} e^{[j\omega(x-y) - c^2\omega^2 t]} d\omega = \frac{1}{c\sqrt{t}} e^{-(x-y)^2/(4c^2 t)} \int_{-\infty}^{\infty} e^{-w^2} dw = \frac{1}{c}\sqrt{\frac{\pi}{t}} e^{-(x-y)^2/(4c^2 t)} \qquad (4.36)$$

Next, substituting (4.36) in (4.35), we finally obtain

$$u(x,t) = \frac{1}{2c\sqrt{\pi t}} \int_{-\infty}^{\infty} f(y) e^{-(x-y)^2/(4c^2 t)} dy \qquad (4.37)$$

∎

Example 4.14 (Potential theory)

The potential of an electric field satisfies Laplace's equation:

$$\nabla^2 u(x,y,z) = \frac{\partial^2 u}{\partial x^2} + \frac{\partial^2 u}{\partial y^2} + \frac{\partial^2 u}{\partial z^2} = u_{xx} + u_{yy} + u_{zz} = 0 \qquad (4.38)$$

Let us find the solution $u(x,y)$ of Laplace's equation for the half plane $y > 0$, when $u(x, 0) = f(x)$ for $-\infty < x < \infty$. Since we have assumed the half plane, the Laplace equation is independent of the z variable. Hence,

$$u_{xx}(x,y) + u_{yy}(x,y) = 0 \qquad (4.39)$$

Applying the FT to (4.39) with respect to x, we obtain

$$\frac{\partial U^2(\omega,y)}{\partial y^2} - \omega^2 U(\omega,y) = 0 \qquad (4.40)$$

The general solution of the above equation is

$$U(\omega,y) = A(\omega)e^{\omega y} + B(\omega)e^{-\omega y} \quad u(x,y) = \text{finite as } y \to \infty \qquad (4.41)$$

For $\omega > 0$, we must set $A(\omega) = 0$. This results in $U(\omega, 0) = B(\omega)$ and, hence, one of the solutions is $U(\omega, y) = U(\omega, 0)e^{-\omega y}$ for $\omega > 0$. Similarly, for $\omega < 0$ we obtain the solution $U(\omega, y) = U(\omega, 0)e^{\omega y}$ for $\omega < 0$. We can combine the two solutions as follows:

$$U(\omega,y) = U(\omega,0)e^{-|\omega|y} \quad -\infty < \omega < \infty \qquad (4.42)$$

Since $u(x, 0) = f(x)$, then we obtain

$$U(\omega, 0) = \mathcal{F}\{u(x, 0)\} = \int\limits_{-\infty}^{\infty} f(x')e^{-j\omega x'}\, dx' \tag{4.43}$$

Introducing the above equation in (4.42), we obtain

$$U(\omega, y) = \left[\int\limits_{-\infty}^{\infty} f(x')e^{-j\omega x'}\, dx'\right] e^{-|\omega|y} \tag{4.44}$$

Therefore, the desired solution is the IFT of (4.44):

$$u(x, y) = \mathcal{F}^{-1}\{U(\omega, y)\} = \frac{1}{2\pi} \int\limits_{-\infty}^{\infty} U(\omega, y)e^{j\omega x}\, d\omega = \frac{1}{2\pi} \int\limits_{-\infty}^{\infty} e^{j\omega x}\left[\int\limits_{-\infty}^{\infty} f(x')e^{-j\omega x'}\, dx'\right] e^{-|\omega|y}\, d\omega$$

$$u(x, y) = \frac{1}{2\pi} \int\limits_{-\infty}^{\infty} f(x')\left[\int\limits_{-\infty}^{\infty} e^{j\omega(x-x')-|\omega|y}\, d\omega\right] dx'$$

$$= \frac{1}{2\pi} \int\limits_{-\infty}^{\infty} f(x')\left[\int\limits_{-\infty}^{0} e^{j\omega(x-x')+\omega y}\, d\omega + \int\limits_{0}^{\infty} e^{j\omega(x-x')-\omega y}\, d\omega\right] dx'$$

$$= \frac{1}{2\pi} \int\limits_{-\infty}^{\infty} f(x')\left[\frac{1}{j(x-x')+y} - \frac{1}{j(x-x')-y}\right] d\omega = \frac{y}{\pi} \int\limits_{-\infty}^{\infty} \frac{f(x')dx'}{(x-x')^2 + y^2} \quad y > 0 \tag{4.45}$$

∎

Example 4.15

If the voltage source is finite of length T_0,

$$f(x) = \begin{cases} T_0 & |x| < b \\ 0 & |x| > b \end{cases}$$

then the solution (see (4.45)) is

$$u(x, y) = \frac{yT_0}{\pi} \int\limits_{-b}^{b} \frac{dx'}{(x'-x)^2 + y^2} = \frac{T_0}{\pi}\left[\tan^{-1}\left(\frac{x+b}{y}\right) - \tan^{-1}\left(\frac{x-b}{y}\right)\right]$$

$$= \frac{T_0}{\pi} \tan^{-1}\left(\frac{2by}{x^2 + y^2 - b^2}\right) \tag{4.46}$$

which was found using the trigonometric relation:

$$\tan(A - B) = \frac{\tan A - \tan B}{1 + \tan A \tan B}$$

If we set $u(x, y) =$ constant, multiply (4.46) by π/T_0 and then taking the tangent of both sides of the equation we obtain

$$x^2 + y^2 - b^2 - \frac{2b}{\tan\left(\dfrac{\text{constant}\pi}{T_0}\right)} y = 0 \quad \text{or} \quad x^2 + y^2 - b^2 - Cby = 0$$

If, for example, we set $b = 1$ and then change the values of the constant C, we produce a set of curves which indicate a constant potential or a constant temperature for heat problems. The following MATLAB® program will produce the desired curve. Let $b = 1$ and $C = 4$, then we write in the command window:

```
≫p1 ='x^2+y^2-1-4*y';
≫ezplot(p1,[-4 4 0 8]);
```

To produce additional iso-potentials you just change the value of C. ∎

4.4 Finite Sine Fourier Transform and Finite Cosine Fourier Transform

If we assume $0 < t < l$, then the finite sine Fourier transform (FFST) and its inverse finite Fourier sine transform IFFST transforms are defined as follows:

$$\mathcal{F}\{f(t)\} \overset{\Delta}{=} \bar{F}_s(p) = \int_0^l f(t) \sin\frac{p\pi t}{l}\, dt$$

$$\mathcal{F}^{-1}\{\bar{F}_s(p)\} = f(t) = \frac{2}{l}\sum_{p=1}^{\infty} \bar{F}_s(p) \sin\frac{p\pi t}{l}$$

(4.47)

The finite cosine Fourier transform (FFCT) and its inverse (IFFCT) are

$$\mathcal{F}\{f(t)\} \overset{\Delta}{=} \bar{F}_c(p) = \int_0^l f(t) \cos\frac{p\pi t}{l}\, dt \quad \text{or}$$

$$\mathcal{F}^{-1}\{\bar{F}_c(p)\} = f(t) = \frac{1}{l}\bar{F}_c(0) + \frac{2}{l}\sum_{p=1}^{\infty} \bar{F}_c(p) \cos\frac{p\pi t}{l}$$

(4.48)

Example 4.16

Let us find the solution of the equation describing the propagation of voltage along an RC transmission line. Specifically, the equation to be solved is given by

$$\frac{\partial V(x,t)}{\partial t} = K\frac{\partial^2 V(x,t)}{\partial x^2} \qquad 0 < x < \pi, t > 0$$

$$\frac{\partial V(x,t)}{\partial x} = 0 \qquad x = 0 \quad \text{and} \quad x = \pi \equiv \text{boundary conditions} \tag{4.49}$$

$$V(x,t) = f(x) \quad t = 0, \quad 0 < x < \pi \equiv \text{initial condition}$$

(distance π was selected to simplify the expressions)
    Taking the FFCT of the given equation, we obtain

$$\frac{1}{K}\int_0^\pi \frac{\partial V}{\partial t}\cos px\,dx = \int_0^\pi \frac{\partial^2 V}{\partial x^2}\cos px\,dx = \left[\frac{\partial V}{\partial x}\cos px\right]_{x=0}^\pi + p\int_0^\pi \frac{\partial V}{\partial x}\cos px$$

$$= 0 + p\left[[V(x,t)\sin px]_{x=0}^\pi - p\int_0^\pi V\cos pxdx\right] = -p^2 \bar{V}_c(p,t) \quad \text{but} \tag{4.50}$$

$$\frac{1}{K}\int_0^\pi \frac{\partial V}{\partial t}\cos px\,dx = \frac{1}{K}\frac{\partial}{\partial t}\int_0^\pi V\cos px\,dx = \frac{1}{K}\frac{\partial \bar{V}_c(p,t)}{\partial t} \quad \text{hence} \quad \frac{\partial \bar{V}_c(p,t)}{\partial t} = -Kp^2\bar{V}_c(p,t)$$

The above equation has the solution

$$\bar{V}_c(p,t) = Ae^{-p^2 Kt} \tag{4.51}$$

Applying the initial condition, we obtain

$$\bar{V}_c(p,0) = A \text{ but } V(x,0) = f(x) \quad \text{and} \quad \text{hence } \bar{V}_c(p,0) = \int_0^\pi f(x)\cos px\,dx = A(p)$$

By the definition of the IFFCT, we find

$$V(x,t) = \frac{1}{\pi}\bar{V}_c(0,t) + \frac{2}{\pi}\sum_{p=1}^\infty \bar{V}_c(x,t)\cos px = \frac{\bar{V}_c(0,t)}{\pi} + \frac{2}{\pi}\sum_{p=1}^\infty A(p)e^{-p^2 Kt}\cos px \tag{4.52}$$

■

Example 4.17

The temperature T in a semi-infinite rod, $0 \leq x < \infty$, is characterized by the PDF

$$\frac{\partial T(x,t)}{\partial t} = K \frac{\partial^2 T(x,t)}{\partial x^2} \tag{4.53}$$

with the following initial conditions: (i) $T = 0$ when $t = 0$, $x > 0$; (ii) $\frac{\partial T}{\partial x} = -a$ when $x = 0$; and (iii) partial derivative of T tends to zero as x tends to infinity. It is desired to find the temperature distribution in the rod. Since the derivative with respect to x is zero, we can apply the FCT to the given equation and, hence,

$$\int_0^\infty \frac{\partial T}{\partial t} \cos px \, dx = K \int_0^\infty \frac{\partial^2 T}{\partial x^2} \cos px \, dx \text{ or } \frac{\partial}{\partial t} \int_0^\infty T \cos px \, dx = \frac{\partial \overline{T}}{\partial t}$$

$$= K \left[\left[\frac{\partial T}{\partial x} \cos px \right]_0^\infty + p \int_0^\infty \frac{\partial T}{\partial x} \cos px \, dx \right]$$

$$= -K \left[\frac{\partial T}{\partial x} \right]_{x=0} + pK \left[(T \sin px)_0^\infty - p \int_0^\infty T \cos px \, dx \right] = Ka - p^2 K \overline{T}_c \left(\lim_{x \to \infty} T = 0 \right);$$

$$\frac{\partial \overline{T}_c}{\partial t} + p^2 K \overline{T}_c = Ka \tag{4.54}$$

The solution of the above ODF is found as follows:

$$e^{p^2 Kt} \frac{\partial \overline{T}_c}{\partial t} + e^{p^2 Kt} p^2 K \overline{T}_c - e^{p^2 Kt} Ka \quad \text{or} \quad \frac{\partial}{\partial t} \left(e^{p^2 Kt} \overline{T}_c \right) = Ka e^{p^2 Kt} \Rightarrow e^{p^2 Kt} \overline{T}_c = Ka \int e^{p^2 Kt} dt + A$$

$$= A + \frac{a}{p^2} e^{p^2 Kt} \Rightarrow \overline{T}_c = \frac{a}{p^2} + Ae^{-p^2 Kt}, T = 0 \text{ as } t \to \infty \Rightarrow \overline{T}_c(x,0) = 0 \Rightarrow 0 = A + \frac{a}{p^2} \Rightarrow A$$

$$= -\frac{a}{p^2}, \text{ now (ii) becomes } \overline{T}_c = \frac{a}{p^2} \left(1 - e^{-p^2 Kt} \right) \Rightarrow T(x,t) = \frac{2}{\pi} \int_0^\infty \frac{a}{p^2} \left(1 - e^{-p^2 Kt} \right) \cos px \, dp$$

$$\tag{4.55}$$

■

Example 4.18

The FFST and the FFCT of $f(t) = 1$ are found as follows:

$$\overline{F}_s(p) = \int_0^l 1 \sin\left(\frac{p\pi t}{l}\right) dt = \frac{l}{p\pi} [1 - (-1)^p] \quad \overline{F}_c(p) = \int_0^l 1 \cos\left(\frac{p\pi t}{l}\right) dt = \begin{cases} l & p = 0 \\ 0 & p \neq 0 \end{cases}$$

$$\tag{4.56}$$

■

Example 4.19

The FFST and the FFCT of $f(t) = t$ are found as follows:

$$\bar{F}_s(p) = \int_0^l t \sin\left(\frac{p\pi t}{l}\right) dt = \frac{(-1)^{p+1} a^2}{p\pi}$$

$$\bar{F}_c(p) = \int_0^l t \cos\left(\frac{p\pi t}{l}\right) dt = \begin{cases} \dfrac{a^2}{2} & p = 0 \\ \left(\dfrac{l}{p\pi}\right)^2 [(-1)^p - 1] & p \neq 0 \end{cases} \tag{4.57}$$

∎

Example 4.20

The FFST and the FFCT of the derivative function are found as follows:

$$F_s\{f'(t)\} = \int_0^l f'(t) \sin\left(\frac{p\pi t}{l}\right) dt = \left[f(t)\sin\left(\frac{p\pi t}{l}\right)\right]_0^l - \frac{p\pi}{l} \int_0^l f(t)\cos\left(\frac{p\pi t}{l}\right) dt$$

$$= -\frac{p\pi}{l} \bar{F}_c(p)$$

$$F_c\{f'(t)\} = \int_0^l f'(t) \cos\left(\frac{p\pi t}{l}\right) dt = \left[f(t)\cos\left(\frac{p\pi t}{l}\right)\right]_0^l + \frac{p\pi}{l} \int_0^l f(t)\sin\left(\frac{p\pi t}{l}\right) dt \tag{4.58}$$

$$= (-1)^p f(l) - f(0) + \frac{p\pi}{l} \bar{F}_s(p)$$

∎

4.5 Two-Dimensional Fourier Transform

A two-dimensional function $f(x, y)$ has a two-dimensional FT $F(u, v)$ connected by the following two-pair relations (Tables 4.3 and 4.4):

$$F(u,v) = \int_{-\infty}^{\infty} \int_{-\infty}^{\infty} f(x,y) e^{-j2\pi(ux+vy)} dx dy \quad f(x,y) = \int_{-\infty}^{\infty} \int_{-\infty}^{\infty} F(u,v) e^{j2\pi(ux+vy)} du dv \tag{4.59}$$

TABLE 4.3 Properties of a Two-Dimensional FT

Properties	$f(x, y)$	$F(u, v)$
Similarity	$f(ax, by)$	$\dfrac{1}{\|ab\|} F\left(\dfrac{u}{a}, \dfrac{v}{b}\right)$
Addition	$f(x, y) + g(x, y)$	$F(u, v) + G(u, v)$
Shift	$f(x - a, y - b)$	$e^{-j2\pi(au+bv)} F(u, v)$
Modulation	$f(x, y) \cos \omega_c x$	$\dfrac{1}{2} F\left(u + \dfrac{\omega_c}{2\pi}, v\right) + \dfrac{1}{2} F\left(u - \dfrac{\omega_c}{2\pi}, v\right)$
Convolution	$f(x, y) * g(x, y)$	$F(u, v) G(u, v)$
Autocorrelation	$f(x, y) * f^*(-x, -y)$	$\|F(u, v)\|^2$
Differentiation	$\left(\dfrac{\partial}{\partial x}\right)^m \left(\dfrac{\partial}{\partial y}\right)^n f(x, y)$	$(2\pi ju)^m (2\pi jv)^n F(u, v)$
	$\dfrac{\partial}{\partial x} f(x, y)$	$2\pi ju F(u, v)$
	$\dfrac{\partial}{\partial y} f(x, y)$	$2\pi jv F(u, v)$
	$\dfrac{\partial^2}{\partial x^2} f(x, y)$	$-4\pi^2 u^2 F(u, v)$
	$\dfrac{\partial^2}{\partial y^2} f(x, y)$	$-4\pi^2 v^2 F(u, v)$
	$\dfrac{\partial^2}{\partial x \partial y} f(x, y)$	$-4\pi^2 uv F(u, v)$
Definite integral	$\int_{-\infty}^{\infty} \int_{-\infty}^{\infty} f(x, y) dx dy = F(0, 0)$	
First moment	$\int_{-\infty}^{\infty} \int_{-\infty}^{\infty} x f(x, y) dx dy = \dfrac{1}{-2\pi j} F_u'(0, 0)$	
Equivalent width	$\dfrac{\int_{-\infty}^{\infty} \int_{-\infty}^{\infty} f(x, y) dx dy}{f(0, 0)}$	$\dfrac{F(0, 0)}{\int_{\infty}^{\infty} \int_{\infty}^{\infty} F(u, v) du dv}$

TABLE 4.4 Two-Dimensional Fourier Transforms

$f(x, y)$	$F(u, v)$
$\cos \pi x \bullet 1$	$\dfrac{1}{2} \delta\left(u - \dfrac{1}{2}, v\right) + \dfrac{1}{2} \delta\left(u + \dfrac{1}{2}, v\right)$
$\delta(x, y)$	1 (in all the u, v-plane)
$\dfrac{\sin x}{x} \dfrac{\sin y}{y}$	$p(x, y)$
$\left(\dfrac{\sin x}{x}\right)^2 \left(\dfrac{\sin y}{y}\right)^2$	$\Lambda(u)\Lambda(v)\left(\Lambda(x) = \begin{cases} 1 - \|x\| & \|x\| < 1 \\ 0 & \text{otherwise} \end{cases}\right)$
$e^{-\pi(x^2 + y^2)}$	$e^{-\pi(u^2 + v^2)}$
$1 \bullet \delta(y)$	$\delta(u) \bullet 1$

Example 4.21

The FT of a unit amplitude two-dimensional pulse, $p_a(x, y)$, is given by

$$F(u, v) = \int_{-a}^{a} \int_{-a}^{a} 1 e^{-j2\pi(ux+vy)} \, dx \, dy = \int_{-a}^{a} e^{-j2\pi ux} \, dx \int_{-a}^{a} e^{-j2\pi vy} \, dy = \frac{1}{-j2\pi u} (e^{-j2\pi ua} - e^{j2\pi ua})$$

$$\times \frac{1}{-j2\pi v} (e^{-j2\pi va} - e^{j2\pi va}) = \frac{\sin(2\pi au)}{2\pi u} \frac{\sin(2\pi av)}{2\pi v} \tag{4.60}$$

Figure E.4.1a shows two ways to find the two-dimensional pulse function—one by multiplying the two-step functions and the other by convolving the two functions. Figure E.4.1b shows the FT $F(u, v)$ of $f(x, y)$. For the plot we used $a = 1$.

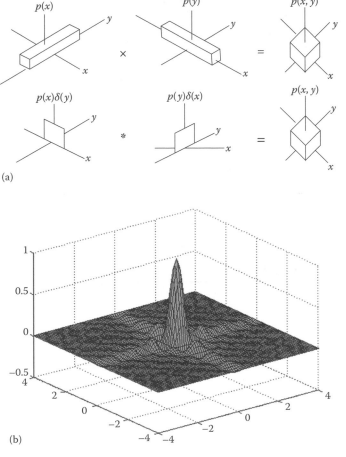

FIGURE E.4.1

For the Figure E.4.1b we used the following MATLAB book program:

```
≫x=-4:0.1:4;x=y;
≫[X,Y]=meshgrid(x,y);
≫F1=sin(2*pi*X)./(2*pi*X+eps);
≫F2=sin(2*pi*Y)./(2*pi*Y+eps);
≫F1(:,41)=1;F2(41,:)=1;%MATLAB for 0/0 gives zero;
≫F=F1.*F2;
≫surf(X,Y,F);
≫colomap([0.7 0.7 0.7]);%gives light gray scale;     ■
```

4.5.1 Two-Dimensional Convolution

The convolution integral of two dimensions is given by

$$f(x,y)^* g(x,y) = \int\limits_{-\infty}^{\infty} \int\limits_{-\infty}^{\infty} f(x',y')g(x-x',y-y')dx'dy' \tag{4.61}$$

Thus, one of the functions is rotated half a turn about its origin (flipped with respect to the y-axis) by reversing the sign of both x and y, displaced, and multiplied with the other function, and the product is then integrated to obtain the value of the convolution integral for that particular displacement x, y.

4.5.2 Two-Dimensional Correlation

The two-dimensional autocorrelation function is found the same way with the exception that the second function is shifted instead of been reversed; thus,

$$f(x,y) \odot g(x,y) = \int\limits_{-\infty}^{\infty} \int\limits_{-\infty}^{\infty} f(x',y')g(x'-x,y'-y)dx'dy' \tag{4.62}$$

If a function can be represented as a convolution of two functions, it is preferred in FTs. For example, the two-dimensional function

$$p(x,y) = \begin{cases} 1 & |x|, |y| < \dfrac{1}{2} \\ 0 & \text{elsewhere} \end{cases} \tag{4.63}$$

may be expressed as a product or as a convolution:

$$p(x,y) = p(x)p(y) = [p(x)\delta(y)]^*[p(y)\delta(x)] \tag{4.64}$$

where, for example, $p(x)$ after the first equal sign is a bar of unit width and infinite length from minus to positive infinity along the y-axis. However, $p(x)$ after the second equal sign is just a one-dimensional pulse function along the x-axis (see Figure E.4.1a).

4.5.3 Theorems of Two-Dimensional Functions

Rotation Theorem:

If $f(x, y)$ is rotated on the (x, y)-plane, then its transform is rotated on the (u, v)-plane through the same angle and in the same sense.

Simple Shear Theorem:

If $f(x, y)$ is subjected to shear then its transform is sheared to the same degree in the perpendicular direction:

$$f(x + by, y) \overset{\mathfrak{F}}{\Leftrightarrow} F(u, v - bu) \qquad (4.65)$$

Affine Theorem:

A function $f(x, y)$ subjected to an affine transformation of its coordinate plane, becomes $f(ax + by + c, dx + ey + f)$ and has the following FT:

$$\frac{1}{|ae - bd|} e^{j2\pi \frac{(ec-bf)u+(af-cd)v}{ae-bd}} F\left[\frac{eu - dv}{ae - bd}, \frac{-bu + av}{ae - bd}\right] \qquad (4.66)$$

5

Sampling of Continuous Signals

The sampling of continuous signals at periodic intervals has become a very important practical, as well as an important mathematical operation. One of the main concerns when signals are sampled is the accuracy with which the sampled signal is represented by its sampled values. Also, what type of a signal is to be used and what the sampling interval should be in order to achieve an optimum recovery of the original signal from the sampled values needs to be determined.

A very important fact in the sampling operation is that all physical engineering systems have frequency response limitations—that is, they respond only to some upper frequency limit, which we call the Nyquist frequency. As a result, the output signal of these systems is band limited. We have already seen that the ordinary house telephone signal has an upper frequency limit of about 4 kHz and the television signals have an upper frequency limit of 4 MHz. A very important consequence of a finite bandwidth signal is that it can be accurately represented by a narrow time-duration sampling sequence, with samples taken at discrete and periodic instants. As already noted, the time space between the samples of one signal can be used to accommodate (**time multiplex**) without interference from the samples of a different signal when transmitted through some transmitting channel. Often the samples are digitized, and this digitization is readily accomplished with an analog-to-digital (A/D) converter, the output being amplitude information in digital form. It is in this form that sampled and digitized signals enter a computer for further processing.

5.1 Fundamentals of Sampling

The values of the function at the sampling points are called **sampled values**. These values are the exact values (infinite precision) of the signal at those corresponding sampling times. The time that separates the sampling points is called the **sampling interval** (T_s), and the reciprocal of the sampling interval is the **sampling frequency**

($F_s = 1/T_s$) or **sampling rate**. The value of any continuous function $f(t)$ at the point nT_s is specified by

$$f(t)\delta(t - nT_s) = f(nT_s)\delta(t - nT_s) \tag{5.1}$$

This relationship is easily proved by integrating both sides of the equation. The sampling interval is chosen to be a real positive number and a constant, and $n = 0, \pm 1, \pm 2, \pm 3, \ldots$. The choice of T_s is critical in recapturing the original signal, a matter to be discussed in detail below. The sampled signal is

$$f_s(t) \overset{\Delta}{=} f(t) \sum_{n=-\infty}^{\infty} \delta(t - nT_s) = f(t)comb_{T_s}(t) = \sum_{n=-\infty}^{\infty} f(nT_s)\delta(t - nT_s) \tag{5.2}$$

where the equality (5.1) was used. The Fourier transform of the above equation is

$$\boxed{F_s(\omega) \overset{\Delta}{=} \mathcal{F}\{f_s(t)\} = \sum_{n=-\infty}^{\infty} f(nT_s)\mathcal{F}\{\delta(t - nT_s)\} = \sum_{n=-\infty}^{\infty} f(nT_s)e^{-j\omega nT_s}} \tag{5.3}$$

where the delta function properties for the Fourier integral were used. Note that the Fourier transform operates on functions with a continuous time t as their independent variable.

Now, consider the Fourier transform of the term $f_s(t) = f(t)comb_{T_s}(t)$, which appears in (5.2) and which is equivalent to the quantity involving the shifted delta function. Using the frequency convolution Fourier property and the Fourier transform of the comb function, we obtain

$$F_s(\omega) = \mathcal{F}\{f(t)comb_{T_s}(t)\} = \frac{1}{2\pi}F(\omega) * \mathcal{F}\{comb_{T_s}(t)\}$$

$$= \frac{1}{2\pi}F(\omega) * \left[\frac{2\pi}{T_s} \sum_{n=-\infty}^{\infty} \delta(\omega - n\omega_s) \right]$$

$$= \frac{1}{T_s} \sum_{n-\infty}^{\infty} \int_{-\infty}^{\infty} F(x)\delta(\omega - x - n\omega_s)dx$$

$$= \frac{1}{T_s} \sum_{n=-\infty}^{\infty} F(\omega - n\omega_s) \tag{5.4}$$

By (5.3) and (5.4),

$$\boxed{F_s(\omega) = \sum_{n=-\infty}^{\infty} f(nT_s)e^{-j\omega nT_s} = \frac{1}{T_s} \sum_{n=-\infty}^{\infty} F(\omega - n\omega_s) \quad \omega_s = \frac{2\pi}{T_s} = 2\pi F_s} \tag{5.5}$$

where

   ω_s is the sampling frequency in rad/s
   F_s is the sampling frequency in s^{-1}

This discussion shows that if we know the Fourier transform of $f(t)$, the Fourier transform of its sampled version is uniquely determined. Furthermore, if we set $\omega = \omega - m\omega_s$ in (5.5), we obtain

$$F_s(\omega - m\omega_s) = \frac{1}{T_s} \sum_{n=-\infty}^{\infty} F[\omega - m\omega_s - n\omega_s] = \frac{1}{T_s} \sum_{n=-\infty}^{\infty} F[\omega - (m+n)\omega_s]$$

$$= \frac{1}{T_s} \sum_{k=-\infty}^{\infty} F[\omega - k\omega_s] = F_s(\omega) \tag{5.6}$$

where we set $n + m = k$. Observe that for any m, as n increases to plus minus infinity the values of k also increase to plus minus infinity.

Note: *The spectrum of the sampled signal $f_s(t)$ is an infinite sum of shifted spectrums of the original signal $f(t)$. The spectrum of the sampled signal is periodic with a period ω_s (see Figure 5.1).*

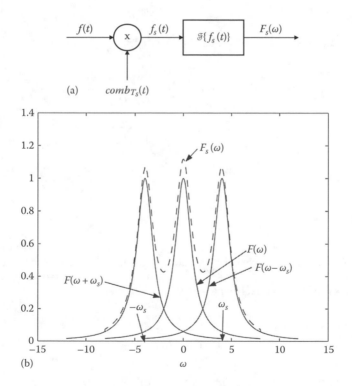

FIGURE 5.1

When the function $f(t)$ is causal (positive time), $f(t) = 0$ *for* $t < 0$; then

$$f_s(t) = \sum_{n=0}^{\infty} f(nT_s)\delta(t - nT_s) \tag{5.7}$$

which can be shown to yield

$$F_s(\omega) = \sum_{n=0}^{\infty} f(nT_s)e^{-j\omega nT_s} = \frac{f(0+)}{2} + \sum_{n=-\infty}^{\infty} F(\omega - n\omega_s) \tag{5.8}$$

Example 5.1

Find the Fourier transform of the sampled functions

(1) $f_s(t) = e^{-|t|}comb_{T_s}(t);$ (2) $f_s(t) = e^{-t}u(t)comb_{T_s}(t)$

SOLUTION

By (5.5) and (5.8) and Table 3.4, we obtain, respectively,

$$\mathscr{F}\{e^{-|t|}comb_{T_s}(t)\} \overset{\Delta}{=} F_s(\omega) = \frac{1}{T_s}\sum_{n=-\infty}^{\infty}\frac{2}{1 + (\omega + n\omega_s)^2} \quad \omega_s = \frac{2\pi}{T_s} \tag{5.9}$$

$$\mathscr{F}\{e^{-t}u(t)comb_{T_s}(t)\} \overset{\Delta}{=} F_s(\omega) = \frac{1}{2} + \frac{1}{T_s}\sum_{n=-\infty}^{\infty}\frac{1}{1 + j(\omega + n\omega_s)} \tag{5.10}$$

∎

Example 5.2

Consider three functions, $f(t)$, $h(t)$, and $g(t)$, with respective frequency characteristics $F(\omega)$, $H(\omega)$, and $G(\omega)$, as shown in Figure 5.2b. Find the maximum sampling interval T_s in order that the function $y(t) = f(t) + h(t)g(t)$ shown in Figure 5.2a is recovered from its sampled version $y_s(t)$ using a low-pass filter.

SOLUTION

The Fourier transform of the sampled function is given by

$$Y_s(\omega) = \mathscr{F}\{f(t)comb_{T_s}(t) + h(t)g(t)comb_{T_s}(t)\}$$

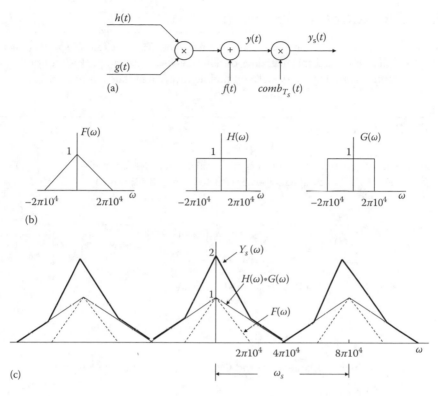

FIGURE 5.2

From the frequency convolution property of the Fourier transform, we find

$$Y_s(\omega) = \frac{1}{2\pi} F(\omega) * \frac{2\pi}{T_s} COMB_{\omega_s}(\omega) + \frac{1}{2\pi} \mathfrak{F}\{h(t)g(t)\} * \frac{2\pi}{T_s} COMB_{\omega_s}(\omega)$$

$$= \frac{1}{T_s} F(\omega) * COMB_{\omega_s}(\omega) + \left(\frac{1}{2\pi}\right)^2 H(\omega) * G(\omega) * \frac{2\pi}{T_s} COMB_{\omega_s}(\omega)$$

The convolution of $F(\omega)$ and $COMB_{\omega_s}(\omega)$ gives us a periodic repetition of the spectrum $F(\omega)$. The convolution $H(\omega) * G(\omega)$ with $COMB_{\omega_s}(\omega)$ gives us a periodic repetition of the spectrum $H(\omega) * G(\omega)$. However, the spectrum width of $H(\omega) * G(\omega)$ is equal to the sum of the spectral widths of $H(\omega)$ and $G(\omega)$; hence, in the present case $\omega_N = \omega_{NH} + \omega_{NG} = 2\pi10^2 + 2\pi10^2 = 4\pi10^2$. The spectrum $Y_s(\omega)$ is shown in Figure 5.2c. We observe that the minimum sampling frequency ω_s, in order that the spectrum of $H(\omega) * G(\omega)$ do not overlap, is $8\pi10^4$ or, equivalently, the maximum $T_s = 2\pi/8\pi10^4 = 0.25 \times 10^{-4}$. Because the spectral width of $H(\omega) * G(\omega)$ is greater than the spectral width of $F(\omega)$, the value of the sampling time is determined by the spectral width of $H(\omega) * G(\omega)$. However, if the spectral width of $F(\omega)$ were greater than the spectral width of $H(\omega) * G(\omega)$, the value of the sampling time must be determined from the spectral width of $F(\omega)$. The spectrums in Figure 5.2c have been normalized to unity.

■

5.2 The Sampling Theorem

We next show that it is possible for a band-limited signal $f(t)$ to be exactly specified by its sampled values provided that the time distance between the sampled values does not exceed a critical sampling interval. The sampling theorem is stated as follows.

THEOREM 5.1

A finite energy function $f(t)$ with a band-limited Fourier spectrum—that is, $F(\omega) = 0$ for $|\omega| \geq \omega_N$—can be completely reconstructed from its sampled values $f(nT_s)$ (see Figure 5.3), with

$$\boxed{f(t) = \sum_{n=-\infty}^{\infty} T_s f(nT_s) \frac{\sin\left[\omega_s(t - nT_s)/2\right]}{\pi(t - nT_s)} \qquad \omega_s = \frac{2\pi}{T_s}} \tag{5.11}$$

provided that the sampling time is selected to satisfy

$$\frac{2\pi}{\omega_s} \triangleq T_s = \frac{\pi}{\omega_N} = \frac{\pi}{2\pi f_N} = \frac{1}{2f_N} \triangleq \frac{T_N}{2} \quad \text{also } \omega_s = 2\omega_N \tag{5.12}$$

where ω_N is the highest frequency of the signal and is known as the **Nyquist frequency**.

Proof: Employ (5.5) and Figure 5.3b to write ($n = 0$)

$$F(\omega) = P_{\omega_s/2}(\omega) T_s F_s(\omega) \tag{5.13}$$

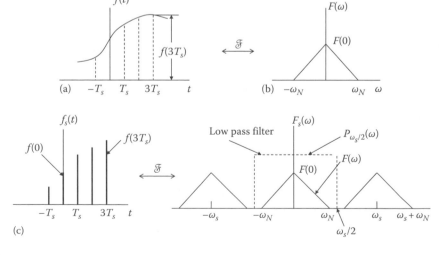

FIGURE 5.3

By (5.3), this equation becomes

$$F(\omega) = P_{\omega_s/2}(\omega)T_s \sum_{n=-\infty}^{\infty} f(nT_s)e^{-j\omega nT_s} \tag{5.14}$$

From which we have that

$$f(t) = \mathscr{F}^{-1}\{F(\omega)\} = \mathscr{F}^{-1}\left\{ P_{\omega_s/2}(\omega)T_s \sum_{n=-\infty}^{\infty} f(nT_s)e^{-j\omega nT_s} \right\}$$

$$= T_s \sum_{n=-\infty}^{\infty} f(nT_s)\mathscr{F}^{-1}\left\{ P_{\omega_s/2}(\omega)e^{-j\omega nT_s} \right\} \tag{5.15}$$

Note that the inverse Fourier transform operates only in function with the independent variable ω. By applying the frequency-shift property of the Fourier transform, it is seen that this equation proves the theorem.

■

This theorem states that no loss of information is incurred through the sampling process if the signal is sampled at a sampling frequency that is at least twice as fast as the highest frequency contained in the signal. Equivalently, the sampling time must be less or equal to one half of the Nyquist time T_N. For band-limiting signals, the sampling process introduces no error since, in theory, we can recover the original continuous time signal from its sampled version. The sinc function in (5.11) is known as the **interpolation function** to indicate that it allows an interpolation between the sampled values to find $f(t)$ for all ts.

For the case when $\omega_s = 2\omega_N$, (5.11) becomes

$$f(t) = \sum_{n=-\infty}^{\infty} f(nT_s)\frac{\sin[\omega_N(t - nT_s)]}{\omega_N(t - nT_s)} \tag{5.16}$$

and the spectrum of the sampled function is periodic with successive replicas of the spectrum of $f(t)$ just touching.

The sampling time

$$\boxed{T_s = \frac{T_N}{2} = \frac{1}{2f_N}} \tag{5.17}$$

is called the **Nyquist interval**. It is the longest time interval that can be used for sampling a band-limited signal while still permitting us to recover the signal without distortion. Figure 5.4 shows how a signal can be reconstructed from its samples using (5.11). Observe that the sinc functions tend to cancel between the sampling times and reinforce at the sampling points. Although we used a Gaussian function that has an infinite spectrum, when we use a unit sampling time and only three-sinc functions

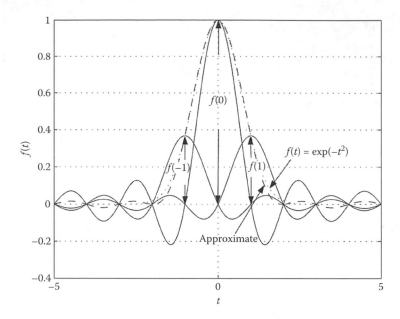

FIGURE 5.4

we observe that the approximation is very good. Of course, for a band-limited function and an infinite number of shifted-sinc functions, the recovery is complete and not approximate.

If we select the sampling time to be larger than half the Nyquist time or, equivalently, the sampling frequency to be less than twice the largest frequency (Nyquist frequency), an overlapping of the spectrums occurs. This overlapping is known as **aliasing**.

The Fourier transform of a sampled function is

$$F_s(\omega) = \mathcal{F}\{f(t)comb_{T_s}(t)\} = \frac{1}{2\pi}F(\omega) * \mathcal{F}\{comb_{T_s}(t)\} \equiv \text{frequency convolution}$$

$$= \frac{1}{2\pi}F(\omega) * \left[\frac{2\pi}{T_s}\sum_{n=-\infty}^{\infty}\delta(\omega - n\omega_s)\right] = \frac{1}{T_s}F(\omega) * \left[\sum_{n=-\infty}^{\infty}\delta(\omega - n\omega_s)\right]$$

$$= \frac{1}{T_s}F(\omega) * COMB_{\omega_s}(\omega) \tag{5.18}$$

If the sampling frequency is not appropriate (is less than twice the Nyquist frequency), then aliasing will take place and the spectrum will look like the one shown in Figure 5.5a. If the sampling time diminishes at least to the value $T_N/2$, then the spectrum of the sampled function will look like the one shown in Figure 5.5b. It is clear from Figure 5.5a that there is no filter available that is capable of extracting the frequency spectrum of the signal without including additional frequencies produced by the shifted spectrums.

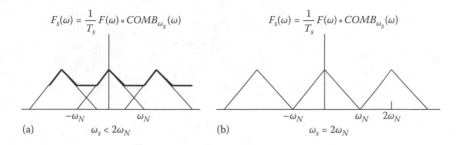

FIGURE 5.5

The effect of aliasing can be used to our advantage in some cases. Suppose that we want to stop optically a repetitive action, for example, the wing undulation of a bee or the turning of a wheel. To accomplish this optical effect, we flash the object with a strobe light. If we adjust the repetition of the strobe flash to equal the wing repetition rate or the wheel turning rate, these events will appear to be stationary. If the strobe frequency is higher than twice that of the periodic phenomenon, the speed of the phenomenon does not appear to change. However, if the strobe flashes are less than twice the frequency of the phenomenon under observation, the repetition slows down; thus, we observe a slow-flying bee or a slow-rotating wheel. This phenomenon is commonly observed in the movies when we see a moving stagecoach with wheels appearing stationary or turning slowly (sometimes backward). In the movies, the sampling rate is 1/20 s because the frame rate of the film is about 20 frames per second.

Sampling the function at twice the largest frequency available in the signal (the Nyquist frequency), the signal spectrums just touch each other and extend from minus infinity to infinity. This situation is shown to the left of Figure 5.6a. To isolate the spectrum of the signal we must multiply the sampled spectrum with an ideal low-pass filter, which is shown in the center of Figure 5.6a. The resulting spectrum of this multiplication is the spectrum of the signal $F(\omega)$. Hence, to recapture the signal intact, we write

$$\mathcal{F}^{-1}\{F(\omega)\} \stackrel{\Delta}{=} f(t) = \mathcal{F}^{-1}\left\{\left[\frac{1}{T_s}F(\omega) * COMB_{\omega_s}(\omega)\right]\frac{T_s}{2\pi}P_{\omega_s/2}(\omega)\right\}$$

$$= \mathcal{F}^{-1}\left\{\frac{1}{T_s}F(\omega) * COMB_{\omega_s}(\omega)\right\} * \frac{T_s}{2\pi}\mathcal{F}^{-1}\{P_{\omega_s/2}(\omega)\}$$

$$= f(t)comb_{T_s}(t) * \frac{T_s}{2\pi}\frac{2\sin(\omega_s t/2)}{t}$$

$$= \frac{T_s}{\pi}\left[\sum_{n=-\infty}^{\infty}f(nT_s)\delta(t - nT_s)\right] * \frac{\sin(\omega_s t/2)}{t}$$

$$= \frac{T_s}{\pi}\sum_{n=-\infty}^{\infty}f(nT_s)\int_{-\infty}^{\infty}\delta(t - x - nT_s)\frac{\sin(\omega_s x/2)}{x}dx$$

$$= T_s\sum_{n=-\infty}^{\infty}f(nT_s)\frac{\sin(\omega_s(t - nT_s)/2)}{\pi(t - nT_s)}$$

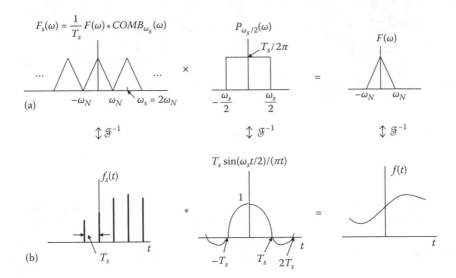

FIGURE 5.6

Example 5.3

Show the aliasing phenomenon by decreasing ω_s or, equivalently, by increasing the sampling time T_s associated with a pure cosine function $f(t) = \cos \omega_0 t$. Use a low-pass filter of bandwidth ω_s in the output.

SOLUTION

The Fourier transform of $f(t)$ is $F(\omega) = \pi\delta(\omega - \omega_0) + \pi\delta(\omega + \omega_0)$. Thus, the Fourier transform of the sampled function $f_s(t)$ is (see also (5.18))

$$F_s(\omega) = \frac{1}{T_s} F(\omega) * COMB_{\omega_s}(\omega)$$

$$= \frac{\pi}{T_s} [\delta(\omega - \omega_0) * COMB_{\omega_s}(\omega) + \delta(\omega + \omega_0) * COMB_{\omega_s}(\omega)]$$

Figure 5.7a shows the spectrum of the sampled function when $\omega_s \gg \omega_0$ or, equivalently, when $T_s \ll T_0 = 2\pi/\omega_0$. Figure 5.7b is the result of the convolution of $\delta(\omega - \omega_0)$ and $COMB_{\omega_s}(\omega)$. Figure 5.7d is the convolution of $\delta(\omega + \omega_0)$ and $COMB_{\omega_s}(\omega)$. By adding the spectrums of Figure 5.7b and d, we obtain the total spectrum of $F_s(\omega)$ as specified by the above equation. The spectrum of the sampled function is shown in Figure 5.7e. If we incorporate a filter with a frequency bandwidth of $\omega_s/2$, we regain our signal since

$$\frac{1}{2\pi} \int_{-\infty}^{\infty} T_s P_{\omega_s}(\omega) e^{j\omega t} d\omega = \frac{1}{2\pi} \int_{-\infty}^{\infty} T_s \left[\frac{\pi}{T_s}\delta(\omega + \omega_0) + \frac{\pi}{T_s}\delta(\omega - \omega_0)\right] e^{j\omega t} d\omega$$

$$= \frac{1}{2}\left[e^{j\omega_0 t} + e^{-j\omega_0 t}\right] = \cos \omega_0 t$$

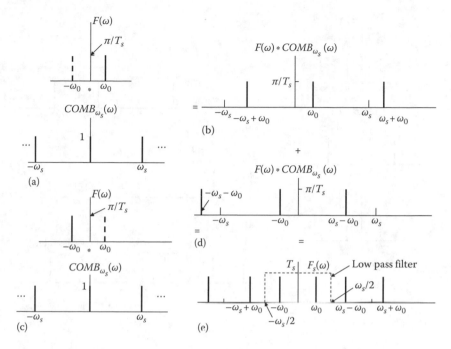

FIGURE 5.7

We follow the same procedure as shown in Figure 5.7 with the difference that, in this case, the sampling frequency was taken less than twice the Nyquist frequency, which, in this case, is ω_0. We observe in Figure 5.8 that the low-pass filter will produce a cosine function but with a different frequency equal to $\omega_s - \omega_0$, which is the alias of ω_0. ∎

Example 5.4

Consider the analog signal $f(t) = e^{-0.5t}\, u(t)$. (1) Find and plot the magnitude of its spectrum. (2) Using MATLAB®, plot the magnitude spectrum of the sampled version of $f(t)$ for $T_s = 0.6$ and $T_s = 0.4$.

Solution

The Fourier transform of $f(t)$ is easily found using the Fourier integral tables and is equal to $F(\omega) = 1/(0.5 + j\omega)$. From (5.3) we find that the Fourier transform of its sampled version is

$$F_s(\omega) = \sum_{n=0}^{\infty} e^{-0.5nT_s} e^{-j\omega nT_s} = \sum_{n=0}^{\infty} e^{-(0.5T_s + j\omega T_s)n} = 1 + e^{-(0.5T_s + j\omega T_s)} + e^{-(0.5T_s + j\omega T_s)2} + \cdots$$

$$= \frac{1}{1 - e^{-(0.5T_s + j\omega T_s)}} = \frac{1}{1 - e^{-0.5T_s} e^{-j\omega T_s}}$$

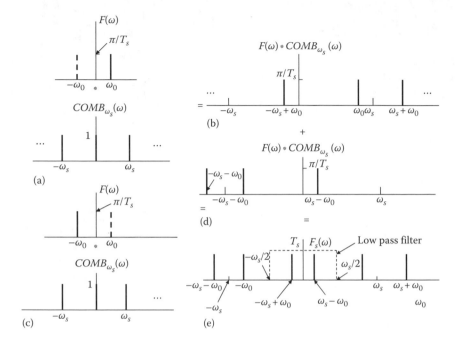

FIGURE 5.8

where the formula for the infinite geometric series was used $1/(1-x) = 1+x+x^2+ x^3+\cdots$. The value of x must be less than one for the series to converge. Figure 5.9a shows the magnitude of the spectrum $F(\omega)$ and Figure 5.9b and c shows the spectrum of the sampled function for $T_s = 0.6$ and $T_s = 0.4$, respectively.

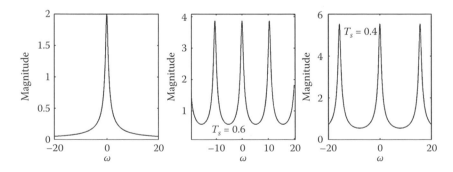

FIGURE 5.9

Book MATLAB m-File: ex5_2_2

```
%ex5_2_2 is the m-file for illustrating Ex 5.4
w=-10:0.01:10;
fw=abs(1./(0.5+j*w));
fs1w=abs(1./(1-exp(-0.5*0.6)*exp(-j*w*0.6)));
fs2w=abs(1./(1-exp(-0.5*0.1)*exp(-j*w*0.1)));
subplot(2,3,1);plot(w,fw,'k');xlabel('\omega');ylabel('magnitude');
subplot(2,3,2);plot(w,fs1w,'k');xlabel('\omega');ylabel('magnitude');
subplot(2,3,3);plot(w,fs2w,'k');xlabel('\omega');ylabel('magnitude');
```

■

6

Discrete-Time Transforms

In the previous chapters, we included discussions on the use of Fourier transform (FT) techniques in continuous time-signal processing and signal-sampling studies. However, in many problems, the signal may be experimentally derived, and analytic functions are not available for the integrations involved. There are two general approaches that might be adopted in such cases. One method calls for approximating the functions and carrying out the integrations by numerical means. The second method, and one that is used extensively, calls for replacing the continuous FT by an equivalent **discrete Fourier transform** (DFT) and then evaluating the DFT using the discrete data. Now, instead of integrations, the direct solution of DFT requires, for each sample, N complex multiplications, N complex additions, and the access of N complex exponentials that appear in the DFT. Hence, with 10^4 samples (a small number in many cases), more than 10^8 mathematical operations are required in the solution. It was the development of the **fast Fourier transform** (FFT), a computational technique that reduces the number of mathematical operations of the DFT to $N \log_2 N$, that made the DFT an extremely useful transform in many fields of science and engineering.

6.1 Discrete-Time Fourier Transform

6.1.1 Approximating the Fourier Transform

The FT of a continuous-time function is given by

$$F(\Omega) = \int_{-\infty}^{\infty} f(t)e^{-j\Omega t} dt \qquad (6.1)$$

where we have used the capital omega to represent the continuous-frequency independent variable in rad/s. We can approximate the above integral by introducing a short time interval T_s such that, essentially, all the radian frequency content of $F(\Omega)$ lies in the interval $-\pi/T_s \le \Omega \le \pi/T_s$. With this approximation, we express the Fourier transform integral (6.1) in the form

$$F(\Omega) = \sum_{n=-\infty}^{\infty} \int_{nT_s}^{nT_s+T_s} f(t)e^{-j\Omega t}\,dt \cong \sum_{n=-\infty}^{\infty} T_s f(nT_s)e^{-j\Omega nT_s} \quad T_s \to 0 \qquad (6.2)$$

If we set

$$f(n) \equiv T_s f(nT_s)$$
$$\omega \equiv \Omega T_s \quad \text{rad/unit} \tag{6.3}$$

in the above equation, we obtain

$$\boxed{\mathscr{F}_{DT}\{f(n)\} \stackrel{\Delta}{=} F(\omega) \stackrel{\Delta}{=} F(e^{j\omega}) = \sum_{n=-\infty}^{\infty} f(n)e^{-j\omega n}} \tag{6.4}$$

which is known as the discrete-time Fourier transform (DTFT).

Note: *The discrete frequency ω is equal to ΩT_s and has the units of radians per unit length, where Ω has units of rad/s. If T_s is unity time, then the discrete frequency and the continuous frequency have the same values but different units. Both frequencies, Ω and ω are continuous independent variables.*

Therefore, the following steps can be taken to approximate the continuous-time FT:

1. Select the time sampling T_s such that $F(\Omega) \cong 0$ for all $|\Omega| > \pi/T_s$.
2. Sample $f(t)$ at times nT_s to obtain $f(nT_s)$.
3. Compute the DTFT using the sequence $\{T_s f(nT_s)\}$.
4. The resulting approximation is then $F(\Omega) \cong F(\omega)$ *for* $-\dfrac{\pi}{T_s} \le \Omega \le \dfrac{\pi}{T_s}$.

If we introduce the value of $\omega + 2\pi$ in (6.4), we obtain

$$F(\omega + 2\pi) \stackrel{\Delta}{=} F(e^{j(\omega+2\pi)}) = F(e^{j\omega}) = F(\omega) \tag{6.5}$$

since $\exp(j2\pi) = 1$. Therefore, the function $F(\omega)$ is periodic with a period of 2π. This property contrasts with the FTs of continuous functions whose range is infinite. Thus, due to periodicity, the frequency range of any discrete-time signal is limited to the range $(-\pi, \pi]$ or $(0, 2\pi]$ for $T_s = 1$; any frequency outside these intervals is equivalent to a frequency within these intervals. If we had included the sampling time, (6.5) would have taken the form

$$F\left(e^{j\left(\Omega + \frac{2\pi}{T_s}\right)T_s}\right) = F\left(e^{j\left(\Omega T_s\right)}\right) \tag{6.6}$$

which indicates that the spectrum is periodic every $2\pi/T_s$.

Because $F(\omega)$ is periodic, it can be expanded in a Fourier series. To accomplish this, multiply both sides of (6.4) by $e^{j\omega m}d\omega$ and integrate the expression within the period. This yields

$$\int_{-\pi}^{\pi} F(e^{j\omega})e^{j\omega m}d\omega = \int_{-\pi}^{\pi} \left[\sum_{n=-\infty}^{\infty} f(n)e^{-jn\omega}\right]e^{jm\omega}d\omega \tag{6.7}$$

But, taking into consideration that the integral of any sinusoidal signal is zero when it is integrated within a period (or multiple period), we obtain

$$\sum_{n=-\infty}^{\infty} f(n)\int_{-\pi}^{\pi} e^{j\omega(m-n)}d\omega = \begin{cases} 2\pi f(n) & m = n \\ 0 & m \neq n \end{cases} \tag{6.8}$$

Then, (6.7) becomes

$$f(n) = \frac{1}{2\pi}\int_{-\pi}^{\pi} F(e^{j\omega})e^{j\omega n}d\omega \tag{6.9}$$

Thus, the DTFT pair of discrete-time signals is

$$\boxed{\begin{aligned} F(e^{j\omega}) &= |F(e^{j\omega})|e^{j\phi(\omega)} = \sum_{n=-\infty}^{\infty} f(n)e^{-j\omega n} \\ f(n) &= \frac{1}{2\pi}\int_{-\pi}^{\pi} F(e^{j\omega})e^{j\omega n}d\omega \end{aligned}} \tag{6.10}$$

Example 6.1

Determine the DTFT of the signal $f(t) = \exp(-t)u(t)$, for $T_s = 1$ and $T_s = 0.1$.

SOLUTION

(a) From (6.10), we write

$$F(e^{j\omega}) = \sum_{n=0}^{\infty} e^{-n}e^{-j\omega n} = \sum_{n=0}^{\infty} e^{-(1+j\omega)n} = \frac{1}{1 - e^{-(1+j\omega)}}$$

since the summation is an infinite geometric series.

(b) Taking into consideration the sampling time in (6.10), we obtain

$$F(e^{j\Omega T_s}) = T_s \sum_{n=0}^{\infty} f(nT_s)e^{-j\Omega T_s n} = 0.1 \sum_{n=0}^{\infty} e^{-nT_s}e^{-j\Omega T_s n}$$

$$= 0.1 \sum_{n=0}^{\infty} e^{-(0.1+j\Omega 0.1)n} = \frac{0.1}{1 - e^{-0.1}e^{-j0.1\Omega}}$$

Since $\{f(n)\}$ and $\{f(nT_s)\}$ are absolutely summable,

$$\sum_{n=0}^{\infty} |e^{-(1+j\omega)n}| = \sum_{n=0}^{\infty} |e^{-n}| = \sum_{n=0}^{\infty} e^{-n} = \frac{1}{1 - e^{-1}} < \infty$$

$$0.1 \sum_{n=0}^{\infty} |e^{-(0.1+j0.1\Omega)n}| = 0.1 \sum_{n=0}^{\infty} |e^{-0.1n}| = \frac{0.1}{1 - e^{-0.1}} < \infty$$

the series converge.

The magnitude and phase of the second case, for example, are given by

$$F(e^{j\Omega 0.1}) = \frac{0.1}{(1 - e^{-0.1}\cos 0.1\Omega) + j(e^{-0.1}\sin 0.1\Omega)}$$

$$= \frac{0.1}{\sqrt{(1 - e^{-0.1}\cos 0.1\Omega)^2 + (e^{-0.1}\sin 0.1\Omega)^2}} e^{-j\tan^{-1}\frac{e^{-0.1}\sin 0.1\Omega}{1 - e^{-0.1}\cos 0.1\Omega}}$$

If we set $\Omega = -\Omega$ in the above equation, we observe that the magnitude of the spectrum is an even function and the phase is an odd function.

Note: *For real functions the amplitude spectrum is **even** and the phase spectrum is an **odd** function. Thus, the representation of the frequency spectrum within the range $0 \le \Omega \le \pi$ will suffice.*

The reader should plot the magnitude of the spectra given above to observe the differences. Furthermore, if we set $T_s = 0.01$, the spectrum will closely approximate the exact one to 100π and its magnitude at zero frequency is $0.01/(1 - \exp(-0.01)) = 1.0050$, which is very close to 1, as it should be. ∎

6.1.2 Symmetry Properties of the DTFT

Sequence	DTFT
Complex signals	
$x(n)$	$X(e^{j\omega})$
$x^*(n)$	$X^*(e^{-j\omega})$
$x^*(-n)$	$X^*(e^{j\omega})$
$x_r(n)$	$X_e(e^{j\omega}) = \frac{1}{2}[X(e^{j\omega}) + X^*(e^{-j\omega})]$
$jx_i(n)$	$X_o(e^{j\omega}) = \frac{1}{2}[X(e^{j\omega}) - X^*(e^{-j\omega})]$
$x_e(n) = \frac{1}{2}[x(n) + x^*(-n)]$	$X_r(e^{j\omega})$
$x_o(n) = \frac{1}{2}[x(n) - x^*(-n)]$	$X_o(e^{j\omega})$
Real signals	
$x(n)$	$X(e^{j\omega}) = X^*(e^{-j\omega})$
$x(n)$	$X_r(e^{j\omega}) = X_r(e^{-j\omega})$
$x(n)$	$X(e^{j\omega})_i = -X_i(e^{-j\omega})$
$x(n)$	$|X(e^{j\omega})| = |X(e^{-j\omega})|$
$x(n)$	angle $X(e^{j\omega}) = -$angle $X(e^{-j\omega})$
$x_e(n) = \frac{1}{2}[x(n) + x(-n)]$	$X_r(e^{j\omega})$
Real and even	Real and even
$x_o(n) = \frac{1}{2}[x(n) - x(-n)]$	$jX_i(e^{j\omega})$
Real and odd	Imaginary and odd

6.2 Summary of DTFT Properties

Linearity	$af(n) + bh(n) \overset{\mathfrak{F}_{DT}}{\leftrightarrow} aF(e^{j\omega}) + bH(e^{j\omega})$
Time shifting	$f(n - m) \overset{\mathfrak{F}_{DT}}{\leftrightarrow} e^{-j\omega m} F(e^{j\omega})$
Time reversal	$f(-n) \overset{\mathfrak{F}_{DT}}{\leftrightarrow} F(e^{-j\omega})$
Convolution	$f(n)^*h(n) \overset{\mathfrak{F}_{DT}}{\leftrightarrow} F(e^{j\omega})H(e^{j\omega})$
Frequency shifting	$e^{j\omega_0 n} f(n) \overset{\mathfrak{F}_{DT}}{\leftrightarrow} F(e^{j(\omega - \omega_0)})$
Modulation	$f(n) \cos \omega_0 n \overset{\mathfrak{F}_{DT}}{\leftrightarrow} \frac{1}{2} F(e^{j(\omega + \omega_0)}) + \frac{1}{2} F(e^{j(\omega - \omega_0)})$
Correlation	$f(n) \odot h(n) \overset{\mathfrak{F}_{DT}}{\leftrightarrow} F(e^{j\omega})H(e^{-j\omega})$
Parseval's formula	$\sum_{n=-\infty}^{\infty} |f(n)|^2 = \frac{1}{2\pi} \int_{-\pi}^{\pi} |F(e^{j\omega})|^2 d\omega$
Time multiplication	$\mathfrak{F}_{DT}\{nx(n)\} = -z \dfrac{dX(z)}{dz}\bigg|_{z=e^{j\omega}}$ (see Chapter 8)
Multiplication on sequences	$\mathfrak{F}_{DT}\{x(n)y(n)\} = \dfrac{1}{2\pi} \int_{-\pi}^{\pi} Y(e^{jx})X(e^{j(\omega - x)})dx = \dfrac{1}{2\pi} Y(e^{j\omega})^*X(e^{j\omega})$

The proofs of these properties are given in Appendix 6.A.1.

Example 6.2

Find the DTFT of the function $g(n) = 0.9^n u(n) \cos 0.1\pi n$.

SOLUTION

We observe that the function $0.9^n u(n)$ is a modulating function and thus,

$$\mathfrak{F}_{DT}\{0.9^n u(n) \cos 0.1\pi n\} = \mathfrak{F}_{DT}\left\{\frac{0.9^n u(n)e^{j0.1\pi n}}{2} + \frac{0.9^n u(n)e^{-j0.1\pi n}}{2}\right\}$$

$$= \frac{1}{2}\sum_{n=0}^{\infty}(0.9e^{-j(\omega-0.1\pi)})^n + \frac{1}{2}\sum_{n=0}^{\infty}(0.9e^{-j(\omega+0.1\pi)})^n$$

$$= \frac{1}{2}\frac{1}{1-0.9e^{-j(\omega-0.1\pi)}} + \frac{1}{2}\frac{1}{1-0.9e^{-j(\omega+0.1\pi)}}$$

$$= \frac{1}{2}G(e^{j(\omega-0.1\pi)}) + \frac{1}{2}G(e^{j(\omega+0.1\pi)})$$

We must always have in mind that the spectrum of discrete signals is infinitely periodic. Therefore, the magnitude of $G(e^{j\omega})$ is a double periodic structure, one shifted to the left by 0.1π and the other to the right by 0.1π. Since in this case $T_s = 1$, the spectrum can be plotted from $-\pi/1$ to $\pi/1$.

Book MATLAB® m-File: ex6_2_1

```
%ex6_2_1 is an m file for illustrating Ex 6.1
w=-pi:0.01:pi;
fw1=abs(0.5*(1./(1-0.9*exp(-j*(w-0.1*pi)))));
fw2=abs(0.5*(1./(1-0.9*exp(-j*(w+0.1*pi)))));
plot(w,fw1,w,fw2);
```
∎

Note: *In contrast to the FT, where we had continuous functions in both the time and the frequency domain, in the case of DTFT we have discrete functions in the time domain and continuous functions in the frequency domain.*

Example 6.3

Verify Parseval's formula using the function

$$F(e^{j\omega}) = \begin{cases} 1 & -\pi < \omega < \pi \\ 0 & \text{otherwise} \end{cases}$$

SOLUTION

The time sequence corresponding to the given $F(e^{j\omega})$ is

$$f(n) = \frac{1}{2\pi}\int_{-\pi}^{\pi} e^{j\omega n}d\omega = \frac{\sin n\pi}{n\pi}$$

But the relations

$$\sum_{n=0}^{\infty}\left(\frac{\sin n\pi}{n\pi}\right)^2 = \frac{1}{\pi^2}\sum_{n=0}^{\infty}\left(\frac{\sin n\pi}{n}\right)^2 = \frac{1}{\pi^2}(\pi^2 + 0 + 0 + \cdots) = 1 \quad \text{and}$$

$$\frac{1}{2\pi}\int_{-\pi}^{\pi} d\omega = 1$$

prove the theorem. ■

6.3 DTFT of Finite Time Sequences

Practical considerations usually dictate that we deal with truncated series. Therefore, the spectrum of a truncated series requires special attention. When we observe a finite number of data points, $f(0), f(1), f(2), \ldots, f(N-1)$, how do we account for the unobserved time-series elements that lie outside the measured interval $0 \le n \le N-1$? We must consider the effect of the missing data since a one-sided FT, for example, requires the entire set of time-series elements $f(n)$ for the interval $0 \le n < \infty$.

The one-sided truncated DTFT is defined by

$$F_N(e^{j\omega}) = \sum_{n=0}^{N-1} f(n)e^{-j\omega n} \tag{6.11}$$

We introduce the DTFT of $f(n)$ in the above expression so that

$$F_N(e^{j\omega}) = \sum_{n=0}^{N-1}\left[\frac{1}{2\pi}\int_{-\pi}^{\pi} F(e^{j\omega'})e^{j\omega' n}d\omega'\right]e^{-j\omega n}$$

$$= \frac{1}{2\pi}\int_{-\pi}^{\pi} F(e^{j\omega'})\sum_{n=0}^{N-1} e^{j(\omega-\omega')n}d\omega' = \frac{1}{2\pi}\int_{-\pi}^{\pi} F(e^{j\omega'})W(e^{j(\omega-\omega')})d\omega'$$

$$= \frac{1}{2\pi}F(e^{j\omega}) * W(e^{j\omega}) \tag{6.12}$$

where

$$W(e^{j\omega}) = \sum_{n=0}^{N-1} e^{-j\omega n} = \frac{1 - (e^{-j\omega})^N}{1 - e^{-j\omega}} = e^{-j\omega(N-1)/2}\frac{\sin(\omega N/2)}{\sin(\omega/2)} \tag{6.13}$$

and the finite geometric series formula was used. The transform function $W(e^{j\omega})$ is the DTFT of the rectangular window, since it is the transform of the time function $w(n) = u(n) - u(n-N) = p_{N/2}\left(n - \frac{N}{2}\right)$, even for N. We observe that with a finite sequence a convolution appears in the frequency domain. From (6.12), we observe that to find the exact spectrum $F(e^{j\omega})$ we require a Fourier-transformed window $W(e^{j\omega})$ equal to a delta function $\delta(\omega)$ in the interval $-\pi \leq \omega \leq \pi$. However, the magnitude of $|W(e^{j\omega})| = \sin(\omega N/2)/\sin(\omega/2)$ has the properties of a delta function and approaches it as $N \to \infty$. Therefore, the longer the observed time-data sequence, the less the distortion in the spectrum of the signal, $F(e^{j\omega})$.

We observe that

$$\cos \omega_0 n = \mathscr{F}_{DT}^{-1}\{\pi\delta(\omega - \omega_0) + \pi\delta(\omega + \omega_0)\} \tag{6.14}$$

and, hence, the DTFT of

$$f(n) = \cos(\omega_0 n) + \cos[(\omega_0 + \Delta\omega_0)n] \tag{6.15}$$

is

$$F(e^{j\omega}) = \pi[\delta(\omega - \omega_0) + \delta(\omega - \omega_0) + \delta(\omega - \omega_0 - \Delta\omega_0) + \delta(\omega + \omega_0 + \Delta\omega_0)] \tag{6.16}$$

The convolution of (6.16) with (6.13) gives the following expression:

$$F_N(e^{j\omega}) = \frac{1}{2}e^{-j(\omega-\omega_0)(N-1)/2}\frac{\sin\dfrac{\omega - \omega_0}{2}N}{\sin\dfrac{\omega - \omega_0}{2}} + \frac{1}{2}e^{-j(\omega+\omega_0)(N-1)/2}\frac{\sin\dfrac{\omega + \omega_0}{2}N}{\sin\dfrac{\omega + \omega_0}{2}}$$

$$+ \frac{1}{2}e^{-j(\omega-\omega_0-\Delta\omega_0)(N-1)/2}\frac{\sin\dfrac{\omega - \omega_0 - \Delta\omega_0}{2}N}{\sin\dfrac{\omega - \omega_0 - \Delta\omega_0}{2}}$$

$$+ \frac{1}{2}e^{-j(\omega+\omega_0+\Delta\omega_0)(N-1)/2}\frac{\sin\dfrac{\omega + \omega_0 + \Delta\omega_0}{2}N}{\sin\dfrac{\omega + \omega_0 + \Delta\omega_0}{2}} \tag{6.17}$$

We observe that the DTFT of two finite sinusoids is made up of four sinc functions, two on the right side and two on the left side. We will plot the magnitude of (6.17) for the different values of $\Delta\omega_0$. This will show us how the value of N affects the resolution capabilities of the DTFT for two sinusoids whose frequency is very close to each other. We shall plot only half of the spectrum for convenience. Figure 6.1a shows the magnitude of the spectrum for two sinusoids with $\omega_0 = 0.2\pi$, $\Delta\omega_0 = 0.1\pi$, and an infinite number of terms. Figure 6.1b shows the magnitude spectrum for the case $\Delta\omega_0 = 0.1\pi \gg \frac{2\pi}{N} = \frac{2\pi}{50}$. Figure 6.1c shows the magnitude spectrum for the case $\Delta\omega_0 = 0.1\pi > \frac{2\pi}{N} = \frac{2\pi}{25}$ and, finally, Figure 6.1d shows the magnitude spectrum for the case $\Delta\omega_0 = 0.1\pi < \frac{2\pi}{N} = \frac{2\pi}{10}$. Since the limit value is $2\pi/N$, then as $N \to \infty$ the value of $2\pi/N$ approaches zero. This

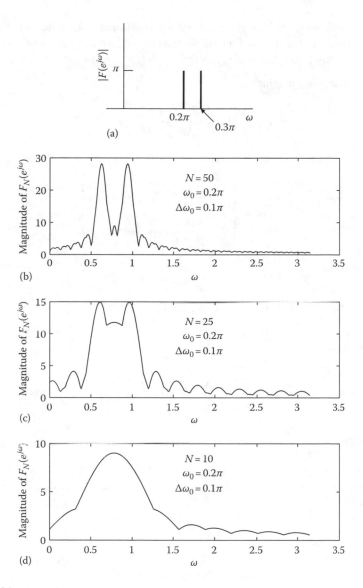

FIGURE 6.1

indicates that we can separate two sinusoids with an infinitesimal difference if we take enough signal values.

6.3.1 Windowing

It turns out that the FT spectrum will depend on the type of window used. It is appropriate that the window functions used are such that the smoothed version of the spectrum resembles the exact spectrum as closely as possible. Typically, it is found that for a given value N, the smoothing effect is directly proportional to the width of the main lobe of the window, and the effects of the ripple decrease as the relative amplitude of the main lobe and the largest side-lobe diverge. A few important discrete windows are

1. von Hann or Hanning

$$w(n) = \frac{1}{2}\left[1 - \cos\left(\frac{2\pi n}{N-1}\right)\right] \quad 0 \le n < N \tag{6.18}$$

2. Bartlett

$$w(n) = \begin{cases} \dfrac{2n}{N-1} & 0 \le n \le \dfrac{N-1}{2} \\[2mm] 2 - \dfrac{2n}{N-1} & \dfrac{N-1}{2} \le n < N \end{cases} \tag{6.19}$$

3. Hamming

$$w(n) = 0.54 - 0.46\cos\left(\frac{2\pi n}{N-1}\right) \quad 0 \le n < N \tag{6.20}$$

4. Blackman

$$w(n) = 0.4 - 0.5\cos\left(\frac{2\pi n}{N-1}\right) + 0.08\cos\left(\frac{4\pi n}{N-1}\right) \quad 0 \le n < N \tag{6.21}$$

Example 6.4

Find the DTFT of the discrete function $f(n) = 0.95^n u(n)$ using $N = 21$ and compare that spectrum with the spectrum using the Hamming window instead of the rectangle one.

SOLUTION

Figure 6.2a shows the spectrum for the rectangular window and Figure 6.2b shows the spectrum using the Hamming window. For both cases, we used $N = 21$. Figure 6.2c is the exact spectrum.

Book MATLAB m-File: ex6_3_1

```
%ex6_3_1 is an m file that illustrates Ex 6.1
N=21;
n=0:N-1;
w=0:.01:pi;
wh=0.54-0.46*cos(2*pi*n'/(N-1));%Nx1 vector;
f1=abs((0.95.^n')'*exp(-j*n'*w));
```

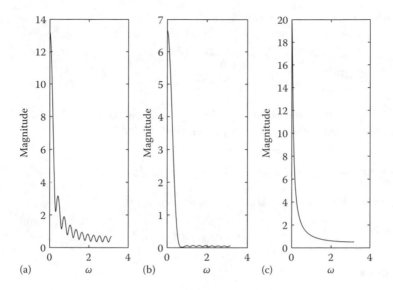

FIGURE 6.2

```
f1s = sum(f1,1);%sums the columns of the Nxlength(w) matrix f1;
f2h = abs((0.95.^n'.*wh)'*exp(-j*n'*w));
f2hs = sum(f2h,1);
subplot(1,3,1);plot(w,f1s,'k');xlabel('\omega');ylabel('magnitude');
subplot(1,3,2);plot(w,f2hs,'k');xlabel('\omega');ylabel('magnitude');
fwe = abs(1./(1-0.95*exp(-j*w)));
subplot(1,3,3);plot(w,fwe,'k');xlabel('\omega');ylabel('magnitude');
```
∎

6.4 Frequency Response of LTI Discrete Systems

In Chapter 1, we introduced discrete systems which are described by difference equations. The general difference equation for the first-order system is given by

$$y(n) = b_0 x(n) + b_1 x(n-1) - a_1 y(n-1) \qquad (6.22)$$

We assume that the DTFT of $y(n)$, $x(n)$, and the system impulse response $h(n)$ all exist. Taking the DTFT of both sides of the above equation, we obtain

$$Y(e^{j\omega}) = b_0 X(e^{j\omega}) + b_1 e^{-j\omega} X(e^{j\omega}) - a_1 e^{-j\omega} Y(e^{j\omega}) \qquad (6.23)$$

from which we write the system function

$$H(e^{j\omega}) = \frac{Y(e^{j\omega})}{X(e^{j\omega})} = \frac{b_0 + b_1 e^{-j\omega}}{1 + a_1 e^{-j\omega}} = b_0 \frac{e^{j\omega} + \frac{b_1}{b_0}}{e^{j\omega} + a_1} \overset{\Delta}{=} b_0 \frac{z - z_1}{z - p_1}\Big|_{z=e^{j\omega}} \tag{6.24}$$

If we set $\omega + 2\pi$ in the above equation we find that $H(e^{j(\omega+2\pi)}) = H(e^{j\omega})$, which indicates that the transfer function is periodic with period 2π ($T_s = 1$).

Example 6.5

Determine the system function of the system specified by

$$y(n) - 0.5y(n-1) = x(n) \tag{6.25}$$

SOLUTION

Comparing (6.25) to (6.22), we observe that the constants are $a_1 = -0.5$, $b_0 = 1$, and $b_1 = 0$. Hence, (6.24) gives the following transfer function:

$$H(e^{j\omega}) = \frac{1}{1 - 0.5e^{-j\omega}} = \frac{e^{j\omega}}{e^{j\omega} - 0.5} \tag{6.26}$$

We know that the DTFT (see Table 6.1) of $0.5^n u(n)$ is

$$\sum_{n=0}^{\infty} 0.5^n e^{-j\omega n} = \frac{1}{1 - 0.5e^{-j\omega}} = H(e^{j\omega})$$

Therefore, the impulse response of the system is

$$h(n) = 0.5^n u(n) \tag{6.27} \quad \blacksquare$$

TABLE 6.1   Some DTFT of Discrete Signals

Time Function $f(n)$	DTFT of $f(n)$
$x(n) = \delta(n)$	$X(e^{j\omega}) = 1$
$x(n) = u(n) - u(n-N-1)$	$X(e^{j\omega}) = e^{-j\frac{\omega}{2}(N-1)} \dfrac{\sin\dfrac{\omega N}{2}}{\sin\dfrac{\omega}{2}}$
$x(n) = \dfrac{\sin \omega_0 n}{\pi n} \quad |\omega_0| < \pi, \; -\infty < n < \infty$	$X(e^{j\omega}) = \begin{cases} 1 & |\omega| < \omega_0 \\ 0 & \omega_0 \leq |\omega| \leq \pi \end{cases}$
$x(n) = a^n \cos n\omega_0 \quad |\omega_0| < \pi$	$X(e^{j\omega}) = \dfrac{1}{2}\dfrac{1}{1 - ae^{-j(\omega-\omega_0)}} + \dfrac{1}{2}\dfrac{1}{1 - ae^{-j(\omega+\omega_0)}}$
$x(n) = a^n u(n)$	$X(e^{j\omega}) = \dfrac{1}{1 - ae^{-j\omega}}$

6.5 Discrete Fourier Transform

We have shown in Chapter 5 that if a time function is sampled uniformly in time, its Fourier spectrum is a periodic function. Therefore, corresponding to any sampled function in the frequency domain, a periodic function exists in the time domain. As a result, the sampled signal values can be related in both domains.

As a practical matter, we are only able to manipulate a certain length of signal. That is, suppose that the data sequence is available only within a finite time window from $n=0$ to $n=N-1$. The transform is discretized for N values by taking samples at the frequencies $2\pi/NT_s$, where T_s is the time interval between sample points in the time domain. Hence, we define the **discrete Fourier transform** (DFT) of a sequence of N samples $\{f(nT_s)\}$ by the relation

$$\boxed{\begin{aligned} F(k\Omega_b) &\stackrel{\Delta}{=} F(e^{jk\Omega_b}) = \mathscr{F}_D\{f(nT_s)\} = T_s \sum_{n=0}^{N-1} f(nT_s)e^{-j\Omega_b T_s kn} \\ \Omega_b &= \frac{2\pi}{NT_s}; \quad 0 \le n \le N-1 \end{aligned}}$$

(6.28)

where

N = number of samples (even number)
T_s = sampling time interval
$(N-1)T_s$ = signal length in the time domain
$\Omega_b = \dfrac{\omega_s}{N} = \dfrac{2\pi}{NT_s}$ = the frequency sampling interval (frequency bin)
$e^{-j\Omega_b T_s}$ = Nth principal root of unity

The inverse DFT (IDFT) is related to the DFT in much the same way as the FT was related to its inverse FT. The IDFT is given by

$$\boxed{\begin{aligned} f(nT_s) &\stackrel{\Delta}{=} \mathscr{F}_D^{-1}\{F(k\Omega_b)\} = \frac{1}{NT_s} \sum_{k=0}^{N-1} F(k\Omega_b)e^{j\Omega_b T_s nk} \\ \Omega_b &= \frac{2\pi}{NT_s}; \quad 0 \le k \le N-1 \end{aligned}}$$

(6.29)

Proof: We write

$$\frac{1}{NT_s} \sum_{k=0}^{N-1} F(k\Omega_b)e^{j\Omega_b T_s nk} = \frac{1}{NT_s} \sum_{k=0}^{N-1} \left[T_s \sum_{m=0}^{N-1} f(mT_s)e^{-j\Omega_b T_s mk} \right] e^{j\Omega_b T_s nk}$$

$$= \frac{1}{N} \sum_{m=0}^{N-1} f(mT_s) \sum_{k=0}^{N-1} e^{-j\Omega_b T_s(m-n)k}$$

However, using the finite number geometric series formula, we obtain

$$\sum_{k=0}^{N-1} e^{-j\Omega_b T_s(m-n)k} = \begin{cases} N & m = n \\ 0 & m \neq n \end{cases}$$

and, hence, the right side of the above equation becomes equal to $f(nT_s)$, as it should. Therefore, (6.29) and (6.28) constitute a pair of the DFT.

Note: *The sequences in the time domain and in the frequency domain are periodic with the period N. This indicates that when we take N terms from a time sequence and use the DFT, we automatically create a periodic-time function and the corresponding periodic-frequency function.*

Note: *In the DFT, both the time domain and the frequency domain are discrete sequences.*

In general, $F(e^{jk\Omega_b})$ is a complex function and can be written in the form

$$F(e^{jk\Omega_b}) = |F(e^{jk\Omega_b})| e^{j\phi(k\Omega_b)} \tag{6.30}$$

where $|F(e^{jk\Omega_b})|$ and $\phi(k\Omega_b)$ are discrete frequency functions. The plots of these functions versus $k\Omega_b$ are referred to as the **amplitude** and the **phase** spectra, respectively, of the signal $f(nT_s)$.

Example 6.6

Find the DFT of the function $f(n) = 0.9^n u(n)$ for $N = 4$.

SOLUTION

Since the sampling time is unity, (6.28) becomes

$$F\left[0\left(\frac{2\pi}{4}\right)\right] = \sum_{n=0}^{4-1} 0.9^n e^{-j\frac{2\pi}{4}0n} = 1 + 0.9 + 0.9^2 + 0.9^3 = 3.4390$$

$$F\left[1\left(\frac{2\pi}{4}\right)\right] = \sum_{n=0}^{4-1} 0.9^n e^{-j\frac{2\pi}{4}1n} = 1 + 0.9e^{-j\frac{2\pi}{4}1 \times 1} + 0.9^2 e^{-j\frac{2\pi}{4}1 \times 2} + 0.9^3 e^{-j\frac{2\pi}{4}1 \times 3}$$

$$= 0.1900 - 0.1710j$$

$$F\left[2\left(\frac{2\pi}{4}\right)\right] = \sum_{n=0}^{4-1} 0.9^n e^{-j\frac{2\pi}{4}2n} = 1 + 0.9e^{-j\frac{2\pi}{4}2 \times 1} + 0.9^2 e^{-j\frac{2\pi}{4}2 \times 2} + 0.9^3 e^{-j\frac{2\pi}{4}2 \times 3}$$

$$= 0.1810 - 0.0000j$$

$$F\left[3\left(\frac{2\pi}{4}\right)\right] = \sum_{n=0}^{4-1} 0.9^n e^{-j\frac{2\pi}{4}3n} = 1 + 0.9e^{-j\frac{2\pi}{4}3 \times 1} + 0.9^2 e^{-j\frac{2\pi}{4}3 \times 2} + 0.9^3 e^{-j\frac{2\pi}{4}3 \times 3}$$

$$= 0.1900 + 0.1710j$$

Note: *The magnitude of the spectrum for $k=3$ and $k=1$ are the same. The frequency $\pi/T_s = \pi/1$ is known as the **fold-over** frequency.*

The amplitude spectrum is 3.4390, 0.2556, 0.1810, and 0.2556 for $k=0$, 1, 2, and 3, respectively. The phase spectrum is 0.0000, -0.7328, 0, and 0.7328 for $k=0$, 1 2 and 3. The phase spectrum is given in radians. Since the frequency bin is $\Omega_b = 2\pi/NT_s = 2\pi/4 \times 1 = \pi/2$, then the amplitude and the frequency spectra are located at 0, $\pi/2$, π (the fold-over frequency) and $3\pi/2$ radians/unit length. ∎

6.6 Summary of DFT Properties

Below we give a summary of the DFT properties. Their proof is given in Appendix 6.A.2. It is convenient to abbreviate the notations of (6.28) and (6.29) by writing

$$F(k\Omega_b) = F(k); \quad f(nT) = f(n); \quad e^{-j2\pi/N} = W \tag{6.31}$$

Using this notation, the DFT pair takes the form

$$
\begin{aligned}
F(k) &= \sum_{n=0}^{N-1} f(n) W^{nk} = \sum_{n=0}^{N-1} f(n) e^{-j2\pi nk/N} \quad k = 0, 1, 2, \ldots, N-1 \\
f(n) &= \frac{1}{N} \sum_{k=0}^{N-1} F(k) W^{-nk} = \frac{1}{N} \sum_{k=0}^{N-1} F(k) e^{j2\pi nk/N} \quad n = 0, 1, 2, \ldots, N-1
\end{aligned}
\tag{6.32}
$$

Periodicity	$x(n+N) = x(n)$ for all n
	$X(k+N) = X(k)$ for all k
Linearity	$af(n) + bh(n) \overset{\mathfrak{I}_D}{\leftrightarrow} aF(k) + bH(k)$
Symmetry	$\dfrac{1}{N} F(n) \overset{\mathfrak{I}_D}{\leftrightarrow} f(-k)$
Time shifting	$f(n-i) \overset{\mathfrak{I}_D}{\leftrightarrow} F(k) e^{-j2\pi ki/N} = F(k) W^{ki}$
Frequency shifting	$f(n) e^{jni} \overset{\mathfrak{I}_D}{\leftrightarrow} F(k-i)$
Time convolution	$y(n) \overset{\Delta}{=} f(n)^* h(n) \overset{\mathfrak{I}_D}{\leftrightarrow} F(k) H(k)$
Frequency convolution	$f(n) h(n) \overset{\mathfrak{I}_D}{\leftrightarrow} \dfrac{1}{N} \sum_{x=0}^{N-1} F(x) H(n-x)$
Parseval's theorem	$\sum_{n=0}^{N-1} |f(n)|^2 = \dfrac{1}{N} \sum_{k=0}^{N-1} |F(k)|^2$
Time reversal	$f(-n) \overset{\mathfrak{I}_D}{\leftrightarrow} F(-k)$
Delta function	$\delta(n) \overset{\mathfrak{I}_D}{\leftrightarrow} 1$
Central ordinate	$f(0) = \dfrac{1}{N} \sum_{k=0}^{N-1} F(k); \quad F(0) = \sum_{n=0}^{N-1} f(n)$

Example 6.7 (Shifting property)

Deduce the DFT of the two sequences shown in Figure 6.3. Observe that $h(n)$ is a shifted function of $f(n)$.

SOLUTION

From (6.28) we obtain, respectively,

$$F\left(k\frac{2\pi}{4}\right) = \sum_{n=0}^{4-1} f(n)e^{-j2\pi kn/4}; \quad H\left(k\frac{2\pi}{4}\right) = \sum_{n=3}^{6} f(n)e^{-j2\pi kn/4}$$

The specific expansions are

$$F\left(0\frac{\pi}{2}\right) = 1\ \cos\left(\frac{\pi}{2}0 \times 0\right) + 2\ \cos\left(\frac{\pi}{2}0 \times 1\right) + 3\ \cos\left(\frac{\pi}{2}0 \times 2\right) + 4\ \cos\left(\frac{\pi}{2}0 \times 3\right)$$
$$- j\left[1\ \sin\left(\frac{\pi}{2}0 \times 0\right) + 2\ \sin\left(\frac{\pi}{2}0 \times 1\right) + 3\ \sin\left(\frac{\pi}{2}0 \times 2\right) + 4\ \sin\left(\frac{\pi}{2}0 \times 3\right)\right]$$
$$= 10 - j0$$

$$F\left(1\frac{\pi}{2}\right) = 1\ \cos\left(\frac{\pi}{2}1 \times 0\right) + 2\ \cos\left(\frac{\pi}{2}1 \times 1\right) + 3\ \cos\left(\frac{\pi}{2}1 \times 2\right) + 4\ \cos\left(\frac{\pi}{2}1 \times 3\right)$$
$$- j\left[1\ \sin\left(\frac{\pi}{2}1 \times 0\right) + 2\ \sin\left(\frac{\pi}{2}1 \times 1\right) + 3\ \sin\left(\frac{\pi}{2}1 \times 2\right) + 4\ \sin\left(\frac{\pi}{2}1 \times 3\right)\right]$$
$$= -2 + j2$$

We can proceed the same way to find the rest of the values. However, we can use the following Book MATLAB m-file to find the values easily.

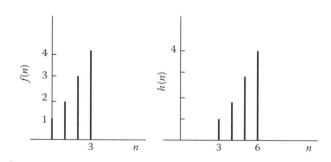

FIGURE 6.3

Book MATLAB m-Files: ex6_6_1, ex6_6_1b

```
%ex6_6_1 is an m-file to find the values of the summations
N=4;
n=0:N−1;
fn= [1 2 3 4];
k=0:N−1;
F=exp(−j*k'*n*2*pi/N)*fn';
subplot(2,1,1);stem(k,abs(F));subplot(2,1,2);
stem(k,angle(F));
%after you call the m-file, type F and return and you will
%be presented with the values of the DFT
```

The values of *F* are:
10.0000
−2.0000 + 2.0000*i*
−2.0000 − 0.0000*i*
−2.0000 − 2.0000*i*

```
%ex6_6_1b is an m-file to find the values of the summations for
% the shifted function h(n);
N=4;
n=3:N+3−1;
fn= [1 2 3 4];
k=0:N−1;
H=exp(−j*k'*n*2*pi/N)*fn';
subplot(2,1,1);stem(k,abs(H));subplot(2,1,2);stem(k,angle(H));
%after you call the m-file, type H and return and you will
%be presented with the values of the DFT
```

If we call the m-file ex6_6_1b, we obtain the DFT of the shifted function $h(n)$. The values are:
10.0000
−2.0000 − 2.0000*i*
2.0000 + 0.0000*i*
−2.0000 + 2.0000*i*

Observe that the magnitude of the spectrum stays the same but the phases have changed. This is the practical verification of the shifting property.   ∎

Example 6.8 (Time convolution)

Consider the two periodic sequences, $f(n) = \{1, -1, 4\}$ and $h(n) = \{0, 1, 3\}$. Verify the time convolution property by showing that for $T_s = 1$,

$$y(2) = \sum_{x=0}^{N-1} f(x)h(2-x) = \mathcal{F}_D\{F(k)H(k)\} = \frac{1}{N}\sum_{k=0}^{N-1} F(k)H(k)e^{j2\pi k2/N}$$

SOLUTION

We first find the summation

$$\sum_{x=0}^{2} f(x)h(2-x) = f(0)h(2) + f(1)h(1) + f(2)h(0) = 1 \times 3 + (-1)1 + 4 \times 0 = 2$$

Next, we obtain the DFT of the sequences $f(n)$ and $h(n)$, which are

$$F(0) = \sum_{n=0}^{2} f(n)e^{-j2\pi 0n/3} = f(0) + f(1) + f(2) = 1 - 1 + 4 = 4$$

$$F(1) = \sum_{n=0}^{2} f(n)e^{-j2\pi 1n/3} = 1 - e^{-j2\pi 1 \times 1/3} + 4e^{-j2\pi 1 \times 2/3}$$

$$F(2) = \sum_{n=0}^{2} f(n)e^{-j2\pi 2n/3} = 1 - e^{-j2\pi 2 \times 1/3} + 4e^{-j2\pi 2 \times 2/3}$$

Similarly, we obtain

$$H(0) = \sum_{n=0}^{2} h(n)e^{-j2\pi 0n/3} = h(0) + h(1) + h(2) = 0 + 1 + 3 = 4$$

$$H(1) = \sum_{n=0}^{2} h(n)e^{-j2\pi 1n/3} = 0 + e^{-j2\pi 1 \times 1/3} + 3e^{-j2\pi 1 \times 2/3}$$

$$H(2) = \sum_{n=0}^{2} h(n)e^{-j2\pi 2n/3} = 0 + e^{-j2\pi 2 \times 1/3} + 3e^{-j2\pi 2 \times 2/3}$$

The second summation given above becomes

$$\frac{1}{3}\sum_{k=0}^{2} F(k)H(k)e^{jk2\pi 2/3} = \frac{1}{3}[F(0)H(0) + F(1)H(1)e^{j4\pi/3} + F(2)H(2)e^{j8\pi/3}]$$

$$= 2.0000 - 0.0000j$$

For convenience, the reader can use MATLAB directly.   ■

Note: *To obtain the circular convolution of the two sequences $f(n)$ and $h(n)$, we follow the following steps: (1) Calculate the DFT of these sequences to obtain $F(k)$ and $H(k)$. (2) Multiply element by element these DFTs to obtain $G(k) = F(k)H(k)$. (3) Calculate the IDFT of $G(k)$. The result is the circular convolution, and for these two given in the example above $g(n) = \{1\ 13\ 2\}$.*

Note: *To obtain the linear convolution of $f(n)$ with N elements and $h(n)$ with M elements, we add zeros at the end of each sequence such that its total length is $Q = N + M - 1$.*

Example 6.9 (Frequency convolution)

Use the values of $F(k)$ and $H(k)$ of Example 6.7 to verify the frequency convolution property.

SOLUTION

Figure 6.4 shows the circular convolution of $F(k)$ and $H(k)$, with the values obtained from Example 6.8. From the frequency convolution property, we obtain the periodic sequence

$$y(n) = \mathcal{F}_{DF}^{-1}\{Y(k)\} = \mathcal{F}_{DF}^{-1}\left[\frac{1}{N}\sum_{i=0}^{N-1}F(i)H(k-i)\right] \quad \text{or} \quad \{y(n)\} = \{f(n)h(n)\} = \{0, -1, 12\}$$

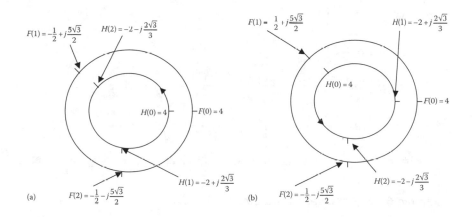

(a)

(b)

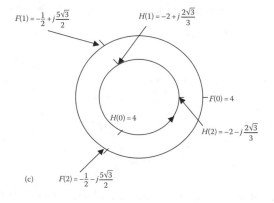

(c)

FIGURE 6.4

Thus, from the results of Figure 6.4, we obtain

$$y(0) = F_D^{-1}\{Y(k)\} = \frac{1}{3} \sum_{k=0}^{N-1} Y(k)e^{jk(2\pi/3)0}$$

$$= \frac{1}{3}\left[11 - \frac{11}{2} - \frac{11}{2} + j13\frac{\sqrt{3}}{2} - j13\frac{\sqrt{3}}{2}\right] = 0$$

$$y(1) = F_D^{-1}\{Y(k)\} = \frac{1}{3}\left[11 + \left(-\frac{11}{2} + j13\frac{\sqrt{3}}{2}\right)\left(\cos\frac{2\pi}{3} + j\sin\frac{2\pi}{3}\right)\right.$$

$$\left. + \left(-\frac{11}{2} - j13\frac{\sqrt{3}}{2}\right)\left(\cos\frac{4\pi}{3} + j\sin\frac{4\pi}{3}\right)\right] = -1$$

$$y(2) = F_D^{-1}\{Y(k)\} = \frac{1}{3}\left[11 + \left(-\frac{11}{2} + j13\frac{\sqrt{3}}{2}\right)\left(\cos\frac{4\pi}{3} + j\sin\frac{4\pi}{3}\right)\right.$$

$$\left. + \left(-\frac{11}{2} - j13\frac{\sqrt{3}}{2}\right)\left(\cos\frac{8\pi}{3} + j\sin\frac{8\pi}{3}\right)\right] = 12$$

From Figure 6.4, we obtain

$$Y(0) = \frac{[F(0)H(0) + F(1)H(2) + F(2)H(1)]}{3} = 11$$

$$Y(1) = \frac{[F(0)H(1) + F(1)H(0) + F(2)H(2)]}{3} = -\frac{11}{2} + j13\frac{\sqrt{3}}{2}$$

$$Y(2) = \frac{[F(0)H(2) + F(1)H(1) + F(2)H(0)]}{3} = -\frac{11}{2} - j13\frac{\sqrt{3}}{2}$$

The results obtained verify the frequency convolution property.   ∎

Example 6.10 (Parseval's theorem)

Verify Parseval's theorem using the sequence $f(n) = \{1, -1, 4\}$.

SOLUTION

We have directly that $\sum_{n=0}^{3-1} f^2(n) = 1 + 1 + 16 = 18$. The values of $F(k)$ for this sequence are given in Example 6.8, so that

$$\frac{1}{3}\sum_{k=0}^{2}|F(k)|^2 = \frac{1}{3}[16 + |-0.5 + j5 \times 0.8660|^2 + |-0.5 - j5 \times 0.8660|^2$$

$$= \frac{1}{3}[16 + 19 + 19] = 18$$ ∎

The above examples elucidated some of the DFT properties. The following two examples will clarify the similarities and the differences between the DFT of the continuous functions and their FT.

Example 6.11

Deduce the DFT of the function shown in Figure 6.5a discretized with $T_s = 1$, and $T_s = 0.5$.

SOLUTION

This signal is a shifted triangle by 2 s. The centered signal is equal to the convolution of two pulses of total width 2 s each: $f(t) = p(t) * p(t)$. From the time convolution property, the FT of $f(t)$ is equal to $F(\omega) = P(\omega)P(\omega)$. But the FT of a pulse is equal to $P(\omega) = \frac{2 \sin \omega}{\omega}$. Hence, the FT of the symmetric signal is $F(\omega) = \frac{4 \sin^2 \omega}{\omega^2}$. Taking into consideration the time-shifting property, the FT of the shifted signal is given by $F(\omega) = e^{-j2\omega} \frac{4 \sin^2 \omega}{\omega^2}$. For the case of $T_s = 1$, we have the following discrete signal $f(nT_s) = \{0\ 1\ 2\ 1\ 0\}$ for times $t = \{0\ 1\ 2\ 3\ 4\}$ seconds, respectively. Let us further define

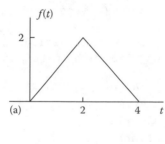

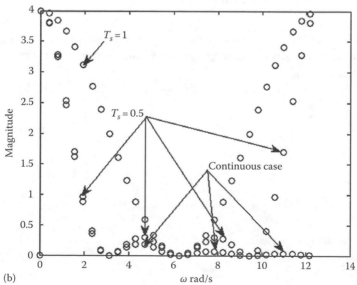

(b)

FIGURE 6.5

$NT_s = 16$. Since we set $T_s = 1$ and $NT_s = 16$, we find that $N = 16$. This value specifies the bins in the frequency domain that are located at every $\Omega_b = 2\pi/NT_s = 2\pi/16 = \pi/8$ rad/s. For the case of $T_s = 0.5$ s, and $NT_s = 16$ we obtain $N = 32$. However, because $NT_s = 16$ the bins at the frequency domain are located every $\pi/8$. But since in this case $N = 32$, the fold-over frequency (the useful frequency extend) is at $(\pi/8)32 = 4\pi$ rad/s. This can be compared with the first case whose fold-over frequency is $(\pi/8)16 = 2\pi$. Comparing the two results we observe that by decreasing the sampling time by 2 we expand the fold-over frequency (the useful frequency) by 2. The spectra can be found following Example 6.6. Here we use MATLAB for convenience.

Book MATLAB m-File: ex6_6_5

```
%ex6_6_5 is an m-file for the Ex 6.5
N1=16;
T1=1;
T2=0.5;
f1= [0 1 2 1 0 zeros(1,27)];
f2= [0 0.5 1 1.5 2 1.5 1 0.5 0 zeros(1,23)];
fd1=fft(f1);
wb1=0:2*pi/(N1*T1):4*pi-2*pi/(N1*T1);
fd2=T2*fft(f2);%the fft must be multiplied by sampling
%time;
ftf=4*(sin(wb1).^2./(wb1+eps).^2);
plot(wb1,abs(fd1),'ko');hold on;
plot(wb1,abs(fd2),'ko'); ...
     hold on;plot(wb1,abs(ftf),'ko');
```

Note: *We observe that by **decreasing** the time sampling, we succeed in extending the useful frequency range (the fold-over frequency) and within that range the approximation becomes better.* ∎

Example 6.12

Deduce the DFT transform of the function

$$f(t) = \begin{cases} 2 & 0 \le t \le 6 \\ 0 & \text{otherwise} \end{cases}$$

for the following three cases:

(a) $T_s = 1.0$, $NT_s = 16$
(b) $T_s = 1.0$, $NT_s = 32$
(c) $T_s = 0.2$, $NT_s = 16$

SOLUTION

We readily find the magnitude of the FT of the continuous time function to be

$$|F(\omega)| = \left| \frac{4 \sin 3\omega}{\omega} \right|$$

Cases (a) and (b): Setting $T_s = 1$ and $NT_s = 16$, we find that $N = 16$. The frequency bins are $\Omega_b = \frac{2\pi}{NT_s} = \frac{2\pi}{16} = \frac{\pi}{8}$ rad/s apart. The fold-over frequency is at $\frac{\pi}{8} \times \frac{N}{2} = \frac{\pi}{8}8 = \pi$ rad/s.

Figure 6.6a shows the absolute value of the difference (error) of the magnitude between the continuous and the DFT case at the same points of the frequency axis. For this case the distance is $\pi/8$. The discrete-time function is $f(n) = [2\ 2\ 2\ 2\ 2\ 2$ zeros $(1,10)]$; Below are nine magnitude values of the DFT spectrum for case (a)

12.0000 9.4713 3.6955 1.3776 2.8284 0.9205 1.5307 1.8840 0

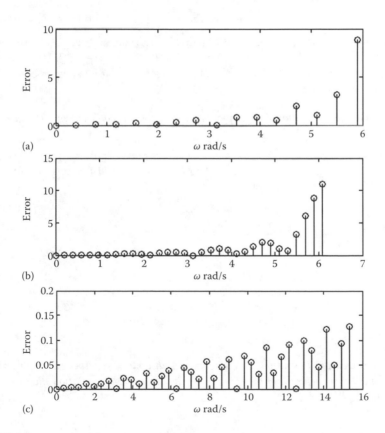

(a)

(b)

(c)

FIGURE 6.6

In case (b), we set $T_s = 1$ and $NT_s = 32$ to obtain $N = 32$ and the frequency bin $\Omega_b = \frac{2\pi}{NT_s} = \frac{2\pi}{32} = \frac{\pi}{16}$ rad/s apart. The fold-over frequency is $\frac{\pi}{16} \times \frac{N}{2} = \frac{\pi}{16} 16 = \pi$, which is identical to that of case (a). The discrete-time function for this case is $f(n) = $ [2 2 2 2 2 2 zeros(1,26)];. Note that in case (a) we added 10 zeros and for case (b) we added 26 zeros so that the numbers N in the two cases have the correct value. Although the fold-over frequency is the same, the number of frequency bins is twice as many as in case (a). Comparing the above nine frequency values with the nine numbers below for case (b),

12.0000   11.3362   9.4713   6.7574   3.6955   0.8277   1.3776   2.6213   2.8284

we observe that new values appeared in the second case between every $\pi/8$ distance. We note that at $\pi/8$ frequency bins the values of the spectrum for both cases are the same.

Note: *By adding zeros to a discrete sequence the accuracy of the spectrum does not increase. However, within any range, the number of points increases and gives us a better overall view of the spectrum.*

Figure 6.6b depicts the error between the discrete case (b) and the exact spectrum.

Case (c): For this case, we set $T_s = 0.2$ and $NT_s = 16$. Hence, the number of elements of the sequence is $N = 16/0.2 = 80$. The frequency bin is $\Omega_b = \frac{2\pi}{16} = \frac{\pi}{8}$. The fold-over frequency is $\Omega_b \frac{N}{2} = \frac{\pi}{8} 40 = 5\pi$ rad/s which is five times the fold-over frequency of cases (a) and (b).

Note: *By decreasing the sampling frequency we extend the useful frequency (fold over) to a larger range.*

The discrete-time function created from the continuous signal is
$f(n0.2) = $ [2 2 2 2 2 2 2 2 2 2 2 2 2 2 2 2 2 2 2 2 2 2 2 2 2 2 2 2 2 2 zeros(1, 50)]. The first nine elements of the DFT magnitude are

12.0000   9.4130   3.6050   1.3023   2.5570   0.7846   1.2116   1.3614   0.0000

The exact first nine frequency values for the continuous case are

12.0000   9.4106   3.6013   1.2993   2.5465   0.7796   1.2004   1.3444   0.0000

If we compare case, (c) with case (a) we observe that the case (c) results are more accurate (they are closer to the continuous case). If we compare case (c) with case (a) for the frequency $3(\pi/8)$ rad/s, we obtain 0.0030 for case (c) and 0.0783 for case (a). Figure 6.6c shows the error between the DFT for case (c) and the exact values.

Note: *By decreasing the sampling frequency, the discrete signal spectrum approximates better to the exact spectrum of the continuous signal. The approximation increases as the time sampling decreases.* ∎

Because the DFT uses a finite number of samples, we must be concerned about the effect that the truncation has on the Fourier spectrum. Specifically, if the signal $f(t)$

extends beyond the total sampling $(N-1)T_s$ time, the resulting frequency spectrum is an appropriation of the exact one. If, for example, we take the DFT of a truncated sinusoidal signal, we find that an error is present in the Fourier spectrum. If N is small and the sampling covers neither a large number of periods nor an integral number of cycles of the signal, a large error may occur, known as the **leakage** problem. Since the truncated signal is equal to $f(t)p_a(t)$, its Fourier spectrum is the convolution of the exact spectrum with the spectrum of the window. Since the truncation pre-assumes a rectangular window, its spectrum consists of a main lobe and side lobes. The side lobes affect the spectra and, sometimes, completely hide weak spectra lines. Since the magnitudes of the side lobes do not decrease with N, it is recommended to use other types of windows that may remedy the problem. However, the main lobe for other windows is wider and we must justify any effect that we may introduce.

Example 6.13

Find the DFT of the exponential function $f(t) = e^{-t}u(t)$ that is truncated at times $t = 0.8$ s and $t = 1.6$ s. Assume the sampling time $T_s = 0.02$ for both cases.

TABLE 6.2 Symmetries of the DFT

N-Point Sequence $x(n)$	DFT $X(k)$
$0 \leq n \leq N-1$	$0 \leq n \leq N-1$
Complex signals	
$x(n)$	$X(k)$
$x^*(n)$	$X^*(N-k)$
$x^*(N-n)$	$X^*(k)$
$x_r(n)$	$X_e(k) = \dfrac{1}{2}[X(k) + X^*(N-k)]$
$jx_i(n)$	$X_o(k) = \dfrac{1}{2}[X(k) - X^*(N-k)]$
$x_e(n) = \dfrac{1}{2}[x(n) + x^*(N-n)]$	$X_r(k)$
$x_o(n) = \dfrac{1}{2}[x(n) - x^*(N-n)]$	$jX_i(k)$
Real signals	
$x(n)$	$X(k) = X^*(N-k)$
$x(n)$	$X_r(k) = X_r(N-k)$
$x(n)$	$X_i(k) = -X_i(N-k)$
$x(n)$	$|X(k)| = |X(N-k)|$
$x(n)$	angle$\{X(k)\}$ = angle$\{X(N-k)\}$
$x_e(n) = \dfrac{1}{2}[x(n) + x(N-n)]$	$X_r(k)$
$x_o(n) = \dfrac{1}{2}[x(n) - x(N-n)]$	$jX_i(k)$

TABLE 6.3   Table of Several Functions DFTs

1. $f(n) = \delta(n - n_0)$ $0 \leq n_0 < N - 1$
   $n_0 = $ integer $0 \leq n < N - 1$
   $\qquad$ $F(k) = e^{-j2\pi n_0 k/N}$ $0 \leq k \leq N - 1$

2. $f(n) = e^{j2\pi nk_0/N}$ $0 < k_0 < N - 1$
   $k_0 = $ integer $0 \leq n \leq N - 1$
   $\qquad$ $F(k) = \delta(k - k_0)$ $0 \leq k \leq N - 1$

3. $f(n) = u(n) - u(n - N)$ $0 \leq n \leq N - 1$
   $\qquad$ $F(k) = \begin{cases} 1 & k = 0 \\ 0 & \text{otherwise} \end{cases}$

4. $f(n) = \cos(2\pi k_0 n/N)$ $0 < k_0 < N - 1$
   $k_0 = $ integer $0 \leq n \leq N - 1$
   $\qquad$ $F(k) = \begin{cases} \dfrac{N}{2}\delta(k - k_0) & k = k_0 \\ \dfrac{N}{2}\delta[k - (N - k_0)] & k = N - k_0 \end{cases}$

5. $f(n) = \cos \pi n$ $0 \leq n \leq N - 1$
   $\qquad$ $F(k) = N\delta\left(k - \dfrac{N}{2}\right)$ $0 \leq k \leq N - 1$

6. $f(n) = \cos\left(\dfrac{\pi n}{N}\right)$ $0 \leq n \leq N - 1$
   $\qquad$ $F(k) = \dfrac{1}{2}\dfrac{1 - \exp[-j(2\pi k - \pi)]}{1 - \exp\left[-j\left(\dfrac{2\pi k}{N} - \dfrac{\pi}{N}\right)\right]}$
   $\qquad + \dfrac{1 - \exp[-j(2\pi k + \pi)]}{1 - \exp\left[-j\left(\dfrac{2\pi k}{N} + \dfrac{\pi}{N}\right)\right]}$ $0 \leq k \leq N - 1$

7. $f(n) = \dfrac{n}{N}$ $0 \leq n \leq N - 1$
   $\qquad$ $F(k) = \begin{cases} \dfrac{N}{2} & k = 0 \\ j\dfrac{\cos(k\pi)\sin(k\pi/N)}{2\sin^2(k\pi/N)} & 0 \leq k \leq N - 1 \end{cases}$

8. $f(n) = \begin{cases} 1 & 0 \leq n \leq m \\ 0 & m \leq n \leq N - 1 \end{cases}$
   $\qquad$ $F(k) = e^{-j\frac{\pi km}{N}}\dfrac{\sin\left(\dfrac{\pi k(m + 1)}{N}\right)}{\sin\left(\dfrac{\pi k}{N}\right)}$ $0 \leq k \leq N - 1$

9. $f(n) = \begin{cases} 1 & 0 \leq n \leq (N/2) - 1 \\ 0 & (N/2) \leq k \leq N - 1 \end{cases}$
   $\qquad$ $F(k) = \exp\left[-j\dfrac{\pi k}{N}(N - 1)\right]\dfrac{\sin\left(\dfrac{\pi k}{2}\right)}{\sin\left(\dfrac{\pi k}{N}\right)}$

10. $f(n) = e^{-an/N}$ $a = $ positive constant
    $0 \leq n \leq N - 1$
    $\qquad$ $F(k) = \dfrac{1 - e^{-a}e^{-j2\pi k}}{1 - e^{-a/N}e^{-j2\pi/N}}$ $0 \leq k \leq N - 1$ $0 \leq k \leq N - 1$

11. $f(n) = 1 - \dfrac{n}{N - 1}$ $0 \leq n \leq N - 1$
    $\qquad$ $F(k) = \dfrac{1 - e^{-j2\pi k}}{1 - e^{-j2\pi k/N}} - \dfrac{1}{N - 1}[e^{-j2\pi k/N}$
    $\qquad [[1 - Ne^{-j2\pi k(N-1)/N} + (N - 1)e^{-j2\pi k}]$
    $\qquad /(1 - e^{-j2\pi k/N})^2]$ $0 \leq k \leq N - 1$

12. $f(n) = ne^{-an}$ $0 \leq n \leq N - 1$
    $a = $ positive constant
    $\qquad$ $F(k) = \dfrac{e^{-a}e^{-j2\pi k/N}}{(1 - e^{-a}e^{-j2\pi k/N})^2}$
    $\qquad \times \left(1 - Ne^{-a(N-1)}e^{-j\frac{2\pi k(N-1)}{N}} + (N - 1)e^{-aN}e^{-\frac{j2\pi k}{N}}\right)$
    $\qquad 0 \leq k \leq N - 1$

13. $f(n) = e^{-an}\cos\left(\dfrac{\pi n}{N}\right)$ $0 \leq n \leq N - 1$
    $\qquad$ $F(k) = \dfrac{1}{2}\dfrac{1 - e^{-aN}e^{-j\pi}e^{-j2\pi k}}{1 - e^{-a}e^{-j\pi/N}e^{-j2\pi k/N}}$
    $\qquad + \dfrac{1}{2}\dfrac{1 - e^{-aN}e^{j\pi}e^{-j2\pi k}}{1 - e^{-a}e^{j\pi/N}e^{-j2\pi k/N}}$
    $\qquad 0 \leq k \leq N - 1$

14. $f(n) = e^{-an}\sin\left(\dfrac{\pi n}{N}\right)$ $0 \leq n \leq N - 1$
    $\qquad$ $F(k) = \dfrac{1}{2j}\dfrac{1 - e^{-aN}e^{j\pi}e^{-j2\pi k}}{1 - e^{-a}e^{j\pi/N}e^{-j2\pi k/N}}$
    $\qquad - \dfrac{1}{2j}\dfrac{1 - e^{-aN}e^{-j\pi}e^{-j2\pi k}}{1 - e^{-a}e^{-j\pi/N}e^{-j2\pi k/N}}$
    $\qquad 0 \leq k \leq N - 1$

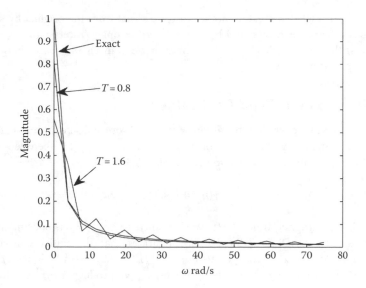

FIGURE 6.7

SOLUTION

The results are shown in Figure 6.7. The discrete spectrum was plotted in the continuous format to produce a better visual picture of the differences. The following Book m-file produces the results shown in the figure.

Book MATLAB m-File: ex6_6_7

```
%ex6_6_7 is an m file for the Ex6.7
t=0:0.02:0.8-.02;
f=[exp(-t) zeros(1,40)];
t1=0:0.02:1.6-0.02;
f1=exp(-t1);
df=0.02*fft(f);
df1=0.02*fft(f1);
w=0:2*pi/(80*0.02):100*pi-(2*pi/(80*0.02));
fcw=1./(1+w);
plot(w(1,1:20),fcw(1,1:20),'k');
hold on;plot(w(1,1:20),abs(df1(1,1:20)),'k');
hold on;plot(w(1,1:20),abs(df(1,1:20)),'k');
```
∎

6.7 Multirate Digital Signal Processing and Spectra

There are many applications where the signal of a given sampling rate must be converted into an equivalent signal with a different sampling rate. In other words, we "sample" digital signals. The sampling can be achieved using the **down-sampler (decimation)**, a device (or process) that creates another discrete signal from the original by skipping a specific number of elements, and by using the **up-sampler** (**interpolation**), a device (or process) that pads with a specific number of zeros in all the spaces between the original discrete-time signal.

6.7.1 Down Sampling (or Decimation)

The down-sampling operation by an **integer** $D > 1$ on a discrete-time signal $x(n)$ consists of keeping every Dth sample of $x(n)$ and removing $D - 1$ in-between samples. Hence, the output of such an operation is (see also Figure 6.8)

$$y(n) = x(nD) \tag{6.33}$$

Since the initial sampling time of $x(t)$ was assumed to be $T_s = 1$ producing $x(n)$, the sampling time of $x(nD)$ is $T_sD = D$. This indicates that the initial sampling rate $\omega_s = 2\pi/T_s = 2\pi$ for $x(n)$ changes to $\omega_{sd} = 2\pi/DT_s = 2\pi/D$, which is $(1/D)$th of that of $x(n)$. By decreasing the sampling frequency, we produce a periodic frequency spectrum whose period is $\omega_{sd} < \omega_s$. Down sampling produces aliasing. Figure 6.9

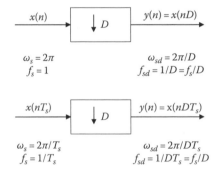

FIGURE 6.8

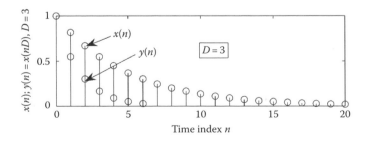

FIGURE 6.9

illustrates the effect of down sampling in the time domain and Figure 6.8 represents the down-sampling operation in a block diagram form.

Book MATLAB function for down sampling:[y, n1] = taadownsampling(x, D)

```
function[y,n1] =taadownsampling(x,D)
        %x=input sequence to be downsampled assuming
        %that Ts=1;
        %D=down-sampling factor;
n1=0:floor(length(x)/D)-1;
        %floor(x) rounds the elements of x to the
        %nearest integers towards minus infinity;
        %if desired, n1 can be used as the x-axis
        %to plot y from the origin;
y=x(1,1:D:D*floor(length(x)/D));
```

Example 6.14

Consider the system shown in Figure 6.10. The sequences that appear in the system are given below.

n:	0	1	2	3	4	5	6
$x(n)$	$x(0)$	$x(1)$	$x(2)$	$x(3)$	$x(4)$	$x(5)$	$x(6)$
$y_1(n)$	$x(-1)$	$x(0)$	$x(1)$	$x(2)$	$x(3)$	$x(4)$	$x(5)$
$y_3(n)$	$x(1)$	$x(2)$	$x(3)$	$x(4)$	$x(5)$	$x(6)$	$x(7)$
$y_2(n)$	$x(-1)$	$x(1)$	$x(3)$	$x(5)$	$x(7)$	$x(9)$	$x(11)$
$y_4(n)$	$x(1)$	$x(4)$	$x(7)$	$x(10)$	$x(13)$	$x(16)$	$x(19)$
$y(n)$	$x(-1)-x(1)$	$x(1)-x(4)$	$x(3)-x(7)$	$x(5)-x(10)$	$x(7)-x(13)$	$x(9)-x(16)$	$x(11)-x(19)$

■

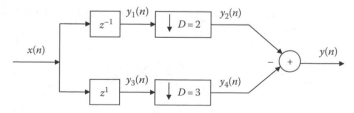

FIGURE 6.10

The sampling rate conversion can also be understood from the point of view of a digital resampling of the same analog signal. If the analog signal $x(t)$ is sampled at the rate $1/T_s$ to generate $x(n)$, the analog signal must be sampled at the rate $1/(DT_s)$ to generate $y(n)$.

6.7.2 Frequency Domain of Down-Sampled Signals

It is instructive to develop the spectra of the input and output sequences of a down sampler. The spectrum of the output signal can be found after being transformed in the DTFT domain. Hence, we write

$$y(e^{j\omega}) = \sum_{n=-\infty}^{\infty} y(n)e^{-j\omega n} = \sum_{n=-\infty}^{\infty} x(nD)e^{-j\omega n} \quad -\infty < n < \infty \tag{6.34}$$

If we set $nD = k$, then $n = k/D$ and (6.34) becomes (n is an integer)

$$Y(e^{j\omega}) = \sum_{k=-\infty}^{\infty} x(k)e^{-j\omega k/D} \quad k = 0, \pm D, \pm 2D, \ldots \tag{6.35}$$

The above equation suggests that we must take the DTFT of the values of $x(n)$ at every D numbers apart so that k/D is an integer. Since only every D value of $x(n)$ is used (6.35) is not in the form of the definition of the DTFT. Therefore, we must write (6.35) in the form

$$Y(e^{j\omega}) = \sum_{n=-\infty}^{\infty} c(n)x(n)e^{-j\omega n/D} \quad n = 0, \pm 1, \pm 2, \ldots \tag{6.36}$$

$$c(n) = \begin{cases} 1 & n = 0, \pm D, \pm 2D, \ldots \\ 0 & \text{otherwise} \end{cases} \tag{6.37}$$

And thus the DTFT applies. The function $x(n)$ is a discrete periodic function, a comb function with impulses every D apart and zero otherwise. It can be represented by the relation

$$c(n) = \frac{1}{D}\sum_{k=0}^{D-1} e^{j2\pi kn/D} = \frac{1}{D}\sum_{k=0}^{D-1} W^{-kn} \quad W = e^{-j2\pi/D} \tag{6.38}$$

which is known as the discrete sampling function. Note that (6.38) is the IDF transform. Introducing (6.38) in (6.36), we obtain

$$Y(e^{j\omega}) = \frac{1}{D}\sum_{k=0}^{D-1}\sum_{n=-\infty}^{\infty} W^{-kn}x(n)e^{-j\omega n/D} = \frac{1}{D}\sum_{k=0}^{D-1}\sum_{n=-\infty}^{\infty} (W^k e^{j\omega/D})^{-n} x(n)$$

$$= \frac{1}{D}\sum_{k=0}^{D-1} X(W^k e^{j\omega/D}) = \frac{1}{D}\sum_{k=0}^{D-1} X(e^{j\frac{\omega-2\pi k}{D}}) \tag{6.39}$$

which indicates that the spectrum of the down-sampled signal is the sum of D, a uniformly shifted and stretched version of the spectrum $X(e^{j\omega})$ and scaled down by D. Graphically, (see Figure 6.11) the above formula is interpreted as follows: (a) stretch $X(e^{j\omega})$ by D to obtain $X(e^{j\omega/D})$, (b) create $D-1$ copies of this stretched version by shifting it uniformly in successive amounts of 2π, and (c) add all these shifted versions to $X(e^{j\omega/D})$ and divide by D.

Let the spectrum of the sequence be $\{x(n)\}$ that is shown at the top of Figure 6.11. Let us further set $D=2$. Then (6.39) becomes

$$Y(e^{j\omega}) = \frac{1}{2}[X(e^{j\omega/2}) + X(e^{-j\pi}e^{j\omega/2})] = \frac{1}{2}[X(e^{j\omega/2}) + X(e^{j(\omega-2\pi)/2})] \qquad (6.40)$$

Because there is a multiplication by $1/D=0.5<1$ of ω, it implies that the function $X(e^{j\omega})$ is stretched by a factor of 2. Figure 6.11 shows the spectrum of a digital signal and the non-aliased spectrum at the output of a decimator with $D=2$. By judiciously assuming the Nyquist frequency of $\{x(n)\}$ to be $\omega_N = \pi/2$ and $D=2$ we avoided the aliasing phenomenon. If, however, the Nyquist frequency was in the range $\pi/2 < \omega < \pi$, a decimator with $D=2$ would produce aliasing at its output as shown in Figure 6.12. In general, aliasing is avoided if the Nyquist frequency of $\{x(n)\}$ is less than π/D.

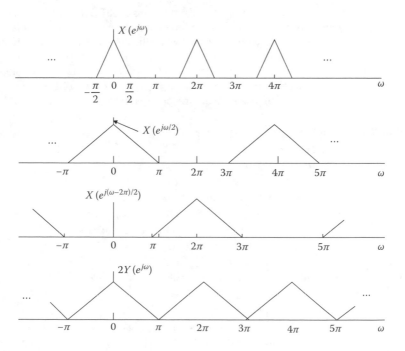

FIGURE 6.11

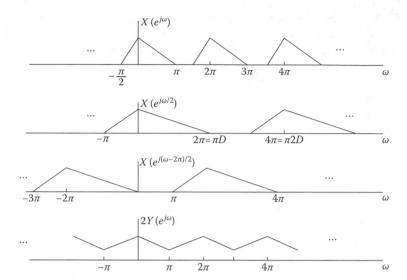

FIGURE 6.12

Example 6.15

Figure 6.13a shows the frequency response of a discrete signal. Figure 6.13b shows the frequency response when the corresponding time signal is down-sampled by a factor of 2. Figure 6.13c shows the frequency response when the time signal is down-sampled by a factor of 4 and, finally, Figure 6.13d shows the spectrum when the time signal is down-sampled by a factor of 6. Observe the distortion of the spectrum in (d) due to aliasing. Observe, also, that the magnitudes in (b) and (c) are 1/2 and 1/4 of the original spectrum, which verifies (6.39). The following program produces Figure 6.13.

Book MATLAB m-File

```
%ex7_7_2 is an m file for the EX 6.15
f = [0 1/8 1/4 1/2 3/4 1]; %normalized frequency bins;
m = [1 1 0 0 0 0]; %magnitude spectrum corresponding
               %to the above frequency bins;
h = fir2(200,f,m); %based on the frequency magnitude m it returns
               %the time function (filter coefficients);
[Hz,w] = freqz(h,1,512); %evaluation of the spectrum;
subplot(2,2,1); plot(w/pi,abs(Hz)); grid on;
xlabel('Normalized frequency'); ylabel('Magnitude');
title('Input spectrum');
D=2; y=h(1,1:D:length(h));
[Yz,w] = freqz(y,1,512); subplot(2,2,2); plot(w/pi,abs(Yz)); grid on;
xlabel('Normalized frequency'); ylabel('Magnitude');
```

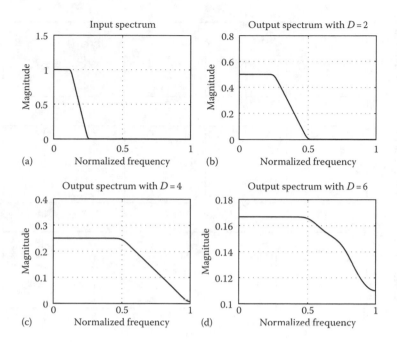

FIGURE 6.13

```
title ('Output spectrum with D = 2');
D1 = 4;y1 = h(1,1:D1:length(h));
[Y1z,w] = freqz(y1,1,512);subplot(2,2,3);plot(w/pi,abs(Y1z));grid on;
xlabel('Normalized frequency');ylabel('Magnitude');
title ('Output spectrum with D = 4');
D2 = 6;y2 = h(1,1:D2:length(h));
[Y2z,w] = freqz(y2,1,512);subplot(2,2,4);plot(w/pi,abs(Y2z));grid on;
xlabel('Normalized frequency');ylabel('Magnitude');
title ('Output spectrum with D = 6');                                ■
```

Note: *To avoid aliasing we must first reduce the bandwidth of the signal by π/D and then down-sample it by a factor D.*

Figure 6.14 graphically shows the relationship among the sampled form of $x(t)$ $(T_s = 1)$, the function $c(n)$, and lastly the output function $y(n)$ of the decimation process with factor $D = 2$. Note how the different functions shown are registered in the time domain.

Let us assume that we have a signal $x(t)$ with a $\pm 200\pi$ rad/s bandwidth. If we sample the signal at $T_s = 0.001$ seconds, the sampled function will occupy the frequency range $\pm 200\pi \times 0.001 = \pm 0.2\pi$ rad. This indicates that we can down-sample up to $D = 5$

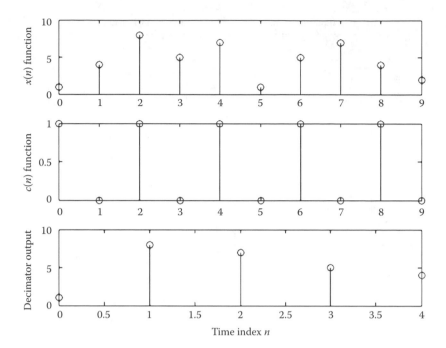

FIGURE 6.14

without aliasing. We can also see that half of the sampling frequency is 1000π rad/s and, hence, the signal bandwidth can be widened up to five times.

6.7.3 Interpolation (Up-Sampling) by a Factor U

Interpolation or up-sampling is the process of increasing the sampling rate of a signal by an **integer** factor $U > 1$ which results in adding $U - 1$ zero samples between two consecutive samples of the input sequence $\{x(n)\}$. Figure 6.15 shows the block-diagram representation of an up-sampling operation (interpolation). Figure 6.16 shows the form of a discrete signal before and after the up-sampler (interpolator) with $U = 3$. The following Book MATLAB will produce the output of an up-sampler.

Book MATLAB function: [y] = taaupsampling(x, U)

```
function[y] =taaupsampling(x,U)
          %y=output of the up-sampler;U=up-sampling factor
          %x=input to the up-sampler;
y1=zeros(1,U*length(x));
y1(1,1:U:length(y1))=x;
y=y1(1,1:length(y1)-U+1);
```

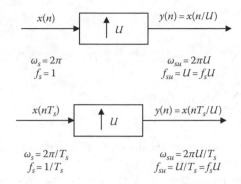

FIGURE 6.15

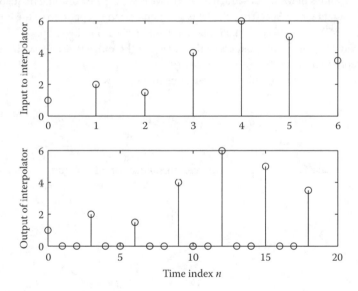

FIGURE 6.16

6.7.4 Frequency Domain Characterization of Up-Sampled Signals

The input–output relationship of the signals of an interpolator is

$$y(n) = \begin{cases} x(n/U) & n = 0, \pm U, \pm 2U, \dots \\ 0 & n \neq \text{ multiple of } U \end{cases} \qquad (6.41)$$

Note: *The sampling rate of y(n) is U times larger than that of the original sequence {x(n)}.*

The DTFT of (6.41) is given by

$$Y(e^{j\omega}) = \sum_{n=-\infty}^{\infty} y(n)e^{-j\omega n} = \sum_{n=-\infty}^{\infty} x(n/U)e^{-j\omega n} = \sum_{m=-\infty}^{\infty} x(m)e^{-j\omega mU}$$

$$= \sum_{n=-\infty}^{\infty} x(m)(e^{j\omega U})^{-m} = X(e^{j\omega U}) \tag{6.42}$$

The above equation states that we take the DTFT of the input sequence to the interpolator and then we substitute every $e^{j\omega}$ with $e^{j\omega U}$.

Let the spectrum of $\{x(n)\}$, given by $X(e^{j\omega})$, be the one shown in Figure 6.17a. The spectrum of the output of the interpolator is given by $X(e^{j\omega U})$. If we set $\omega = \omega + (2\pi k/U)$ in $Y(e^{j\omega})$, we find that the output is periodic for $k = \pm U, \pm 2U, \ldots$. Furthermore, since $U > 1$ it implies that the ω − axis is **compressed** by a factor of U. By adding $U - 1$ zeros between the values of the input sequence $\{x(n)\}$, we increase the sampling frequency which results in a signal whose spectrum $Y(e^{j\omega})$ is a U-fold periodic repetition of the input signal spectrum $X(e^{j\omega})$. This effect on the input spectrum is shown in Figure 6.17b.

Since only frequency components of this $\{x(n)\}$ in the range $0 \leq \omega \leq \pi/U$ are unique for the reproduction of the sequence $\{x(n)\}$, the images above $\omega = \pi/U$ should be eliminated using a low-pass filter with ideal characteristics:

$$H(e^{j\omega}) = \begin{cases} A & 0 \leq \omega \leq \pi/U \\ 0 & \text{otherwise} \end{cases} \tag{6.43}$$

where A is a scale factor to be used for the normalization of the output sequence $\{y(n)\}$ which is U.

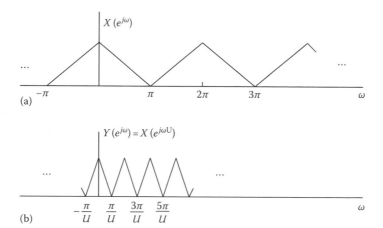

FIGURE 6.17

Example 6.16

Let the input spectrum to an interpolator with an up-sampling factor $U = 4$ be as that shown in Figure 6.18a. The Book MATLAB m-file given below produces the spectrum which is shown in Figure 6.18b.

Book MATLAB m file

```
%ex6_7_3 an m file for the EX6.16
freq= [0 0.2 0.3 0.4 1];%frequency range must run from 0 to 1;
mag = [1 1 0.5 0 0];
x=fir2(59,freq,mag);%given the magnitude vector of a spectrum
                    %and its corresponding frequency vector,
                    %fir2(N,freq,mag) produces N+1 filter
                    %coefficients (time function);
[xw,w] =freqz(x,1,512);%input frequency spectrum;
[y] =ssupsampling(x,6);
[yw,w] =freqz(y,1,512);
subplot(2,1,1);plot(w/pi,abs(xw),'k');ylabel('Magnitude');
title('Input spectrum to upsampler');
subplot(2,1,2);plot(w/pi,abs(yw),'k');xlabel('Normalized frequency');
ylabel('Magnitude');title('Output spectrum of the upsampler');    ■
```

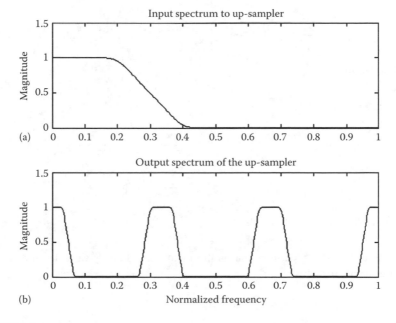

FIGURE 6.18

Appendix

6.A.1 Proofs of DTFT Properties

Linearity

Proof

$$\sum_{n=-\infty}^{\infty} [af(n) + bh(n)]e^{-j\omega n} = a\sum_{n=-\infty}^{\infty} f(n)e^{-j\omega n} + b\sum_{n=-\infty}^{\infty} h(n)e^{-j\omega n} = aF(e^{j\omega}) + bH(e^{j\omega})$$

Time shifting

Proof

$$F_{sh}(e^{j\omega}) = \sum_{n=-\infty}^{\infty} f(n-m)e^{-j\omega n} = e^{-j\omega m}\sum_{k=-\infty}^{\infty} f(k)e^{-j\omega k} = e^{-j\omega m}F(e^{j\omega})$$

where we set $n - m = k$. Observe that k is a dummy variable.

Time reversal

Proof

$$\sum_{n=-\infty}^{\infty} f(-n)e^{-j\omega n} = \sum_{m=\infty}^{-\infty} f(m)e^{-j(-\omega)m} = F(e^{-j\omega})$$

Time convolution

Proof

$$\sum_{n=-\infty}^{\infty} \left(\sum_{m=-\infty}^{\infty} f(m)h(n-m) \right) e^{-j\omega n} = \sum_{m=-\infty}^{\infty} f(m)\sum_{n=-\infty}^{\infty} h(n-m)e^{-j\omega n}$$

$$= \sum_{m=-\infty}^{\infty} f(m)e^{-j\omega m}\sum_{k=-\infty}^{\infty} h(k)e^{-j\omega k}$$

$$= F(e^{j\omega})H(e^{j\omega})$$

where we set $n - m = k$. Observe that both m and k are dummy variables and can take any other name, for example, n.

Frequency shifting

Proof

$$\sum_{n=-\infty}^{\infty} e^{\pm j\omega_0 n} f(n) e^{-j\omega n} = \sum_{n=-\infty}^{\infty} f(n) e^{-j(\omega \mp \omega_0)n} = F(e^{j(\omega \mp \omega_0)})$$

Time multiplication

$$\sum_{n=-\infty}^{\infty} nf(n) e^{-j\omega n} = -\sum_{n=-\infty}^{\infty} e^{j\omega} f(n) \frac{d(e^{j\omega})^{-n}}{d(e^{j\omega})} = -e^{j\omega} \frac{d}{d(e^{j\omega})} \sum_{n=-\infty}^{\infty} f(n) e^{-j\omega n}$$

$$= -e^{j\omega} \frac{d}{d(e^{j\omega})} F(e^{j\omega})$$

Example 6.A.1

Find the DTFT of $y(n) = na^n$, $a < 1$ and $0 \le n < \infty$.

SOLUTION

The DTFT of $f(n) = a^n$ for $n = 0, 1, 2, \ldots$ is equal to

$$F(e^{j\omega}) = \sum_{n=0}^{\infty} a^n e^{-j\omega n} = \sum_{n=0}^{\infty} (ae^{-j\omega})^n = \frac{1}{1 - ae^{-j\omega}} = \frac{e^{j\omega}}{e^{j\omega} - a}$$

and, hence, $-e^{j\omega} \dfrac{d}{d(e^{j\omega})} \left(\dfrac{e^{j\omega}}{e^{j\omega} - a} \right) = a \dfrac{e^{j\omega}}{(e^{j\omega} - a)^2}$. ∎

Modulation

Proof

$$\sum_{n=-\infty}^{\infty} f(n) \cos \omega_0 n e^{-j\omega n} = \sum_{n=-\infty}^{\infty} \left(\frac{f(n) e^{j\omega_0 n}}{2} + \frac{f(n) e^{-j\omega_0 n}}{2} \right) e^{-j\omega n}$$

$$= \frac{1}{2} \sum_{n=-\infty}^{\infty} f(n) e^{-j(\omega - \omega_0)n} + \frac{1}{2} \sum_{n=-\infty}^{\infty} f(n) e^{-j(\omega + \omega_0)n}$$

$$= \frac{1}{2} F(e^{j(\omega - \omega_0)}) + \frac{1}{2} F(e^{j(\omega + \omega_0)})$$

Correlation

Proof

$$\sum_{n=-\infty}^{\infty}\left(\sum_{m=-\infty}^{\infty} f(m)h(m-n)\right)e^{-j\omega n} = \sum_{n=-\infty}^{\infty}\left(\sum_{m=-\infty}^{\infty} f(m)h[-(n-m)]\right)e^{-j\omega n}$$

$$= \sum_{m=-\infty}^{\infty} f(m)\sum_{n=-\infty}^{\infty} h[-(n-m)]e^{-j\omega n} = \sum_{m=-\infty}^{\infty} f(m)e^{j\omega m}H(e^{-j\omega}) = F(e^{j\omega})H(e^{-j\omega})$$

where in the last step we used the time reversal and shifting properties.

Parseval's formula

Proof: We start from the right-hand side of the identity. Hence,

$$\frac{1}{2\pi}\int_{-\pi}^{\pi}\left[\sum_{n=-\infty}^{\infty} f(n)e^{-j\omega n}\right]F^*(e^{j\omega})d\omega = \sum_{n=-\infty}^{\infty} f(n)\frac{1}{2\pi}\int_{-\pi}^{\pi} F^*(e^{j\omega})e^{-j\omega n}d\omega$$

$$= \sum_{n=-\infty}^{\infty} f(n)\left[\frac{1}{2\pi}\int_{-\pi}^{\pi} F^*(e^{j\omega})e^{j\omega n}d\omega\right]^* = \sum_{n=-\infty}^{\infty} f(n)f^*(n) = \sum_{n=-\infty}^{\infty} |f(n)|^2$$

6.A.2 Proofs of DFT Properties

(**Note:** *For simplicity, we have deleted the constant $2\pi/N$ in the exponent.*)

Linearity

Proof: See the linearity proof in Appendix 6.A.1.

Symmetry

Proof: Set $n=-n$ in the IDFT. Hence, $f(-n)=\dfrac{1}{N}\sum_{n=0}^{N-1} F(k)e^{jk(-n)}$. Next, interchange the parameters n and k to find, $f(-k)=\dfrac{1}{N}\sum_{n=0}^{N-1} F(n)e^{-jkn} = \text{DFT}\left\{\dfrac{1}{N}F(n)\right\}$.

Time shifting

Proof: Substitute $m=n-i$ into   the   IDFT   so   that   the   equation $f(m)=\dfrac{1}{N}\sum_{k=0}^{N-1} F(k)e^{jkm}$ becomes

$$f(n-i) = \frac{1}{N}\sum_{k=0}^{N-1} F(k)e^{j(n-i)k} = \frac{1}{N}\sum_{k=0}^{N-1} [F(k)e^{-jik}]e^{jnk} = \text{IDFT}\{F(k)e^{-jik}\}.$$

Frequency shifting

Proof: We write the DFT in the form $F(m) = \sum_{n=0}^{N-1} f(n)e^{-jmn}$. Next, we set $m = k - i$ in the expression to find

$$F(k - i) = \sum_{k=0}^{N-1} f(n)e^{-j(k-i)n} = \sum_{k=0}^{N-1} [f(n)e^{jin}]e^{-jnk} = \text{DFT}\{f(n)e^{jin}\}$$

Time convolution

Proof: Start with the function

$$\sum_{i=0}^{N-1} f(i)h(n - i) = \sum_{i=0}^{N-1} \frac{1}{N} \sum_{k=0}^{N-1} F(k)e^{jik} \times \frac{1}{N} \sum_{m=0}^{N-1} H(m)e^{j(n-i)m}$$

This is rearranged to

$$= \frac{1}{N} \sum_{k=0}^{N-1} \sum_{m=0}^{N-1} F(k)H(m)e^{jmn} \left[\frac{1}{N} \sum_{i=0}^{N-1} e^{jik}e^{-jim} \right]$$

But, using the finite geometric series formula, we can prove that

$$\frac{1}{N} \sum_{i=0}^{N-1} e^{jik}e^{-jim} = \begin{cases} n & n = m \\ 0 & n \neq m \end{cases}$$

Hence, for $n = m$ in the second sum, we find finally

$$y(n) = \sum_{i=0}^{N-1} f(i)h(n - i) = \frac{1}{N} \sum_{k=0}^{N-1} F(k)H(k)e^{jnk} \triangleq \text{IDFT}\{F(k)H(k)\}$$

Since we have shown that the DFT automatically supposes periodic sequences in the time and the frequency domain, the periodic convolution for the two specific signals, $f(n) = \{1\ 2\ 3\}$ and $h(n) = \{1\ 1\ 0\}$, are shown in Figure 6.A.1.

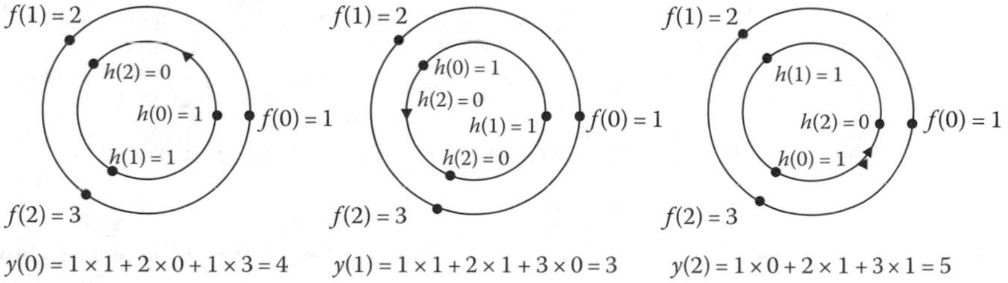

$y(0) = 1 \times 1 + 2 \times 0 + 1 \times 3 = 4$ $\quad$ $y(1) = 1 \times 1 + 2 \times 1 + 3 \times 0 = 3$ $\quad$ $y(2) = 1 \times 0 + 2 \times 1 + 3 \times 1 = 5$

FIGURE 6.A.1

Another approach to the evaluation of the cyclic convolution is to cast the convolution form in the matrix form. For the case of the periodic sequences $f(n)$ and $h(n)$ each of the three elements, as in Figure 6.A.1, we write

$$y(0) = f(0)h(0) + f(1)h(2) + f(2)h(1)$$
$$y(1) = f(0)h(1) + f(1)h(0) + f(2)h(2)$$
$$y(2) = f(0)h(2) + f(1)h(1) + f(2)h(0)$$

This set can be written in the matrix form

$$[y]^T = [f(0) \quad f(1) \quad f(2)] \begin{bmatrix} h(0) & h(1) & h(2) \\ h(2) & h(0) & h(1) \\ h(1) & h(2) & h(0) \end{bmatrix}$$

For this specific example

$$[y(0) \quad y(1) \quad y(2)] = [1 \quad 2 \quad 3] \begin{bmatrix} 1 & 1 & 0 \\ 0 & 1 & 1 \\ 1 & 0 & 1 \end{bmatrix} = [4 \quad 3 \quad 5]$$

To produce a linear convolution for this case, each sequence must have $N = 3 + 3 - 1 = 5$ elements. But since each sequence has only three elements, we pad them with zeros. Hence, for $f(n) = \{1\ 2\ 3\ 0\ 0\}$ and $h(n) = \{1\ 1\ 1\ 0\ 0\}$, the result is $f * h = \{0\ 1\ 2\ 1\ 12\ 0\ 0\ 0\ 0\}$.

Frequency convolution

Proof: Substitute known forms into

$$\sum_{i=0}^{N-1} F(i)H(k-i) = \sum_{i=0}^{N-1} \left[\sum_{m=0}^{N-1} f(m)e^{-jmi} \right] \left[\sum_{n=0}^{N-1} h(n)e^{-jn(k-i)} \right]$$

$$= \sum_{m=0}^{N-1}\sum_{n=0}^{N-1} f(m)h(n)e^{-jkn} \left[\sum_{i=0}^{N-1} e^{-jmi}e^{jni} \right]$$

The bracketed term is the orthogonality relationship and is equal to N if $m = k$ and the zero if m is different from k. Therefore,

$$\sum_{i=0}^{N-1} F(i)H(k-i) = N\sum_{n=0}^{N-1} f(n)h(n)e^{-jkn} \triangleq N \times \text{DFT}\{f(n)h(n)\}$$

from which the property is verified if we take the IDFT of both sides of the above equation. Because $F(k)$ and $H(k)$ are periodic, their convolution is a circular convolution in the frequency domain.

Parseval's theorem

Using the equation which characterizes the frequency convolution property, we obtain

$$\sum_{n=0}^{N-1} f(n)f(n)e^{-jkn} = \frac{1}{N}\sum_{i=0}^{N-1} F(i)F(k-i)$$

If we set $k=0$ in this expression, we find

$$\sum_{n=0}^{N-1} f^2(n) = \frac{1}{N}\sum_{i=0}^{N-1} F(i)F(-i) = \frac{1}{N}\sum_{i=0}^{N-1} |F(i)|^2$$

From this result we may **define a discrete energy spectral density** or a **periodogram spectral estimate**:

$$S(k) = |F(k)|^2 \quad 0 \le k \le N-1$$

Time reversal

Setting $n=-n$ in the IDFT, we find that $f(-n) = \frac{1}{N}\sum_{k=0}^{N-1} F(k)e^{-jnk}$. Next, set $k=-m$ on the right side. Then, $f(-n) = \frac{1}{N}\sum_{m=0}^{-(N-1)} F(-m)e^{jnm}$. Because of the periodic nature of $F(-m)$ and $\exp(jmn)$, the sum over $-(N-1)$ to 0 and $(N-1)$ to 0 is the same. Thus,

$$f(-n) = \frac{1}{N}\sum_{m=0}^{N-1} F(-m)e^{jnm} \triangleq \text{IDFT}\{F(-m)\}$$

which verifies the time-reversal property.

6.A.3 Fast Fourier Transform

6.A.3.1 Decimation in Time Procedure

The DFT of a sequence $\{x(n)\}$ is given by

$$F_N(k) = \sum_{n=0}^{N-1} f(n)W_N^{kn}, \quad W_N = e^{-j2\pi/N}, \quad k=0,1,\ldots,N-1 \tag{6.A.1}$$

Let us also assume that $N=2^r$ for r integer. This indicates that N is always an even number. Because of N being even, we can decompose the summation of the above equation into two terms, one with even indices and the other with odd indices. Hence,

$$F_N(k) = \sum_{\text{even}} f(n)W_N^{kn} + \sum_{\text{odd}} f(n)W_N^{kn} \tag{6.A.2}$$

We can write the even indices as $n = 2m$ and the odd indices as $n = 2m + 1$, with $m = 0$, $1, \ldots, (N/2) - 1$. Therefore, (6.A.2) becomes

$$F_N(k) = \sum_{m=0}^{(N/2)-1} f(2m) W_N^{2mk} + W_N^k \sum_{m=0}^{(N/2)-1} f(2m) W_N^{2mk} \qquad (6.A.3)$$

In the second summation, we used the identity $W_N^{(2m+1)k} = W_N^k W_N^{2mk}$. Observe that

(1) $W_N^2 = e^{-j2\pi/(N/2)} = W_{N/2}$, (2) $W_N^N = e^{-j2\pi} = 1$, and (3) $W_N^{N/2} = e^{-j\pi} = -1$. Furthermore, the two summations in (6.A.3) are DFT's themselves: the DFT of the even samples $x(0), x(2), \ldots, x(N-2)$ and the DFT of the odd samples $x(1), x(3), \ldots, x(N-1)$. Therefore, we write

$$F_N(k) = F_{N/2}^{(e)}(k) + W_N F_{N/2}^{(o)}(k) \quad k = 0, 1, \ldots, N-1 \qquad (6.A.4)$$

where

$$F_{N/2}^{(e)}(k) = \text{DFT}\{x(0), x(2), \ldots, x(N-2)\} = \sum_{m=0}^{(N/2)-1} f(2m) W_{N/2}^{km} \qquad (6.A.5)$$

$$F_{N/2}^{(o)}(k) = \text{DFT}\{x(1), x(3), \ldots, x(N-1)\} = \sum_{m=0}^{(N/2)-1} f(2m+1) W_{N/2}^{km} \qquad (6.A.6)$$

We observe that the DFT of the N-point sequence has been decomposed into two DFT's of $N/2$ samples each.

For example, when $N = 4 = 2^2$ we need to compute $F_4(0), F_4(1), F_4(2), F_4(3)$, on the basis of $F_2^{(e)}(0), F_2^{(e)}(1)$ and $F_2^{(o)}(0), F_2^{(o)}(1)$. Hence, for $k = 0,1$ we write

$$F_4(k) = F_2^{(e)}(k) + W_4^k F_2^{(o)}(k) \quad k = 0, 1 \qquad (6.A.7)$$

For the two indices, $k = 2, 3$, we use the fact that the DFT of a 2-point sequence is periodic with a period 2. Hence, for $k = 0, 1$, we find

$$F_4(k+2) = F_2^{(e)}(k) + W_4^{2+k} F_2^{(o)}(k) \quad k = 0, 1 \qquad (6.A.8)$$

because $F_2(k) = F_2(k+2)$. Because $W_4^2 = (e^{-j2\pi/4})^2 = e^{-j\pi} = -1$, (6.A.7) and (6.A.8) become

$$F_4(k) = F_2^{(e)}(k) + W_N^k F_2^{(o)}(k) \quad k = 0, 1$$
$$F_4(k+2) = F_2^{(e)}(k) - W_N^k F_2^{(o)}(k) \quad k = 0, 1 \qquad (6.A.9)$$

The above equation can be generalized for any sequence of even number N samples in the form

$$F_N(k) = F^{(e)}_{N/2}(k) + W^k_N F^{(o)}_{N/2}(k) \quad k = 0, 1, \ldots, (N/2) - 1$$
$$F_N(k + 2) = F^{(e)}_{N/2}(k) - W^k_N F^{(o)}_{N/2}(k) \quad k = 0, 1, \ldots, (N/2) - 1$$

(6.A.10)

The above equation is known as a **butterfly**. Figure 6.A.2 depicts (6.A.10). Figure 6.A.3 shows the FFT for $N = 8$.

Next, the question is how efficient the FFT is. Let us look at the case when $N = 8$. In Figure 6.A.2, we see that the first stage is a two 4-order DFT. Each one of these DFTs can be split in two 2-order FFTs. As we can easily see, a 2-order DFT is just a butterfly, since $W_2 = e^{-j2\pi/2} = -1$ and $W^k_2 = (-1)^k = \pm 1$ when $k = 0, 1$. For $N = 8$, we find that we obtain $3 = \log_2(8)$ layers of butterflies with $8/2$ butterflies to be computed in each layer. Since each butterfly involves one multiplication and two additions we require about $(8/2)\log_2(8)$ complex multiplications and $8\log_2(8)$ complex additions. Hence, the complexity of the DFT is $O(8\log_2(8))$. If we substitute N for 8, we find that the FFT complexity is of the order of $N\log_2(N)$. This must be compared with the N^2

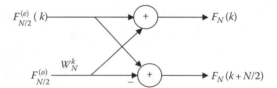

FIGURE 6.A.2

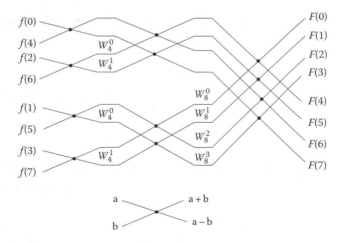

FIGURE 6.A.3

operations that must be performed by the DFT. For $N = 1024 = 2^{10}$, the DFT needs $(1024)^2 = 1,048,576$ operations whereas the FFT needs only $1024 \times 10 = 10240$ operations.

Observe that to obtain the unscrambled output from the FFT operation, we must scramble the input as shown in Figure 6.A.3. To accomplish it we first write the numbers, in this case 0–7, in their 3-bit representation. Next, we reverse the bits and we thus find, the scrambled input. This process is shown below:

$000 = 0,\ 001 = 1,\ 010 = 2,\ 011 = 3,\ 100 = 4,\ 101 = 5,\ 110 = 6,\ 111 = 7.$
$000 = 0,\ 100 = 4,\ 010 = 2,\ 110 = 6,\ 001 = 1,\ 101 = 5,\ 011 = 3,\ 111 = 7.$

7

Laplace Transform

The use of the Fourier transform (FT) in systems analysis involves the decomposition of the excitation function, $f(t)$, into a function, $F(\omega)$, over an infinite band of frequencies. The excitation function, $F(\omega)$, together with the appropriate system function, $H(\omega)$, leads, through the inverse Fourier transform, to the response of the system to the prescribed excitation or excitations. Despite its general importance in systems functions, the Fourier integral is not generally useful in determining the transient response of networks (systems).

The discussion of the FT showed that we could represent a function $f(t)$ by a continuous sum of weighted exponential functions of the form $f(t)e^{j\omega t}$. The values of the exponential are restricted along the imaginary axis of the complex plane as omega varies from minus infinity to infinity. This restriction proves to be undesirable in many cases. We can remove this restriction by representing $f(t)$ by a continuous sum of weighted damped exponential functions of the form $f(t)e^{-st}$, where $s = \sigma + j\omega$ with some real constant σ. The choice of s moves the values of the exponential function from the $j\omega$-axis to a parallel line off the $j\omega$-axis in the complex plane. The Laplace transform (LT) is well adapted to linear time domain systems analysis. Another feature of the LT is that it automatically provides for initial conditions in the systems.

7.1 One-Sided Laplace Transform

In systems problems, it is usually possible to restrict considerations to positive-time function. The reason is that the response of physical systems can be determined for all $t \geq 0$ from a knowledge of the input for $t \geq 0$ and the initial conditions. The one-sided LT is defined by

$$\boxed{F(s) = \int_0^\infty f(t)e^{-st}dt \triangleq \mathscr{L}\{f(t)\}} \qquad (7.1)$$

provided that the function is defined in the range of integration and that the integral exists for all values of s greater than a specific value s_0.

In our studies, we will consider piecewise continuous functions (a function is piecewise continuous on an interval if the interval can be subdivided into a finite number of subintervals, in each of which the function is continuous and has finite left- and right-hand limits) and those functions for which

$$\lim_{t \to \infty} f(t)e^{-ct} = 0, \quad c = \text{real constant} \tag{7.2}$$

Functions of this type are known as functions of **exponential order** c. Also, from the expression

$$\int_0^\infty |f(t)e^{-st}|dt = \int_0^\infty |f(t)e^{-\sigma t}e^{-j\omega t}|dt = \int_0^\infty |f(t)||e^{-\sigma t}||e^{-j\omega t}|dt$$

$$= \int_0^\infty |f(t)|e^{-\sigma t}dt = \int_0^\infty |f(t)|e^{-ct}e^{-(\sigma-c)t}dt$$

we observe that, if $f(t)$ is of exponential order, the integral converges for $\text{Re}\{s\} = \sigma > c$ (c is the abscissa of convergence) since the integrand goes to zero as $t \to \infty$. The abscissa of convergence may be positive, negative, or zero depending on the function. The importance of this result is that a finite number of infinite discontinuities are permissible so long as they have finite areas under them. In addition, the convergence is also uniform, which permits us to alter the order of integration in multiple integrals without affecting the results. The restriction in this equation, namely, $\text{Re}\{s\} = \sigma > c$, indicates that when we wish to find the inverse LT, we must choose an appropriate path of integration in the complex plane. By doing so, it is guaranteed that the time function so obtained is unique. The left-hand infinite space with boundary, the abscissa of convergence, is known as the **region of convergence** (ROC).

Example 7.1

Find the LT of the unit step function $f(t) = u(t)$ and establish the region of convergence.

SOLUTION

By (7.1), we find

$$U(s) = \int_0^\infty u(t)e^{-st}dt = \int_0^\infty e^{-st}dt = -\frac{e^{-st}}{s}\Big|_0^\infty = \frac{1}{s} \tag{7.3}$$

The ROC is found from the expression $|e^{-\sigma t}e^{-j\omega t}| = |e^{-\sigma t}| < \infty$, which is true only if $\sigma > 0$. ∎

Example 7.2

Find the LT of $f(t) = e^{-2t}u(t)$ and establish the ROC.

SOLUTION

The LT is

$$F(s) = \int_0^\infty e^{-2t}e^{-st}\,dt = \int_0^\infty e^{-(s+2)t}\,dt = \frac{1}{-(s+2)}e^{-(s+2)t}\Big|_0^\infty = \frac{1}{s+2} \tag{7.4}$$

and the ROC is found from

$$\left|e^{-(\sigma+j\omega+2)t}\right| = \left|e^{-(\sigma+2)t}\right|\left|e^{-j\omega t}\right| = e^{-(\sigma+2)t}$$

which results in $\sigma > -2$. Figure 7.1a shows the ROC for the function $f(t) = u(t)$ and Figure 7.1b for the function $f(t) = e^{-2t}u(t)$. These are found in Examples 7.1 and 7.2, respectively. ∎

Example 7.3

Find the LT of the delta function and the ROC.

SOLUTION

The LT is

$$\Delta(s) = \int_0^\infty \delta(t)e^{-st}\,dt = e^{-s0}\int_{0-}^{0+} \delta(t)\,dt = 1 \tag{7.5}$$

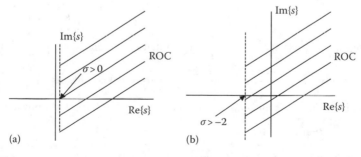

(a)   (b)

FIGURE 7.1

and the ROC is found from

$$\int_0^\infty |\delta(t)e^{-st}|dt = \int_0^\infty |\delta(t)e^{-s0}|dt = \int_0^\infty \delta(t)dt = 1 < \infty$$

which indicates that the result is independent of σ and, hence, the ROC is the entire s-plane.   ∎

Example 7.4

Find the LT of the function $f(t) = 2u(t) + e^{-t}u(t)$.

SOLUTION

From (7.1), we find

$$F(s) = \int_0^\infty [u(t) + e^{-t}]e^{-st}dt = \int_0^\infty e^{-st}dt + \int_0^\infty e^{-(s+1)t}dt = \frac{1}{s} + \frac{1}{s+1}$$

The ROC is $\sigma > 0$. Observe that the ROC, which is accepted, is the region of the s-plane where the two ROCs overlap.   ∎

7.2 Summary of the Laplace Transform Properties

1. *Linearity* $af(t) + bh(t) \overset{\mathscr{L}}{\leftrightarrow} aF(s) + bF(s)$

2. *Time derivative* $\dfrac{df(t)}{dt} \overset{\mathscr{L}}{\leftrightarrow} sF(s) - f(0)$

$\dfrac{d^nf(t)}{dt^n} \overset{\mathscr{L}}{\leftrightarrow} s^nF(s) - s^{n-1}f(0) - s^{n-2}f^{(1)}(0)$

$\qquad\qquad - \cdots - f^{(n-1)}(0), \quad f^{(m)}(0) \overset{\Delta}{=} \dfrac{d^mf(t)}{dt^m}\Big|_{t=0}$

3. *Integral* $\int_0^t f(x)dx \overset{\mathscr{L}}{\leftrightarrow} \dfrac{F(s)}{s}$ zero initial conditions

$\int_{-\infty}^t f(x)dx \overset{\mathscr{L}}{\leftrightarrow} \dfrac{F(s)}{s} + \dfrac{\int_{-\infty}^0 f(t)dt}{s}$ nonzero initial conditions

4. *Multiplication by exponential* $e^{-at}f(t) \overset{\mathscr{L}}{\leftrightarrow} F(s+a), \quad a = $ positive constant

5. *Multiplication by t* $tf(t) \overset{\mathscr{L}}{\leftrightarrow} -\dfrac{dF(s)}{ds}$

6. *Time shifting* $f(t-a)u(t-a) \overset{\mathscr{L}}{\leftrightarrow} e^{-as}F(s), \quad a > 0$

7. *Complex frequency shift* $\quad e^{s_0 t} f(t) \quad \overset{\mathcal{L}}{\leftrightarrow} \quad F(s - s_0)$

8. *Scaling* $\quad f(at) \quad \overset{\mathcal{L}}{\leftrightarrow} \quad \dfrac{1}{a} F\left(\dfrac{s}{a}\right), \quad a > 0$

9. *Time convolution* $\quad f(t) * h(t) \quad \overset{\mathcal{L}}{\leftrightarrow} \quad F(s)H(s)$

10. *Initial value* $\quad \lim_{t \to 0} f(t) = \lim_{s \to \infty} sF(s)$ provided that no delta function exists at $t = 0$

11. *Final value* $\quad \lim_{t \to \infty} f(t) = \lim_{s \to 0} sF(s)$ provided that $sF(s)$ is analytic on the $j\omega$-axis and in the right half of the s-plane

Example 7.5 (Time derivative)

Find the LT of the differential equation characterizing an *RL* series circuit with a voltage $v(t)$ as input.

SOLUTION

From the KVL, we obtain the differential equation $L\dfrac{di(t)}{dt} + Ri(t) = v(t)$. Using the linearity property, we obtain

$$\mathcal{L}\left\{ L\frac{di(t)}{dt} + Ri(t) \right\} = \mathcal{L}\{v(t)\}$$

or

$$L\mathcal{L}\left\{ \frac{di(t)}{dt} \right\} + R\mathcal{L}\{i(t)\} = \mathcal{L}\{v(t)\}$$

or

$$LsI(s) - Li(0) + RI(s) = V(s) \qquad \blacksquare$$

Example 7.6 (Integral)

Find the LT of the integrodifferential equation of a series *RC* circuit with a voltage input. There is an initial voltage, $v_C(0)$, across the capacitor.

SOLUTION

From the KVL, we obtain the following equation: $Ri(t) + 1/C \displaystyle\int_{-\infty}^{t} i(x)dx = v(t)$. Applying the linearity property, we can write the LT of the above equation as

$$R\mathcal{L}\{i(t)\} + \mathcal{L}\left\{ \frac{1}{C} \int_{-\infty}^{t} i(x)dx \right\} = \mathcal{L}\{v(t)\} \quad \text{or} \quad RI(s) + \frac{1}{C}\frac{I(s)}{s} + \frac{1}{C}\frac{\displaystyle\int_{-\infty}^{0} i(x)dx}{s} = V(s)$$

From the integration $\int_{-\infty}^{0} i(x)dx = q(0)$, we obtain the charge accumulated at the initial time $t=0$. But $q(0)/C$ is the initial voltage across the capacitor and hence, the transformed equation becomes $\left(R + \dfrac{1}{Cs}\right)I(s) = -\dfrac{v_C(0)}{s} + V(s)$. ∎

Example 7.7 (Multiplication by an exponential)

Find the LT of the function $g(t) = e^{-at}\cos\omega_0 t\, u(t)$.

SOLUTION

To find the transformation, we need to find the LT of the cosine function. Therefore, we write

$$\mathcal{L}\{\cos\omega_0 t\, u(t)\} = \int_0^\infty \left[\frac{e^{-j\omega_0 t} + e^{j\omega_0 t}}{2}\right]e^{-st}dt = \frac{1}{2}\int_0^\infty e^{-(j\omega_0+s)t}dt + \frac{1}{2}\int_0^\infty e^{-(-j\omega_0+s)t}dt$$

$$= \frac{1}{2}\frac{1}{-(j\omega_0+s)}e^{-(j\omega_0+s)t}\Big|_0^\infty + \frac{1}{2}\frac{1}{-(-j\omega_0+s)}e^{-(-j\omega_0+s)t}\Big|_0^\infty$$

$$= \frac{1}{2}\frac{1}{(j\omega_0+s)} + \frac{1}{2}\frac{1}{(-j\omega_0+s)} = \frac{s}{s^2+\omega_0^2}$$

Therefore, the answer is $G(s) = \dfrac{s+a}{(s+a)^2+\omega_0^2}$. ∎

Example 7.8 (Multiplication by t)

Find the LT of the function $g(t) = t\sin\omega_0 t\, u(t)$. From the property, we only need to find the LT of the sine signal. Hence, following the procedure of Example 7.7, we find

$$F(s) = \int_0^\infty \frac{e^{j\omega_0 t} - e^{-j\omega_0 t}}{2j}e^{-st}dt = \frac{\omega_0}{s^2+\omega_0^2}$$

Hence,

$$G(s) = -\frac{d}{ds}\left(\frac{\omega_0}{s^2+\omega_0^2}\right) = \frac{2s\omega_0}{(s^2+\omega_0^2)^2}$$ ∎

Example 7.9 (Time shifting)

Find the LT of the function $f_s(t) = (t - a)u(t - a)$ $a > 0$.

SOLUTION

We first must find the LT of the un-shifted function, which is

$$F(s) = \int_0^\infty t e^{-st} dt = \frac{1}{-s} \int_0^\infty t \frac{de^{-st}}{dt} dt = \frac{1}{-s} \int_0^\infty t d(e^{-st}) = \frac{1}{-s} \left[t e^{-st} \Big|_0^\infty - \int_0^\infty e^{-st} dt \right] = \frac{1}{s^2}$$

Hence, applying the time-shifting property, we obtain $F_s(s) = e^{-sa} F(s) = e^{-sa} \frac{1}{s^2}$. ∎

Example 7.10 (Frequency shift)

Find the LT of the function $g(t) = e^{-as} f(t) = e^{-as} \cos \omega_0 t\, u(t)$.

SOLUTION

We have already found earlier the LT of the cosine function. Hence, the solution is
$$G(s) = \frac{s + a}{(s + a)^2 + \omega_0^2}.$$ ∎

Example 7.11 (Scaling)

Verify the LT of the function $f(\omega_0 t) = \cos \omega_0 t\, u(t)$ by using the scaling properties and the LT of the function $f(t) = \cos t$.

SOLUTION

The LT of the signal $f(t)$ is equal to $\frac{s}{s^2 + 1}$. Applying the scaling property, we find that
$$\frac{1}{\omega_0} F\left(\frac{s}{\omega_0}\right) = \frac{1}{\omega_0} \frac{s/\omega_0}{(s/\omega_0)^2 + 1} = \frac{s}{s^2 + \omega_0^2}.$$ ∎

Example 7.12 (Time convolution)

Find the LT of the convolution of the following two functions: $f(t) = tu(t)$, and $h(t) = \cos t\, u(t)$.

SOLUTION

$$\mathscr{L}\{f(t) * h(t)\} = F(s)H(s) = \frac{1}{s^2} \frac{s}{s^2 + 1} = \frac{1}{s} \frac{1}{s^2 + 1}.$$ ∎

Example 7.13 (Initial value)

Apply the initial value property to the following functions:

$$F(s) = \frac{s}{s^2 + 3}, \quad H(s) = \frac{s^2 + s + 3}{s^2 + 3}$$

SOLUTION

(a) $\lim_{s \to \infty} sF(s) = \lim_{s \to \infty} \frac{s^2}{s^2 + 3} = \lim_{s \to \infty} \frac{1}{1 + (3/s^2)} = 1$

(b) If we expand the division of $H(s)$, we obtain $H(s) = 1 + \frac{s}{s^2 + 3}$. This shows the presence of an impulse at $t = 0$ and, therefore, we cannot find the initial value of the time function.   ∎

Example 7.14 (Final value)

Apply the final value property to the following two functions: (a) $F(s) = \frac{s + a}{(s + a)^2 + b^2}$ and (b) $H(s) = \frac{s}{s^2 + b}$.

SOLUTION

(a) $\lim_{s \to 0} sF(s) = \lim_{s \to 0} \frac{s(s + a)}{(s + a)^2 + b^2} = 0$. (b) In this case, the property is not applicable because the function has singularities on the imaginary axis at $s = \pm jb$.   ∎

7.3 Systems Analysis: Transfer Functions of LTI Systems

The transfer function or system function, $H(s)$, of a LTI system, an almost essential entity in system analysis, is defined as the ration of the LT of the output to the LT of the input. Thus, if the input is $f(t)$ and the output is $y(t)$, then

$$\boxed{H(s) = \frac{Y(s)}{F(s)} = \text{transfer function or system function}} \qquad (7.6)$$

The output may be a voltage or a current anywhere in the system, and $H(s)$ is then appropriate to the selected output for a specified input. That is, $H(s)$ may be an impedance, an admittance, or a transfer entity in the given problem. The transfer function $H(s)$ describes the properties of the system alone. That is, the system is assumed to be in its quiescent state (zero state); hence, the initial conditions are assumed zero.

The output time function is given by

$$y(t) = \mathcal{L}^{-1}\{H(s)F(s)\} \tag{7.7}$$

If the input is a delta function, $f(t) = \delta(t)$, the output in the s-domain is $H(s)$ since the LT of the delta function is 1. The impulse response of the system is found by taking the ILT of $H(s)$. Hence, we find

$$h(t) = \mathcal{L}^{-1}\{H(s)\} \tag{7.8}$$

The above relation shows that the impulse response of a system is the ILT of the transfer function $H(s)$, and the transfer function is equal to the LT of the system's impulse response.

An important feature of the LT method is that it is not necessary to isolate and identify the system transfer function since $H(s)$ appears automatically through the transform of the differential equation and is included in the mathematical operations. The situation changes considerably in those cases when the system consists of a number of subsystems that are interconnected to form the completed system. Next, one must take due account of whether the subsystems are interconnected in cascade, parallel, or feedback configuration.

The following examples illustrate how we can find the transfer function of a system from its time domain representation. Often, it is convenient to draw a block-diagram representation of the system and then use the properties of block-diagram reduction as given in Chapter 1.

Example 7.15

Determine the transfer function of the system shown in Figure 7.2a.

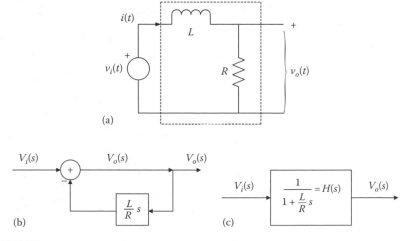

(a)

(b)

(c)

FIGURE 7.2

SOLUTION

The differential equation describing the system is

$$L\frac{di(t)}{dt} + Ri(t) = v_i(t),$$

which we write as

$$\frac{L}{R}\frac{dv_o(t)}{dt} + v_o(t) = v_i(t); \quad v_o(t) = Ri(t)$$

The LT of both sides of the differential equation, invoking the linearity and differentiation properties of LT plus zero initial conditions (remember that the transfer function is defined only with zero initial conditions), is

$$\frac{L}{R}sV_o(s) + V_o(s) = V_i(s); \quad H(s) = \frac{V_o(s)}{V_i(s)} = \frac{R/L}{s + (R/L)}$$

To obtain the block-diagram representation of the system, the above equation is written in the following form:

$$\text{(a) } V_o(s) = V_i(s) - \frac{L}{R}sV_o(s); \quad \text{(b) } V_o(s) = \frac{1}{1 + \frac{L}{R}s}V_i(s)$$

The first equation is represented in block-diagram form in Figure 7.2b, and the second equation is represented in block-diagram form in Figure 7.2c.   ∎

Example 7.16

Find the transfer function for the system shown in Figure 7.3a.

SOLUTION

The differential equation describing the system is

$$v_i(t) = Ri(t) + \frac{1}{C}\int_0^t i(x)dx \quad \text{(zero initial conditions)}$$

The above equation is multiplied and divided by R to become

$$V_i(s) = V_o(s) + \frac{V_o(s)}{RCs} \quad \text{or} \quad H(s) = \frac{V_o(s)}{V_i(s)} = \frac{s}{s + \frac{1}{RC}}$$

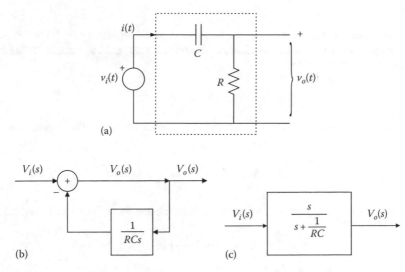

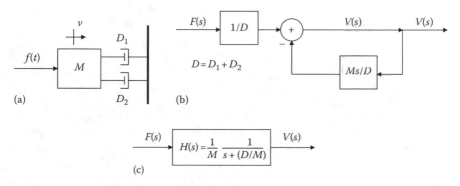

FIGURE 7.3

The transfer function is shown in block-diagram form in Figure 7.3c. Further, to obtain the block-diagram format, we write the transformed equation in the form

$$V_o(s) = V_i(s) - \frac{1}{RCs} V_o(s)$$

The block-diagram representation of the above equation is shown in Figure 7.3b. ■

Example 7.17 (Mechanical system)

Determine the transfer function, $H(s) = V(s)/F(s)$, of the system shown in Figure 7.4a.

FIGURE 7.4

SOLUTION

To find the differential equation describing the system, we add algebraically the forces that are as follows: the inertia force, the two frictional forces, and the input force. Since the forces are collinear, we obtain

$$M\frac{dv(t)}{dt} + D_1v(t) + D_2v(t) = f(t)$$

The LT of the above equation and its transfer function, assuming zero initial conditions, are

$$MsV(s) + (D_1 + D_2)V(s) = F(s), \quad H(s) = \frac{1}{M}\frac{1}{s + \dfrac{D}{M}} \quad D = D_1 + D_2$$

To obtain the block-diagram representation, we write the transformed equation in the form

$$V(s) = \frac{1}{D}F(s) - \frac{Ms}{D}V(s)$$

The block diagram for the above equation is shown in Figure 7.4b and the transfer function is shown in Figure 7.4c   ■

Example 7.18 (Mechanical system)

Find the transfer function $H(s) = V_1(s)/F(s)$ of the system shown in Figure 7.5.

SOLUTION

We first develop the network equivalent diagram. It is shown in Figure 7.5b, where the velocities v_1 and v_2 are specified relative to ground as a fixed frame of reference. Observe that the force moves with velocity v_1; hence, this source is connected between ground and level v_1 as shown in Figure 7.5c. Observe also that the mass M_1 moves with velocity v_1 and is connected, therefore, between v_1 and v_g. The damper D_1 moves with relative velocities specified by v_1 and v_2; hence, it is connected between these two velocities. Lastly, since both M_2 and D_2 move with velocity v_2, they are connected in parallel between v_2 and v_g. A rearrangement of the resulting geometry yields Figure 7.5d, a familiar circuit configuration for the mechanical system.

Since the sum of the forces at each node must be zero, we obtain

$$M_1\frac{dv_1(t)}{dt} + D_1[v_1(t) - v_2(t)] - f(t) = 0 \qquad \text{(a)}$$

$$M_2\frac{dv_2(t)}{dt} + D_1[v_2(t) - v_1(t)] + D_2v_2(t) = 0 \quad \text{(b)}$$

(7.9)

In writing these equations, we assumed that non-source terms were pointing away from the nodes and were assumed positive. This selection is arbitrary and must be kept consistent for all nodes. The LT of these equations is

$$(M_1s + D_1)V_1(s) - D_1V_2(s) = F(s) \qquad \text{(a)}$$

$$-D_1V_1(s) + (M_2s + D_1 + D_2)V_2(s) = 0 \quad \text{(b)}$$

(7.10)

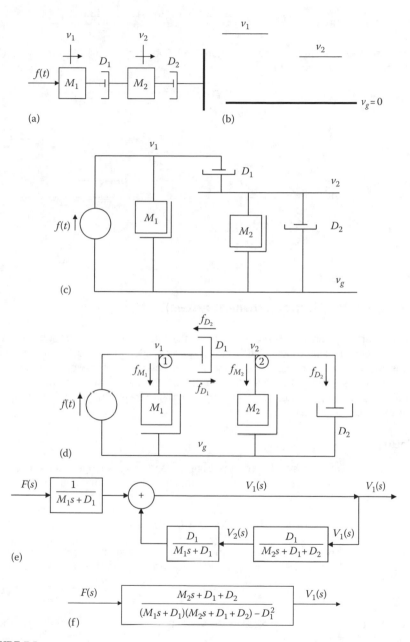

FIGURE 7.5

To draw the block diagram of the system, we rearrange these equations to

$$V_1(s) = \frac{D_1}{M_1 s + D_1} V_2(s) + \frac{1}{M_1 s + D_1} F(s), \quad V_2(s) = \frac{D_1}{M_2 s + D_1 + D_2} V_1(s)$$

These equations are readily seen in Figure 7.5e. Substituting $V_2(s)$ from the second equation in the first, we obtain

$$V_1(s) = \frac{D_1}{M_1 s + D_1} \frac{D_1}{M_2 s + D_1 + D_2} V_1(s) + \frac{1}{M_1 s + D_1} F(s)$$

or

$$\frac{V_1(s)}{F(s)} = \frac{\dfrac{1}{M_1 s + D_1}}{1 - \dfrac{D_1^2}{M_1 s + D_1} \dfrac{1}{M_2 s + D_1 + D_2}} = \frac{1}{(M_1 s + D_1) - \dfrac{D_1^2}{M_2 s + D_1 + D_2}}$$

$$= \frac{M_2 s + D_1 + D_2}{(M_1 s + D_1)(M_2 s + D_1 + D_2)}$$

The final configuration is shown in Figure 7.5f.   ■

Example 7.19 (Electromechanical system)

Find the transfer function $H(s) = \Omega(s)/E(s)$ of the rotational electromechanical transducer shown in Figure 7.6a. Mechanical and air friction damping are taken into consideration by the damping constant D. The movement of the cylinder–pointer combination is restrained by a spring with spring constant K_s. The moment of inertia of the coil assembly is J. There are N turns in the coil (in this example, e's define voltages and v's define velocities).

SOLUTION

Because there are N turns on the coil, there are $2N$ conductors of length l perpendicular to the magnetic field that are at a distance a from the center of rotation. Therefore, the electrical torque is

$$\mathfrak{T}_e = f_t a = (2NBli)a = (2NBla)i = K_e i \qquad (7.11)$$

where K_e is a constant depending on the physical and geometrical properties of the apparatus. The spring develops an equal and opposite torque, which is written as follows:

$$\mathfrak{T}_s = K_s \theta \quad K_s = (\text{N-m})/\text{degree} \qquad (7.12)$$

Owing to the movement of the coil in the magnetic field, a voltage is generated in the coil. The voltage is given by

$$e_m = 2NBlv = (2NBla)\frac{d\theta}{dt} = K_m \frac{d\theta}{dt} = K_m \omega \quad \omega = \text{angular velocity} \qquad (7.13)$$

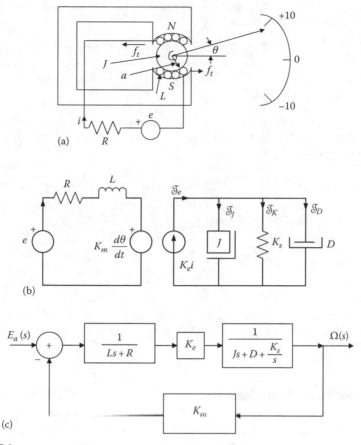

FIGURE 7.6

From (7.11) and (7.13), we observe that K_e and K_m are equal. From the equivalent circuit representation of the system shown in Figure 7.6b, we obtain the equations

$$L\frac{di}{dt} + Ri + K_m\omega = e \qquad \text{Kirchhoff voltage law} \quad \text{(a)}$$

$$J\frac{d\omega}{dt} + D\omega + K_s\int \omega dt = K_e i \quad \text{D'Alembert's principle} \quad \text{(b)}$$

(7.14)

The LT of these equations yields

$$(Ls + R)I(s) + K_m\Omega(s) = E(s) \qquad \text{(a)}$$

$$\left(Js + D + \frac{K_s}{s}\right)\Omega(s) - K_e I(s) = 0 \quad \text{(b)}$$

(7.15)

Substituting the value of $I(s)$ from the second of these equations in the first and solving for the ratio $\Omega(s)/E(s)$, we obtain

$$H(s) \triangleq \frac{\Omega(s)}{E(s)} = \frac{K_e}{(Ls+R)\left(Js+D+\dfrac{K_s}{s}\right) + K_eK_m} \tag{7.16}$$

A block-diagram representation of this system is shown in Figure 7.6c. To obtain the block diagram, we first observe that (7.15a) gives the error signal (the signal just after the summer) that is equal to

$$E(s) - K_m\Omega(s) = (Ls+R)I(s)$$

From (7.15b), we find the relationship

$$\Omega(s) = K_e \frac{1}{Ls+R} \frac{1}{Js+D+\dfrac{K_s}{s}} \{(Ls+R)I(s)\},$$

which indicates that the error signal must be multiplied by the three front factors shown in the above equation to obtain the output. These are shown as a combination of three systems in series. ∎

Example 7.20 (Bioengineering)

Determine the transfer function $H(s) \triangleq \Theta(s)/\mathfrak{I}(s)$ for the mechanical system (pendulum) shown in Figure 7.7a. Draw a block diagram of the system and use block-diagram reductions (see Chapter 1) to deduce the transfer function.

SOLUTION

By an application of D'Alembert's principle, which requires that the algebraic sum of torques must be zero at a node, we write

$$\mathfrak{I} = \mathfrak{I}_g + \mathfrak{I}_D + \mathfrak{I}_J \tag{7.17}$$

where

$\mathfrak{I}$ = input torque

$\mathfrak{I}_g$ = gravity torque = $Mgl\sin\theta$

$\mathfrak{I}_D$ = frictional torque = $D\omega = D\dfrac{d\theta}{dt}$

$\mathfrak{I}_J$ = inertial torque = $D\dfrac{d\omega}{dt} = J\dfrac{d^2\theta}{dt^2}$

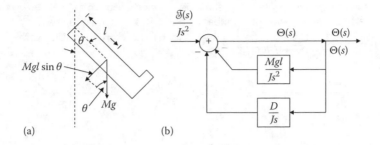

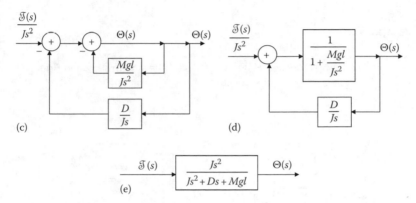

FIGURE 7.7

We have characterized the problem as a bioengineering one because we think that the physical presentation shown in Figure 7.7a is an idealized stiff human limb with the idea to assess the passive control of the locomotive action as an initial step. Therefore, the equation that describes the system is (see Figure 7.7a)

$$J\frac{d^2\theta(t)}{dt^2} + D\frac{d\theta(t)}{dt} + Mgl \sin \theta(t) = \Im(t) \tag{7.18}$$

This equation is nonlinear owing to the presence of the $\sin \theta(t)$ in the differential equation. However, if we assume small deflections to be less than 20–30°, we can approximate the sine function with $\theta(t)$. Under these conditions, the above equation becomes

$$J\frac{d^2\theta(t)}{dt^2} + D\frac{d\theta(t)}{dt} + Mgl\,\theta(t) = \Im(t) \tag{7.19}$$

The LT of this equation is

$$Js^2\Theta(s) + Ds\Theta(s) + Mgl\Theta(s) = \Im(s) \tag{7.20}$$

To obtain the block-diagram representation, we write the above equation in the form

$$\Theta(s) = -\frac{D}{Js}\Theta(s) - \frac{Mgl}{Js^2}\Theta(s) + \frac{1}{Js^2}\Im(s)$$

This result is shown in Figure 7.7b. Rearrange the block diagram as in Figure 7.7c and then reduce the innermost feedback loop. This inner loop has the value $1/(1 + Mgl/Js^2)$. The block diagram simplifies to that shown in Figure 7.7d. Now, reduce this feedback loop to obtain finally

$$\frac{\Theta(s)}{\Im(s)} = \frac{1/(1 + Mgl/Js^2)}{1 + \dfrac{D}{Js^2}\dfrac{1}{Js}\dfrac{1}{(1 + Mgl/Js^2)}} \quad \text{or} \quad \frac{\Theta(s)}{\Im(s)} \overset{\Delta}{=} H(s) = \frac{1}{Js^2 + Ds + Mgl}$$

The transfer function is shown in Figure 7.7e.   ∎

Example 7.21

Determine the transfer function $H(s) \overset{\Delta}{=} V_o(s)/V_i(s)$ of the system shown in Figure 7.8.

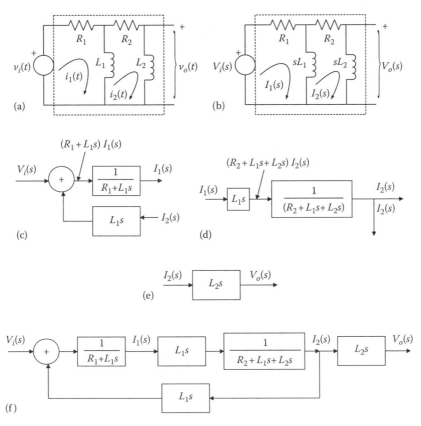

FIGURE 7.8

SOLUTION

We have seen that in the LT operation, the operator d/dt in a differential equation is replaced by s and the operator $\int dt$ is replaced by $1/s$ in problems with zero initial conditions (see previous examples). Therefore, we can write Kirchhoff's voltage law in Laplace form by direct reference to Figure 7.8b. The equations are

$$
\begin{aligned}
(R_1 + L_1 s)I_1(s) - L_1 s I_2(s) &= V_i(s) && \text{(a)} \\
-L_1 s I_1(s) + (R_2 + L_1 s + L_2 s)I_2(s) &= 0 && \text{(b)} \\
L_2 s I_2(s) &= V_o(s) && \text{(c)}
\end{aligned}
\qquad (7.21)
$$

These equations are shown in Figure 7.8c through e, respectively. When the parts are combined, the resulting block diagram is that shown in Figure 7.8f. Using the rules of block reduction (see Chapter 1), the transfer function is easily determined. ■

7.4 Inverse Laplace Transform

As already discussed, the LT is the integral that converts $F(s)$ into the equivalent $f(t)$. To perform the inverse transformation requires that the integration be performed in the complex plane along a path, which ensures that $f(t)$ is unique. Since this type of integration is beyond the level of this text, we will concentrate in the one-to-one correspondence between the direct and the inverse transforms, as expressed by the pair

$$
F(s) = \mathcal{L}\{f(t)\} \quad f(t) = \mathcal{L}^{-1}\{F(s)\} \qquad (7.22)
$$

Consequently, for this level of mathematical knowledge we must refer to Table 7.1 to write $f(t)$ appropriate to a given $\mathcal{L}^{-1}\{F(s)\}$. The following examples illustrate some of the usual methods used in finding inverse LTs.

Example 7.22 (Separate roots)

Find the inverse LT of the function

$$
F(s) = \frac{s-3}{s^2 + 5s + 6} \qquad (7.23)
$$

SOLUTION

Observe that the denominator can be factored in the form $(s+2)(s+3)$. Thus, $F(s)$ is written in partial fraction form:

$$
F(s) = \frac{s-3}{(s+2)(s+3)} = \frac{A}{s+2} + \frac{B}{s+3} \qquad (7.24)
$$

TABLE 7.1   Table of Elementary LT Pairs

Entry No.	$f(t)$	$F(s) = \int_0^\infty f(t)e^{-st}dt$
1	$\delta(t)$	1
2	$u(t)$	$\dfrac{1}{s}$
3	$t^n \quad n = 1, 2, 3, \ldots$	$\dfrac{n!}{s^{n+1}}$
4	e^{-at}	$\dfrac{1}{s+a}$
5	$t^n e^{-at}$	$\dfrac{n!}{(s+a)^{n+1}}$
6	$\sin \omega t$	$\dfrac{\omega}{s^2 + \omega^2}$
7	$\cos \omega t$	$\dfrac{s}{s^2 + \omega^2}$
8	$e^{-at} \sin \omega t$	$\dfrac{\omega}{(s+a)^2 + \omega^2}$
9	$e^{-at} \cos \omega t$	$\dfrac{s+a}{(s+a)^2 + \omega^2}$
10	$\sinh \omega t$	$\dfrac{\omega}{s^2 - \omega^2}$
11	$\cosh \omega t$	$\dfrac{s}{s^2 - \omega^2}$
12	$t \sin \omega t$	$\dfrac{2\omega s}{(s^2 + \omega^2)^2}$
13	$t \cos \omega t$	$\dfrac{s^2 - \omega^2}{(s^2 + \omega^2)^2}$
14	$\dfrac{at + e^{-at} - 1}{a^2}$	$\dfrac{1}{s^2(a + s)}$
15	$\dfrac{bat + (b - a)e^{-at} + (a - b)}{a^2}$	$\dfrac{b + s}{s(a + s)^2}$
16	$\dfrac{(b^2 + \omega^2)^{1/2} \sin(\omega t + \phi)}{\omega}$ $\phi = \tan^{-1}\dfrac{\omega}{b}$	$\dfrac{s + b}{s^2 + \omega^2}$
17	$\dfrac{b - (b^2 + \omega^2)^{1/2} \cos(\omega t + \phi)}{\omega^2}$ $\phi = \tan^{-1}\dfrac{\omega}{b}$	$\dfrac{s + b}{s(s^2 + \omega^2)}$
18	$\dfrac{(a^2 + b^2 + c^2 - 2ac)^{1/2} e^{-at} \sin(bt + \phi)}{b}$ $\phi = \tan^{-1}\dfrac{b}{c - a}$	$\dfrac{s + c}{(s + a)^2 + b^2)}$
19	$\dfrac{b + (a^2 + b^2)^{1/2} e^{-at} \sin(bt - \phi)}{b(a^2 + b^2)}$ $\phi = -\tan^{-1}\dfrac{b}{a}$	$\dfrac{1}{s[(s + a)^2 + b^2)]}$

Note: A more extensive table is at the end of the chapter.

where A and B are constants to be determined. To evaluate A, multiply both sides of (7.24) by $(s+2)$ and set $s=-2$. The result is

$$\frac{s-3}{(s+2)(s+3)}(s+2)\bigg|_{s=-2} = A + \frac{(s+2)B}{s+3}\bigg|_{s=-2} \quad \text{or} \quad A = -5$$

We proceed in the same manner to reduce the constant B. By multiplying both sides by $(s+3)$ and setting $s=-3$, we obtain $B=6$. Hence, the inverse LT is given by

$$f(t) = \mathcal{L}^{-1}\{F(s)\} = -5\mathcal{L}^{-1}\left\{\frac{1}{s+2}\right\} + 6\mathcal{L}^{-1}\left\{\frac{1}{s+3}\right\} = -5e^{-2t} + 6e^{-3t} \quad t \geq 0$$

where Table 7.1, entry 4, was used.

MATLAB® function residue

We can find the partial fraction expansion of the rational functions by invoking the residue command of MATLAB: $[r,p,k]=\text{residue}(A,B)$, where B and A are row vectors specifying the coefficients of the numerator and denominator polynomials in descending powers of s. The residues are returned in the column vector r, the poles in column vector p, and the direct terms in row vector k. If, for example, $k=[1\ -5]$ we add in the expression the function $s-5$. For the above case, we write at the command window the following:

$$\gg [r,p,k,] = \text{residue}([0\ 1\ -3],[1\ 5\ 6]);$$

where
  $r=$ residues in column form
  $p=$ poles in column form
  $k=$ some function of s or constant

The results of our example above are

r	p	k
6	-3	$[\,]$
-5	-2	

With these results, we write $F(s) = \dfrac{6}{s-(-3)} + \dfrac{-5}{s-(-2)} + 0$. This result is identical to the one found above. ∎

Example 7.23

Find the inverse LT of the function

$$F(s) = \frac{s+1}{[(s+2)^2 + 1](s+3)}$$

SOLUTION

This function is written in the form

$$F(s) = \frac{A}{s+3} + \frac{Bs+C}{[(s+2)^2+1]} = \frac{s+1}{[(s+2)^2+1]}$$

The value of A is evaluated by multiplying both sides of the equation by $s+3$ and then setting $s=-3$. The result is

$$A = (s+3)F(s) = \frac{-3+1}{(-3+2)^2+1} = -1$$

To evaluate B and C, combine the two equations

$$\frac{-1[(s+2)^2+1] + (s+3)(Bs+C)}{[(s+2)^2+1](s+3)} = \frac{s+1}{[(s+2)^2+1](s+3)}$$

From which it follows that

$$-(s^2+4s+5) + Bs^2 + (C+3B)s + 3C = s+1$$

Combine like-powered terms of s to write

$$(-1+B)s^2 + (-4+C+3B)s + (-5+3C) = s+1$$

Equating the coefficients of equal power of s, we then have $-1+B=0$; $-4+C+3B=1$; $-5+3C=1$. From these equations, we obtain $B=1, C=2$. Hence, the function is written in the equivalent form

$$F(s) = \frac{-1}{s+3} + \frac{s+2}{(s+2)^2+1}$$

Now using Table 7.1, the result is $f(t) = -e^{3t} + e^{-2t}\cos t$ $t \geq 0$.

Using the residue MATLAB function, $\mathrm{[r,p,k]=residue([0\ 0\ 1\ 1],[1\ 7\ 17}$ $\mathrm{15])}$, we obtain

r	p	k
−1.0000	−3.0000	
0.5000 − 0.0000i	−2.0000 + 1.0000i	
0.5000 + 0.0000i	−2.0000 − 1.0000i	[ ]

Therefore, the function becomes

$$F(s) = \frac{-1}{s-(-3)} + \frac{0.5}{s-(-2+j)} + \frac{0.5}{s-(-2-j)} + 0$$

or $f(t) = -e^{-3t} + 0.5e^{-2t+jt} + 0.5e^{-2t-jt} = -e^{-3t} + 0.5e^{-2t}(e^{jt}+e^{-jt}) = -e^{-3t} + e^{-2t}\cos t,$

which is identical to the value found above.

■

In many cases, $F(s)$ is the quotient of two polynomials with real coefficients. If the numerator polynomial is of the same or higher degree than the denominator polynomial, we must first divide the numerator polynomial by the denominator polynomial, the division carried forward until the numerator polynomial is of degree one less than the denominator polynomial. This procedure results in a polynomial of s plus a **proper function**. The proper function can be expanded into a partial fraction expansion. The result of such an expansion is an expression of the form

$$F'(s) = B_0 + B_1 s + B_2 s^2 + \cdots + \frac{A_1}{s - s_1} + \frac{A_2}{s - s_2} + \frac{A_3}{s - s_3} + \cdots + \frac{A_{p1}}{s - s_p}$$

$$+ \frac{A_{p2}}{(s - s_p)^2} + \frac{A_{p3}}{(s - s_p)^3} + \cdots + \frac{A_{pr}}{(s - s_p)^r} \tag{7.25}$$

This expression has been written in a form to show three types of terms:

1. Polynomial
2. Simple partial fraction including all terms with distinct roots
3. Partial fraction appropriate to multiple roots

To find the constants A_1, A_2, A_3, ..., the polynomial terms are removed, leaving the proper fraction

$$F(s) = F'(s) - (B_0 + B_1 s + B_3 s^2 + \cdots) \tag{7.26}$$

where $F(s)$ is the partial fraction expansion containing all the A's. To find the constants, A_k, which in complex variable terminology are the residues of the function $F(s)$ at the simple poles s_k, it is only necessary to note that as $s \rightarrow s_k$, the term $A_k/(s - s_k)$ will become large compared with all other terms. In the limit,

$$\boxed{A_k = \lim_{s \to s_k} [(s - s_k)F(s)]} \tag{7.27}$$

Therefore, for each simple pole, upon taking the inverse transform, the result will be a simple exponential of the form

$$\mathcal{L}^{-1}\left\{\frac{A_k}{s - s_k}\right\} = A_k e^{s_k t} \tag{7.28}$$

Note also that since $F(s)$ contains only real coefficients, if s_k is a complex pole with A_k, there exists also a conjugate pole s_k^* with residue A_k^*. For such complex poles,

$$\mathcal{L}^{-1}\left\{\frac{A_k}{s - s_k} + \frac{A_k^*}{s - s_k^*}\right\} = A_k e^{s_k t} + A_k^* e^{s_k^* t}$$

These terms can be combined in the following way:

$$\begin{aligned}
\text{response} &= (a_k + jb_k)e^{(\sigma_k + j\omega_k)t} + (a_k - jb_k)e^{(\sigma_k - j\omega_k)t} \\
&= e^{\sigma_k t}[(a_k + jb_k)(\cos\omega_k t + j\sin\omega_k t) + (a_k - jb_k)(\cos\omega_k t - j\sin\omega_k t)] \\
&= 2e^{\sigma_k t}(a_k \cos\omega_k t - b_k \sin\omega_k t) = 2|A_k|e^{\sigma_k t}\cos(\omega_k t + \theta_k) \qquad (7.29)
\end{aligned}$$

$$|A_k| = \sqrt{a_k^2 + b_k^2}, \quad \theta = \tan^{-1}\left(\frac{b_k}{a_k}\right)$$

When the proper fraction contains a multiple pole of order r, the coefficients in the partial fraction expansion, which are involved in the terms

$$\frac{A_{p1}}{(s - s_p)} + \frac{A_{p2}}{(s - s_p)^2} + \frac{A_{p3}}{(s - s_p)^3} + \cdots + \frac{A_{pr}}{(s - s_p)^r}$$

must be evaluated. A simple application of (7.27) is not adequate. Now, the procedure is to multiply both sides of (7.26) by $(s - s_p)^r$, which gives

$$\begin{aligned}
(s - s_p)^r F(s) &= (s - s_p)^r \left(\frac{A_1}{s - s_1} + \frac{A_2}{s - s_2} + \cdots + \frac{A_k}{s - s_k}\right) + A_{p1}(s - s_p)^{r-1} \\
&\quad + A_{p2}(s - s_p)^{r-2} + \cdots + A_{p(r-1)}(s - s_p) + A_{pr} \qquad (7.30)
\end{aligned}$$

In the limit as $s = s_p$, all terms on the right vanish with the exception of A_{pr}. Suppose we know that this equation is differentiated once with respect to s. The constant A_{pr} will vanish in the differentiation, but $A_{p(r-1)}$ will be determined by setting $s = s_p$. This procedure is continued to find each of the coefficients A_{pk}. Specifically, this procedure is quantified by

$$\boxed{A_{pk} = \frac{1}{(r - k)!}\frac{d^{r-k}}{ds^{r-k}}[F(s)(s - s_p)^r]\bigg|_{s=s_p} \quad k = 1, 2, \ldots, r \quad 0! = 1} \qquad (7.31)$$

Example 7.24

Find the inverse transform of the following function:

$$F'(s) = \frac{s^3 + 2s^2 + 3s + 1}{s^2(s + 1)}$$

SOLUTION

This fraction is not a proper one. The numerator polynomial is divided by the denominator polynomial by simple long division. The result is

$$F'(s) = 1 + \frac{s^2 + 3s + 1}{s^2(s + 1)}$$

The proper fraction is expanded into partial fraction form:

$$F(s) = \frac{s^2 + 3s + 1}{s^2(s+1)} = \frac{A_{11}}{s} + \frac{A_{12}}{s^2} + \frac{A_2}{s+1}$$

The value of A_2 is deduced by using (7.27):

$$A_2 = [(s+1)F(s)]_{s=-1} = \frac{s^2 + 3s + 1}{s^2(s+1)}\bigg|_{s=-1} = -1$$

To find A_{11} and A_{12}, we proceed as specified by (7.31):

$$A_{12} = s^2 F(s)\big|_{s=0} = \frac{s^2 + 3s + 1}{s+1}\bigg|_{s=0} = 1$$

$$A_{11} = \frac{1}{(2-1)!}\left[\frac{d^{2-1}}{ds^{2-1}} s^2 F(s)\right]_{s=0} = \frac{d}{ds}\left(\frac{s^2+3s+1}{s+1}\right)\bigg|_{s=0} = \frac{s^2+2s+2}{(s+1)^2}\bigg|_{s=0} = 2$$

Therefore,

$$F'(s) = 1 - \frac{1}{s+1} + 2\frac{1}{s} + \frac{1}{s^2} \quad \text{or} \quad f'(t) = \delta(t) - e^{-t} + 2 + t \quad \text{for } t \ge 0$$

where Table 7.1 was used. ■

We can use MATLAB function $[\text{r,p,k}] = \text{residue (b,a)}$ for multiple roots. For example, to invert the function

$$F(s) = \frac{1}{s(s+1)^4} = \frac{1}{s^5 + 4s^4 + 6s^3 + 4s^2 + s}$$

we proceed as follows: At the command sign $\gg$, we write
$[\text{r,p,k}] = \text{residue}([0\ 0\ 0\ 0\ 0\ 1], [1\ 4\ 6\ 4\ 1\ 0])$; At the enter command, we find the following values:

r	p	k
−1.0000	−1.0000	[ ]
−1.0000	−1.0000	
−1.0000	−1.0000	
−1.0000	−1.0000	
1.0000	0	

Based on these values, we write

$$F(s) = \frac{-1}{s - (-1)} + \frac{-1}{[s - (-1)]^2} + \frac{-1}{[s - (-1)]^3} + \frac{-1}{[s - (-1)]^4} + \frac{1}{s - (0)}$$

Its inverse LT, consulting Table 7.1, is

$$f(t) = -e^{-t} - \frac{te^{-t}}{1!} - \frac{t^2 e^{-t}}{2!} - \frac{t^3 e^{-t}}{3!} + u(t)$$

7.5 Problem Solving with Laplace Transform

7.5.1 Ordinary Differential Equations

It is instructive to present several examples and their solutions using the LT method.

Example 7.25

Study the changes in the time of the current in an *RL* series circuit with initial condition $i(0) = $ constant and input voltage source $v(t)$.

SOLUTION

The differential equation describing the system (KVL) is

$$L \frac{di(t)}{dt} + Ri(t) = v(t) \tag{7.32}$$

With the help of the LT properties (see Section 7.2), we find that the LT of the above equation and the unknown $I(s)$ is

$$LsI(s) - Li(0) + RI(s) = V(s) \quad \text{or} \quad I(s) = \frac{L}{Ls + R} i(0) + \frac{1}{Ls + R} V(s) \overset{\Delta}{=} I_{zi}(s) + I_{zs}(s) \tag{7.33}$$

where

$$I_{zi}(s) = \frac{L}{Ls + R} i(0) = \textbf{zero-input solution} \text{ (dependent on initial conditions} \tag{7.34}$$
$$\text{and independent of inputs)}$$

and

$$I_{zs}(s) = \frac{1}{Ls + R} V(s) = \textbf{zero-state solution} \text{ (dependent on inputs and} \tag{7.35}$$
$$\text{independent of initial conditions)}$$

Therefore, the total solution in the transform and time domains is

$$I(s) = I_{zi}(s) + I_{zs}(s) \quad \text{or} \quad i(t) = i_{zi}(t) + i_{zs}(t) \tag{7.36}$$

Impulse response: To find the impulse response of the system, we set the initial condition equal to zero and a delta function as the input voltage. This implies that the zero-input solution is zero. Therefore, the **system function** for this problem, which is the zero-state solution in transform domain, and the **impulse response** are

$$H(s) = \frac{1}{L}\frac{1}{s+(R/L)}; \quad h(t) = \mathcal{L}^{-1}\left\{\frac{1}{L}\frac{1}{s+(R/L)}\right\} = \frac{1}{L}e^{-(R/L)t}$$

Superposition: Let us assume that $v(t) = v_1(t) + v_2(t)$. Then the zero-state solution will be

$$I_{zs}(s) = \frac{1}{Ls+R}[V_1(s)+V_2(s)] = \frac{1}{Ls+R}V_1(s) + \frac{1}{Ls+R}V_2(s) = I_{zs1}(s) + I_{zs2}(s) \quad (7.37)$$

Note: *The input voltages affect only the zero state part of the solution.*

Time invariance: Suppose that the source $v(t)$ is shifted by t_0. Then the zero-state solution is

$$I_{0zs}(s) = \frac{1}{Ls+R}\mathcal{L}\{v(t-t_0)u(t-t_0)\} = \frac{1}{Ls+R}V(s)e^{-st_0} = I_{zs}(s)e^{-st_0}$$

This indicates that the current due to a shifted source is of the same form but shifted by t_0 due to $\exp(-st_0)$. Hence,

$$i_{0zs}(t) = i_{zs}(t-t_0)u(t-t_0) \quad (7.38)$$

Note: *A delay of the input results in an equal delay of the zero-state solution.*

Step response: If $v(t) = u(t)$ is a unit step signal applied at $t=0$ to a system with zero initial conditions, then the resulting zero-state response is called **step response**. This step response will be denoted by $u_{sr}(t)$. Since the LT of $u(t)$ is $1/s$, then (7.35) yields

$$I_{zs}(s) = \frac{1}{Ls+R}\frac{1}{s} = \frac{1}{R}\frac{1}{s} - \frac{1}{R}\frac{1}{s+(R/L)}$$

Hence,

$$i_{zs}(t) \triangleq u_{sr}(t) = \frac{1}{R}(1-e^{-(R/L)t})u(t) \quad (7.39)$$

Let, for example, the values of the circuit elements be $R=2, L=1$. We assume a voltage input of the form shown in Figure 7.9a and zero initial condition. The input voltage is

$$v(t) = 2u(t) + u(t-1) - 2u(t-2)$$

Therefore (superposition),

$$i_{zs}(t) = 2u_{sr}(t) + u_{sr}(t-1) - 2u_{sr}(t-2)$$

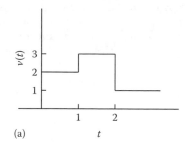

(a)

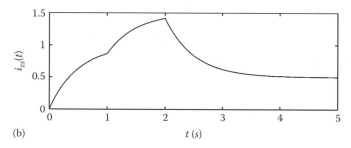

(b)

FIGURE 7.9

But

$$u_{sr}(t) = \frac{1}{R}(1 - e^{-(R/L)t})u(t) = \frac{1}{2}(1 - e^{-2t})u(t)$$

and, hence, the total current is

$$i_{zs}(t) = \begin{cases} (1 - e^{-2t}) & t \geq 0 \\ \frac{1}{2}(1 - e^{-2(t-1)}) & 1 \leq t < \infty \\ -(1 - e^{-2(t-2)}) & 2 \leq t < \infty \end{cases}$$

This current is plotted in Figure 7.9b. If, however, we had also assumed initial condition $i(0)$, then the general LT format would have been as follows:

$$I(s) = \underbrace{\frac{1}{R}\left(\frac{1}{s} - \frac{1}{s + (R/L)}\right)}_{\text{zero–state response}} + \underbrace{\frac{1}{s + (R/L)}i(0)}_{\text{zero–input response}}$$

Using Table 7.1, the inverse transform of the above equation is

$$i(t) = \underbrace{\frac{1}{R}(1 - e^{-(R/L)t})}_{\text{zero-state response}} + \underbrace{i(0)e^{-(R/L)t}}_{\text{zero-input response}} = \underbrace{\left[i(0) - \frac{1}{R}\right]e^{-(R/L)t}}_{\text{transient response}} + \underbrace{\frac{1}{R}}_{\substack{\text{steady–state} \\ \text{response}}} \quad t \geq 0 \quad \blacksquare$$

Example 7.26

Find the current in an initially relaxed *RL* series circuit when the input is a pulse shown in Figure 7.10a.

SOLUTION

The differential equation describing the system is

$$L\frac{di(t)}{dt} + Ri(t) = v(t)$$

The input signal is decomposed into two unit step functions, as shown in Figure 7.10:

$$v(t) = u(t) - u(t - 1)$$

Because the system is LTI, we can straightaway use the superposition property and the time-invariant properties discussed above. Hence, we write

$$i_{zs}(t) = u_{sr}(t) - u_{sr}(t - 1)$$

But $u_{sr}(t)$ is given by (7.39) and, thus, the current is given by

$$i_{zs}(t) = \frac{1}{R}(1 - e^{-(R/L)t})u(t) - \frac{1}{R}(1 - e^{-(R/L)(t-1)})u(t - 1)$$

For $L = 1$ and $R = 2$, the current is shown in Figure 7.10c. ∎

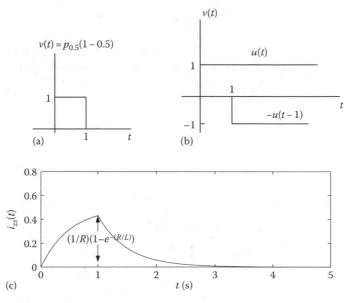

(a)

(b)

(c)

FIGURE 7.10

Example 7.27

A force is applied to a relaxed mechanical system of negligible mass, as shown in Figure 7.11a. Find the velocity of the system.

SOLUTION

From Figure 7.11b, we write

$$Dv + K \int_0^t v(x)dx = au(t)$$

Now, define a new variable $y(t) = \int_0^t v(x)dx$; this equation takes the form

$$D\frac{dy(t)}{dt} + Ky = au(t)$$

Observe that the solution is equal to $au_{sr}(t)$, where $u_{sr}(t)$ is the step response solution. Therefore, the LT of the above equation and its inverse are (see Example 7.25)

$$Y(s) = \frac{a}{D}\frac{1}{[s+(K/D)]s}; \quad y(t) = \frac{a}{K}(1 - e^{-(K/D)t}) \quad t \geq 0$$

Hence, the velocity is given by

$$v(t) \triangleq \frac{dy(t)}{dt} = \frac{a}{D}e^{-(K/D)t} \quad t \geq 0$$

Next, we approach this problem from a different point of view, namely, the use of convolution. This approach requires that we find the impulse response of the system in the time and LT domains, which are ($h(t)$ has the same units as the velocity)

$$Dh(t) + K \int_0^t h(x)dx = \delta(t); \quad DH(s) + \frac{K}{s}H(s) = 1$$

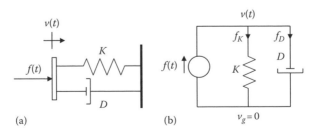

(a)　　　　　(b)

FIGURE 7.11

from which the transfer function and its impulse response are

$$H(s) = \frac{1}{D} - \frac{K}{D^2}\frac{1}{s + (K/D)}; \quad h(t) = \frac{1}{D}\delta(t) - \frac{K}{D^2}e^{-(K/D)t} \quad t \geq 0$$

For a step function input $au(t)$, the output is

$$v(t) = au(t) * h(t) = \frac{a}{D}\int_{-\infty}^{\infty}\delta(x)u(t-x)dx - \frac{Ka}{D^2}\int_{-\infty}^{\infty}e^{-(K/D)x}u(x)u(t-x)dx$$

$$= \frac{a}{D}u(t) - \frac{Ka}{D^2}\int_{0}^{t}e^{-(K/D)x}dx = \frac{a}{D}u(t) + \frac{a}{D}(e^{-(K/D)t} - 1)u(t) = \frac{a}{D}e^{-(K/D)t}u(t)$$

This result is precisely what we found earlier using a different approach. This makes sense if we remember that the derivative of a step function is a delta function.

We could have also taken the LT of the defining equation, which yields

$$DV(s) + K\frac{V(s)}{s} = a\frac{1}{s}$$

By solving for the unknown $V(s)$ and taking the inverse LT, we obtain, as before, the desired solution:

$$V(s) = \frac{a}{D}\frac{1}{s + (K/D)}; \quad v(t) = \frac{a}{D}e^{-(K/D)t} \quad t \geq 0 \qquad \blacksquare$$

Example 7.28

Refer to Figure 7.12a, which shows the switching of an inductor into a circuit with an initial current. Prior to switching, the circuit inductance is L_1; after switching, the total circuit inductance is $L_1 + L_2$. The switching occurs at $t = 0$. Determine the current in the circuit for $t \geq 0$.

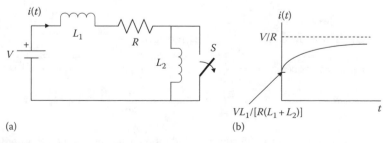

(a)

(b)

FIGURE 7.12

SOLUTION

The current just before switching is

$$i(0-) = \frac{V}{R}$$

To find the current after switching at $t=0+$, we employ the law of conservation of flux linkages. We can write over the switching period,

$$L_1 i(0-) = (L_1 + L_2)i(0+) \quad \text{or} \quad i(0+) = \frac{L_1}{(L_1 + L_2)}i(0-) = \frac{L_1}{(L_1 + L_2)}\frac{V}{R}$$

The differential equation that governs the circuit response, after the switch S is closed, is

$$(L_1 + L_2)\frac{di(t)}{dt} + Ri(t) = Vu(t)$$

The LT of this equation yields

$$I(s) = \frac{V}{R}\left[\frac{1}{s} - \frac{1}{s + R/(L_1 + L_2)}\right] + \frac{1}{s + R/(L_1 + L_2)}i(0+)$$

Include the value of $i(0+)$ in this expression and then take the inverse LT. The result is

$$i(t) = \frac{V}{R}\left(1 - \frac{L_2}{L_1 + L_2}e^{-Rt/(L_1 + L_2)}\right) \quad t \geq 0$$

The form of the current variation is shown in Figure 7.12b.   ∎

Example 7.29 (Second-order systems—the series RLC circuit)

Let us analyze a series RLC circuit driven by a voltage source, $v_i(t)$, as shown in Figure 7.13a. Specifically we will be interested in the step response of the system. The output is the voltage across the capacitor. In our study, the following important parameters are needed:

Critical resistance: $R_c = 2\sqrt{\dfrac{L}{C}}$ Damping: $\alpha = \dfrac{R}{2L}$

Resonant frequency: $\omega_r = \dfrac{1}{\sqrt{LC}}$ Natural frequency: $\beta = \sqrt{\omega_r^2 - \alpha^2}$

Q-factor: $Q = \dfrac{\omega_r}{2\alpha}$ Damping ratio: $\varsigma = \dfrac{\alpha}{\omega_r} = \dfrac{R}{R_c}$

which are used to determine the current $i(t)$ and the voltage across the capacitor $v(t)$.

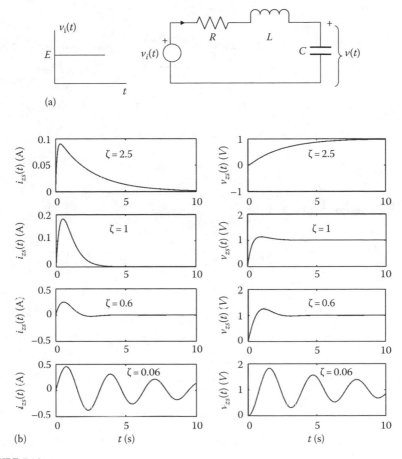

FIGURE 7.13

Let us assume that there exist an initial current $i(0)$ and an initial voltage across the capacitor $v(0)$. The following equation (KVL) characterizes the system:

$$Ri(t) + L\frac{di(t)}{dt} + \frac{1}{C}\int_0^t i(x)dx + v(0) = v_i(t) \tag{7.40}$$

Taking the transform of both sides, we obtain

$$RI(s) + LsI(s) - Li(0) + \frac{I(s)}{Cs} + \frac{v(0)}{s} = V_i(s)$$

Therefore,

$$I(s) = \underbrace{\frac{V_i(s)}{R + Ls + (1/Cs)}}_{\text{zero-state response}} + \underbrace{\frac{Li(0) - (v(0)/s)}{R + Ls + (1/Cs)}}_{\text{zero-input response}} \overset{\Delta}{=} I_{zs}(s) + I_{zi}(s) \tag{7.41}$$

The voltage $v(t)$ can be expressed in terms of the current $i(t)$. Since $i(t) = C(dv(t)/dt)$, we conclude that $I(s) = C[sV(s) - v(0)]$.

The zero-state response $V_{zs}(s)$ of the voltage $V(s)$ is, therefore, given by

$$V_{zs}(s) = I_{zs}(s)\frac{1}{Cs} = \frac{V(s)}{LCs^2 + RCs + 1} \tag{7.42}$$

Step response: If $v_i(t) = Eu(t)$, then $V_i(s) = E/s$; hence,

$$I_{zs}(s) = \frac{E/s}{R + Ls + (1/Cs)} = \frac{E/L}{s^2 + 2\alpha s + \omega_r^2}; \quad V_{zs}(s) = \frac{E\omega_r^2}{s(s^2 + 2\alpha s + \omega_r^2)} \tag{7.43}$$

To obtain the inverse LT of these quantities, we must investigate their poles, which are

$$p_{1,2} = -\frac{R}{2L} \pm \sqrt{\left(\frac{R}{2L}\right)^2 - \frac{1}{LC}} = -\alpha \pm \omega_r\sqrt{\left(\frac{R}{R_c}\right)^2 - 1} \tag{7.44}$$

Based on the above equation, the following three cases may be considered:

1. If $R > R_c$, then the roots are real and the time functions are

$$i_{zs}(t) = \frac{E}{L(p_1 - p_2)}(e^{p_1 t} - e^{p_2 t}); \quad v_{zs}(t) = E + \frac{E\omega_r^2}{(p_1 - p_2)}\left(\frac{1}{p_1}e^{p_1 t} - \frac{1}{p_2}e^{p_2 t}\right) \tag{7.45}$$

2. If $R = R_c$, then $p_1 = p_2 = -\alpha = -\omega_r$ and

$$i_{zs}(t) = \frac{E}{L}te^{-\alpha t}; \quad v_{zs}(t) = E - E(1 + \omega_r t)e^{-\alpha t} \tag{7.46}$$

3. If $R < R_c$, then the roots are complex: $p_{1,2} = -\alpha \pm j\beta$ and the time functions are

$$i_{zs}(t) = \frac{E}{\beta L}e^{-\alpha t}\sin\beta t; \quad v_{zs}(t) = E - Ee^{-\alpha t}\left(\cos\beta t + \frac{\alpha}{\beta}\sin\beta t\right) \tag{7.47}$$

We note that if $R \ll R_c$, then $\alpha \ll \beta \cong \omega_r$, and (7.47) yields

$$i_{zs}(t) \cong E\sqrt{\frac{C}{L}}e^{-\alpha t}\sin\omega_r t; \quad v_{zs}(t) \cong E - Ee^{-\alpha t}\cos\omega_r t \tag{7.48}$$

It is instructive to find out how the current and the voltage behave when the damping ratio ς varies. Let us assume the following circuit element values: $v_i(t) = u(t)$, $L = 1$, $C = 0.25$. The four cases, which we investigate, are shown in Figure 7.13b from top to bottom. The parameters for these four cases are

a) $R = 10, \varsigma = 2.5, \omega_r = 2, \alpha = 5$, roots: $-9.5826, -0.4174$

b) $R = 4, \varsigma = 1.0, \omega_r = 2, \alpha = 2$, roots: $-2.0000, -2.0000$

c) $R = 2.4, \varsigma = 0.6, \omega_r = 2, \alpha = 1.2$, roots: $-1.2000 + j1.6000, -1.2000 - j1.6000$

d) $R = 0.24, \varsigma = 0.06, \omega_r = 2, \alpha = 0.12$, roots: $-0.1200 + j1.9964, -0.1200 - j1.9964$

There exists the MATLAB function `step(num,den)` where "num" is a vector containing the coefficients of the numerator and "den" is a vector containing the coefficients of the denominator. The variable, which we want to find, must be given in a Laplace-transformed rational fraction form. To plot the output, we write

`>>sr = step (num, den) ; plot (sr) ; .` ∎

Example 7.30

Find the velocity of the system shown in Figure 7.14a when the applied force is $f(t) = \exp(-t)u(t)$. Use the LT method and assume zero initial conditions. Solve the same problem by means of the convolution technique. The input is the force and the output is the velocity.

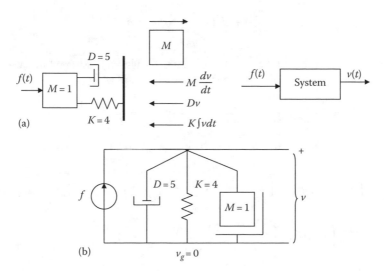

FIGURE 7.14

SOLUTION

From Figure 7.14b, we write the controlling equation:

$$M\frac{dv(t)}{dt} + Dv(t) + K\int_0^t v(x)dx = f(t) \quad \text{or} \quad \frac{dv(t)}{dt} + 5v(t) + 4\int_0^t v(x)dx = e^{-t}u(t)$$

LT these equations and then solve for $V(s)$. This yields

$$H(s) = \frac{s}{Ms^2 + Ds + K}; \quad V(s) = \frac{s}{(s+1)(s^2 + 5s + 4)} = \frac{s}{(s+1)^2(s+4)}$$

Write the expression in the form

$$V(s) = \frac{A}{s+4} + \frac{B}{s+1} + \frac{C}{(s+1)^2}$$

where

$$A = \frac{s}{(s+1)^2}\bigg|_{s=-4} = -\frac{4}{9}; \quad B = \frac{1}{1!}\frac{d}{ds}\left(\frac{s}{s+4}\right)\bigg|_{s=-1} = \frac{4}{9}; \quad C = \frac{s}{s+4}\bigg|_{s=-1} = -\frac{1}{3}$$

The inverse transform of $V(s)$ is given by

$$v(t) = -\frac{4}{9}e^{-4t} + \frac{4}{9}e^{-t} - \frac{1}{3}te^{-t} \quad t \geq 0$$

To find the aero state solution, $v(t)$, by the use of the convolution integral, we must first find the impulse response of the system, $h(t)$. The quantity is specified by

$$\frac{dh(t)}{dt} + 5h(t) + 4\int_0^t h(x)dx = \delta(t)$$

Because we want to find the impulse response, the system is assumed to have zero initial conditions. The LT of the equation yields and its inverse are

$$H(s) = \frac{s}{s^2 + 5s + 4} = \frac{4}{3}\frac{1}{s+4} - \frac{1}{3}\frac{1}{s+1}; \quad h(t) = \frac{4}{3}e^{-4t} - \frac{1}{3}e^{-t} \quad t \geq 0$$

Therefore, the output of the system to the input $\exp(-t)u(t)$ is written as

$$v(t) = \int_{-\infty}^{\infty} h(x)f(t-x)dx = \int_{-\infty}^{\infty} e^{-(t-x)}u(t-x)\left(\frac{4}{3}e^{-4x} - \frac{1}{3}e^{-x}\right)u(x)dx$$

$$= e^{-t}\left(\frac{4}{3}\int_0^t e^{-3x}dx - \frac{1}{3}\int_0^t dx\right) = e^{-t}\left(\frac{4}{3}\left(\frac{1}{-3}\right)e^{-3x}\bigg|_0^t - \frac{1}{3}t\right) = -\frac{4}{9}e^{-4t} + \frac{4}{9}e^{-t} - \frac{1}{3}te^{-t} \quad t \geq 0$$

This result is identical with that found using the LT technique, as it should.   ∎

Example 7.31 (Environmental engineering)

Let us assume that at time $t = 0$, there are a certain number of some species, say $n(0)$, all of the same age. For convenience, let us classify this age as the zero age. At future time t, there are $n_1(t)$ of these members still in the population. Therefore, $n_1(t)$ and $n(0)$ are connected through a **survival function**, $f(t)$, as follows

$$n_1(t) = n(0)f(t)$$

Let us assume that at time, t_1, we placed m of zero age. At any future time $t > t_1$, the number of these individuals will be

$$m(t) = m(t_1)f(t - t_1) \quad t > t_1$$

Consider that there is a **replacement rate** $r(t)$ of the age-zero individuals. In the time interval from τ to $\tau + \Delta\tau$, $r(\tau_1)\Delta\tau$ individuals are placed in the population. τ_1 must be in the range $\tau \le \tau_1 \le \tau + \Delta\tau$. The survival law dictates that $r(\tau_1)\Delta\tau f(t - \tau_1)$ individuals will still be present in the population at time t. We, therefore, can think of splitting the interval $[0, t]$ into subintervals of length $\Delta\tau$ and adding up the number of survivors for each interval. At the limit, as $\Delta\tau$ approaches zero, the summation becomes an integral. Hence, the number of species at time t is

$$n(t) = n(0)f(t) + \int_0^t r(\tau)f(t - \tau)d\tau \tag{7.49}$$

Let us assume that in 2005, the population of deer was 55,000, and their survival function was the exponential function $\exp(-t)$. Suppose we need to determine a rate function $r(t)$ such that the population is linear function of time, that is,

$$n(t) = n(0) + at$$

Solution

Here $n(0) = 55,000$ and a is a constant. Then $r(t)$ must satisfy the equation

$$n(0) + at = n(0)e^{-t} + \int_0^t r(x)e^{-(t-x)}dx \quad t \ge 0 \tag{7.50}$$

Taking the LT of both sides of the above equation, we find that

$$\frac{n(0)}{s} + \frac{a}{s^2} = \left[\frac{n(0)}{s+1} + R(s)\frac{1}{s+1}\right] \quad \text{or} \quad R(s) = \frac{(n(0) + a)s + a}{s^2} = \frac{n(0) + a}{s} + \frac{a}{s^2}$$

Therefore, the rate function is $r(t) = n(0) + a + at$ ∎

Note: *Observe that the LT was used in (7.49), even if it was not a differential equation.*

Example 7.32 (Systems of differential equations)

Find the voltage $v_2(t)$ for a delta function (impulse response, $h(t)$) and step function (step response) inputs to the system shown in Figure 7.15a. Assuming zero initial conditions, we find the zero-state response for the step function.

SOLUTION

Using the node equation law, we write

$$\frac{v_1(t)}{R} + C\frac{dv_1(t)}{dt} + i(t) = i_i(t); \quad \frac{v_2(t)}{R} + C\frac{dv_2(t)}{dt} - i(t) = 0; \quad v_1(t) - v_2(t) = L\frac{di(t)}{dt} \quad (7.51)$$

Taking transforms of both sides and setting $G = 1/R$, we obtain

$$GV_1(s) + CsV_1(s) + I(s) = I_i(s)$$
$$GV_2(s) + CsV_2(s) - I() = 0 \quad (7.52)$$
$$V_1(s) - V_2(s) - LsI(s) = 0$$

The desired output is

$$V_2(s) = \frac{I_i(s)}{LC^2s^3 + 2LCGs^2 + (LG^2 + 2C)s + 2G} \quad (7.53)$$

Let us assume the following values: $L = 1$, $R = 4$, $C = 1$. Then (7.53) becomes

$$V_2(s) = I_i(s)\frac{1}{s^3 + 0.5s^2 + 2.0625s + 0.5}$$

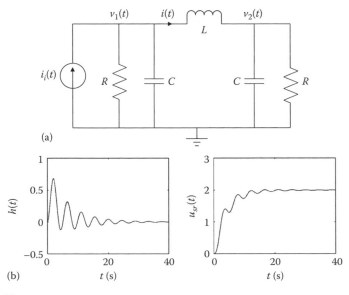

(a)

(b)

FIGURE 7.15

For the impulse response $I_i(s) = 1$ and for step response is equal to $1/s$. To obtain the time domain of these two responses, we used the following MATLAB programs:

```
>>t=0:0.05:40;
>>h=impulse([0 0 0 1],[1 0.5 2.0625 0.5],t]);
>>subplot(2,2,1);plot(t,h);xlabel('t s');ylabel('h(t)');
```

For the step response we used the following commands:

```
>>t=0:0.05:40;
>>sr=step([0 0 0 1],[1 0.5 2.0625 0.5],t]);
```

The outputs are shown in Figure 7.15b. ∎

7.5.2 Partial Differential Equations

The gamma function

If n is positive integer quantity, then the **gamma function** is defined by

$$\Gamma(n) = \int_0^\infty x^{n-1} e^{-x} dx \tag{7.54}$$

Integrating by parts the function $\Gamma(n+1) = \int_0^\infty x^n e^{-x} dx$ for $n > 0$, we easily find the relationship

$$\Gamma(n+1) = n \int_0^\infty x^{n-1} e^{-x} dx = n\Gamma(n) \tag{7.55}$$

The gamma function for half-integral arguments is found as follows:

$$\Gamma\left(\frac{1}{2}\right) = \int_0^\infty x^{-\frac{1}{2}} e^{-x} dx = 2 \int_0^\infty e^{-y^2} dy = \sqrt{\pi} \tag{7.56}$$

where we set $x = y^2$ and $dx = 2y\, dy$, and we used the mathematical tables for the last integral. Based on (7.55) and (7.56), we obtain

$$\Gamma\left(\frac{3}{2}\right) = \frac{1}{2}\Gamma\left(\frac{1}{2}\right) = \frac{\sqrt{\pi}}{2}, \ \Gamma\left(\frac{5}{2}\right) = \frac{3}{2}\Gamma\left(\frac{3}{2}\right) = \frac{3\sqrt{\pi}}{2^2}, \ \Gamma\left(\frac{n+1}{2}\right) = \frac{1 \cdot 3 \cdot 5 \cdots (n-1)}{2^{(n/2)}} \sqrt{\pi} \tag{7.57}$$

For $m > -1$, the LT of t^m is

$$\mathcal{L}\{t^m\} = \int_0^\infty t^m e^{-st} dt = \int_{x/s=0}^{x/s=\infty} \frac{x^m e^{-x}}{s^{m+1}} dx = \frac{1}{s^{m+1}} \int_0^\infty x^m e^{-x} dx = \frac{\Gamma(m+1)}{s^{m+1}} \tag{7.58}$$

where we set $x = st$.

Error function
The **error** and **co-error** functions are defined as follows:

$$\text{erf}(x) = \frac{2}{\sqrt{\pi}} \int_0^x e^{-x^2} dx, \quad \text{erfc}(x) = \frac{2}{\sqrt{\pi}} \int_x^\infty e^{-x^2} dx = 1 - \text{erf}(x) \tag{7.59}$$

By (7.58) and (7.57), we obtain

$$\mathcal{L}^{-1}\left\{\frac{1}{\sqrt{\pi}}\right\} = \frac{t^{-1/2}}{\Gamma(1/2)} = \frac{1}{\sqrt{\pi t}} \tag{7.60}$$

But, from the LT tables we observe that $\mathcal{L}^{-1}\{1/(s-1)\} = e^t$ and, therefore, using the convolution property of the LT, we find that

$$\mathcal{L}^{-1}\left\{\frac{1}{\sqrt{s}(s-1)}\right\} = \int_0^t f(x)h(t-x)dx = \int_0^t \frac{1}{\sqrt{\pi x}} e^{(t-x)} dx = \frac{e^t}{\sqrt{\pi}} \int_0^t \frac{e^x}{\sqrt{x}} dx$$

$$= \frac{e^t}{\sqrt{\pi}} \int_{y=0}^{y=\sqrt{t}} 2e^{-y^2} dy = e^t \text{erf}\left(\sqrt{t}\right) \tag{7.61}$$

where we set $x = y^2$ and $dx = 2ydy$.

Since the LT of $e^{-t}f(t)$ is $F(s+1)$, then (7.61) becomes

$$\mathcal{L}^{-1}\left\{\frac{1}{s\sqrt{s+1}}\right\} = \text{erf}\left(\sqrt{t}\right) \tag{7.62}$$

By somewhat similar procedure, transformations, and substitutions, we find the relation

$$\mathcal{L}^{-1}\left\{\frac{1}{s} e^{-k\sqrt{s}}\right\} = \text{erfc}\left(\frac{k}{2\sqrt{t}}\right) \quad k > 0 \tag{7.63}$$

Transmission line
If R and L are the resistance (in ohms) and the inductance (in Henry) per unit length of an electrical transmission line, and C and G are the capacitance (in farads) and the conductance (in mhos) across a transmission line per unit length, then the equations which characterize the voltage and current transmitted along the line are

$$L\frac{\partial i(x,t)}{\partial t} + Ri(x,t) = -\frac{\partial v(x,t)}{\partial x}$$
$$C\frac{\partial v(x,t)}{\partial t} + Gv(x,t) = -\frac{\partial i(x,t)}{\partial x} \tag{7.64}$$

These are **simultaneous partial differential equations**, and the independent variables are x and t.

Taking the LT with respect to time t of the above equations and applying the LT property of a derivative, we obtain

$$(Ls + R)I(x,s) = -\frac{\partial V(x,s)}{\partial x} + Li(x,0)$$

$$(Cs + G)V(x,s) = -\frac{\partial I(x,s)}{\partial x} + Cv(x,0)$$

(7.65)

where $I(x,s) = \int_0^\infty i(x,t)e^{-st}dt$, $V(x,s) = \int_0^\infty v(x,t)e^{-st}dt$.

Next, multiplying the first equation of (7.65) by $(Cs + G)$ and differentiating the second equation of (7.65) with respect to x, and adding the results, we obtain the ordinary differential equation

$$\frac{\partial^2 I(x,s)}{\partial x^2} - [Ls + R][Cs + G]I(x,s) = C\frac{\partial v(x,0)}{\partial x} - L[Cs + G]i(x,0)$$

(7.66)

Similarly, we find

$$\frac{\partial^2 V(x,s)}{\partial x^2} - [Ls + R][Cs + G]V(x,s) = L\frac{\partial i(x,0)}{\partial x} - C[Ls + R]v(x,0)$$

(7.67)

Example 7.33 (Semi-infinite transmission line)

It is desired to find the propagation of a signal in an infinitely long line with the initial voltage $v(0, t) = f(t)$ across the input end of the line. We assume that no initial voltage or current is present in the line ($i(x, 0) = 0$ and $v(x, 0) = 0$).

SOLUTION

Introducing the initial conditions in (7.66) and (7.67), we obtain

$$\frac{\partial^2 I(x,s)}{\partial x^2} - [Ls + R][Cs + G]I(x,s) = 0$$

$$\frac{\partial^2 V(x,s)}{\partial x^2} - [Ls + R][Cs + G]V(x,s) = 0$$

(7.68)

The solutions (homogeneous solutions) to the above equations are

$$I(x,s) = A_1 e^{-ax} + B_1 e^{ax} \quad V(x,s) = A_2 e^{-ax} + B_2 e^{ax} \quad a = \sqrt{(Ls + R)(Cs + G)}$$

(7.69)

The A's and B's are functions of s and are to be determined from the conditions imposed by the problem at hand. For example, let us determine A_2 and B_2 from the conditions of the problem.

Because we must have a finite-energy signal, it implies that the voltage at $x = \infty$ must be equal to zero. That forces us to set

$$B_2 = 0 \tag{7.70}$$

Next, with $x = 0$ and the help of (7.69) and (7.70), it follows that

$$V(0, s) = A_2 \tag{7.71}$$

Using the above results and remembering the condition $v(0, t) = f(t)$, we find

$$V(0, s) = \mathcal{L}\{v(0, t)\} = \mathcal{L}\{f(t)\} = F(s) \tag{7.72}$$

Therefore, the second equation of (7.69) becomes

$$V(x, s) = F(s)e^{-ax} = F(s)e^{-\sqrt{(Ls+R)(Cs+G)}\,x} \tag{7.73}$$

Special case I (No attenuation and no distortion case)
If we assume no dissipation in the transmission line, that is,

$$R = G = 0 \tag{7.74}$$

then (7.73) becomes

$$V(x, s) = F(s)e^{-\sqrt{LC}\,x} \tag{7.75}$$

Taking into consideration the time shift of the LT, we obtain

$$v(x, t) = \mathcal{L}^{-1}\{V(x, s)\} = \mathcal{L}^{-1}\{F(s)e^{-\sqrt{LC}\,x}\} = f(t - \sqrt{LC}\,x) \tag{7.76}$$

Thus, the voltage at the point $x = x$ is exactly the same as the voltage at $x = 0$ except that there is a time lag of the amount $\sqrt{LC}\,x$. In other words, signals are propagated along the line with the velocity $1/\sqrt{LC}$ without attenuation or distortion.

Special case II (Distortionless case)
Another case in which the shape of the signal is not changing, but a decrease in magnitude is found in a transmission line with the following characteristic:

$$\frac{R}{L} = \frac{G}{C} \tag{7.77}$$

For this case, (7.73) becomes

$$V(x, s) = F(s)e^{-x\sqrt{\frac{C}{L}(Ls+R)}} = F(s)e^{-x\sqrt{\frac{C}{L}R}}e^{-x\sqrt{LC}\,s} \tag{7.78}$$

The ILT of the above equation gives

$$v(x,t) = \mathcal{L}^{-1}\{V(x,s)\} = \mathcal{L}^{-1}\left\{F(s)e^{-x\sqrt{\frac{C}{L}}R}e^{-x\sqrt{LC}s}\right\} = e^{-x\sqrt{\frac{C}{L}}R}\mathcal{L}^{-1}\{F(s)e^{-x\sqrt{LC}s}\}$$

$$= e^{-x\sqrt{\frac{C}{L}}R}f(t - x\sqrt{LC}) \tag{7.79}$$

Therefore, in this case signals propagate with a velocity $1/\sqrt{LC}$, do not change their waveforms but are attenuated by the factor $\exp(-x\sqrt{C/L}\,R)$. ∎

Example 7.34 (Semi-infinite transmission line $L = G = 0$ or heat conduction)

Introducing the values $L = G = 0$ in (7.64), we obtain the following two equations:

$$Ri(x,t) = -\frac{\partial v(x,t)}{\partial x} \tag{a}$$

$$C\frac{\partial v(x,t)}{\partial t} = -\frac{\partial i(x,t)}{\partial x} \tag{b}$$

Taking the derivative with respect to x of equation (a) above, and taking the derivative with respect to t of equation (b) above and combining the results, we obtain

$$\frac{\partial^2 v(x,t)}{\partial x^2} = RC\frac{\partial i(x,t)}{\partial t}$$

$$\left(\frac{\partial^2 u(x,t)}{\partial x^2} = \frac{1}{a^2}\frac{\partial u(x,t)}{\partial t} \equiv \text{heat equation, } a^2 = K/\sigma\delta = \left\{\begin{array}{l} K = \text{thermal conductivity} \\ \sigma = \text{specific heat} \\ \delta = \text{density} \end{array}\right.\right) \tag{7.80}$$

Consider the following boundary and initial condition problems:

$$\frac{\partial^2 v(x,t)}{\partial x^2} = RC\frac{\partial i(x,t)}{\partial t} \quad 0 < x < \infty, t > 0$$

B.C.: $\quad v(0,t) = f(t) \quad v(x,t) \to 0 \text{ as } x \to \infty \tag{7.81}$

I.C.: $\quad v(x,0) = 0 \quad 0 < x < \infty$

By applying the LT to the PDE and boundary conditions, we obtain

$$\frac{\partial^2 V(x,s)}{\partial x^2} - (sRC)V(x,s) = 0 \quad 0 < x < \infty$$

B.C.: $\quad V(0,s) = F(s) \quad V(x,s) \to 0 \text{ as } x \to \infty \tag{7.82}$

where $V(x,s) = \mathcal{L}\{v(x,t); t \to s\}$ and $F(s) = \mathcal{L}\{f(t)\}$. The general solution of this second-order DE is

$$V(x, s) = A(s)e^{x\sqrt{RCs}} + B(s)e^{-x\sqrt{RCs}} \tag{7.83}$$

where $A(s)$ and $B(s)$ are arbitrary functions of s. However, to satisfy the condition $V(x, s) \to 0$ as $x \to \infty$, we must choose $A(s) = 0$. The remaining boundary conditions demand that $B(s) = F(s)$, and thus

$$V(x, s) = F(s)e^{-x\sqrt{sRC}} \tag{7.84}$$

From the extended table at the end of the chapter, we have

$$\mathcal{L}^{-1}\{e^{-x\sqrt{RC}\sqrt{s}}; s \to t\} = \frac{x\sqrt{RC}}{2\sqrt{\pi}\, t^{3/2}} e^{-x^2 RC/4t} \tag{7.85}$$

Therefore, using the convolution property, we find

$$v(x, t) = \frac{x\sqrt{RC}}{2\sqrt{\pi}} \int_0^t \frac{f(\tau)}{(t-\tau)^{3/2}} \exp\left[-\frac{x^2 RC}{4(t-\tau)}\right] d\tau \tag{7.86}$$

An alternate form of the solution of (7.86) can be derived by making a change in variable $z = x\sqrt{RC}/2\sqrt{t - \tau}$, which leads to

$$v(x, t) = \frac{2}{\sqrt{\pi}} \int_{x\sqrt{RC}/2\sqrt{t}}^{\infty} f(t - x^2 RC/4z^2)e^{-z^2} dz \tag{7.87}$$

Constant input voltage
When the voltage at the input to the line is constant $v(0, t) = V_0$, we obtain from (7.63)

$$v(x, t) = \frac{2}{\sqrt{\pi}} V_0 \int_{x\sqrt{RC}/2\sqrt{t}}^{\infty} e^{-z^2} dz = V_0 \mathrm{erfc}(x\sqrt{RC}/2\sqrt{t}) \tag{7.88}$$

where $\mathrm{erfc}(x)$ is the complementary error function.   ∎

Example 7.35 (Heat diffusion)

Let us have a finite rod of length a, and a constant source of heat is applied at its end. Therefore, the problem is stated as follows:

$$\frac{\partial u(x, t)}{\partial t} = \kappa \frac{\partial^2 u(x, t)}{\partial x^2} \quad 0 < x < a, t > 0$$

I.C.: $u(x, 0) = 0 \quad 0 < x < a$

B.C.: $u(x, t) = U_0 \quad x = a, t > 0$ $\tag{7.89}$

B.C.: $\dfrac{\partial u(x, t)}{\partial x} = 0 \quad x = 0, t > 0$

SOLUTION

An application of the LT of $u(x, t)$ with respect to t gives

$$\frac{\partial^2 U(x, s)}{\partial x^2} - \frac{s}{\kappa} U(x, s) = 0 \qquad 0 < x < a$$

$$\text{B.C.: } U(a, s) = \frac{U_0}{s} \qquad \left(\frac{\partial U(x, s)}{\partial x}\right)_{x=0} = 0 \tag{7.90}$$

The solution of this system is

$$U(x, s) = \frac{U_0 \cosh\left(x\sqrt{\frac{s}{\kappa}}\right)}{s \cosh\left(a\sqrt{\frac{s}{\kappa}}\right)} \tag{7.91}$$

Substituting $\alpha = \sqrt{s/\kappa}$ in the above equation, we obtain

$$\mathcal{L}^{-1}\{U(x, s)\} = u(x, t) = \mathcal{L}^{-1}\left\{\frac{U_0}{s}\left[e^{-\beta(a-x)} + e^{-\beta(a+x)}\right]\sum_{n=0}^{\infty}(-1)^n e^{-2n\beta a}\right\}$$

$$= \mathcal{L}^{-1}\left\{\frac{U_0}{s}\left[\sum_{n=0}^{\infty}(-1)^n e^{-\beta[(2n+1)a-x]} + \sum_{n=0}^{\infty}(-1)^n e^{-\beta[(2n+1)a+x]}\right]\right\}$$

$$= U_0\left\{\sum_{n=0}^{\infty}(-1)^2\left[\text{erfc}\left(\frac{(2n+1)a-x}{2\sqrt{\kappa t}}\right) + \text{erfc}\left(\frac{(2n+1)a+x}{2\sqrt{\kappa t}}\right)\right]\right\} \tag{7.92}$$

where we used (7.63) (see also Table 7.5). ∎

Example 7.36 (General form of transmission line)

If we eliminate $i(x, t)$ or $v(x, t)$ from equations (7.65), both $i(x, t)$ and $v(x, t)$ satisfy the same equation of the form

$$\frac{1}{c^2}\frac{\partial^2 u(x, t)}{\partial t^2} - \frac{\partial^2 u(x, t)}{\partial x^2} + a\frac{\partial u(x, t)}{\partial t} + bu(x, t) = 0 \tag{7.93}$$

$$c^2 = \frac{1}{LC} \qquad a = LG + RC \qquad b = RG$$

The above equation can also be written in the form

$$\frac{\partial^2 u(x, t)}{\partial t^2} = c^2\frac{\partial^2 u(x, t)}{\partial x^2} - (p+q)\frac{\partial u(x, t)}{\partial t} - pqu(x, t) \tag{7.94}$$

$$ac^2 = \frac{R}{C} + \frac{G}{C} = p+q \qquad a = LG + RC \qquad b = RG \qquad bc^2 = pq$$

Case I: $R = G = 0$

If $R = G = 0$, then (7.93) becomes the **wave equation**

$$\frac{\partial^2 u(x, t)}{\partial t^2} = c^2 \frac{\partial^2 u(x, t)}{\partial x^2} \tag{7.95}$$

B.C.:

$$
\begin{aligned}
u(x, t) &= Af(t) \quad \text{at } x = 0, t \geq 0 \\
u(x, t) &\to 0 \quad \text{as } x \to \infty, t \geq 0
\end{aligned} \tag{7.96}
$$

I.C.:

$$u(x, t) = \frac{\partial u(x, t)}{\partial t} = 0 \quad \text{at } t = 0 \quad \text{for } 0 \leq x < \infty \tag{7.97}$$

An application of LT of $u(x, t)$ with respect to t gives

$$
\begin{aligned}
\frac{d^2 U(x, s)}{dx^2} - \frac{s^2}{c^2} U(x, s) &= 0 \quad 0 \leq x < \infty \\
U(x, s) &= AF(s) \quad x = 0 \\
U(x, s) &\to 0 \qquad x \to \infty
\end{aligned} \tag{7.98}
$$

The solution of the above ordinary differential equation is

$$U(x, s) = AF(s)e^{-\frac{x}{c}s} \tag{7.99}$$

Remembering the shift property, the inverse gives the solution

$$u(x, t) = \begin{cases} Af\left(t - \dfrac{x}{c}\right) & t > \dfrac{x}{c} \\[2mm] 0 & t > \dfrac{x}{c} \end{cases} \tag{7.100}$$

Case II: $L = G = 0$ (an *RC* line)

This type of transmission line reduces to classical diffusion equation

$$\frac{\partial u(x, t)}{\partial t} = k \frac{\partial^2 u(x, t)}{x^2} \quad k = \frac{1}{RC} \tag{7.101}$$

Assuming the following boundary and initial conditions for the voltage in the line

B.C.:

$$
\begin{aligned}
u(0, t) &= V_0 \quad t > 0 \\
u(x, t) &\to 0 \quad x \to \infty, \quad t > 0
\end{aligned} \tag{7.102}
$$

I.C.:

$$u(x, 0) = 0 \quad x > 0 \tag{7.103}$$

and taking the LT with respect to time of (7.101), we obtain

$$U(x, s) = Ae^{-x\sqrt{\frac{s}{k}}} + Be^{x\sqrt{\frac{s}{k}}} \tag{7.104}$$

A and B are integrating constants. To obtain a bounded solution, we must set $B = 0$. Using $U(0,s) = V_0$ (see (7.102)), we obtain the solution

$$U(x, s) = V_0 e^{-x\sqrt{\frac{s}{k}}} \tag{7.105}$$

The inverse LT of the above equation gives us the voltage distribution in space and time in the transmission line, and it is given by (see Table 7.2)

$$u(x, t) = V_0 \mathrm{erfc}\left(\frac{x}{2\sqrt{kt}}\right) \tag{7.106}$$

TABLE 7.2   Some LT of Functions Corresponding to Error and Co-Error Functions

$$\mathscr{L}^{-1}\left\{\frac{1}{s\sqrt{s+1}}\right\} = \int_0^t 1 \cdot \frac{e^{-x}}{\sqrt{\pi x}}\, dx$$

$$\mathrm{erf}(x) = \frac{2}{\sqrt{\pi}} \sum_{n=0}^{\infty} \frac{(-1)^n x^{2n+1}}{(2n+1)n!}$$

$$\mathscr{L}\left\{\mathrm{erfc}\left(\frac{k}{\sqrt{t}}\right)\right\} = \frac{1}{s} e^{-2k\sqrt{s}} \quad k > 0 \quad s > 0$$

$$\mathscr{L}^{-1}\left\{\frac{\sinh(x\sqrt{s})}{s \sinh\sqrt{s}}\right\} = \sum_{n=0}^{\infty}\left[\mathrm{erfc}\left(\frac{1-x+2n}{2\sqrt{t}}\right) - \mathrm{erfc}\left(\frac{1+x+2n}{2\sqrt{t}}\right)\right]$$

$$\mathscr{L}\{t^{-1/2}\mathrm{erf}(\sqrt{t})\} = \frac{2}{\sqrt{\pi s}}\mathrm{Arctan}\frac{1}{\sqrt{s}} \quad s > 0$$

$$\mathscr{L}^{-1}\left\{\frac{1}{1+\sqrt{1+s}}\right\} = \frac{e^{-t}}{\sqrt{\pi t}} - \mathrm{erfc}(\sqrt{t})$$

$$\mathscr{L}^{-1}\left\{\frac{1}{(s-1)\sqrt{s}}\right\} = e^t \mathrm{erf}(\sqrt{t})$$

$$\mathscr{L}^{-1}\left\{\frac{1}{\sqrt{s}(\sqrt{s}+1)}\right\} = e^t \mathrm{erfc}(\sqrt{t})$$

$$\mathscr{L}^{-1}\left\{\frac{1}{\sqrt{s}+1}\right\} = \frac{1}{\sqrt{\pi t}} - e^t \mathrm{erfc}(\sqrt{t})$$

The current in the transmission line is given by

$$i(x,t) = -\frac{1}{R}\left(\frac{\partial u(x,t)}{x}\right) = \frac{V_0}{R}\frac{1}{\sqrt{\pi kt}}e^{-\frac{x^2}{4kt}} \tag{7.107}$$

■

Example 7.37 (First-order PDE)

Let us solve the following first-order PDE:

$$\frac{\partial u(x,t)}{\partial t} + x\frac{\partial u(x,t)}{\partial x} = x \quad x > 0, t > 0 \tag{7.108}$$

with the boundary and initial conditions

$$\begin{aligned} u(x,0) &= 0 \quad \text{for } x > 0 \\ u(0,t) &= 0 \quad \text{for } t > 0 \end{aligned} \tag{7.109}$$

We take the LT with respect to t to obtain

$$sU(x,s) + x\frac{dU(x,s)}{dx} = \frac{x}{s} \quad U(0,s) = 0 \tag{7.110}$$

Using the integrating factor x^s, we write the above equation in the form

$$\frac{d}{dx}(U(x,s)x^s) = \frac{x^s}{s} \tag{7.111}$$

Integrating both sides, we obtain

$$U(x,s) = \frac{x}{s(s+1)} + Ax^{-s} = x\left(\frac{1}{s} - \frac{1}{s+1}\right) + Ax^{-s} \tag{7.112}$$

Since $U(0,s) = 0$, we must set the constant of integration A equal to zero to have bounded solution. The ILT of the remaining equation gives the desired solution

$$u(x,t) = x(1 - e^{-t}) \tag{7.113}$$

■

Example 7.38 (First-order PDE)

Let us find the solution of the equation

$$x\frac{\partial u(x,t)}{\partial t} + \frac{\partial u(x,t)}{\partial x} = x \quad x > 0, t > 0 \tag{7.114}$$

having the same boundary and initial conditions as the equation in the previous example. Applying the LT with respect to *t*, we obtain

$$\frac{dU(x,s)}{dx} + xsU(x,s) = \frac{x}{s} \tag{7.115}$$

Using the integrating factor $x^2 s/2$ gives the solution

$$U(x,s) = \frac{1}{s^2} + A\exp\left(-\frac{1}{2}x^2 s\right)$$

where Λ is an integrating constant. Since $U(0,s) = 0$, $A = -1/s^2$ and, hence, the solution is

$$U(x,s) = \frac{1}{s^2}\left(1 - e^{-\frac{1}{2}x^2 s}\right) \tag{7.116}$$

The *s* in the exponent indicates time shift, and since the ILT of $1/s^2$ is *t*, the solution becomes

$$u(x,t) = t - \left(t - \frac{1}{2}x^2\right)H\left(t - \frac{x^2}{2}\right) \quad \text{or}$$

$$u(x,t) = \begin{cases} t & 2t < x^2 \\ \frac{1}{2}x^2 & 2t > x^2 \end{cases} \tag{7.117}$$

In this example, the function $H(\cdot)$ is the unit step function. ∎

7.6 Frequency Response of LTI Systems

The frequency response of systems is defined when the initial conditions are zero; hence, we address only the zero-state response. We know that the output of an analog system is given by the convolution of the input and the impulse response of the system:

$$\underbrace{g(t)}_{\text{output}} = \underbrace{f(t)}_{\text{input}} * \underbrace{h(t)}_{\substack{\text{impulse}\\\text{response}}} \tag{7.118}$$

The LT of both sides of (7.118) gives (see Section 7.2)

$$G(s) = F(s)H(s) \tag{7.119}$$

The time and transformed representations of a system are shown diagrammatically in Figure 7.16. Because $h(t)$ is the inverse LT of $H(s)$, then

$$h(t) = \mathcal{L}^{-1}\{H(s)\} = \mathcal{L}^{-1}\left\{\frac{G(s)}{F(s)}\right\} \tag{7.120}$$

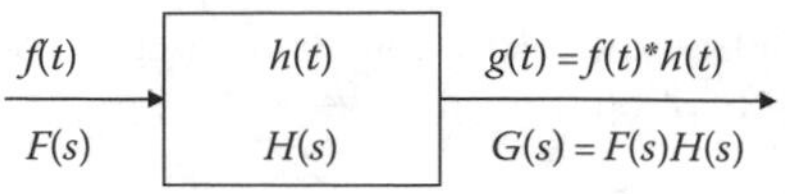

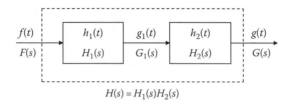

FIGURE 7.16

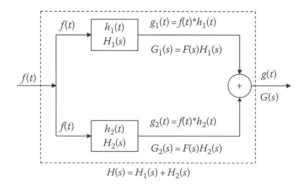

$$H(s) = H_1(s)H_2(s)$$

FIGURE 7.17

When two systems are connected in cascade, as shown in Figure 7.17, we obtain the following equations:

$$g_1(t) = f(t) * h_1(t) \quad G_1(s) = F(s)H_1(s); \quad g(t) = g_1(t) * h_2(t) \quad G(s) = G_1(s)H_2(s)$$

Eliminating $G_1(s)$ from the transformed equations above, we find

$$G(s) = H_1(s)H_2(s)F(s)$$

which shows that the combined transfer function is

$$\boxed{H(s) = H_1(s)H_2(s)} \tag{7.121}$$

Therefore, if n systems are connected in series, their total transfer function is

$$H(s) = H_1(s)H_2(s)H_3(s)\cdots H_n(s) \tag{7.122}$$

If two systems are connected in parallel, as shown in Figure 7.18, we have

$$g(t) = g_1(t) + g_2(t) = f(t) * h_1(t) + f(t) * h_2(t)$$

FIGURE 7.18

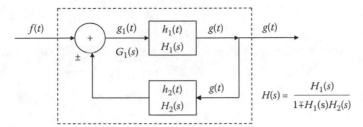

FIGURE 7.19

from which

$$G(s) = F(s)H_1(s) + F(s)H_2(s) = F(s)[H_1(s) + H_2(s)] \qquad (7.123)$$

This equation shows that the transfer function of two systems connected in parallel is given by

$$\boxed{H(s) = H_1(s) + H_2(s)} \qquad (7.124)$$

For systems with feedback connection, as shown in Figure 7.19, we write

$$g_1(t) = f(t) \pm g(t) * h_2(t) \quad \text{and} \quad g(t) = g_1(t) * h_1(t) = f(t) * h_1(t) \pm g(t) * h_2(t) * h_1(t)$$

The LT of the second equation above gives

$$G(s) = F(s)H_1(s) \pm G(s)H_1(s)H_2(s)$$

from which

$$\boxed{G(s) = F(s)\frac{H_1(s)}{1 \mp H_1(s)H_2(s)}} \qquad (7.125)$$

and

$$\boxed{H(s) = \frac{G(s)}{F(s)} = \frac{H_1(s)}{1 \mp H_1(s)H_2(s)}} \qquad (7.126)$$

These results are consistent with those discussed for block-diagram configurations (see Chapter 1).

Example 7.39

Determine the transfer function, $I_2(s)/V(s)$, for the system shown in Figure 7.20a. Also, find the magnitude and the phase response functions.

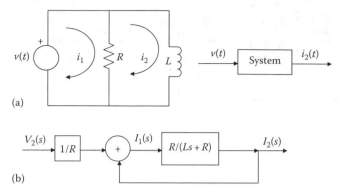

(a)

(b)

FIGURE 7.20

Solution

The differential equations describing the system are

$$Ri_1(t) - Ri_2(t) = v(t) \quad L\frac{di_2(t)}{dt} + Ri_2(t) - Ri_1(t) = 0$$

from which

$$I_1(s) = \frac{1}{R}V(s) + I_2(s) \quad I_2(s) = I_1(s)\frac{R}{Ls + R}$$

These two equations are shown in block-diagram representation in Figure 7.20b. By block-diagram simplification or by eliminating $I_1(s)$ from these equations, we obtain

$$H(s) = \frac{I_2(s)}{V(s)} = \frac{1}{Ls} \quad \text{or} \quad H(j\omega) = \frac{1}{j\omega L}$$

As a consequence,

$$|H(j\omega)| = \frac{1}{\omega L} \quad \phi(\omega) = -\frac{\pi}{2} = \text{constant} \qquad \blacksquare$$

Example 7.40

Deduce an expression for the current (voltage input–current output) in the circuit shown in Figure 7.21, if a cosine is applied with $\omega = 5$.

Solution

Taking the LT of the KVL equation that describes the system, we find the transfer function to be equal to

$$H(s) = \frac{1}{s + 2}$$

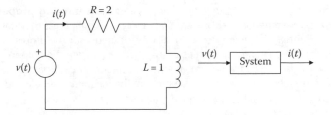

FIGURE 7.21

The output of the system, if the input is a complex exponential function, is

$$i(t) = \text{Re}\{e^{j\omega t} * h(t)\} = \text{Re}\left\{ \int_{-\infty}^{\infty} h(x)e^{j\omega(t-x)}dx \right\} = \text{Re}\left\{ e^{j\omega t} \int_{-\infty}^{\infty} h(x)e^{-j\omega x}dx \right\}$$

$$= \text{Re}\{e^{j\omega t} H(j\omega)\}$$

Note: *For linear operations (integrations, differentiations, etc.), we can exchange those operations with the Re{} or Im{} operations.*

Based on our results above, we write the output as follows:

$$i(t) = \text{Re}\left\{ \frac{1}{j5+2} e^{j5t} \right\} = \text{Re}\left\{ \frac{1}{\sqrt{5^2 + 2^2} e^{j \tan^{-1}(5/2)}} e^{j5t} \right\} = \frac{1}{\sqrt{29}} \cos [5t - \tan^{-1}(5/2)]$$

When there is a sum of exponential functions inputs

$$f(t) = a_1 e^{j\omega_1 t} + a_2 e^{j\omega_2 t} + \cdots + e^{j\omega_n t} a_n \qquad (7.127)$$

the output function becomes

$$g(t) = a_1 H(j\omega_1)e^{j\omega_1 t} + a_2 H(j\omega_2)e^{j\omega_2 t} + \cdots + a_n H(j\omega_n)e^{j\omega_n t} \qquad (7.128)$$

■

Example 7.41

Repeat Example 7.40 for an input voltage $v(t) = 2 + 5\cos 3t + 6\sin 6t$.

SOLUTION

For the given circuit $H(j\omega) = \dfrac{1}{j\omega + 2}$ and, thus, we write

$$i(t) = \text{Re}\left\{ 2\frac{1}{j0+2}e^{j0t} \right\} + \text{Re}\left\{ 5\frac{1}{j3+2}e^{j3t} \right\} + \text{Im}\left\{ 6\frac{1}{j6+2}e^{j6t} \right\}$$

$$= 1 + \frac{5}{\sqrt{13}} \cos\left(3t - \tan^{-1}\frac{3}{2} \right) + \frac{6}{\sqrt{40}} \cos\left(6t - \tan^{-1}\frac{6}{2} \right)$$

A basic and very important property of any system is its **filtering properties** that define how different frequencies are attenuated and phase shifted as they pass through the system. Further, the energy of a signal is associated with the amplitude of each harmonic. Thus, the filtering properties of a system will dictate the amount of energy that is to be transferred by the system and the percentage for each particular frequency component. Therefore, we generally prefer to plot the magnitude $|H(j\omega)|$ and the phase $\phi(\omega) = \tan^{-1}[H_i(\omega)/H_r(\omega)]$ versus frequency ω, where $H_i(\omega)$ and $H_r(\omega)$ are the imaginary and real components of $H(j\omega)$. ∎

Example 7.42

Deduce and plot the frequency characteristics (filtering) of the circuit shown in Figure 7.22a.

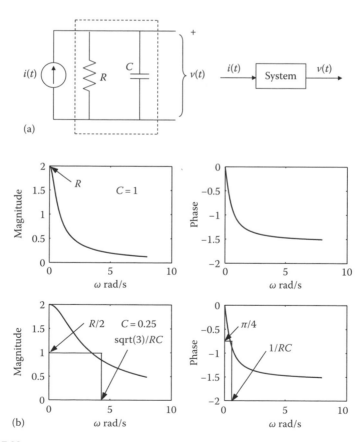

(a)

(b)

FIGURE 7.22

SOLUTION

Applying the KCL and taking the LT of the differential equation, we obtain

$$C\frac{dv(t)}{dt} + \frac{v(t)}{R} = i(t) \quad H(s) = \frac{V(s)}{I(s)} = \frac{1}{C}\frac{1}{s + (1/RC)}$$

The magnitude and phase spectra are obtained by setting $s = j\omega$. These are

$$|H(j\omega)| = \frac{1}{C}\frac{1}{\sqrt{\omega^2 + (1/RC)^2}} \quad \phi(\omega) = -\tan^{-1}(\omega RC)$$

The magnitude and phase characteristics are graphed in Figure 7.22b. ∎

Example 7.43

Deduce and plot the frequency characteristics of the system shown in Figure 7.23a.

SOLUTION

From an inspection of the circuit, we write the following two KCL equations:

$$C\frac{dv_c(t)}{dt} + \frac{1}{L}\int [v_c(t) - v_o(t)]dt = i(t)$$

$$\frac{1}{L}\int [v_o(t) - v_c(t)]dt + \frac{v_o(t)}{R} = 0$$

The LT of these equations are

$$CsV_c(s) + \frac{1}{Ls}[V_c(s) - V_o(s)] = I(s); \quad \frac{1}{Ls}[V_o(s) - V_c(s)] + \frac{V_o(s)}{R} = 0$$

Eliminate $V_c(s)$ from these last equations and solve for $H(s) \triangleq V_o(s)/I(s)$. The result is

$$H(s) \triangleq \frac{V_o(s)}{I(s)} = \frac{R/LC}{s^2 + (R/L)s + 1/LC}$$

Therefore,

$$|H(j\omega)| = \frac{1/LC}{\left[(-\omega^2 + 1/LC)^2 + \left(\frac{R}{L}\right)^2 \omega^2\right]^{1/2}}; \quad \phi(\omega) = -\tan^{-1}\left(\frac{(R/L)\omega}{1/LC - \omega^2}\right)$$

The above characteristics are plotted in Figure 7.23b for $L=R=1$ and $C=0.2$. These plots can be done by two different approaches.

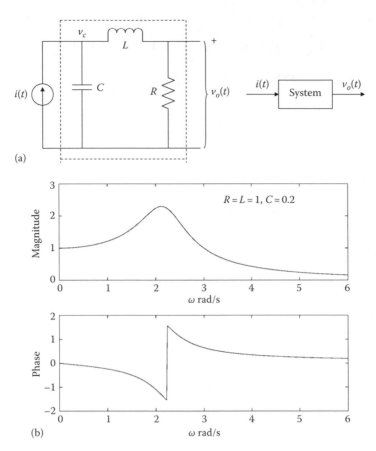

FIGURE 7.23

(a) **Book MATLAB calculations**

```
>>w = 0:0.02:6;
>>r = 1; l = 1; c = 0.2;
>>H = (r/(l*c))./sqrt((−w.^2+1/(l*c))^2+(r/l)^2*w.^2);
>>phw = −atan((r/l)*w./(1/(l*c)−w.^2));
>>subplot(2,1,1);plot(w,H);subplot(2,1,2);plot(w,phw);
```

(b) **MATLAB functions**

```
>>w = 0:0.02:6;
>>r = 1; l = 1; c = 0.2;
>>num = [0 0 r/(l*c)]; den[1 r/l 1/(l*c)];%numerator and
                                    %denominator coefficients;
>>H = freqs(num,den,w);%the function freqs() is used for
                    %Laplace transform rational functions;
>>phw = angle(H);
>>subplot(2,1,1);plot(w,abs(H));
subplot(2,1,2);plot(w,phw);%note that we used the function
                                    %abs();   ∎
```

Example 7.44

A signal $2 + 3\sin 5t$ is the input to the system of Example 7.43. Find the output $v_o(t)$.

SOLUTION

The output is given by

$$v_o(t) = \text{Re}\{2H(j0)e^{j0t}\} + \text{Im}\{3H(j5)e^{j5t}\} = \text{Re}\{2|H(j0)|e^{j\phi(0)}e^{j0t}\} + \text{Im}\{3|H(j5)|e^{j\phi(5)}e^{j5t}\}$$

Therefore, we obtain

$$v_o(t) = 2\frac{R/LC}{[(1/LC)^2 + (R/L)^2 0]^{1/2}} + 3\frac{R/LC}{[(-25 + 1/LC)^2 25]^{1/2}} \sin\left(5t - \tan^{-1}\left[\frac{5R/L}{1/LC - 25}\right]\right) \quad \blacksquare$$

7.7 Pole Location and the Stability of LTI Systems

The physical idea of stability is closely related to a bounded system response to a sudden disturbance or input. If the system is displaced slightly from its equilibrium state, several different behaviors are possible. If the system remains near the equilibrium state, the system is said to be **stable**. If the system tends to return to the equilibrium state or tends to a bounded or limited state, it is said to be **asymptotically stable**. Here, it should be noted that the stability can be examined by studying a system either through its impulse response $h(t)$ or through its Laplace-transformed system $H(s)$.

Let us assume that we can expand the transfer function in terms of its roots as follows:

$$H(s) = \frac{A_1}{s - s_1} + \frac{A_2}{s - s_2} + \cdots + \frac{A_k}{s - s_k} + \cdots + \frac{A_n}{s - s_n} \quad (7.129)$$

The time response for an applied impulse due to kth pole will be of the form $A_k e^{s_k t}$. Thus, the nature of the response will depend on the location of the roots s_k in the s-plane. Because the controlling differential equation that describes the system has real coefficients, the roots are either real or, if complex, will occur in complex conjugate pairs. Three general cases exist that depend intimately on the **order** and **location** of the poles s_k in the s-plane. These are

1. The point representing s_k lies to the left of the imaginary axis in the s-plane.
2. The point representing s_k lies on the $j\omega$-axis.
3. The point representing s_k lies to the right of the imaginary axis in the s-plane.

We examine each of these alternatives for both simple-order and higher order poles.

Simple-order poles

Case I: The root is a real number $s_k = \sigma_k$, and it is located on the negative real axis of the s-plane. The response due to this root will be of the form

$$\text{response} = A_k e^{\sigma_k t} \quad \sigma_k < 0 \tag{7.130}$$

This expression indicates that after a lapse time, the response will be vanishing small.

For the case when a pair of complex conjugate roots exists, the response is given by

$$\text{response} = A_k e^{s_k t} + A_k^* e^{s_k^* t} \tag{7.131}$$

The response terms can be combined, noting that $A_k = a + jb$ and $s_k = \sigma_k + j\omega_k$:

$$\text{response} = (a + jb)e^{(\sigma_k + j\omega_k)t} + (a - jb)e^{(\sigma_k - j\omega_k)t}$$

or

$$\text{response} = 2\sqrt{a^2 + b^2}\, e^{\sigma_k t} \cos(\omega_k t + \phi_k), \quad \phi_k = \tan^{-1}(b/a), \quad \sigma_k < 0 \tag{7.132}$$

This response is a damped sinusoid, and it ultimately decays to zero.

Case II: The point representing s_k lies in the imaginary axis. This condition is a special case of Case I, but now $\sigma_k = 0$. The response for complex conjugate poles (see 7.132) is

$$\text{response} = 2\sqrt{a^2 + b^2} \cos(\omega_k t + \phi_k), \quad \phi_k = \tan^{-1}(b/a), \quad \sigma_k = 0 \tag{7.133}$$

Observe that there is no damping, and the response is thus a sustained oscillatory function. Such a system has a bounded response to a bounded input, and the system is defined as **stable** even though it is oscillatory.

Case III: The point representing s_k lies in the right half of the s-plane. The response function will be of the form

$$\text{response} = A_k e^{\sigma_k t} \quad \sigma_k > 0 \tag{7.134}$$

for real roots and will be of the form

$$\text{response} = 2\sqrt{a^2 + b^2}\, e^{\sigma_k t} \cos(\omega_k t + \phi_k), \quad \phi_k = \tan^{-1}(b/a), \quad \sigma_k > 0 \tag{7.135}$$

for complex conjugate roots. Because both functions increase with time without limit even for bounded inputs, the system for which these functions are roots is said to be **unstable**.

Multiple-order poles

We now examine the situation when multiple-order poles exist. The following cases are examined.

Case I: Multiple real poles exist in the left-half of the s-plane. As previously discussed, a second-order real pole (two repeated roots) gives rise to the response function

$$\text{response} = (A_{k1} + A_{k2}t)e^{\sigma_k t} \quad \sigma_k < 0 \tag{7.136}$$

Because σ_k is negative, and because the exponential decreases faster than the linearly increasing time, the response eventually becomes zero. The system with such poles is stable.

Case II: Multiple poles exist on the imaginary axis. The response function is made up of the responses due to each pair of poles and it is

$$\text{response} = (A_{k1} + A_{k2}t)e^{j\omega_k t} + (A_{k1}^* + A_{k2}^*t)e^{-j\omega_k t} \tag{7.137}$$

Following the procedure discussed above for the simple complex poles, this result can be written as

$$\text{response} = 2\sqrt{a^2 + b^2} \cos(\omega_k t + \phi_k) + 2\sqrt{c^2 + d^2}\, t \cos(\omega_k t + \theta_k)$$
$$\phi_k = \tan^{-1}(b/a) \quad \theta_k = \tan^{-1}(d/c) \tag{7.138}$$

Because the second term is oscillatory and increases linearly with time, the system is unstable.

Case III: Multiple roots exist in the right-hand of the s-plane. For a double real root, for example, the solution in this case will be

$$\text{response} = (A_{k1} + A_{k2}t)e^{\sigma_k t} \quad \sigma_k > 0 \tag{7.139}$$

For complex roots, the solution will be

$$\text{response} = e^{\sigma_k t}[2\sqrt{a^2 + b^2} \cos(\omega_k t + \phi_k) + 2\sqrt{c^2 + d^2}\, t \cos(\omega_k t + \theta_k)] \tag{7.140}$$

In both cases, owing to the exponential factor, the response increases with time and the system is unstable.

Note:

1. *A system with simple poles is unstable if one or more poles of its transfer function appears in the right half of the s-plane.*
2. *A system whose transfer function has simple poles is stable when all of the poles are in the left-half of the s-plane and on its boundary (imaginary axis).*

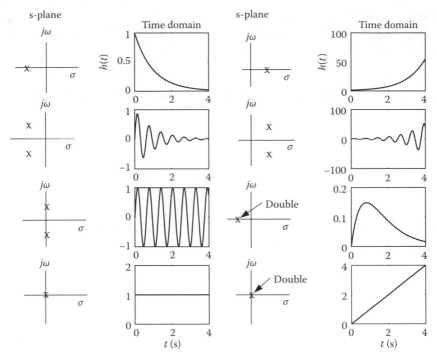

FIGURE 7.24

3. *A system with multiple poles is unstable if one or more of its poles appear on the imaginary axis or in the right half of the s-plane.*
4. *When all of the multiple poles of the system are confined to the left-hand s-plane, the system is stable.*

The impulse response of a system and the location of its poles are shown in Figure 7.24.

7.8 Feedback for Linear Systems

The feedback concept is one of the most important engineering discoveries of the last century. In general, the introduction of feedback is to adjust the output so that it coincides with the desired one. The following are some of the features that can be achieved by using feedback:

1. Stabilization of the system is possible
2. The system output sensitivity may be reduced due to the system parameter variations
3. Reduction of input disturbance to the system is possible
4. Feedback improves system transient response
5. Feedback improves the steady-state output

Cascade stabilization of systems

For cascaded systems, it is possible, by selecting the appropriate transfer functions, to cancel zeros or poles and thus produce stabilization. It has been found in Chapter 1 that the combined transfer function of two systems in cascade is given by

$H(s) = H_1(s)H_2(s)$. If the first system transfer function is of the form $H_1(s) = s/(s - 1.2)$, we observe that the pole is on the right side of the s-plane, which indicates a non-stable system. Hence if we add in series a second system with the transfer function of the form $H_2(s) = (s - 1.2)/s$, the combined transfer function will be $H(s) = 1$. Hence, we succeeded to create a stable system because the ROC is the whole s-plane. In practice, however, most of the times we cannot create a system with a pole having the exact value that is needed for stabilization. If the first system, for example, is of the form $H_1(s) = s/(s - 1.2001)$, then the total transfer function is

$$H(s) = H_1(s)H_2(s) = \frac{s}{s - 1.2001} \frac{s - 1.2}{s} = \frac{s - 1.2}{s - 1.2001}$$

and indicates an unstable combined system.

In communications, for example, a signal leaves a transmitter (antenna), propagates through space, and reaches the receiver (cell phone). During transmissions, the information-carrying signal is distorted by the transfer function of the space it traveled known as **channel**. One remedy would be to create a system in front of the receiver that would be the reciprocal of the space transfer function. This system is known as an **equalizer** and acts as a **compensator** that reduces the distortion or completely eliminates it. Let, for example, the channel transfer function be of the form

$$H_1(s) = \frac{s}{s + 0.5}$$

The equalizer must have a transfer function of the form

$$H_2(s) = \{H_1(s)\}^{-1} = \frac{s + 0.5}{s} = 1 + 0.5\frac{1}{s}$$

The impulse response of the above filter is $h_2(t) = \delta(t) + 0.5u(t)$. This filter is stable and can serve as an equalizer.

Most of the times, the functional form of the communication channel is not known. One way to circumvent this difficulty, especially in slow-changing channels, is to transmit a known signal and, then, design the appropriate inverse filter that eliminates the channel disturbance. The sequence of the known signals is called the **training sequence**.

Parallel composition

The transfer function of two systems in parallel is given by

$$H(s) = H_1(s) + H_2(s)$$

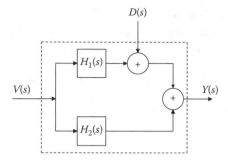

FIGURE 7.25

The transfer function of the system shown in Figure 7.25 is

$$Y(s) = D(s) + [H_1(s) + H_2(s)]V(s)$$

where $D(s)$ is the transformed desired signal and $V(s)$ is the transformed noise that must be eliminated. If we can create a filter such that $H_2(s) = -H_1(s)$, then the desired signal is detected. This type of process is known as **noise cancellation**. One of the important tasks of engineers is to build **noise cancellers** that suppress or remove unwanted signals (noise). One common case is when a person speaks through microphone, as a person in a convention center or a pilot in cockpit, and noise is also introduced (the noise of the attendees or the engine noise). Figure 7.26a shows a physical setup of the situation discussed, and Figure 7.26b shows the diagrammatic representation of the system. From the figure, we obtain

$$V(s)H_1(s) + D(s) + V(s)[H_2(s)H_3(s)] = Y(s)$$

or we can write

$$Y(s) = V(s)H_1(s) + D(s) + V(s)H_2(s)\left(-\frac{H_1(s)}{H_2(s)}\right) = D(s)$$

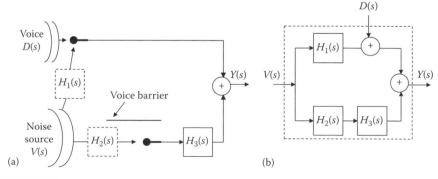

FIGURE 7.26

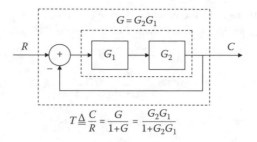

$$T \triangleq \frac{C}{R} = \frac{G}{1+G} = \frac{G_2 G_1}{1+G_2 G_1}$$

FIGURE 7.27

From the above results, we observe that if we introduce in the second parallel branch a system with the transfer function $-H_1(s)/H_2(s)$, we are able to eliminate the noise and receive the desired signal. A common use of such a scheme is in the telephone systems. At the receiving and sending ends, the two-wire transmission changes into a four-wire transmission by device known as the **hybrid**. The imperfection of the hybrid devices create echoes that we, sometimes, encounter during our conversation.

Feedback stabilization

Figure 7.27 shows a simple negative feedback system with unity return. Let the **open loop transfer function** be of the form

$$G(s) = G_1(s)G_2(s) = \frac{1}{(s - 0.6)(s + 1.2)}$$

This system is unstable due to the pole at $+0.6$. This pole creates in time domain an exponential function of the form $e^{0.6t}$, which becomes unbounded as $t \to \infty$. If the open system is modified to a negative feedback system, as shown in Figure 7.27, the total transfer function (**closed loop transfer function**) is

$$T = \frac{C}{R} = \frac{1}{s^2 + 0.6s + 0.28}$$

The poles of this new system are $-0.3000 + j0.4359$; $-0.3000 - j0.4359$. Because the real part of both poles is negative, it indicates a sinusoidal decaying function in the time domain. Observe also that the closed loop poles are different from those of the open loop.

Note: *An unstable system may become stable by introducing a negative feedback path.*

Sensitivity in feedback

The input–output relation for a unit feedback system (see Figure 7.27) is

$$C(s) = \frac{G(s)}{1 + G(s)} R(s)$$

If we assume that there is an incremental change of the open-loop transfer function by $\Delta G(s)$, there will be an incremental change of the output. Therefore, we will have the following relation:

$$C(s) + \Delta C(s) = \frac{G(s) + \Delta G(s)}{1 + G(s) + \Delta G(s)} R(s)$$

Proceeding to rearrange the above equation, we find that

$$\Delta C(s) = \frac{G(s) + \Delta G(s)}{1 + G(s) + \Delta G(s)} R(s) - C(s) = \frac{G(s) + \Delta G(s)}{1 + G(s) + \Delta G(s)} R(s) - \frac{G(s)}{1 + G(s)} R(s)$$

$$= \frac{\Delta G(s)}{[1 + G(s) + \Delta G(s)][1 + G(s)]} R(s)$$

Since, in general, $|\Delta G(s)| \ll |G(s)|$, the last expression can be approximated as follows:

$$\Delta C(s) \cong \frac{\Delta G(s)}{[1 + G(s)][1 + G(s)]} R(s) = \frac{G(s)\Delta G(s)}{G(s)[1 + G(s)][1 + G(s)]} R(s) = \frac{\Delta G(s)}{G(s)} \frac{1}{1 + G(s)} C(s)$$

We can now write the above equation in the form

$$\frac{\Delta C(s)}{C(s)} = \frac{1}{1 + G(s)} \frac{\Delta G(s)}{G(s)} = S(s) \frac{\Delta G(s)}{G(s)} \qquad (7.141)$$

where $S(s)$ is called the system **output sensitivity function**. Since $|S(s)| = |1/[1 + G(s)]| < 1$, it indicates that the variation of the open-loop transfer function due to the perturbation of its parameter will reduce the perturbation of the output by the factor of $|S(s)| < 1$.

Rejection of disturbance using feedback

Figure 7.28 shows a feedback configuration that completely eliminates the disturbance $V(s)$. Based on the design shown and referring to systems block-diagram transformations Table 1.3, we obtain

$$C(s) = V(s) + G(s)[R(s) - C(s)]$$

The above equation can be written in the form

$$C(s) = \frac{G(s)}{1 + G(s)} R(s) + \frac{1}{1 + G(s)} V(s) = \frac{G(s)}{1 + G(s)} R(s) + S(s)V(s) \qquad (7.142)$$

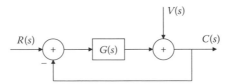

FIGURE 7.28

It is apparent that for $|S(s)V(s)| \cong 0$ in the frequency range of $V(s)$, the feedback configuration is able to diminish and eliminate the disturbance.

Step response
The voltage across the capacitor of an RLC series circuit to a step input is given by (7.43) and is given also here with the following substitutions: $E = 1, \omega_r^2 = K, 2\alpha = 6$.

$$V_{zs} = \frac{1}{s} \frac{K}{s^2 + 6s + K}$$

The second factor of the above equation is produced by a feedback system shown in Figure 7.29a. The voltage responses for $K = 5, 15$, and 60 are shown in Figure 7.29b. We observe that we can shorten the rise time with the drawback in creating overshoots. In this case, the **percent overshoot** is about 25%. The **rise time** is defined as the time required for the response to rise from 10% to 90% of the steady-state value. From Figure 7.29b, we can categorize the step responses in three main categories. For this case, we have (a) $K < 5$ is said to be an **over-damped** case, (b) $K = 15$ is said to be **critically damped,** and (c) $K > 15$ is said to be **under-damped**.

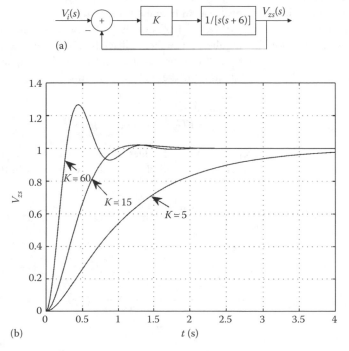

(a)

(b)

FIGURE 7.29

Book MATLAB m-File: step7_8_1

```
%step7_8_1 is an m file to produce one of the
%curves of Fig 7.29b;n=numerator coefficient vector;
%d=denomenator coefficient vector;H(s)=15/(s^2+6s+15);
n=15;d=[1 6 15];
sys=tf(n,d);
[y,t]=step(sys,4);%4 is the desired end of time axis;
plot(t,y);
text(3,1.2,'a)');%this sets the text 'a)'at point (3,1.2);
```

Proportional controllers

Let a voltage source be attached to an inductor. The relationship is well known to be equal to

$$L\frac{di(t)}{dt} = v(t) \quad \text{or} \quad \frac{di(t)}{dt} = \frac{1}{L}v(t) \quad \text{also} \quad I(s) = \frac{1}{Ls}V(s) \quad \text{or} \quad H_2(s) = \frac{I(s)}{V(s)} = \frac{1}{Ls}$$

where

$i(t)$ is the current in the circuit at time t

$v(t)$ is the voltage source

L is the inductance

Based on the above relationships, we can create a unity feedback system made of the **plant** and the **controller**. The feedback configuration is shown in Figure 7.30 where G_2 is called the plant and G_1 is called the controller. This particular feedback controller is known as the **proportional controller**. Since the feedback loop transfers the output I to summation point, it is apparent that the **error signal** E will be zero if the input is equal to the output. The total transfer function for the feedback system is

$$T(s) = \frac{KG_2(s)}{1 + KG_2(s)} = \frac{K/L}{s + (K/L)} \tag{7.143}$$

with a pole at $-K/L$. To obtain the range of values of the gain K must take to create a bounded feedback system we can vary K and plot the values of the root in the s-plane.

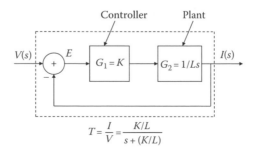

FIGURE 7.30

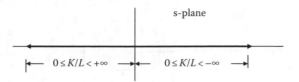

FIGURE 7.31

For the present case, as K vary from $-\infty$ to $+\infty$ the pole varies from $+\infty$ to $-\infty$. The trace of the values of the roots is known as the **root locus**. Figure 7.31 shows the root locus for this case. If the system had additional roots, each root would have produced its own root locus (**branch**).

If the input is a step function, the output in LT form and in time domain is

$$I(s) = \frac{K/L}{s + (K/L)} \frac{1}{s} = -\frac{1}{s + (K/L)} + \frac{1}{s} \quad \text{or} \quad i(t) = u(t) - e^{-(K/L)t} u(t)$$

We observe that at time infinity the current is equal to its input. The first term is known as the **steady-state response** $i_{ss}(t)$. Hence, the **error signal (tracking error)**, $i(t) - u(t) = e(t) = -\exp[-(K/L)T]$, decreases faster the larger the K values are.

Example 7.45

The position of the mass shown in Figure 7.32a is found by solving the equation

$$M\frac{dv(t)}{dt} + Dv(t) = M\frac{d^2x(t)}{dt^2} + D\frac{dx(t)}{dt} = Af(t)$$

where
  M is the mass
  D is the damping factor
  A is a constant

The transfer function of this system is

$$G_2(s) \triangleq \frac{X(s)}{F(s)} = \frac{A}{Ms^2 + Ds} = \frac{A/M}{s[s + (D/M)]}$$

The step response of this system is unstable due to the double root at $s = 0$. If we use a proportional controller with unit feedback loop, as shown in Figure 7.30, the total transfer function is

$$T(s) \triangleq \frac{X(s)}{F(s)} = \frac{KG_2(s)}{1 + KG_2(s)} = \frac{KA}{Ms^2 + Ds + KA},$$

$$s_{1,2} = -\frac{D}{2M} \pm \sqrt{\frac{D^2}{4M^2} - \frac{KA}{M}} \equiv \text{roots of the characteristic equation } Ms^2 + Ds + KA$$

If K is positive, both roots have negative real parts and the system is stable. If, $K < [D^2/(4MA)]$, both poles are real. If, however, $K > [D^2/(4MA)]$

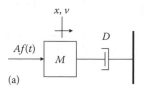

(a)

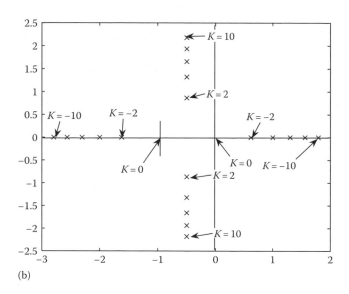

(b)

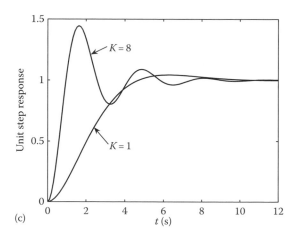

FIGURE 7.32

the two roots are complex conjugate, which are

$$s_{1,2} = -\frac{D}{2M} \pm j\sqrt{\frac{KA}{M} - \frac{D^2}{4M^2}}$$

If we set $A/M = 0.5$, $D/M = 1$, the transfer function becomes

$$T(s) = \frac{0.5\,K}{s^2 + 0.2s + 0.5\,K}$$

Figure 7.32b shows the roots for different values of the gain K, and Figure 7.32c shows the unit step response of the feedback system.

Book MATLAB m-File Name ex7_8_8

```
%m-file for plotting the roots of the total
%transfer function of the Ex 7.45; name:ex7_8_8;
for k=1:5
z(:,k)=roots([1 1 0.5*2*k]);
end;
plot(real(z(1,:)),imag(z(1,:)),'xk')%'xk'=plot black x's;
hold on;
plot(real(z(2,:)),imag(z(2,:)),'xk');
for m=1:5
z1(:,k)=roots([1 1 -0.5*2*k]);
end;
plot(real(z1(1,:)),imag(z1(1,:)),'xk');
hold on;
plot(real(z1(2,:)),imag(z1(2,:)),'xk');                    ■
```

We must mention that proportional controller does not always produce a closed-loop stability.

Proportional integral differential controllers (PID controllers)

Before we proceed with an example, we must mention that a proportional controller is just a constant, as it was developed above. If the controller has an integrator, its transform domain representation is proportional to $1/s$. And similarly the differentiator is proportional to s. This type of controller is, basically, a generalization of a proportional controller.

Example 7.46

The KVL of a series RLC-relaxed circuit and the equivalent differential equation with respect to charge are

$$L\frac{di(t)}{dt} + Ri(t) + \frac{1}{C}\int i(t)dt = v(t) \quad or \quad L\frac{d^2q(t)}{dt^2} + R\frac{dq(t)}{dt} + \frac{1}{C}q(t) = v(t) \qquad (7.144)$$

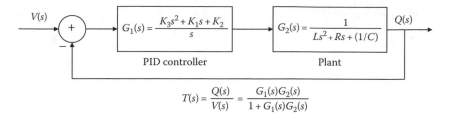

$$T(s) = \frac{Q(s)}{V(s)} = \frac{G_1(s)G_2(s)}{1 + G_1(s)G_2(s)}$$

FIGURE 7.33

The LT of the second equation and the transfer function are

$$Ls^2Q(s) + RsQ(s) + (1/C)Q(s) = V(s); \qquad G_2(s) \triangleq \frac{Q(s)}{V(s)} = \frac{1}{Ls^2 + Rs + (1/C)} \qquad (7.145)$$

If, for example, we set $L = R = 1$ and $1/C = 1.3$ in (7.145) we will observe that the transient of a step input disappears to about 10 units of time, which may be very slow for some applications. In addition, the overshoot may also be objectionable.

To alleviate these objections, we can use a **PID** controller, which is of the form

$$G_1(s) = K_1 + \frac{K_2}{s} + K_3 s = \frac{K_3 s^2 + K_1 s + K_2}{s} \qquad (7.146)$$

The feedback controller is shown in Figure 7.33. The configuration indicates that by selecting the values of K_i's, we can create the following individual controllers: (a) Proportional controller, $K_3 = K_2 = 0$; (b) **integral** controller, $K_3 = K_1 = 0$; and (c) **derivative** controller, $K_1 = K_2 = 0$. ■

Example 7.47

To evaluate the PID controller, we set $L = R = 1$ and $1/C = 1.30$. The total transfer function (closed-loop transfer function) is

$$T(s) = \frac{G_1(s)G_2(s)}{1 + G_1(s)G_2(s)} = \frac{K_3 s^2 + K_1 s + K_2}{s^3 + (1 + K_3)s^2 + (1.3 + K_1)s + K_2} \qquad (7.147)$$

Setting $K_1 = 8$ and $K_2 = K_3 = 0$, the step response of the P controller is given by

$$Q_P(s) = \frac{1}{s} \frac{8}{s^2 + s + 9.3} \qquad (7.148)$$

Setting $K_1 = K_3 = 8$ and $K_2 = 0$, the step response of the PD controller is given by

$$Q_{PD}(s) = \frac{1}{s} \frac{8s + 8}{s^2 + 9s + 9.3} \qquad (7.149)$$

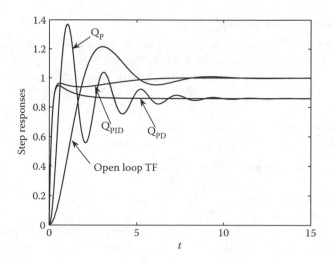

FIGURE 7.34

Finally, with $K_1 = K_3 = 8$ and $K_2 = 4$ the step response of the PID controller is given by

$$Q_{PID}(s) = \frac{1}{s} \frac{8s^2 + 8s + 4}{s^3 + 9s^2 + 9.3s + 4} \tag{7.150}$$

Figure 7.34 shows the step responses for the above three cases and the open loop response with $L = R = 1$ and $(1/C) = 1.3$. ∎

7.9 Bode Plots

Bode plots represent the magnitude and phase versus frequency based on 10 base log–log scales. The data are plotted as follows:

$$\text{decibels (dB)} = 20 \log |H(j\omega)| \quad \text{versus} \quad \log \omega, \quad \log(\cdot) \overset{\Delta}{=} \log_{10}(\cdot) \tag{7.151}$$

$$\theta(\omega) \quad \text{versus} \quad \log(\omega) \tag{7.152}$$

Bode plots are extensively used in feedback control studies since they are one of the several techniques to specify if a system is stable or not.

Bode plots of constants

Based on (7.151) and, due to the fact that $C = Ce^{j0}$ *and* $-C = Ce^{j\pi} = Ce^{-j\pi}$, we obtain

$$dB = 20 \log |C| = \begin{cases} \text{positive number} & |C| > 1 \\ \text{negative number} & |C < 1| \end{cases}$$

$$\theta = \begin{cases} 0 & C > 1 \\ \pi = -\pi & C < 1 \end{cases} \tag{7.153}$$

For example, the number -0.5 gives $20 \log(0.5) = -6.0206$ and $\theta = \pi$ or $-\pi$. This means that the plot will show a straight line at height -6.0206.

Bode diagram for differentiator

The frequency transfer function, amplitude, and phase of a differentiator are

$$H(j\omega) = j\omega = \omega e^{j\frac{\pi}{2}}; \quad |H(j\omega)|_{dB} = 10\log\sqrt{(j\omega)(-j\omega)} = 20\log\omega; \quad \theta(\omega) = \frac{\pi}{2}$$

$$(7.154)$$

Since $s = j\omega$, the transfer function of a differentiator is $H(s) = s/1$. Based on the Book MATLAB function given below, Figure 7.35 was found.

Book MATLAB m-File: bode_differentiator

```
%m-file: bode_differentiator
num= [1 0];den= [0 1];
sys=tf(num,den);
w=logspace(-2,3,100);%(-2,3,100)=create 100 points
   %along x-axis from 10^{-2}=0.01 to 10^{3}=1000
   %in a log scale;
[ma,ph] =bode(sys,w);
mag=reshape(ma,[100,1]);%since ma=1x1x100 it reshapes
                       %to 100 by 1 vector;
phg=reshape(ph,[100,1]);
subplot(2,1,1);semilogx(w,20*log10(mag),'k');
xlabel('\omega rad/s');ylabel('Magnitude (dB)');
subplot(2,1,2);semilogx(w,phg,'k');
xlabel('\omega rad/s');ylabel('Phase (deg)');
```

Bode diagram for an integrator

The frequency transfer function, amplitude, and phase of an integrator are $(\log(1) = 0)$.

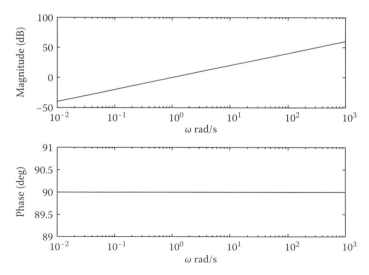

FIGURE 7.35

$$H(j\omega) = \frac{1}{j\omega}, \quad |H(j\omega)|_{dB} = 20\log\frac{1}{\omega} = -20\log\omega, \quad \theta(\omega) = -\frac{\pi}{2} \tag{7.155}$$

If we plot the Bode diagram of an integrator, the magnitude line has an opposite inclination to that of the differentiator.

Bode diagram for a real pole

Let the transfer function of a system be given by

$$H(s) = \frac{p}{p+s} \quad \text{or} \quad H(j\omega) = \frac{p}{p+j\omega} = \frac{1}{1+j\frac{\omega}{p}} = \frac{1}{\sqrt{1+\left(\frac{\omega}{p}\right)^2}}e^{-j\tan^-\left(\frac{\omega}{p}\right)} = |H(j\omega)|e^{j\theta(\omega)} \tag{7.156}$$

The magnitude and phase of the above transfer function for a Bode plot are

$$|H(j\omega)| = 20\log\frac{1}{\sqrt{1+\left(\frac{\omega}{p}\right)^2}} = 20\log(1) - 20\log\sqrt{1+\left(\frac{\omega}{p}\right)^2} = -20\log\sqrt{1+\left(\frac{\omega}{p}\right)^2} \tag{7.157}$$

$$\theta(\omega) = -\tan^{-1}\left(\frac{\omega}{p}\right)$$

For $\omega \ll p$, the magnitude is equal to about $-20\log(\sim1)=0$, a constant. For the same case, the angle is equal to $-\tan^{-1}(\sim0) = \sim0$. On the other hand, if $\omega \gg p$ the magnitude is equal to about $-20\log(\omega/p) = -20\log(\omega) + 20\log(p)$ which is a straight line with negative slope of -20 dB/decade. The Bode plots are shown in Figure 7.36 as curves, and the approximations are shown as straight lines for the particular case when $p = 1$.

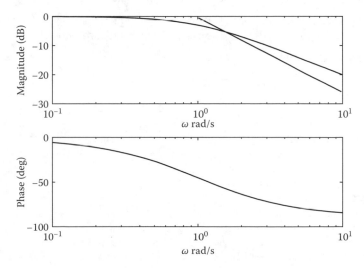

FIGURE 7.36

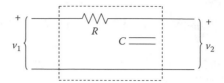

FIGURE 7.37

For the case of a transfer function having a real zero, we expect the curves to be reflected with respect to frequency axis since the zero is the reciprocal of the pole.

Example 7.48

Find the transfer function for the system shown in Figure 7.37. Plot the s-plane of zeros and poles, the locus and the Bode plots.

SOLUTION

The voltage ratio system function is

$$\frac{V_2(s)}{V_1(s)} \triangleq H(s) = \frac{\frac{1}{Cs}}{R + \frac{1}{Cs}} = \frac{1}{1 + RCs} = \frac{\frac{1}{RC}}{s + \frac{1}{RC}}, \quad H(j\omega) = \frac{\frac{1}{RC}}{j\omega + \frac{1}{RC}}$$

There are three critical ranges to be examined:

1. For low frequencies $H(j\omega) \to 1$, with phase angle approximately zero
2. For high frequencies $H(j\omega) \to 1/(j\omega RC)$, with angle approximately $-90°$
3. For frequency $\omega = 1/RC, H(j\omega) = 1/(1 + j1) = 0.707\angle -45°$

The appropriate figures are given in Figure 7.38a through c.
The following Book MATLAB m-files can be used to produce Figure 7.38c.

Book MATLAB m-File

```
%m-file for Ex7.48:ex7_9_1
num = [0 1];den = [1 1];
sys = tf(num,den);
bode(sys, 'k',{0.1 100});
```

Book MATLAB m-File

```
%another way to find the Body plots;
%m file: ex7_9_1a
num = [0 1]; den = [1 1];
w = 0.1:0.01:100;
[mag,pha,w] = bode(num,den,w);
subplot(2,1,1);semilogx(w,20*log10(mag),'k');
xlabel('\omega rad/s');ylabel('Magnitude (dB)');
subplot(2,1,2);semilogx(w,pha,'k');
xlabel('\omega rad/s');ylabel('Phase (deg)');
```

■

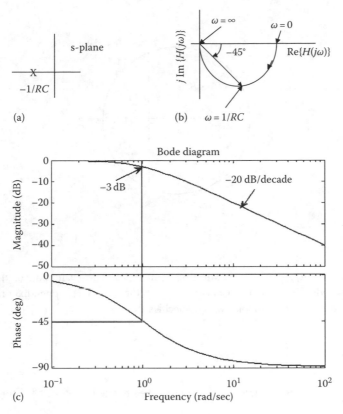

FIGURE 7.38

*7.10 Inversion Integral

When the Laplace $F(s)$ of a one-sided time function is known, the corresponding time function is found using the inversion formula:

$$f(t) = \mathcal{L}^{-1}\{F(s)\} = \frac{1}{2\pi j} \int_{\sigma-j\infty}^{\sigma+j\infty} F(s)e^{st}\,ds \tag{7.158}$$

This equation applies equally well to both the one-sided and two-sided LTs.

The line of integration is restricted to values of σ for which the direct Laplace formula converges (it is drawn in the ROC). As we will see later in this chapter, for the two-sided LT the ROC must be specified so that we obtain the time function uniquely. For the one-sided transform, the ROC is given by σ, where σ is the abscissa of absolute convergence.

The path of integration in (7.158) is usually taken as shown in Figure 7.39, and consists of the straight line ABC displaced to the right of the origin by σ and extending in the

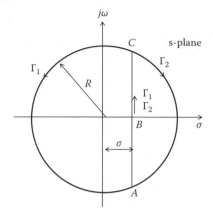

FIGURE 7.39

limit from $-j\infty$ to $+j\infty$ with connecting semicircles. The evaluation of the integral usually proceeds by using the Cauchy integral theorem (see appendix on complex variables at the end of the book), which specifies that

$$f(t) = \frac{1}{2\pi j} \lim_{R \to \infty} \oint_{\Gamma_1} F(s)e^{st}\,ds$$

$$= \sum [\text{residues of } F(s)e^{st} \text{ at the singularities to the left of } ABC], \quad \text{for} \quad t > 0$$

(7.159)

The integration is taken counterclockwise.

Example 7.49

A one-sided signal ($t > 0$) has the following LT: $\omega/(s^2 + \omega^2)$. The inversion is written in a form that shows the poles of the integrand

$$f(t) = \frac{1}{2\pi j} \oint \frac{\omega\, e^{st}}{(s + j\omega)(s - j\omega)}\,ds$$

Since the time function is one sided for $t > 0$, the path of integration must be Γ_1 in Figure 7.39. Next, we evaluate the residues

$$\text{Res}\left[(s - j\omega)\frac{\omega e^{st}}{(s^2 + \omega^2)}\right]_{s=j\omega} = \frac{\omega e^{st}}{(s + j\omega)}\Bigg|_{s=j\omega} = \frac{e^{j\omega}}{2j}$$

$$\text{Res}\left[(s + j\omega)\frac{\omega e^{st}}{(s^2 + \omega^2)}\right]_{s=-j\omega} = \frac{\omega e^{st}}{(s - j\omega)}\Bigg|_{s=-j\omega} = \frac{e^{-j\omega}}{-2j}$$

Therefore,

$$f(t) = \sum \text{Res} = \frac{e^{j\omega} - e^{-j\omega}}{2j} = \sin \omega t$$

■

Example 7.50

To find the inverse LT of the function $F(s) = 1/\sqrt{s}$, we must observe that the function is a double-valued one because of the square root operation. That is, if s is represented in polar form by $re^{j\theta}$, then $re^{j(\theta+2\pi)}$ is a second accepted representation, and $\sqrt{s} = \sqrt{re^{j(\theta+2\pi)}} = -\sqrt{re^{j\theta}}$, thus showing two different values for $\sqrt{s}$. But a double-valued function is not analytic and requires a special procedure in its solution.

The procedure is to make the function analytic by restricting the angle of s to the range $-\pi < \theta < \pi$ and by excluding the point at $s = 0$. This is done by constructing a **branch cut** along the negative axis, as shown in Figure 7.40. The end of the branch cut, which is the origin in this case, is called the **branch point**. Since the branch cut can never be crossed, this essentially ensures that $F(s)$ is single-valued function. Therefore, the inversion integral (7.158) becomes and, hence, for $t > 0$, we have

$$f(t) = \lim_{R \to \infty} \frac{1}{2\pi j} \int_{GAB} F(s)e^{st} ds = \frac{1}{2\pi j} \int_{\sigma-j\infty}^{\sigma+j\infty} F(s)e^{st} ds$$

$$= -\frac{1}{2\pi j} \left[\int_{BC} + \int_{\Gamma_2} + \int_{-l} + \int_{\gamma} + \int_{l+} + \int_{\Gamma_3} + \int_{FG} \right] \tag{7.160}$$

which does not include any singularity.

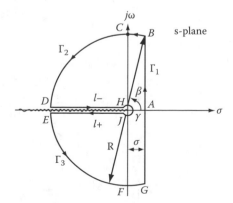

FIGURE 7.40

First we will show that for $t > 0$ the integrals over the contours BC and CD vanish as $R \to \infty$, from which $\int_{\Gamma_2} = \int_{\Gamma_3} = \int_{BC} = \int_{FG} = 0$. Note from Figure 7.40 that $\beta = \cos^{-1}(\sigma/R)$, so that the integral over the arc BC is, since $|e^{j\theta}| = 1$,

$$|I| \leq \int_{BC} \left| \frac{e^{\sigma t} e^{j\omega t}}{R^{1/2} e^{j\theta/2}} jR\, e^{j\theta}\, d\theta \right| = e^{\sigma t} R^{1/2} \int_{\beta}^{\pi/2} d\theta = e^{\sigma t} R^{1/2} \left(\frac{\pi}{2} - \cos^{-1} \frac{\sigma}{R} \right) = e^{\sigma t} R^{1/2} \sin^{-1} \frac{\sigma}{R}$$

But for small arguments $\sin^{-1} \frac{\sigma}{R} = \frac{\sigma}{R'}$, and in the limit as $R \to \infty, I \to 0$. By similar approach, we find that the integral over CD is zero. Thus, the integrals over the contours Γ_2 and Γ_3 are also zero as $R \to \infty$.

For evaluating the integral over γ, let $s = re^{j\theta} = r(\cos\theta + j\sin\theta)$ and

$$\int_{\gamma} F(s)e^{st}\, ds = \int_{\pi}^{-\pi} \frac{e^{t(\cos\theta + j\sin\theta)}}{\sqrt{re^{j\theta/2}}} jre^{j\theta}\, d\theta = 0 \quad \text{as } r \to 0 \tag{7.161}$$

The remaining integrals in (7.160) are written as

$$f(t) = -\frac{1}{2\pi j} \left[\int_{I-} F(s)e^{st}\, ds + \int_{I+} F(s)e^{st}\, ds \right] \tag{7.162}$$

Along the path $I-$, let $s = -u$; $\sqrt{s} = j\sqrt{u}$, and $ds = -du$, where u and $\sqrt{u}$ are real positive quantities. Then,

$$\int_{I-} F(s)e^{st}\, ds = -\int_{-\infty}^{0} \frac{e^{-ut}}{j\sqrt{u}}\, du = \frac{1}{j} \int_{0}^{\infty} \frac{e^{-ut}}{\sqrt{u}}\, du$$

Along the path $I+, s = -u, \sqrt{s} = -j\sqrt{u}$ (not $+j\sqrt{u}$), and $ds = -du$. Then,

$$\int_{I+} F(s)e^{st}\, ds = -\int_{0}^{\infty} \frac{e^{-ut}}{-j\sqrt{u}}\, du = \frac{1}{j} \int_{0}^{\infty} \frac{e^{-ut}}{\sqrt{u}}\, du$$

Combine these results to find that

$$f(t) = -\frac{1}{2\pi j} \left[\frac{2}{j} \int_{0}^{\infty} \frac{e^{-ut}}{\sqrt{u}}\, du \right] = \frac{1}{\pi} \int_{0}^{\infty} \frac{e^{-ut}}{\sqrt{u}}\, du$$

which is a standard form integral listed in most handbooks of mathematical tables, with the result

$$f(t) = \frac{1}{\pi}\sqrt{\frac{\pi}{t}} = \frac{1}{\sqrt{\pi t}} \quad t > 0 \tag{7.163}$$

■

Example 7.51

The *RC* transmission line equation with its initial and boundary conditions for this example are (see also (7.101))

$$\frac{\partial^2 v(t,x)}{\partial x^2} - RC\frac{\partial v(t,x)}{\partial t} \tag{7.164}$$

B.C.

$$v(t,0) = a \cos \omega t \quad t > 0 \tag{7.165}$$

I.C.

$$v(0,x) = 0 \quad x > 0 \tag{7.166}$$

Taking the LT of the above equations, we obtain the following set:

$$\frac{d^2 V(s,x)}{dx^2} - RCsV(s,x) = 0 \quad x > 0 \tag{7.167}$$

$$V(s,0) = \frac{as}{s^2 + \omega^2} \quad x = 0 \tag{7.168}$$

$$V(0,x) = 0 \quad t = 0 \tag{7.169}$$

The characteristic equation of (7.167) is

$$D^2 - RCs = 0 \quad \text{or} \quad D = \pm\sqrt{RC}\sqrt{s}$$

Therefore, the solution is

$$V(s,x) = Ae^{\sqrt{RC}\sqrt{s}x} + Be^{-\sqrt{RC}\sqrt{s}x} \tag{7.170}$$

Since the voltage at infinity must be zero, $V(s,\infty) = 0$, constant A is set equal to zero. Hence,

$$V(s,x) = Be^{-\sqrt{RC}\sqrt{s}x} \tag{7.171}$$

From (7.168) and (7.171), we obtain

$$V(s,x) = \frac{as}{s^2 + \omega^2} e^{-\sqrt{RC}\sqrt{s}x} \tag{7.172}$$

Therefore, the voltage in the transmission line is found by using the inverse LT

$$v(t,x) = \frac{a}{2\pi j} \int_{\sigma-j\infty}^{\sigma+j\infty} e^{st-\sqrt{RC}\sqrt{s}x} \frac{s}{s^2 + \omega^2} ds \tag{7.173}$$

The above integral is found by applying the Cauchy theorem around the contour in a counterclockwise sense as shown in Figure 7.41. From (7.173), we observe that there exist a branch point at the origin and two poles at $s = \pm j\omega$. The application of the theorem gives the following relation:

$$\frac{a}{2\pi j} \oint = \frac{a}{2\pi j} \left[\int_{AB} + \int_{BB'} + \int_{B'F} + \int_{GA} + \int_{C_1} + \int_{C_3} + \int_{FE} + \int_{ED} + \int_{DG} \right] = 0 \tag{7.174}$$

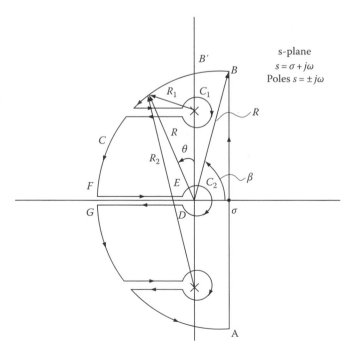

FIGURE 7.41

We will show below that the integral BB', $B'F$, and GA are all equal to zero. From Figure 7.41, we obtain the relations

$$\beta = \cos^{-1}\frac{\sigma}{R}, \quad R_1 = \left|Re^{j\theta} - j\omega\right| \quad \pi \geq \theta \geq \beta, \quad R_2 = \left|Re^{j\theta} + j\omega\right| \quad \pi \geq \theta \geq \beta$$

We also have the following relations:

$$\frac{R}{R_2} < 1 \quad (1), \quad R_1 \geq R - \omega \quad (2), \quad \frac{R}{R_1} \leq \frac{R}{R-\omega} < 2 \quad \text{if } R > 2\omega \quad (3)$$

Multiply (1) and (3) to find

$$\frac{R^2}{R_1 R_2} < 2 \quad \text{if } R > 2\omega \quad (4)$$

Over BB'

We proceed to find the value of the integral BB'. Hence,

$$|I_{BB'}| \leq \int_{BB'} |e^{st}| \left|e^{-\sqrt{RC}\sqrt{sx}}\right| \frac{|s|}{|s^2+\omega^2|}|ds| < \int_{BB'} \left|e^{\sigma t + j\omega t}\right| \frac{\left|Re^{j\theta}\right|}{|Re^{j\theta} - j\omega||Re^{j\theta} + j\omega|} d(Re^{j\theta})$$

$$= \int_{BB'} e^{\sigma t} \frac{R^2}{R_1 R_2} d\theta = e^{\sigma t} \frac{R^2}{R_1 R_2} \int_{\beta}^{\pi/2} d\theta < 2e^{\sigma t}\left(\frac{\pi}{2} - \beta\right) = 2e^{\sigma t}\sin\frac{\sigma}{R} \quad (\text{see (4)})$$

As $R \to \infty$, the last expression approaches zero and, hence, $|I_{BB'}| \to 0$.

Over $B'F$

We, next, proceed to find the value of the integral over the contour $B'F$. Following the same procedure as above, we obtain

$$|I_{B'F}| < \int_{B'F} |e^{st}| \frac{R^2}{R_1 R_2} d\theta = \frac{R^2}{R_1 R_2}\int_{B'F} |e^{st}| d\theta = \frac{R^2}{R_1 R_2}\int_{B'F} \left|e^{(R\cos\theta + j\sin\theta)t}\right| d\theta = \frac{R^2}{R_1 R_2}\int_{B'F} \left|e^{Rt\cos\theta}\right| d\theta$$

Next, we make the following substitutions:

$$\theta = \frac{\pi}{2} + \varphi, \Rightarrow \cos\theta = -\sin\varphi, \text{ also } \theta = \frac{\pi}{2} \Rightarrow \varphi = 0, \quad \theta = \pi \Rightarrow \varphi = \frac{\pi}{2}, \text{ and } d\theta = d\varphi$$

Therefore, the above inequality becomes (see also 4)

$$|I_{B'F}| < \frac{R^2}{R_1 R_2}\int_{B'F} e^{Rt\cos\theta} d\theta < 2\int_0^{\pi/2} e^{-Rt\sin\varphi} d\varphi < 2\int_0^{\pi/2} e^{-2Rt\varphi/\pi} d\varphi < \frac{\pi}{Rt}, \Rightarrow \lim_{R\to\infty}|I_{B'F}| = 0$$

From (7.174), we obtain

$$\frac{a}{2\pi j}\int_{\sigma-j\infty}^{\sigma+j\infty} e^{st-\sqrt{RC}\sqrt{s}x}\frac{s}{s^2+\omega^2}\,ds = \frac{a}{2\pi j} \text{ (value around the branch point counterclockwise}$$

and branch lines opposite to figure since we eliminated the $(-)$signe $+2\pi j\sum \text{Res})$

$$=\underbrace{\frac{a}{2\pi j}\ \text{(branch point)}}_{I_b} + \underbrace{a\frac{e^{st-\sqrt{RC}\sqrt{s}x}s}{(s+j\omega)}\Big|}_{I_1}{\Big|_{s=j\omega=\omega e^{j\pi/2}}}\ \ \underbrace{+a\frac{e^{st-\sqrt{RC}\sqrt{s}x}s}{(s-j\omega)}\Big|}_{I_2}{\Big|_{s=-j\omega=\omega e^{-j\pi/2}}}$$

$$I_1 = a\frac{e^{j\omega t-\sqrt{RC}\sqrt{e^{j\pi/2}}\sqrt{\omega}x}j\omega}{2j\omega} = \frac{a}{2}e^{j\omega t}e^{-\sqrt{RC}\sqrt{\omega}x}\left(\cos\frac{\pi}{4}+j\sin\frac{\pi}{4}\right)$$

$$I_2 = a\frac{e^{-j\omega t-\sqrt{RC}\sqrt{e^{j\pi/2}}\sqrt{\omega}x}(-j\omega)}{-2j\omega} = \frac{a}{2}e^{-j\omega t}e^{-\sqrt{RC}\sqrt{\omega}x}\left(\cos\frac{\pi}{4}-j\sin\frac{\pi}{4}\right)$$

Evaluation of I_b

On GD (we have reversed the integration path)

$$s=\rho e^{-j\pi}\Rightarrow s=-R\Rightarrow\rho=R \text{ and } s=-\varepsilon\Rightarrow\rho=\varepsilon, \text{ also } \sqrt{-\rho}=\sqrt{\rho e^{-j\pi}}=\sqrt{\rho}e^{-j\frac{\pi}{2}}=-j\sqrt{\rho}$$

$$\lim_{\substack{\varepsilon\to 0\\ R\to-\infty}}\frac{a}{2\pi j}\int_{+R}^{+\varepsilon} e^{-\rho t}e^{-\sqrt{RC}\sqrt{\rho}(-j)x}\frac{\rho e^{-j\pi}}{\rho^2 e^{-j2\pi}+\omega^2}e^{-j\pi}\,d\rho = \lim_{\substack{\varepsilon\to 0\\ R\to-\infty}}\frac{a}{2\pi j}\int_{+R}^{+\varepsilon} e^{-\rho t}e^{j\sqrt{RC}\sqrt{\rho}x}\frac{\rho}{\rho^2+\omega^2}\,d\rho$$

$$=-\frac{a}{2\pi j}\int_0^{R} e^{-\rho t}e^{j\sqrt{RC}\sqrt{\rho}x}\frac{\rho}{\rho^2+\omega^2}\,d\rho$$

On EF

$$s=\rho e^{j\pi}=-\rho\Rightarrow s=-R \text{ and } \rho=+R, \ \ s=-\varepsilon\Rightarrow\rho=\varepsilon, \ \ ds=-d\rho=e^{j\pi}d\rho$$

$$\lim_{\substack{\varepsilon\to 0\\ R\to-\infty}}\frac{a}{2\pi j}\int_{\varepsilon}^{R} e^{-\rho t}e^{-\sqrt{RC}j\sqrt{\rho}x}\frac{\rho e^{j\pi}}{\rho^2 e^{j2\pi}+\omega^2}e^{j\pi}\,d\rho = \frac{a}{2\pi j}\int_{\varepsilon}^{R} e^{-\rho t}e^{-j\sqrt{RC}\sqrt{\rho}x}\frac{\rho}{\rho^2+\omega^2}\,d\rho$$

Around the circle on the origin gives zero. Therefore, putting together the above relations we find that the voltage at any time and any point along the transmission line is

$$v(t,x) = \frac{a}{2} e^{j\omega t} e^{-\sqrt{RC\omega}x/\sqrt{2}} e^{-j\sqrt{RC\omega}x/\sqrt{2}} + \frac{a}{2} e^{-j\omega t} e^{-\sqrt{RC\omega}x/\sqrt{2}} e^{j\sqrt{RC\omega}x/\sqrt{2}} + I_b$$

$$= a e^{-\sqrt{RC\omega/2}x} \cos\left(\omega t - \sqrt{RC\omega/2}x\right) + \frac{a}{2\pi j} \int_0^\infty e^{-\rho t} \frac{\rho}{\rho^2 + \omega^2} \left(e^{-j\sqrt{RC\rho}x} - e^{j\sqrt{RC\rho}x}\right) d\rho$$

$$= a e^{-\sqrt{RC\omega/2}x} \cos\left(\omega t - \sqrt{RC\omega/2}x\right) - \frac{a}{\pi} \int_0^\infty e^{-\rho t} \sin\left(\sqrt{RC\rho}x\right) \frac{\rho}{\rho^2 + \omega^2} d\rho \qquad ■$$

Example 7.52

It is desired to find the inverse transform of the function

$$F(s) = \frac{1}{s(1 + e^{-s})}, \qquad (7.175)$$

which has infinite number of poles. The integrand in the inversion integral has the simple poles

$$s = 0, \quad \text{and} \quad s = jn\pi \quad n = \pm 1, \pm 3, + \cdots \text{(odd values)} \qquad (7.176)$$

These are illustrated in Figure 7.42. This means that the function $e^{st}/[s(1 + e^{-s})]$ is analytic in the s plane except at the simple poles $s = 0$ and $s = jn\pi$. Hence, the integral is specified in terms of the residues in the various poles. We thus have

$$\text{For } s = 0 \text{ Res} \left\{ \frac{s e^{st}}{s(s + e^{-s})} \right\} \bigg|_{s=0} = \frac{1}{2}, \quad \text{For } s = jn\pi \text{ Res} \left\{ \frac{(s - jn\pi)e^{st}}{s(s + e^{-s})} \right\} \bigg|_{s=jn\pi} = \frac{0}{0} \qquad (7.177)$$

The problem we now face in this evaluation is that

$$\text{Re } s \left\{ (s - a) \frac{n(s)}{d(s)} \right\} \bigg|_{s=a} = \frac{0}{0},$$

where the roots of $d(s)$ are such that $s = a$ cannot be factored. However, the appendix on complex variables develops this situation and the results are

$$\frac{d[d(s)]}{ds} \bigg|_{s=a} = \lim_{s \to a} \frac{d(s) - d(a)}{s - a} = \lim_{s \to a} \frac{d(s)}{s - a},$$

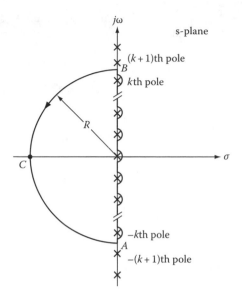

FIGURE 7.42

since $d(a) = 0$. Combine this expression with the above equation to obtain

$$\text{Res}\left\{(s-a)\frac{n(s)}{d(s)}\right\}\Bigg|_{s=a} = \frac{n(s)}{\dfrac{d}{ds}[d(s)]}\Bigg|_{s=a} \tag{7.178}$$

We use (7.178) in evaluating (7.177) to find

$$\text{Re } s\left\{\frac{e^{st}}{s\dfrac{d}{ds}(1+e^{-s})}\right\}\Bigg|_{s=a} = \frac{e^{jn\pi t}}{jn\pi} \quad (n \text{ odd})$$

Adding all the residues, we obtain

$$f(t) = \frac{1}{2} + \sum_{\substack{n=-\infty \\ n \text{ odd}}}^{\infty} \frac{e^{jn\pi t}}{jn\pi}$$

The above equation can be written as follows:

$$f(t) = \frac{1}{2} + \left[\cdots + \frac{e^{-j3\pi t}}{-j3\pi} + \frac{e^{-j\pi t}}{-j\pi} + \frac{e^{j\pi t}}{j\pi} + \frac{e^{j3\pi t}}{j3\pi} + \cdots\right] = \frac{1}{2} + \sum_{\substack{n=1 \\ n \text{ odd}}}^{\infty} \frac{2j \sin n\pi t}{jn\pi}$$

which we write, finally as

$$f(t) = \frac{1}{2} + \frac{2}{\pi} \sum_{k=1}^{\infty} \frac{\sin(2k-1)\pi t}{2k-1} \tag{7.179}$$

As a second approach to a solution to this problem, we will use the contour integration. We choose the path shown in Figure 7.42 that includes semicircular hooks around each pole, the vertical connecting line from hook to hook, and semicircular path at $R \to \infty$. Thus, we have

$$f(t) = \frac{1}{2\pi j} \oint \frac{e^{st}}{s(s+e^{-s})} ds$$

$$= \frac{1}{2\pi j} \left[\underbrace{\int}_{\substack{BCA \\ I_1}} + \underbrace{\int}_{\substack{\text{vertical connecting lines} \\ I_2}} + \underbrace{\sum \int}_{\substack{\text{hooks} \\ I_3}} - \sum \text{Res} \right] \tag{7.180}$$

Next, we consider the several integrals:

Integral I_1 By setting $s = re^{j\theta}$ and taking into consideration that $\cos\theta = -\cos\theta$ for $\theta > \pi/2$, the integral $I_1 \to 0$ as $r \to \infty$.

Integral I_2 Along the y-axis, $s = jy$ and

$$I_2 = j \int_{\substack{-\infty \\ r \to 0}}^{\infty} \frac{e^{jyt}}{jy(1+e^{-jy})} dy$$

Note that the integrand is an odd function, hence $I_2 = 0$.

Integral I_3 Consider a typical hook at $s = jn\pi$. Since

$$\lim_{\substack{r \to 0 \\ s \to jn\pi}} \left[\frac{(s-jn\pi)e^{st}}{s(1+e^{-s})} \right] = \frac{0}{0}$$

we evaluate this expression with the help of (7.178). The result is the expression $e^{jn\pi t}/jn\pi$. Thus, for all poles

$$I_3 = \frac{1}{2\pi j} \int_{\substack{-\pi/2 \\ r \to 0 \\ s \to jn\pi}}^{\pi/2} \frac{e^{st}}{s(1+e^{-s})} ds = \frac{j\pi}{2\pi j} \left[\sum_{\substack{n=-\infty \\ n \text{ odd}}}^{\infty} \frac{e^{jn\pi t}}{jn\pi} + \frac{1}{2} \right] = \frac{1}{2} \left[\frac{2}{\pi} \sum_{\substack{n=-\infty \\ n \text{ odd}}}^{\infty} \frac{\sin n\pi t}{n} + \frac{1}{2} \right]$$

Finally, the residues enclosed within the contour are

$$\text{Res} \frac{e^{st}}{s(1+e^{-s})} = \frac{1}{2} + \sum_{\substack{n=-\infty \\ n \text{ odd}}}^{\infty} \frac{e^{jn\pi t}}{jn\pi} = \frac{2}{\pi} \sum_{\substack{n=-\infty \\ n \text{ odd}}}^{\infty} \frac{\sin n\pi t}{n} + \frac{1}{2},$$

which is seen to be twice the value around the hooks. Then when all terms are included in (7.180)

$$f(t) = \frac{2}{\pi} \sum_{\substack{n=-\infty \\ n \text{ odd}}}^{\infty} \frac{\sin n\pi t}{n} + \frac{1}{2} = \frac{1}{2} + \frac{2}{\pi} \sum_{k=1}^{\infty} \frac{\sin(2k-1)\pi t}{2k-1} \qquad (7.181)$$

■

*7.11 Complex Integration and the Bilateral Laplace Transform

We have discussed the fact that the region of absolute convergence of the unilateral LT is the region to the left of the abscissa of convergence. This is not true for the bilateral LT: the ROC must be specified to invert a function $F(s)$ obtained using the bilateral transform. This requirement is necessary because different time signals might have the same LT, but different ROC.

To establish the ROC, we write the LT in the form

$$F_{II}(s) = \int_0^\infty e^{-st} f(t)dt + \int_{-\infty}^0 e^{-st} f(t)dt \qquad (7.182)$$

If the function $f(t)$ is of exponential order ($e^{\sigma_1 t}$), then the ROC for $t > 0$ is $\mathrm{Re}\{s\} > \sigma_1$. If the function $f(t)$ for $t < 0$ is of exponential order ($e^{\sigma_2 t}$), then the ROC is $\mathrm{Re}\{s\} < \sigma_2$. Hence, the function $F_2(s)$ exists and is analytic in the vertical strip defined by

$$\sigma_1 < \mathrm{Re}\{s\} < \sigma_2 \qquad (7.183)$$

Provided, of course, that $\sigma_1 < \sigma_2$. If the inequality is reversed, no ROC would exist and the inversion process could not be performed. This ROC is shown in Figure 7.43.

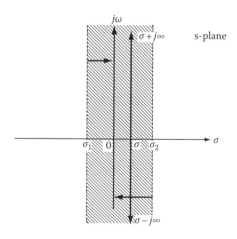

FIGURE 7.43

Example 7.53

It is desired to find the bilateral LT of the signals (a) $f(t) = e^{-at}u(t)$ and (b) $g(t) = -e^{-at}u(-t)$ and to specify their regions of convergence.

Applying the basic definition of the transform, we obtain

(a)

$$F_{//}(s) = \int_{-\infty}^{\infty} e^{-at}u(t)e^{-st}dt = \int_{0}^{\infty} e^{-(s+a)t}dt = \frac{1}{s+a}$$

and its ROC is

$$\text{Re}\{s\} = -a$$

(b)

$$G_{//}(s) = \int_{-\infty}^{\infty} -e^{-at}u(-t)e^{-st}dt = -\int_{-\infty}^{0} e^{-(s+a)t}dt = \frac{1}{s+a}$$

and the ROC is

$$\text{Re}\{s\} < -a$$

Clearly, the knowledge of the ROC is necessary to find the time function unambiguously. ∎

Example 7.54

The inverse LT of the function

$$F_{//}(s) = \frac{3}{(s-4)(s+1)(s+2)} \qquad -2 < \text{Re}\{s\} < -1$$

is found with the help of the Figure 7.44.

For $t > 0$, we close the contour to the left, and we obtain

$$f(t) = \frac{3e^{st}}{(s-4)(s+1)}\bigg|_{s=-2} = \frac{1}{2}e^{-2t} \quad t > 0$$

For $t < 0$, the contour closes t to the right, and now

$$f(t) = \frac{3e^{st}}{(s-4)(s+2)}\bigg|_{s=-1} + \frac{3e^{st}}{(s+2)(s+1)}\bigg|_{s=4} = -\frac{3}{5}e^{-t} + \frac{1}{10}e^{4t} \quad t < 0$$

∎

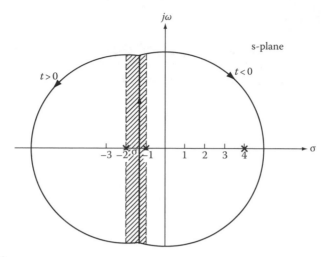

FIGURE 7.44

*7.12 State Space and State Equations

A state space representation of the dynamics of linear time-invariant system, both continuous time and discrete time, is an important description. In this representation, systems are described by a set of equations that describe unique relations among the input, the output, and the state of the system instead of a system description in terms of input–output relationships. The most important features are (a) This type of description has been extensively investigated and is important in both analysis and synthesis in controlled systems. (b) The technique can be extended to time-varying and nonlinear systems. (c) The state variables are part of the solution and are often important factors in the solution. (d) The form is compact in its representation and is suitable for analog and digital computer solution. (e) The form of solution is common to all systems. (f) General system characteristics can be discussed and developed from the state equations.

The state space model of a continuous-time linear system is represented by n linear first-order differential equations. In matrix form, this is given by

$$\boxed{\begin{aligned} \dot{x}(t) &\triangleq \frac{dx(t)}{dt} = Ax(t) + Bw(t) \quad x(0) = x_0 \\ y(t) &= Cx(t) + Dw(t) \end{aligned}}$$

(7.184)

where the dimensions of the matrices, the state vector, and the derivatives of the state vector, the input vector, and the output vector are given below:

$$A^{n \times n}, B^{n \times r}, C^{p \times n}, D^{p \times r}, x(t) = \begin{bmatrix} x_1(t) \\ x_2(t) \\ \vdots \\ x_n(t) \end{bmatrix}, \dot{x}(t) = \begin{bmatrix} \dot{x}_1(t) \\ \dot{x}_2(t) \\ \vdots \\ \dot{x}_n(t) \end{bmatrix}, w(t) = \begin{bmatrix} w_1(t) \\ w_2(t) \\ \vdots \\ w_r(t) \end{bmatrix}, y(t) = \begin{bmatrix} y_1(t) \\ y_2(t) \\ \vdots \\ y_p(t) \end{bmatrix}$$

Matrix A describes the **internal** behavior of the system, while matrices B, C, and D represent connections between the external world and the system. If there are no direct paths between inputs and outputs, which is often the case, matrix D is zero.

In one-dimensional rigid-body mechanics (Newtonian), we know that an (output of the system) of the body for $t > t_0$ is uniquely defined if we know the **position** and its **velocity** at $t = t_0$. Hence, the position and velocity may be used as state variables. For electrical systems, assuming that we do not have loops made up exclusively of voltage sources and capacitors or nodes made up exclusively of current sources and inductors, we can proceed in the selection of the state variables as follows:

1. Currents (through-variables) associated with inductor-type elements (inductors springs).
2. Voltage (across-variables) associated with capacitor-type elements (capacitors, mass elements).
3. Dissipative-type elements do not specify independent state variables.
4. When closed loops of capacitors or junctions of inductors exist, not all state variables chosen according to Rules 1 and 2 are independent.

Example 7.55

It is desired to find the state model representation for the system shown in Figure 7.45a. Therefore, the KVL around the loop yields the expression

$$\frac{di(t)}{dt} = -\frac{R}{L}i(t) + \frac{1}{L}v(t)$$

$$v_o(t) = Ri(t)$$

(7.185)

If we set the current through the inductor as the state variable with $x = i$, $y = v_o$, $a = -R/L$, $b = 1/L$, $c = R$, and $v = w$ in these equations, the resulting equations are the

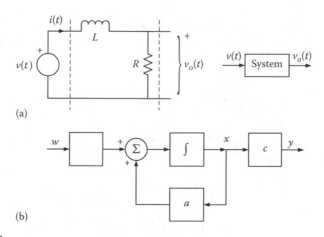

(a)

(b)

FIGURE 7.45

standard form of (7.184). Here, of course, since the system is of the first order, the vector functions become simple functions and the matrices become constant numbers. The transformed equations are

$$\dot{x} = ax + bw$$
$$y = cx$$

(7.186)

And it is the state variable description of the system. Figure 7.45b shows the implicit feedback structure of the state equations, and this is apparent from the block diagram representation.   ∎

7.12.1 State Equations in Phase Variable Form

Let us assume that the input–output relationship of a system is described by an nth-order ordinary differential equation. It is readily possible to convert these higher order differential equations into **normal** form, a set of first-order differential equations. This form expressed in state matrix form is known as the **phase variable** form.

Example 7.56

We must deduce the following differential equation

$$4\frac{d^2y}{dt^2} + 3\frac{dy}{dt} - 2y = 2\frac{dw}{dt} + 5w$$

(7.187)

to its normal form, where y is the output and w is the input to the system. We next write

$$4\frac{d^2\eta}{dt^2} + 3\frac{d\eta}{dt} - 2\eta = w$$

(7.188)

The response now is η under the excitation w. Therefore, the response to the input $d^r w/dt^r$ will be $d^r\eta/dt^r$ (due to linearity). This conclusion follows by the straightforward differentiation of (7.188), with the result

$$4\frac{d^2}{dt^2}\left(\frac{d^r\eta}{dt^r}\right) + 3\frac{d}{dt}\left(\frac{d^r\eta}{dt^r}\right) - 2\left(\frac{d^r\eta}{dt^r}\right) = \frac{d^r w}{dt^r}$$

(7.189)

Next, we multiply (7.188) by 5 and multiply the derivative of the same equation by 2. The two resulting equations are

$$4 \times 5\frac{d^2\eta}{dt^2} + 3 \times 5\frac{d\eta}{dt} - 2 \times 5\eta = 5w$$
$$4 \times 2\frac{d^2}{dt^2}\left(\frac{d\eta}{dt}\right) + 3 \times 2\frac{d}{dt}\left(\frac{d\eta}{dt}\right) - 2 \times 2\left(\frac{d\eta}{dt}\right) = 2\frac{dw}{dt}$$

(7.190)

Add these two equations to obtain

$$4\frac{d^2}{dt^2}\left(2\frac{d\eta}{dt}+5\eta\right)+3\frac{d}{dt}\left(2\frac{d\eta}{dt}+5\eta\right)-2\left(2\frac{d\eta}{dt}+5\eta\right)=2\frac{dw}{dt}+5w \quad (7.191)$$

By comparing (7.191) and (7.187), we see that

$$y=2\frac{d\eta}{dt}+5\eta \quad (7.192)$$

We now write in (7.192) and (7.188),

$$x_1=\eta$$
$$x_2=\dot{x}_1=\frac{d\eta}{dt}$$
$$\dot{x}_2=\frac{d^2\eta}{dt^2}=-\frac{3}{4}\frac{d\eta}{dt}+\frac{2\eta}{4}+\frac{w}{4}$$

The above equations attain the form

$$\dot{x}_1=x_2$$
$$\dot{x}_2=-\frac{3}{4}x_2+\frac{2}{4}x_1+\frac{1}{4}w \quad (7.193)$$
$$y=2x_2+5x_1$$

In matrix form, the above equations are

$$\begin{bmatrix}\dot{x}_1\\\dot{x}_1\end{bmatrix}=\begin{bmatrix}0 & 1\\\frac{2}{4} & -\frac{3}{4}\end{bmatrix}\begin{bmatrix}x_1\\x_1\end{bmatrix}+\begin{bmatrix}0\\\frac{1}{4}\end{bmatrix}w$$

$$y=[5 \quad 2]\begin{bmatrix}x_1\\x_1\end{bmatrix} \quad (7.194)$$

The block-diagram representation of (7.193) is given in Figure 7.46. When (7.194) is written in the matrix form

$$\dot{\mathbf{x}}=\mathbf{A}\mathbf{x}+\mathbf{B}w$$
$$y=\mathbf{C}\mathbf{x} \quad (7.195)$$

∎

Example 7.57

Consider the simple RLC electric circuit shown in Figure 7.47. Applying the basic circuit laws for voltage and currents, we obtain

$$v(t)=L\frac{di_1(t)}{dt}+Ri_1(t)+v_o(t) \quad (7.196)$$

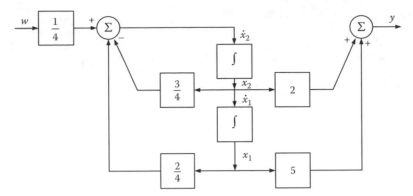

FIGURE 7.46

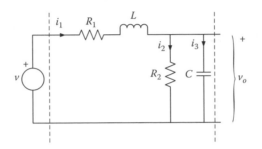

FIGURE 7.47

$$v_o(t) = R_2 i_2(t) = \frac{1}{C}\int\limits_0^t i_3(\tau)d\tau + v_C(0), \quad i_3(t) = C\frac{dv_o(t)}{dt} \tag{7.197}$$

$$i_1(t) = i_2(t) + i_3(t) \tag{7.198}$$

Referring to (7.197) and (7.198), we obtain

$$i_1(t) = \frac{1}{R_2}v_o(t) + C\frac{dv_o(t)}{dt} \tag{7.199}$$

Taking the derivative of (7.199) and combining (7.196) and (7.199), we obtain the desired second-order differential equation:

$$\frac{d^2v_o(t)}{dt^2} + \left(\frac{L+R_1R_2C}{R_2LC}\right)\frac{dv_o(t)}{dt} + \left(\frac{R_1+R_2}{R_2LC}\right)v_o(t) = \frac{1}{LC}v(t) \tag{7.200}$$

This equation relates the input and the output of the system, and represents a mathematical model of the given circuit. In order to solve this equation for the voltage output, we must know the initial conditions $v_o(0)$ and $dv_o(0)/dt$. For electrical circuits, the initial conditions are usually specified in terms of capacitor voltages and inductor currents. Hence, in this example, $v_o(0)$ and $dv_o(0)/dt$ should be expressed in terms of $v_C(0)$ and $i_1(0)$.

Next, introduce the following change of variables:

$$x_1(t) = v_o(t) \Rightarrow \frac{dx_1(t)}{dt} = \frac{dv_o(t)}{dt} = x_2(t)$$

$$x_2(t) = \frac{dv_o(t)}{dt}$$

$$w(t) = v(t)$$

$$y(t) \overset{\Delta}{=} v_o(t) \Rightarrow y(t) = x_1(t)$$

(7.201)

Combining the above variables with (7.200), we obtain

$$\frac{dx_2(t)}{dt} + \left(\frac{L + R_1 R_2 C}{R_2 LC}\right) x_2(t) + \left(\frac{R_1 + R_2}{R_2 LC}\right) x_1(t) = \frac{1}{LC} w(t)$$

(7.202)

The first equation of (7.201) and (7.202) can be put in the matrix form as

$$\begin{bmatrix} \dot{x}_1(t) \\ \dot{x}_2(t) \end{bmatrix} = \begin{bmatrix} 0 & 1 \\ -\dfrac{R_1 + R_2}{R_2 LC} & -\dfrac{L + R_1 R_2 C}{R_2 LC} \end{bmatrix} \begin{bmatrix} x_1(t) \\ x_2(t) \end{bmatrix} + \begin{bmatrix} 0 \\ \dfrac{1}{LC} \end{bmatrix} w(t)$$

(7.203)

The last equation of (7.201), in matrix form, is written as

$$y(t) = \begin{bmatrix} 1 & 0 \end{bmatrix} \begin{bmatrix} x_1(t) \\ x_2(t) \end{bmatrix}$$

(7.204)

The last two equations correspond to the state space model with

$$A = \begin{bmatrix} 0 & 1 \\ -\dfrac{R_1 + R_2}{R_2 LC} & -\dfrac{L + R_1 R_2 C}{R_2 LC} \end{bmatrix}, \quad B = \begin{bmatrix} 0 \\ \dfrac{1}{LC} \end{bmatrix}, \quad C = \begin{bmatrix} 1 & 0 \end{bmatrix}, \quad D = [0] \quad (7.205)$$

∎

Consider a general nth-order model represented by an nth-order differential equation

$$\frac{d^n y(t)}{dt^n} + a_{n-1} \frac{d^{n-1} y(t)}{dt^{n-1}} + \cdots + a_1 \frac{dy(t)}{dt} + a_0 y(t)$$
$$= b_n \frac{d^n w(t)}{dt^n} + b_{n-1} \frac{d^{n-1} w(t)}{dt^{n-1}} + \cdots + b_1 \frac{dw(t)}{dt} + b_0 w(t)$$

(7.206)

We assume that all initial conditions for this differential equation, that is, $y(0^-), dy(0^-)/dt, \ldots, d^{n-1}(y0^-)/dt^{n-1}$, are equal to zero.

To proceed, we first rewrite the above equation with only a single input as follows:

$$\frac{d^n y(t)}{dt^n} + a_{n-1}\frac{d^{n-1} y(t)}{dt^{n-1}} + \cdots + a_1 \frac{dy(t)}{dt} + a_0 y(t) = w(t) \qquad (7.207)$$

We, next, introduce the following change of variables:

$$x_1(t) = y(t), \quad x_2(t) = \frac{dy(t)}{dt}, \quad x_3(t) = \frac{d^2 y(t)}{dt^2}, \ldots, \quad x_n(t) = \frac{d^{n-1} y(t)}{dt^{n-1}} \qquad (7.208)$$

which, after taking derivatives, leads to

$$\frac{dx_1(t)}{dt} = \dot{x}_1(t) = \frac{dy(t)}{dt} = x_2(t)$$

$$\frac{dx_2(t)}{dt} = \dot{x}_2(t) = \frac{d^2 y(t)}{dt^2} = x_3(t)$$

$$\frac{dx_3(t)}{dt} = \dot{x}_3(t) = \frac{d^3 y(t)}{dt^3} = x_4(t)$$

$$\vdots$$

$$\frac{dx_n(t)}{dt} = \dot{x}_n(t) = \frac{d^n y(t)}{dt^n}$$

$$= -a_0 y(t) - a_1 \frac{dy(t)}{dt} - a_2 \frac{d^2 y(t)}{dt^2} - \cdots - a_{n-1}\frac{d^{n-1} y(t)}{dt^{n-1}} + w(t)$$

$$= -a_0 x_1(t) - a_1 x_2(t) - a_2 x_3(t) - \cdots - a_{n-1} x_n(t) + w(t) \qquad (7.209)$$

The state space in matrix form is

$$\begin{bmatrix} \dot{x}_1(t) \\ \dot{x}_2(t) \\ \vdots \\ \vdots \\ \dot{x}_{n-1}(t) \\ \dot{x}_n(t) \end{bmatrix} = \begin{bmatrix} 0 & 1 & 0 & \cdots & \cdots & 0 \\ 0 & 0 & 1 & 0 & \cdots & 0 \\ & & & \vdots & & \\ & & & \vdots & & \\ 0 & 0 & \cdots & \cdots & 0 & 1 \\ -a_0 & -a_1 & -a_2 & \cdots & \cdots & -a_{n-1} \end{bmatrix} \begin{bmatrix} x_1(t) \\ x_2(t) \\ \vdots \\ \vdots \\ x_{n-1}(t) \\ x_n(t) \end{bmatrix} + \begin{bmatrix} 0 \\ 0 \\ \vdots \\ \vdots \\ 0 \\ 1 \end{bmatrix} w(t)$$

$$(7.210)$$

and the corresponding output is obtained from (7.208) as

$$y(t) = \begin{bmatrix} 1 & 0 & 0 & \cdots & 0 \end{bmatrix} \begin{bmatrix} x_1(t) \\ x_2(t) \\ \vdots \\ \vdots \\ x_{n-1}(t) \\ x_n(t) \end{bmatrix} \qquad (7.211)$$

The general state space form in (7.210) and (7.211) is known as the **phase variable canonical form**.

In order to extend this approach to the general case defined by (7.206), which includes derivatives with respect to the input, we proceed by setting an auxiliary equation:

$$\frac{d^n \eta(t)}{dt^n} + a_{n-1}\frac{d^{n-1}\eta(t)}{dt^{n-1}} + \cdots + a_1\frac{d\eta(t)}{dt} + a_0\eta(t) = w(t) \qquad (7.212)$$

with the change of variables

$$x_1(t) = \eta(t), \quad x_2(t) = \frac{d\eta(t)}{dt}, \quad x_3(t) = \frac{d^2\eta(t)}{dt^2}, \ldots, \quad x_n(t) = \frac{d^{n-1}\eta(t)}{dt^{n-1}} \qquad (7.213)$$

and then apply the superposition principle to (7.206) and (7.212). Since $\eta(t)$ is the response of (7.212), by superposition principle the response of (7.206) is given by

$$y(t) = b_0\eta(t) + b_1\frac{d\eta(t)}{dt} + b_2\frac{d^2\eta(t)}{dt^2} + \cdots + b_n\frac{d^n\eta(t)}{dt^n} \qquad (7.214)$$

Equations 7.213 produce the state space equations in the form already given by (7.210). The output equation can be obtained by eliminating $\frac{d^n\eta(t)}{dt^n}$ from (7.214), using (7.212):

$$\frac{d^n\eta(t)}{dt^n} = w(t) - a_{n-1}x_n - \cdots - a_1x_2(t) - a_0x_1(t)$$

This leads to the output equation

$$y(t) = \left[(b_0 - a_0b_n) \quad (b_1 - a_1b_n) \quad \cdots \quad (b_{n-1} - a_{n-1}b_n) \right] \begin{bmatrix} x_1(t) \\ x_2(t) \\ \vdots \\ x_n(t) \end{bmatrix} + b_n w(t) \quad (7.215)$$

Example 7.58

Consider an LTI system that is represented by the differential equation

$$y^{(6)}(t) + 5y^{(5)}(t) - 2y^{(4)}(t) + y^{(2)}(t) - 5y^{(1)}(t) + 3y(t) = 7w^{(3)}(t) + w^{(1)}(t) + 4w(t)$$

where the exponent stands for the order of the derivative. According to (7.210) and (7.211), we write

$$
A = \begin{bmatrix} 0 & 1 & 0 & 0 & 0 & 0 \\ 0 & 0 & 1 & 0 & 0 & 0 \\ 0 & 0 & 0 & 1 & 0 & 0 \\ 0 & 0 & 0 & 0 & 1 & 0 \\ 0 & 0 & 0 & 0 & 0 & 0 \\ -3 & 5 & -1 & 0 & 2 & -5 \end{bmatrix}, \quad B = \begin{bmatrix} 0 \\ 0 \\ 0 \\ 0 \\ 0 \\ 1 \end{bmatrix}, \quad C = [4 \ 1 \ 0 \ 7 \ 0 \ 0], \quad D = [0] \qquad \blacksquare
$$

The transfer function of (7.206), which is obtained using LT, is

$$
H(s) = \frac{b_n s^n + b_{n-1} s^{n-1} + \cdots + b_1 s + b_0}{s^n + a_{n-1} s^{n-1} + \cdots + a_1 s + a_0} \tag{7.216}
$$

Example 7.59

Let the transfer function of a system be

$$
H(s) = \frac{20}{s^3 + 21 s^2 + 24 s + 80} = \frac{0 s^3 + 0 s^2 + 0 s + 20}{s^3 + 21 s^2 + 24 s + 80}
$$

This transfer function corresponds to the following differential equation:

$$
y^{(3)}(t) + 21 y^{(2)}(t) + 24 y^{(1)}(t) + 80 y(t) = 20
$$

Based on (7.210) and (7.215), the state space phase variable canonical form is given by

$$
A = \begin{bmatrix} 0 & 1 & 0 \\ 0 & 0 & 1 \\ -80 & -20 & -21 \end{bmatrix}, \quad B = \begin{bmatrix} 0 \\ 0 \\ 1 \end{bmatrix}, \quad C = [20 \ 0 \ 0], \quad D = [0]
$$

MATLAB has built-in function called **tf2ss** (transfer function to state space), which produces the state space form from the coefficients of the transfer function. For this example, we write in the command window:

```
>>num= [0 0 0 20];
>>den= [1 21 24 80];
>> [A,B,C,D] =tf2ss (num, den);
```

The resulting matrices are

$$
A = \begin{bmatrix} -21 & -24 & -80 \\ 1 & 0 & 0 \\ 0 & 1 & 0 \end{bmatrix}, \quad B = \begin{bmatrix} 1 \\ 0 \\ 0 \end{bmatrix}, \quad C = [0 \ 0 \ 20], \quad D = [0]
$$

Also, the MATLAB function **ss2tf** (state space to transfer function) produces the system transfer function given the state space matrices. For this case, we write

```
>>[num,den] =ss2tf(A,B,C,D);
```
The results are num= [0 0.0000 0.0000 20.0000], den= [1.0000 21.0000 20.0000 80.0000] ∎

Note that the space state form obtained using the MATLAB command differs from the space state form derived above and indicates that the space state form is not unique. To produce the MATLAB version, we make the following substitutions:

$$x_n(t) = y(t), \quad x_{n-1}(t) = \frac{dy(t)}{dt}, \quad x_{n-2}(t) = \frac{d^2y(t)}{dt^2}, \dots, \quad x_1(t) = \frac{d^{n-1}y(t)}{dt^{n-1}} \quad (7.217)$$

Example 7.60

A space state representation of a system is given by the matrices

$$\begin{bmatrix} \dot{x}_1(t) \\ \dot{x}_2(t) \\ \dot{x}_3(t) \end{bmatrix} = \begin{bmatrix} 0 & 1 & 0 \\ 0 & 0 & 1 \\ -1 & -4 & -2 \end{bmatrix} \begin{bmatrix} x_1(t) \\ x_2(t) \\ x_3(t) \end{bmatrix} + \begin{bmatrix} 0 \\ 0 \\ 1 \end{bmatrix} w(t); \quad y(t) = \begin{bmatrix} 0 & 3 & 1 \end{bmatrix} \begin{bmatrix} x_1(t) \\ x_2(t) \\ x_3(t) \end{bmatrix}.$$

The block diagram representing the above phase variable form is shown in Figure 7.48. ∎

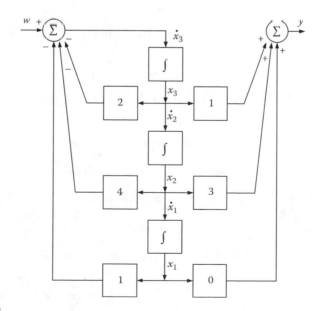

FIGURE 7.48

7.12.2 Time Response Using State Space Representation

The solution of the state space equations (7.184) can be either obtained in the time domain or in the frequency domain using the LT technique.

Time domain solution

Let us start with the scalar case

$$\dot{x}(t) = ax(t) + bw(t) \tag{7.218}$$

with given initial condition $x(0)$. The homogeneous solution of the equation $\dot{x}(t) - ax(t) = 0$ is $x(t) = x(0)e^{at}$. To proceed, we multiply the above equation by the negative exponent of the homogeneous solution and write the results in the form

$$\frac{d}{dt}[e^{-at}x(t)] = be^{at}w(t) \tag{7.219}$$

Integrate both sides of this equation from 0 to t to obtain

$$x(t) = \underbrace{e^{at}x(0)}_{\substack{\text{zero-input} \\ \text{response}}} + \underbrace{\int_0^t e^{a(t-\tau)}bw(\tau)d\tau}_{\substack{\text{zero-state} \\ \text{response}}} \tag{7.220}$$

The exponential term can be expanded in Taylor's series about $x(0) = 0$ as

$$e^{at} = 1 + at + \frac{1}{2!}a^2t^2 + \frac{1}{3!}a^3t^3 + \cdots = \sum_{i=0}^{\infty} \frac{1}{i!}(at)^i \tag{7.221}$$

Analogously, we write the solution for a general nth-order matrix state equation as follows:

$$x(t) = e^{At}x(0) + \int_0^t e^{A(t-\tau)}d\tau \tag{7.222}$$

Let us first consider the homogeneous system without an input

$$\dot{x}(t) = Ax(t) \quad x(0) = \text{initial condition} \tag{7.223}$$

By analogy with the scalar case, the homogeneous solution is

$$x(t) = e^{At}x(0) \tag{7.224}$$

The matrix exponential is defined using the Taylor series expansion as

$$e^{At} = I + At + \frac{1}{2!}A^2t^2 + \frac{1}{3!}A^3t^3 + \cdots = \sum_{i=0}^{\infty} \frac{1}{i!}A^it^i \tag{7.225}$$

Next, we find the derivative of the matrix exponential as follows:

$$\frac{de^{At}}{dt} = \frac{d}{dt}\left(I + At + \frac{1}{2!}A^2t^2 + \cdots\right) = A + \frac{2}{2!}A^2t + \frac{3}{3!}A^3t^2 + \cdots$$

$$= \left(I + \frac{1}{1!}At + \frac{1}{2!}A^2t^2 + \cdots\right)A = e^{At}A$$

Next, substitute (7.224) in differential equation (7.223) to find

$$\dot{x}(t) = \frac{d}{dt}x(t) = \frac{d}{dt}e^{At}x(0) = Ae^{At}x(0) = Ax(t)$$

The above result indicates that the matrix differential equation (7.223) is satisfied, and hence $x(t) = e^{At}x(0)$ is its solution. Note that at $t = 0$, we have $x(0) = e^{A0}x(0) = Ix(0) = x(0)$, which shows that the initial conditions are also satisfied.

The matrix e^{At} is known as the **state transition** or **fundamental matrix**. This matrix relates the system state at time t to that at time zero, and is denoted by

$$\Phi(t) = e^{At} = I + At + A^2\frac{t^2}{2!} + \cdots = \sum_{i=0}^{\infty} A^i \frac{t^i}{i!} \tag{7.226}$$

Properties of $\Phi(t)$

The solution (7.224) can now be written in the form

$$x(t) = \Phi(t)x(0)$$

Property 1

$$\Phi(0) = I$$

Property 2

$$\Phi^{-1}(t) = \Phi(-t) \Rightarrow \Phi(t) \text{ is nonsingular for every } t$$

Property 3

$$\Phi(t_2 - t_0) = \Phi(t_2 - t_1)\Phi(t_1 - t_0)$$

Property 4

$$\Phi^i(t) = \Phi(it) \quad i = \text{integer}$$

Property 5

$$\frac{d}{dt}\Phi(t) = A\Phi(t) \Leftrightarrow \frac{d}{dt}e^{At} = e^{At}A = Ae^{At}$$

The proofs are straightforward. Property 5 has already been established.

The state transition matrix $\boldsymbol{\Phi}(t)$ can be found using several methods. The first is already given in (7.226), the second is based on the use of the LT, and another is based on the Cayley–Hamilton theorem. These last two approaches will be given in subsequent sections.

In the case when the input $w(t)$ is present in the system (forced response), then the equation is

$$\dot{x}(t) = Ax(t) + Bw(t) \quad x(0) = \text{initial condition} \tag{7.227}$$

Multiply each term in this equation by the factor e^{-At}, a procedure that is suggested by the method of variation of parameters. The result is

$$e^{-At}\dot{x}(t) = Ae^{-At}x(t) + e^{-At}Bw(t)$$

Rearrange this equation to the form

$$e^{-At}\dot{x}(t) - e^{-At}Ax(t) = e^{-At}Bw(t)$$

which is

$$\frac{d}{dt}(e^{-At}x(t)) = e^{-At}Bw(t)$$

Multiply by dt and integrate over the interval $t_0 \leq \tau \leq t$. The result is

$$\int_{t_0}^{t} \frac{d}{d\tau}(e^{-A\tau}x(\tau))d\tau = \int_{t_0}^{t} e^{-A\tau}Bw(\tau)d\tau$$

This expression yields

$$e^{At}x(t) - e^{At_0}x(t_0) = \int_{t_0}^{t} e^{A(t-\tau)}Bw(\tau)d\tau$$

Now, multiply all the terms with the positive exponential function, we obtain

$$x(t) = e^{A(t-t_0)}x(t_0) + \int_{t_0}^{t} e^{A(t-\tau)}Bw(\tau)d\tau \tag{7.228}$$

When written in terms of the fundamental matrix, the above expression becomes

$$x(t) = \boldsymbol{\Phi}(t-t_0)x(t_0) + \int_{t_0}^{t} \boldsymbol{\Phi}(t-\tau)Bw(\tau)d\tau \tag{7.229}$$

From the output equation of the state space formulation and (7.228), we obtain

$$y(t) = Ce^{A(t-t_0)}x(t_0) + \int_{t_0}^{t} e^{A(t-\tau)}Bw(\tau)d\tau + Dw(t) \qquad (7.230)$$

Example 7.61

For the system given below

$$\dot{x}(t) = \begin{bmatrix} -1 & 0 & 0 \\ 0 & -2 & 0 \\ 0 & 0 & -3 \end{bmatrix} x(t) + \begin{bmatrix} 1 \\ 1 \\ 1 \end{bmatrix} w(t), \quad y(t) = \begin{bmatrix} 6 & -6 & 1 \end{bmatrix} x(t), \quad t \geq 0$$

we will first find the transition matrix, next we will find the state variable, and then the output. Therefore,

$$\Phi(t) = e^{\begin{bmatrix} -1 & 0 & 0 \\ 0 & -2 & 0 \\ 0 & 0 & -3 \end{bmatrix} t} = \begin{bmatrix} 1 & 0 & 0 \\ 0 & 1 & 0 \\ 0 & 0 & 1 \end{bmatrix} + \begin{bmatrix} 1 & 0 & 0 \\ 0 & -2 & 0 \\ 0 & 0 & -3 \end{bmatrix} t + \frac{1}{2!} \begin{bmatrix} -1 & 0 & 0 \\ 0 & -2 & 0 \\ 0 & 0 & -3 \end{bmatrix}^2 t^2 + \frac{1}{3!} \begin{bmatrix} -1 & 0 & 0 \\ 0 & -2 & 0 \\ 0 & 0 & -3 \end{bmatrix}^3 t^3 + \cdots$$

$$= \begin{bmatrix} 1 & 0 & 0 \\ 0 & 1 & 0 \\ 0 & 0 & 1 \end{bmatrix} + \begin{bmatrix} -t & 0 & 0 \\ 0 & -2t & 0 \\ 0 & 0 & -3t \end{bmatrix} + \begin{bmatrix} \frac{1}{2!}(-1)^2 t^2 & 0 & 0 \\ 0 & \frac{1}{2!}(-2)^2 t^2 & 0 \\ 0 & 0 & \frac{1}{2!}(-3)t^2 \end{bmatrix} + \begin{bmatrix} \frac{1}{3!}(-1)^3 t^3 & 0 & 0 \\ 0 & \frac{1}{3!}(-2)^3 t^3 & 0 \\ 0 & 0 & \frac{1}{3!}(-3)t^3 \end{bmatrix} + \cdots$$

$$= \begin{bmatrix} 1+(-1)t+\frac{1}{2!}(-1)^2t^2+\frac{1}{3!}(-1)^3t^3\cdots & 0 & 0 \\ 0 & 1+(-2)t+\frac{1}{2!}(-2)^2t^2+\frac{1}{3!}(-2)^3t^3\cdots & 0 \\ 0 & 0 & 1+(-3)t+\frac{1}{3!}(-3)^2t^2+\frac{1}{3!}(-3)^3t^3\cdots \end{bmatrix} = \begin{bmatrix} e^{-t} & 0 & 0 \\ 0 & e^{-2t} & 0 \\ 0 & 0 & e^{-3t} \end{bmatrix}$$

At any particular time t, the MATLAB function `expm(A*t)` will evaluate the transition matrix given values at each of its elements.

The state variables to a unit step response is given by

$$x(t) = \int_0^t \Phi(t-\tau)Bw(\tau)d\tau = \int_0^t \begin{bmatrix} e^{-(t-\tau)} & 0 & 0 \\ 0 & e^{-2(t-\tau)} & 0 \\ 0 & 0 & e^{-3(t-\tau)} \end{bmatrix} \begin{bmatrix} 1 \\ 1 \\ 1 \end{bmatrix} 1 d\tau = \int_0^t \begin{bmatrix} e^{-(t-\tau)} \\ e^{-2(t-\tau)} \\ e^{-3(t-\tau)} \end{bmatrix} d\tau$$

$$= \begin{bmatrix} 1 - e^{-t} \\ 0.5(1 - e^{-2t}) \\ 0.333(1 - e^{-3t}) \end{bmatrix}$$

The output is

$$y(t) = \begin{bmatrix} 3 & -3 & 1 \end{bmatrix} \begin{bmatrix} 1 - e^{-t} \\ 0.5(1 - e^{-2t}) \\ 0.333(1 - e^{-3t}) \end{bmatrix} = 3(1 - e^{-t}) - 1.5(1 - e^{-2t}) + 0.333(1 - e^{-3t})$$

For the step response, the initial conditions are set equal to zero and MATLAB uses the following function: `[y,x,t] = step(A,B,C,D)`. The plot of the state variables is accomplished by the command `plot(t,x)`. ∎

7.12.3 Solution Using the Laplace Transform

The time trajectory (solution) of the state vector $x(t)$ can also be found using the LT method.

The LT applied to the state equation (7.184) gives

$$sX(s) - x(0) = AX(s) + BW(s) \quad \text{or} \quad (sI - A)X(s) = x(0) + BW(s)$$

where I is the identity matrix of order n. Since we are dealing with matrices, we must use their inverses, and thus the above matrix becomes

$$X(s) = (sI - A)^{-1}x(0) + (sI - A)^{-1}BW(s) \tag{7.231}$$

Comparing (7.231) and (7.229) with $t_0 = 0$, we conclude that the Laplace of the transition matrix is equal to $(sI - A)^{-1}$. Therefore, we write

$$\Phi(s) = (sI - A)^{-1} = \frac{1}{\det(sI - A)} adj(sI - A) = \mathcal{L}\{\Phi(t)\}$$

or

$$\Phi(t) = \mathcal{L}^{-1}\{\Phi(s)\} = \mathcal{L}^{-1}\{(sI - A)^{-1}\} \tag{7.232}$$

Therefore, the time form of the state vector is obtained by applying the inverse LT of the following equation:

$$X(s) = \Phi(s)x(0) + \Phi(s)BW(s) \tag{7.233}$$

Since the last term is a product in the s-domain, it becomes a convolution integral in the time domain (see LT properties). Hence,

$$x(t) = x_{zi}(t) + x_{zs}(t) = e^{At}x(0) + \int_0^t e^{A(t-\tau)}Bw(\tau)d\tau \tag{7.234}$$

where $x_{zi}(t)$ and $x_{zs}(t)$ are the zero-input and the zero-state components of the system state response.

Once the state vector $\boldsymbol{x}(t)$ is determined, the system output $\boldsymbol{y}(t)$ is simply obtained by substituting $\boldsymbol{x}(t)$ into (7.230), hence,

$$y(t) = y_{zi}(t) + y_{zs}(t) = Ce^{At}x(0) + C \int_0^t e^{A(t-\tau)}Bw(\tau)d\tau + Dw(t) \qquad (7.235)$$

Where $y_{zi}(t)$ and $y_{zs}(t)$ are the zero-input and zero-state components of the output $\boldsymbol{y}(t)$ of the system. The LT of (7.235) becomes

$$Y(s) = Y_{zi}(s) + Y_{zs}(s) = C\boldsymbol{\Phi}(s)x(0) + [C\boldsymbol{\Phi}(s)B + D]W(s) \qquad (7.236)$$

Example 7.62

Let the state space matrices be

$$A = \begin{bmatrix} 0 & 1 \\ -6 & -5 \end{bmatrix}, \quad B = \begin{bmatrix} 0 \\ 1 \end{bmatrix}, \quad C = [1 \ \ 0], \quad D = [0]$$

Then, the state transition matrix is

$$\boldsymbol{\Phi}(s) = (sI - A)^{-1} = \left[\begin{bmatrix} s & 0 \\ 0 & s \end{bmatrix} - \begin{bmatrix} 0 & 1 \\ -6 & -5 \end{bmatrix} \right]^{-1} = \begin{bmatrix} s & -1 \\ 6 & s+5 \end{bmatrix}^{-1}$$

$$= \frac{1}{s(s+5)+6} \begin{bmatrix} s+5 & -6 \\ 1 & s \end{bmatrix}^{T} = \begin{bmatrix} \dfrac{s+5}{(s+2)(s+3)} & \dfrac{1}{(s+2)(s+3)} \\ \dfrac{-6}{(s+2)(s+3)} & \dfrac{s}{(s+2)(s+3)} \end{bmatrix}$$

This implies that

$$\boldsymbol{\Phi}(t) = e^{At} = \mathcal{L}^{-1}\{\boldsymbol{\Phi}(s)\} = \mathcal{L}^{-1}\left\{ \begin{bmatrix} \dfrac{3}{s+2} - \dfrac{2}{s+3} & \dfrac{1}{s+2} - \dfrac{1}{s+3} \\ \dfrac{-6}{s+2} + \dfrac{6}{s+3} & \dfrac{-2}{s+2} + \dfrac{3}{s+3} \end{bmatrix} \right\}$$

$$= \begin{bmatrix} 3e^{-2t} - 2e^{-3t} & e^{-2t} - e^{-3t} \\ -6e^{-2t} + 6e^{-3t} & -2e^{-2t} + 3e^{-3t} \end{bmatrix}$$

If the system input and the initial condition are

$$w(t) = e^{-4t}u(t) \quad \mathbf{x}(0) = \begin{bmatrix} 1 & 0 \end{bmatrix}^{\mathsf{T}}$$

Then, the state variable response is obtained from (7.234)

$$\mathbf{x}(t) = \mathbf{x}_{zi}(t) + \mathbf{x}_{zs}(t) = \begin{bmatrix} x_1(t) \\ x_2(t) \end{bmatrix} = e^{\mathbf{A}t} \begin{bmatrix} 1 \\ 0 \end{bmatrix} + e^{\mathbf{A}t} \int_0^t e^{-\mathbf{A}\tau} \begin{bmatrix} 0 \\ 1 \end{bmatrix} e^{-4t} d\tau$$

$$= \begin{bmatrix} 3e^{-2t} - 2e^{-3t} \\ -6e^{-2t} + 6e^{-3t} \end{bmatrix} + e^{\mathbf{A}t} \int_0^t \begin{bmatrix} e^{2\tau} - e^{-3\tau} \\ -2e^{2\tau} + 3e^{3\tau} \end{bmatrix} e^{-4\tau} d\tau$$

$$= \begin{bmatrix} 3e^{-2t} - 2e^{-3t} \\ -6e^{-2t} + 6e^{-3t} \end{bmatrix} + \begin{bmatrix} \frac{1}{2}e^{-2t} - e^{-3t} + \frac{1}{2}e^{-4t} \\ -e^{-2t} + 3e^{-3t} - 2e^{-4t} \end{bmatrix} = \begin{bmatrix} \frac{7}{2}e^{-2t} - 3e^{-3t} + \frac{1}{2}e^{-4t} \\ -7e^{-2t} + 9e^{-3t} - 2e^{-4t} \end{bmatrix} \quad t \geq 0$$

The system output is obtained as follows:

$$y(t) = \mathbf{C}\mathbf{x}(t) = \begin{bmatrix} 1 & 0 \end{bmatrix} \mathbf{x}(t) = x_1(t) = \frac{7}{2}e^{-2t} - 3e^{-3t} + \frac{1}{2}e^{-4t} \quad t \geq 0 \qquad \blacksquare$$

The state of the system and its output can also be obtained using the transform domain. We use the formulas (7.233) and (7.236) by applying the inverse LT. Therefore, to show this procedure, we write

$$\mathbf{X}(s) = \boldsymbol{\Phi}(s)\mathbf{x}(0) + \boldsymbol{\Phi}(s)\mathbf{B}W(s) = \boldsymbol{\Phi}(s) \begin{bmatrix} 1 \\ 0 \end{bmatrix} + \boldsymbol{\Phi}(s) \begin{bmatrix} 0 \\ 1 \end{bmatrix} \frac{1}{s+4}$$

$$= \begin{bmatrix} \dfrac{s+5}{(s+2)(s+3)} \\ \dfrac{-6}{(s+2)(s+3)} \end{bmatrix} + \begin{bmatrix} \dfrac{1}{(s+2)(s+3)(s+4)} \\ \dfrac{s}{(s+2)(s+3)(s+4)} \end{bmatrix} = \begin{bmatrix} X_1(s) \\ X_2(s) \end{bmatrix}$$

$$Y(s) = \mathbf{C}\mathbf{X}(s) = \begin{bmatrix} 1 & 0 \end{bmatrix} \mathbf{X}(s) = X_1(s)$$

The time domain solution is found from the above equation to be

$$x(t) = \mathcal{L}^{-1}\{X(s)\} = \left\{ \begin{bmatrix} \dfrac{3}{(s+2)} - \dfrac{2}{s+2} \\ \dfrac{-6}{(s+2)} + \dfrac{6}{s+3} \end{bmatrix} + \begin{bmatrix} \dfrac{0.5}{(s+2)} - \dfrac{1}{s+3} + \dfrac{0.5}{s+4} \\ -\dfrac{1}{(s+2)} + \dfrac{3}{s+3} - \dfrac{2}{s+4} \end{bmatrix} \right\}$$

$$= \begin{bmatrix} \dfrac{7}{2}e^{-2t} - 3e^{-3t} + \dfrac{1}{2}e^{-4t} \\ -7e^{-2t} + 9e^{-3t} - 2e^{-4t} \end{bmatrix} = \begin{bmatrix} x_1(t) \\ x_2(t) \end{bmatrix}$$

$$y(t) = \begin{bmatrix} 1 & 0 \end{bmatrix} x(t) = x_1(t)$$

Book MATLAB file: state_response_7_12_3_1

```
%Book MATLAB file: state_response_7_12_3_1
A=[0 1;-6 -5];B=[0;1];C=[1 0];D=[0];x0=[1 0];
t=0:0.1:4;
w=exp(-4*t);
[y,x]=lsim(A,B,C,D,w,t,x0);%lsim is a MATLAB function;
subplot(2,1,1);plot(t,x,'k');grid;
xlabel('Time (s)');ylabel('System state response');
subplot(2,1,2);plot(x(:,1),x(:,2),'k');
xlabel('x_1(t)');ylabel('x_2(t)');
```

The above script file produces Figure 7.49.

7.12.4 State Space Transfer Function

The transfer function is defined with zero initial conditions, and it is the ratio of the output to the input of the system in the s-domain. From (7.236), we obtain the transfer function

$$H(s) = \frac{Y(s)}{W(s)} = C\boldsymbol{\Phi}(s)\boldsymbol{B} + \boldsymbol{D} = C(sI - A)^{-1}\boldsymbol{B} + \boldsymbol{D} \tag{7.237}$$

MATLAB has a special function that gives the numerator and denominator factors. The function is

```
≫[num,den]=ss2tf(A,B,C,D);
```

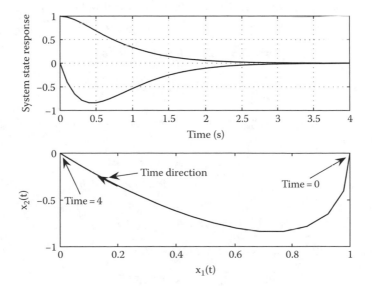

FIGURE 7.49

7.12.5 Impulse and Step Response

One of the most useful outputs of a system is when it is excited by an impulse or step response. Therefore, for a MIMO system with zero initial conditions, we have

$$x(t) = \int_0^t e^{A(t-\tau)} Bw(\tau) d\tau$$

$$y(t) = Cx(t) + Dw(t)$$

(7.238)

Let the input function have r components, each one is zero besides the jth component. Then, we write

$$w^{(j)} = [0 \quad 0 \cdots 0 \quad w_j(t) \quad 0 \cdots 0]^T$$

(7.239)

Based on the above definition, we also have

$$Bw^{(j)}(t) = w_j(t) b_j \quad B = [b_1 \quad b_2 \cdots b_{j-1} \quad b_j \quad b_{j+1} \cdots b_{r-1} \quad b_r]$$

(7.240)

The state and output responses, due to the jth component of the input signal, are

$$x^{(j)}(t) = \int_0^t e^{A(t-\tau)} b_j w_j(\tau) d\tau$$

$$y^{(j)}(t) = Cx^{(j)}(t) + Dw^{(j)}(t) = Cx^{(j)}(t) + d_j w_j(t)$$

(7.241)

where d_j is the jth column of the matrix D.

If $w_j(t)$ is an impulse, then the state variables and output will be

$$x^{(j)}(t) = \int_0^t e^{A(t-\tau)} b_j \delta(\tau) d\tau = e^{At} b_j$$

$$y^{(j)}(t) = Ce^{At} b_j + d_j \delta(t) \triangleq h^{(j)}(t)$$

(7.242)

In the same way, if $w_j(t) = u(t)$, with all the other inputs equal to zero, then the system variables and system output are

$$x_{step}^{(j)}(t) = \int_0^t e^{A(t-\tau)} b_j u(\tau) d\tau = \int_0^t e^{A(t-\tau)} d\tau b_j$$

$$y_{step}^{(j)}(t) = Cx_{step}^{(j)}(t) + d_j u(t)$$

(7.243)

Taking the derivative of both sides of (7.243), we obtain

$$h^{(t)}(t) = \frac{d}{dt} y_{step}^{(j)}(t)$$

(7.244)

The differentiation in (7.244) must be performed component-wise; that is, the derivative operator must be applied to every component of the output vector.

The impulse and step responses of MIMO systems can be found using the following MATLAB statements:

```
>>t = ti:dt:tf;
>> [y,x] = impulse(A,B,C,D,in,t);%in=number of inputs to the
>>                                %system;
>>                        %ti=initial time; dt=sampling time;
>> [y,x] = step(A,B,C,D,in,t);   %tf=final time; t=response
                                 %time interval;
```

TABLE 7.3 LT Pairs

$F(s)$		$f(t)$
1	s^n	$\delta^{(n)}(t)$ n^{th} derivative of the delta function
2	s	$\dfrac{d\delta(t)}{dt}$
3	1	$\delta(t)$
4	$\dfrac{1}{s}$	1
5	$\dfrac{1}{s^2}$	t
6	$\dfrac{1}{s^n}\,(n = 1, 2, \ldots)$	$\dfrac{t^{n-1}}{(n-1)!}$
7	$\dfrac{1}{\sqrt{s}}$	$\dfrac{1}{\sqrt{\pi t}}$
8	$s^{-3/2}$	$2\sqrt{\dfrac{t}{\pi}}$
9	$s^{-[n+(1/2)]}\,(n = 1, 2, \ldots)$	$\dfrac{2^n t^{n-(1/2)}}{1 \cdot 3 \cdot 5 \cdots (2n-1)\sqrt{\pi}}$
10	$\dfrac{\Gamma(k)}{s^k}\,(k \geq 0)$	t^{k-1}
11	$\dfrac{1}{s-a}$	e^{at}
12	$\dfrac{1}{(s-a)^2}$	te^{at}
13	$\dfrac{1}{(s-a)^n}\,(n = 1, 2, \ldots)$	$\dfrac{1}{(n-1)!}t^{n-1}e^{at}$
14	$\dfrac{\Gamma(k)}{(s-a)^k}\,(k \geq 0)$	$t^{k-1}e^{at}$
15	$\dfrac{1}{(s-a)(s-b)}$	$\dfrac{1}{(a-b)}(e^{at} - e^{bt})$
16	$\dfrac{s}{(s-a)(s-b)}$	$\dfrac{1}{(a-b)}(ae^{at} - be^{bt})$
17	$\dfrac{1}{(s-a)(s-b)(s-c)}$	$-\dfrac{(b-c)e^{at} + (c-a)e^{bt} + (a-b)e^{ct}}{(a-b)(b-c)(c-a)}$

(continued)

TABLE 7.3 (continued)   LT Pairs

$F(s)$	$f(t)$
18 $\dfrac{1}{(s+a)}$	e^{-at} valid for complex a
19 $\dfrac{1}{s(s+a)}$	$\dfrac{1}{a}(1-e^{-at})$
20 $\dfrac{1}{s^2(s+a)}$	$\dfrac{1}{a^2}(e^{-at}+at-1)$
21 $\dfrac{1}{s^3(s+a)}$	$\dfrac{1}{a^2}\left[\dfrac{1}{a}-t+\dfrac{at^2}{2}-\dfrac{1}{a}e^{-at}\right]$
22 $\dfrac{1}{(s+a)(s+b)}$	$\dfrac{1}{(b-a)}(e^{-at}-e^{-bt})$
23 $\dfrac{1}{s(s+a)(s+b)}$	$\dfrac{1}{ab}\left[1+\dfrac{1}{(a-b)}(be^{-at}-ae^{-bt})\right]$
24 $\dfrac{1}{s^2(s+a)(s+b)}$	$\dfrac{1}{(ab)^2}\left[\dfrac{1}{(a-b)}(a^2e^{-bt}-b^2e^{-at})+abt-a-b\right]$
25 $\dfrac{1}{s^3(s+a)(s+b)}$	$\dfrac{1}{(ab)}\left[\dfrac{a^3-b^3}{(ab)^2(a-b)}+\dfrac{1}{2}t^2-\dfrac{(a+b)}{ab}t+\dfrac{1}{(a-b)}\left(\dfrac{b}{a^2}e^{-at}-\dfrac{a}{b^2}e^{-bt}\right)\right]$
26 $\dfrac{1}{(s+a)(s+b)(s+c)}$	$\dfrac{1}{(b-a)(c-a)}e^{-at}+\dfrac{1}{(a-b)(c-b)}e^{-bt}+\dfrac{1}{(a-c)(b-c)}e^{-ct}$
27 $\dfrac{1}{s(s+a)(s+b)(s+c)}$	$\dfrac{1}{abc}-\dfrac{1}{a(b-a)(c-a)}e^{-at}-\dfrac{1}{b(a-b)(c-b)}e^{-bt}-\dfrac{1}{c(a-c)(b-c)}e^{-ct}$
28 $\dfrac{1}{s^2(s+a)(s+b)(s+c)}$	$\begin{cases}\dfrac{ab(ct-1)-ac-bc}{(abc)^2}+\dfrac{1}{a^2(b-a)(c-a)}e^{-at}\\[2mm]+\dfrac{1}{b^2(a-b)(c-b)}e^{-bt}+\dfrac{1}{c^2(a-c)(b-c)}e^{-ct}\end{cases}$
29 $\dfrac{1}{s^3(s+a)(s+b)(s+c)}$	$\begin{cases}\dfrac{1}{(abc)^3}[(ab+ac+bc)^2-abc(a+b+c)]-\dfrac{ab+ac+bc}{(abc)^2}t+\dfrac{1}{2abc}t^2\\[2mm]-\dfrac{1}{a^3(b-a)(c-a)}e^{-at}-\dfrac{1}{b^3(a-b)(c-b)}e^{-bt}-\dfrac{1}{c^3(a-c)(b-c)}e^{-ct}\end{cases}$
30 $\dfrac{1}{s^2+a^2}$	$\dfrac{1}{a}\sin at$
31 $\dfrac{s}{s^2+a^2}$	$\cos at$
32 $\dfrac{1}{s^2-a^2}$	$\dfrac{1}{a}\sinh at$
33 $\dfrac{s}{s^2-a^2}$	$\cosh at$
34 $\dfrac{1}{s(s^2+a^2)}$	$\dfrac{1}{a^2}(1-\cos at)$
35 $\dfrac{1}{s^2(s^2+a^2)}$	$\dfrac{1}{a^3}(at-\sin at)$
36 $\dfrac{1}{(s^2+a^2)^2}$	$\dfrac{1}{2a^3}(\sin at-at\cos at)$
37 $\dfrac{s}{(s^2+a^2)^2}$	$\dfrac{t}{2a}\sin at$
38 $\dfrac{s^2}{(s^2+a^2)^2}$	$\dfrac{1}{2a}(\sin at+at\cos at)$

TABLE 7.3 (continued) LT Pairs

$F(s)$	$f(t)$
39 $\dfrac{s^2 - a^2}{(s^2 + a^2)^2}$	$t \cos at$
40 $\dfrac{s}{(s^2 + a^2)(s^2 + b^2)}\,(a^2 \neq b^2)$	$\dfrac{\cos at - \cos bt}{b^2 - a^2}$
41 $\dfrac{1}{(s - a)^2 + b^2}$	$\dfrac{1}{b}e^{at}\sin bt$
42 $\dfrac{s - a}{(s - a)^2 + b^2}$	$e^{at}\cos bt$
43 $\dfrac{1}{[(s + a)^2 + b^2]^n}$	$\dfrac{-e^{-at}}{4^{n-1}b^{2n}}\sum_{r=1}^{n}\binom{2n - r - 1}{n - 1}(-2t)^{r-1}\dfrac{d^r}{dt^r}[\cos(bt)]$
44 $\dfrac{s}{[(s + a)^2 + b^2]^n}$	$\begin{cases}\dfrac{e^{-at}}{4^{n-1}b^{2n}}\left\{\sum_{r=1}^{n}\binom{2n - r - 1}{n - 1}(-2t)^{r-1}\dfrac{d^r}{dt^r}[a\cos(bt) + b\sin(bt)]\right.\\[2mm] \left.-2b\sum_{r=1}^{n-1}r\binom{2n - r - 2}{n - 1}(-2t)^{r-1}\dfrac{d^r}{dt^r}[\sin(bt)]\right\}\end{cases}$
45 $\dfrac{3a^2}{s^3 + a^3}$	$e^{-at} - e^{(at)/2}\left(\cos\dfrac{at\sqrt{3}}{2}\quad \sqrt{3}\sin\dfrac{at\sqrt{3}}{2}\right)$
46 $\dfrac{4a^3}{s^4 + 4a^4}$	$\sin at \cosh at - \cos at \sinh at$
47 $\dfrac{s}{s^4 + 4a^4}$	$\dfrac{1}{2a^2}(\sin at\ \sinh at)$
48 $\dfrac{1}{s^4 - a^4}$	$\dfrac{1}{2a^3}(\sinh at - \sin at)$
49 $\dfrac{s}{s^4 - a^4}$	$\dfrac{1}{2a^2}(\cosh at - \cos at)$
50 $\dfrac{8a^3 s^2}{(s^2 + a^2)^3}$	$(1 + a^2 t^2)\sin at - \cos at$
51 $\dfrac{1}{s}\left(\dfrac{s - 1}{s}\right)^n$	$L_n(t) = \dfrac{e^t}{n!}\dfrac{d^n}{dt^n}(t^n e^{-t})\ [L_n(t)$ is the Laguerre polynomial of degree $n]$
52 $\dfrac{1}{(s + a)^n}$	$\dfrac{t^{(n-1)}e^{-at}}{(n - 1)!}$ where n is a positive integer
53 $\dfrac{1}{s(s + a)^2}$	$\dfrac{1}{a^2}[1 - e^{-at} - ate^{-at}]$
54 $\dfrac{1}{s^2(s + a)^2}$	$\dfrac{1}{a^3}[at - 2 + ate^{-at} + 2e^{-at}]$
55 $\dfrac{1}{s(s + a)^3}$	$\dfrac{1}{a^3}\left[1 - \left(\dfrac{1}{2}a^2 t^2 + at + 1\right)e^{-at}\right]$
56 $\dfrac{1}{(s + a)(s + b)^2}$	$\dfrac{1}{(a - b)^2}\{e^{-at} + [(a - b)t - 1]e^{-bt}\}$
57 $\dfrac{1}{s(s + a)(s + b)^2}$	$\dfrac{1}{ab^2} - \dfrac{1}{a(a - b)^2}e^{-at} - \left[\dfrac{1}{b(a - b)}t + \dfrac{a - 2b}{b^2(a - b)^2}\right]e^{-bt}$
58 $\dfrac{1}{s^2(s + a)(s + b)^2}$	$\dfrac{1}{a^2(a - b)^2}e^{-at} + \dfrac{1}{ab^2}\left(t - \dfrac{1}{a} - \dfrac{2}{b}\right) + \left[\dfrac{1}{b^2(a - b)}t + \dfrac{2(a - b) - b}{b^3(a - b)^2}\right]e^{-bt}$

(continued)

TABLE 7.3 (continued) LT Pairs

	$F(s)$	$f(t)$
59	$\dfrac{1}{(s+a)(s+b)(s+c)^2}$	$\begin{cases} \left[\dfrac{1}{(c-b)(c-a)}t + \dfrac{2c-a-b}{(c-a)^2(c-b)^2}\right]e^{-ct} \\[2mm] + \dfrac{1}{(b-a)(c-a)^2}e^{-at} + \dfrac{1}{(a-b)(c-b)^2}e^{-bt} \end{cases}$
60	$\dfrac{1}{(s+a)(s^2+\omega^2)}$	$\dfrac{1}{a^2+\omega^2}e^{-at} + \dfrac{1}{\omega\sqrt{a^2+\omega^2}}\sin(\omega t - \phi); \quad \phi = \tan^{-1}\left(\dfrac{\omega}{a}\right)$
61	$\dfrac{1}{s(s+a)(s^2+\omega^2)}$	$\dfrac{1}{a\omega^2} - \dfrac{1}{a^2+\omega^2}\left(\dfrac{1}{\omega}\sin\omega t + \dfrac{a}{\omega^2}\cos\omega t + \dfrac{1}{a}e^{-at}\right)$
62	$\dfrac{1}{s^2(s+a)(s^2+\omega^2)}$	$\begin{cases} \dfrac{1}{a\omega^2}t - \dfrac{1}{a^2\omega^2} + \dfrac{1}{a^2(a^2+\omega^2)}e^{-at} \\[2mm] + \dfrac{1}{\omega^3\sqrt{a^2+\omega^2}}\cos(\omega t + \phi); \quad \phi = \tan^{-1}\left(\dfrac{a}{\omega}\right) \end{cases}$
63	$\dfrac{1}{[(s+a)^2+\omega^2]^2}$	$\dfrac{1}{2\omega^3}e^{-at}[\sin\omega t - \omega t\cos\omega t]$
64	$\dfrac{1}{s^2-a^2}$	$\dfrac{1}{a}\sinh at$
65	$\dfrac{1}{s^2(s^2-a^2)}$	$\dfrac{1}{a^3}\sinh at - \dfrac{1}{a^2}t$
66	$\dfrac{1}{s^3(s^2-a^2)}$	$\dfrac{1}{a^4}(\cosh at - 1) - \dfrac{1}{2a^2}t^2$
67	$\dfrac{1}{s^3+a^3}$	$\dfrac{1}{3a^2}\left[e^{-at} - e^{\frac{a}{2}t}\left(\cos\dfrac{\sqrt{3}}{2}at - \sqrt{3}\sin\dfrac{\sqrt{3}}{2}at\right)\right]$
68	$\dfrac{1}{s^4+4a^4}$	$\dfrac{1}{4a^3}(\sin at\cosh at - \cos at\sinh at)$
69	$\dfrac{1}{s^4-a^4}$	$\dfrac{1}{2a^3}(\sinh at - \sin at)$
70	$\dfrac{1}{[(s+a)^2-\omega^2]}$	$\dfrac{1}{\omega}e^{-at}\sinh\omega t$
71	$\dfrac{s+a}{s[(s+a)^2+\omega^2]}$	$\begin{cases} \dfrac{a}{b^2+\omega^2} - \dfrac{1}{\omega} + \sqrt{\dfrac{(a-b)^2+\omega^2}{b^2+\omega^2}}e^{-bt}\sin(\omega t + \phi); \\[3mm] \phi = \tan^{-1}\left(\dfrac{\omega}{b}\right) + \tan^{-1}\left(\dfrac{\omega}{a-b}\right) \end{cases}$
72	$\dfrac{s+a}{s^2[(s+b)^2+\omega^2]}$	$\begin{cases} \dfrac{1}{b^2+\omega^2}[1+at] - \dfrac{2ab}{(b^2+\omega^2)^2} + \sqrt{\dfrac{(a-b)^2+\omega^2}{\omega(b^2+\omega^2)}}e^{-bt}\sin(\omega t + \phi) \\[3mm] \phi = \tan^{-1}\left(\dfrac{\omega}{a-b}\right) + 2\tan^{-1}\left(\dfrac{\omega}{b}\right). \end{cases}$
73	$\dfrac{s+a}{(s+c)[(s+b)^2+\omega^2]}$	$\begin{cases} \dfrac{a-c}{(c-b)^2+\omega^2}e^{-ct} + \dfrac{1}{\omega} + \sqrt{\dfrac{(a-b)^2+\omega^2}{(c-b)^2+\omega^2}}e^{-bt}\sin(\omega t + \phi) \\[3mm] \phi = \tan^{-1}\left(\dfrac{\omega}{a-b}\right) - \tan^{-1}\left(\dfrac{\omega}{c-b}\right) \end{cases}$
74	$\dfrac{s+a}{s(s+c)[(s+b)^2+\omega^2]}$	$\begin{cases} \dfrac{a}{c(b^2+\omega^2)} + \dfrac{(c-a)}{c[(b-c)^2+\omega^2]}e^{-ct} \\[3mm] - \dfrac{1}{\omega\sqrt{b^2+\omega^2}}\sqrt{\dfrac{(a-b)^2+\omega^2}{(b-c)^2+\omega^2}}e^{-bt}\sin(\omega t + \phi) \\[3mm] \phi = \tan^{-1}\left(\dfrac{\omega}{b}\right) + \tan^{-1}\left(\dfrac{\omega}{a-b}\right) - \tan^{-1}\left(\dfrac{\omega}{c-b}\right) \end{cases}$

TABLE 7.3 (continued) LT Pairs

$F(s)$	$f(t)$
75 $\dfrac{s+a}{s^2(s+b)^3}$	$\dfrac{a}{b^3}t+\dfrac{b-3a}{b^4}+\left[\dfrac{3a-b}{b^4}+\dfrac{a-b}{2b^2}t^2+\dfrac{2a-b}{b^3}t\right]e^{-bt}$
76 $\dfrac{s+a}{(s+c)(s+b)^3}$	$\dfrac{a-c}{(b-c)^3}e^{-ct}+\left[\dfrac{a-b}{2(c-b)}t^2+\dfrac{c-a}{(c-b)^2}t+\dfrac{a-c}{(c-b)^3}\right]e^{-bt}$
77 $\dfrac{s^2}{(s+a)(s+b)(s+c)}$	$\dfrac{a^2}{(b-a)(c-a)}e^{-at}+\dfrac{b^2}{(a-b)(c-b)}e^{-bt}+\dfrac{c^2}{(a-c)(b-c)}e^{-ct}$
78 $\dfrac{s^2}{(s+a)(s+b)^2}$	$\dfrac{a^2}{(b-a)^2}e^{-at}+\left[\dfrac{b^2}{(a-b)}t+\dfrac{b^2-2ab}{(a-b)^2}\right]e^{-bt}$
79 $\dfrac{s^2}{(s+a)^3}$	$\left[2-2at+\dfrac{a^2}{2}t^2\right]e^{-at}$
80 $\dfrac{s^2}{(s+a)(s^2+\omega^2)}$	$\dfrac{a^2}{(a^2+\omega^2)}e^{-at}-\dfrac{\omega}{\sqrt{a^2+\omega^2}}\sin(\omega t+\phi);\ \phi=\tan^{-1}\left(\dfrac{\omega}{a}\right)$
81 $\dfrac{s^2}{(s+a)^2(s^2+\omega^2)}$	$\begin{cases}\left[\dfrac{a^2}{(a^2+\omega^2)}t-\dfrac{2a\omega^2}{(a^2+\omega^2)^2}\right]e^{-at}-\dfrac{\omega}{(a^2+\omega^2)}\sin(\omega t+\phi);\\[2mm]\phi=-2\tan^{-1}\left(\dfrac{\omega}{a}\right)\end{cases}$
82 $\dfrac{s^2}{(s+a)(s+b)(s^2+\omega^2)}$	$\begin{cases}\dfrac{a^2}{(b-a)(a^2+\omega^2)}e^{-at}+\dfrac{b^2}{(a-b)(b^2+\omega^2)}e^{-bt}\\[2mm]-\dfrac{\omega}{\sqrt{(a^2+\omega^2)(b^2+\omega^2)}}\sin(\omega t+\phi);\ \phi=-\left[\tan^{-1}\left(\dfrac{\omega}{a}\right)+\tan^{-1}\left(\dfrac{\omega}{b}\right)\right]\end{cases}$
83 $\dfrac{s^2}{(s^2+a^2)(s^2+\omega^2)}$	$-\dfrac{a}{(\omega^2-a^2)}\sin(at)-\dfrac{\omega}{(a^2-\omega^2)}\sin(\omega t)$
84 $\dfrac{s^2}{(s^2+\omega^2)^2}$	$\dfrac{1}{2\omega}(\sin\omega t+\omega t\cos\omega t)$
85 $\dfrac{s^2}{(s+a)[(s+b)^2+\omega^2]}$	$\begin{cases}\dfrac{a^2}{(a-b)^2+\omega^2}e^{-at}+\dfrac{1}{\omega}\sqrt{\dfrac{(b^2-\omega^2)^2+4b^2\omega^2}{(a-b)^2+\omega^2}}\,e^{-bt}\sin(\omega t+\phi)\\[3mm]\phi=\tan^{-1}\left(\dfrac{-2b\omega}{b^2-\omega^2}\right)-\tan^{-1}\left(\dfrac{\omega}{a-b}\right)\end{cases}$
86 $\dfrac{s^2}{(s+a)^2[(s+b)^2+\omega^2]}$	$\begin{cases}\dfrac{a^2}{(a-b)^2+\omega^2}te^{-at}-2\left[\dfrac{a[(b-a)^2+\omega^2]+a^2(b-a)}{[(b-a)^2+\omega^2]^2}\right]e^{-at}\\[3mm]+\dfrac{\sqrt{(b^2-\omega^2)^2+4b^2\omega^2}}{\omega[(a-b)^2+\omega^2]}e^{-bt}\sin(\omega t+\phi)\\[3mm]\phi=\tan^{-1}\left(\dfrac{-2b\omega}{b^2-\omega^2}\right)-2\tan^{-1}\left(\dfrac{\omega}{a-b}\right)\end{cases}$
87 $\dfrac{s^2+a}{s^2(s+b)}$	$\dfrac{b^2+a}{b^2}e^{-bt}+\dfrac{a}{b}t-\dfrac{a}{b^2}$
88 $\dfrac{s^2+a}{s^3(s+b)}$	$\dfrac{a}{2b}t^2-\dfrac{a}{b^2}t+\dfrac{1}{b^3}\left[b^2+a-(a+b^2)e^{-bt}\right]$
89 $\dfrac{s^2+a}{s(s+b)(s+c)}$	$\dfrac{a}{bc}+\dfrac{(b^2+a)}{b(b-c)}e^{-bt}-\dfrac{(c^2+a)}{c(b-c)}e^{-ct}$

(continued)

TABLE 7.3 (continued)   LT Pairs

$F(s)$	$f(t)$
90 $\dfrac{s^2 + a}{s^2(s+b)(s+c)}$	$\dfrac{b^2 + a}{b^2(c-b)}e^{-bt} + \dfrac{c^2 + a}{c^2(b-c)}e^{-ct} + \dfrac{a}{bc}t - \dfrac{a(b+c)}{b^2 c^2}$
91 $\dfrac{s^2 + a}{(s+b)(s+c)(s+d)}$	$\dfrac{b^2 + a}{(c-b)(d-b)}e^{-bt} + \dfrac{c^2 + a}{(b-c)(d-c)}e^{-ct} + \dfrac{d^2 + a}{(b-d)(c-d)}e^{-dt}$
92 $\dfrac{s^2 + a}{s(s+b)(s+c)(s+d)}$	$\dfrac{a}{bcd} + \dfrac{b^2 + a}{b(b-c)(d-b)}e^{-bt} + \dfrac{c^2 + a}{c(b-c)(c-d)}e^{-ct} + \dfrac{d^2 + a}{d(b-d)(d-c)}e^{-dt}$
93 $\dfrac{s^2 + a}{s^2(s+b)(s+c)(s+d)}$	$\begin{cases} \dfrac{a}{bcd}t - \dfrac{a}{b^2 c^2 d^2}(bc + cd + db) + \dfrac{b^2 + a}{b^2(b-c)(b-d)}e^{-bt} \\[2mm] + \dfrac{c^2 + a}{c^2(c-b)(c-d)}e^{-ct} + \dfrac{d^2 + a}{d^2(d-b)(d-c)}e^{-dt} \end{cases}$
94 $\dfrac{s^2 + a}{(s^2 + \omega^2)^2}$	$\dfrac{1}{2\omega^3}(a + \omega^2)\sin\omega t - \dfrac{1}{2\omega^2}(a - \omega^2)t\cos\omega t$
95 $\dfrac{s^2 - \omega^2}{(s^2 + \omega^2)^2}$	$t\cos\omega t$
96 $\dfrac{s^2 + a}{s(s^2 + \omega^2)^2}$	$\dfrac{a}{\omega^4} - \dfrac{(a - \omega^2)}{2\omega^3}t\sin\omega t - \dfrac{a}{\omega^4}\cos\omega t$
97 $\dfrac{s(s+a)}{(s+b)(s+c)^2}$	$\dfrac{b^2 - ab}{(c-b)^2}e^{-bt} + \left[\dfrac{c^2 - ac}{b-c}t + \dfrac{c^2 - 2bc + ab}{(b-c)^2}\right]e^{-ct}$
98 $\dfrac{s(s+a)}{(s+b)(s+c)(s+d)^2}$	$\begin{cases} \dfrac{b^2 - ab}{(c-b)(d-b)^2}e^{-bt} + \dfrac{c^2 - ac}{(b-c)(d-c)^2}e^{-ct} + \dfrac{d^2 - ad}{(b-d)(c-d)}te^{-dt} \\[2mm] + \dfrac{a(bc - d^2) + d(db + dc - 2bc)}{(b-d)^2(c-d)^2}e^{-dt} \end{cases}$
99 $\dfrac{s^2 + a_1 s + a_0}{s^2(s+b)}$	$\dfrac{b^2 - a_1 b + a_0}{b^2}e^{-bt} + \dfrac{a_0}{b}t + \dfrac{a_1 b - a_0}{b^2}$
100 $\dfrac{s^2 + a_1 s + a_0}{s^3(s+b)}$	$\dfrac{a_1 b - b^2 - a_0}{b^3}e^{-bt} + \dfrac{a_0}{2b}t^2 + \dfrac{a_1 b - a_0}{b^2}t + \dfrac{b^2 - a_1 b + a_0}{b^3}$
101 $\dfrac{s^2 + a_1 s + a_0}{s(s+b)(s+c)}$	$\dfrac{a_0}{bc} + \dfrac{b^2 - a_1 b + a_0}{b(b-c)}e^{-bt} + \dfrac{c^2 - a_1 c + a_0}{c(c-b)}e^{-ct}$
102 $\dfrac{s^2 + a_1 s + a_0}{s^2(s+b)(s+c)}$	$\dfrac{a_0}{bc}t + \dfrac{a_1 bc - a_0(b+c)}{b^2 c^2} + \dfrac{b^2 - a_1 b + a_0}{b^2(c-b)}e^{-bt} + \dfrac{c^2 - a_1 c + a_0}{c^2(b-c)}e^{-ct}$
103 $\dfrac{s^2 + a_1 s + a_0}{(s+b)(s+c)(s+d)}$	$\dfrac{b^2 - a_1 b + a_0}{(c-b)(d-b)}e^{-bt} + \dfrac{c^2 - a_1 c + a_0}{(b-c)(d-c)}e^{-ct} + \dfrac{d^2 - a_1 d + a_0}{(b-d)(c-d)}e^{-dt}$
104 $\dfrac{s^2 + a_1 s + a_0}{s(s+b)(s+c)(s+d)}$	$\dfrac{a_0}{bcd} - \dfrac{b^2 - a_1 b + a_0}{b(c-b)(d-b)}e^{-bt} - \dfrac{c^2 - a_1 c + a_0}{c(b-c)(d-c)}e^{-ct} - \dfrac{d^2 - a_1 d + a_0}{d(b-d)(c-d)}e^{-dt}$
105 $\dfrac{s^2 + a_1 s + a_0}{s(s+b)^2}$	$\dfrac{a_0}{b^2} - \dfrac{b^2 - a_1 b + a_0}{b}te^{-bt} + \dfrac{b^2 - a_0}{b^2}e^{-bt}$
106 $\dfrac{s^2 + a_1 s + a_0}{s^2(s+b)^2}$	$\dfrac{a_0}{b^2}t + \dfrac{a_1 b - 2a_0}{b^3} + \dfrac{b^2 - a_1 b + a_0}{b^2}t\varepsilon^{-bt} + \dfrac{2a_0 - a_1 b}{b^3}e^{-bt}$

TABLE 7.3 (continued) LT Pairs

$F(s)$	$f(t)$
107 $\dfrac{s^2 + a_1 s + a_0}{(s+b)(s+c)^2}$	$\dfrac{b^2 - a_1 b + a_0}{(c-b)^2} e^{-bt} + \dfrac{c^2 - a_1 c + a_0}{(b-c)} te^{-ct} + \dfrac{c^2 - 2bc + a_1 b - a_0}{(b-c)^2} e^{-ct}$

108 $\dfrac{s^3}{(s+b)(s+c)(s+d)^2}$

$$\begin{cases} \dfrac{b^3}{(b-c)(d-b)^2} e^{-bt} + \dfrac{c^3}{(c-b)(d-c)^2} e^{-ct} + \dfrac{d^3}{(d-b)(c-d)} te^{-dt} \\[4mm] + \dfrac{d^2[d^2 - 2d(b+c) + 3bc]}{(b-d)^2(c-d)^2} e^{-dt} \end{cases}$$

109 $\dfrac{s^3}{(s+b)(s+c)(s+d)(s+f)^2}$

$$\begin{cases} \dfrac{b^3}{(b-c)(d-b)(f-b)^2} e^{-bt} + \dfrac{c^3}{(c-b)(d-c)(f-c)^2} e^{-ct} \\[4mm] + \dfrac{d^3}{(d-b)(c-d)(f-d)^2} e^{-dt} + \dfrac{f^3}{(f-b)(c-f)(d-f)} te^{-ft} \\[4mm] + \left[\dfrac{3f^2}{(b-f)(c-f)(d-f)} \right. \\[4mm] \left. + \dfrac{f^3[(b-f)(c-f) + (b-f)(d-f) + (c-f)(d-f)]}{(b-f)^2(c-f)^2(d-f)^2} \right]\varepsilon^{-dt} \end{cases}$$

110 $\dfrac{s^3}{(s+b)^2(s+c)^2}$

$$-\dfrac{b^3}{(c-b)^2} te^{-bt} + \dfrac{b^2(3c-b)}{(c-b)^3} e^{-bt} - \dfrac{c^3}{(b-c)^2} te^{-ct} + \dfrac{c^2(3b-c)}{(b-c)^3} e^{-ct}$$

111 $\dfrac{s^3}{(s+d)(s+b)^2(s+c)^2}$

$$\begin{cases} -\dfrac{d^3}{(b-d)^2(c-d)^2} e^{-dt} + \dfrac{b^3}{(c-b)^2(b-d)} te^{-bt} \\[4mm] + \left[\dfrac{3b^2}{(c-b)^2(d-b)} + \dfrac{b^3(c+2d-3b)}{(c-b)^3(d-b)^2}\right] e^{-bt} + \dfrac{c^3}{(b-c)^2(c-d)} te^{-ct} \\[4mm] + \left[\dfrac{3c^2}{(b-c)^2(d-c)} + \dfrac{c^3(b+2d-3c)}{(b-c)^3(d-c)^2}\right] e^{-ct} \end{cases}$$

112 $\dfrac{s^3}{(s+b)(s+c)(s^2+\omega^2)}$

$$\begin{cases} \dfrac{b^3}{(b-c)(b^2+\omega^2)} e^{-bt} + \dfrac{c^3}{(c-b)(c^2+\omega^2)} e^{-ct} \\[4mm] - \dfrac{\omega^2}{\sqrt{(b^2+\omega^2)(c^2+\omega^2)}} \sin(\omega t + \phi) \\[4mm] \phi = \tan^{-1}\left(\dfrac{c}{\omega}\right) - \tan^{-1}\left(\dfrac{\omega}{b}\right) \end{cases}$$

113 $\dfrac{s^3}{(s+b)(s+c)(s+d)(s^2+\omega^2)}$

$$\begin{cases} \dfrac{b^3}{(b-c)(d-b)(b^2+\omega^2)} e^{-bt} + \dfrac{c^3}{(c-b)(d-c)(c^2+\omega^2)} e^{-ct} \\[4mm] + \dfrac{d^3}{(d-b)(c-d)(d^2+\omega^2)} e^{-dt} \\[4mm] - \dfrac{\omega^2}{\sqrt{(b^2+\omega^2)(c^2+\omega^2)(d^2+\omega^2)}} \cos(\omega t - \phi) \\[4mm] \phi = \tan^{-1}\left(\dfrac{\omega}{b}\right) + \tan^{-1}\left(\dfrac{\omega}{c}\right) + \tan^{-1}\left(\dfrac{\omega}{d}\right) \end{cases}$$

(continued)

TABLE 7.3 (continued) LT Pairs

$F(s)$	$f(t)$
114 $\dfrac{s^3}{(s+b)^2(s^2+\omega^2)}$	$\begin{cases} -\dfrac{b^3}{b^2+\omega^2}\,t\,e^{-bt} + \dfrac{b^2(b^2+3\omega^2)}{(b^2+\omega^2)^2}e^{-bt} - \dfrac{\omega^2}{(b^2+\omega^2)}\sin(\omega t+\phi) \\[3mm] \phi = \tan^{-1}\left(\dfrac{b}{\omega}\right) - \tan^{-1}\left(\dfrac{\omega}{b}\right) \end{cases}$
115 $\dfrac{s^3}{s^4+4\omega^4}$	$\cos(\omega t)\cosh(\omega t)$
116 $\dfrac{s^3}{s^4-\omega^4}$	$\dfrac{1}{2}[\cosh(\omega t)+\cos(\omega t)]$
117 $\dfrac{s^3+a_2s^2+a_1s+a_0}{s^2(s+b)(s+c)}$	$\begin{cases} \dfrac{a_0}{bc}t - \dfrac{a_0(b+c)-a_1bc}{b^2c^2} + \dfrac{-b^3+a_2b^2-a_1b+a_0}{b^2(c-b)}e^{-bt} \\[3mm] + \dfrac{-c^3+a_2c^2-a_1c+a_0}{c^2(b-c)}e^{-ct} \end{cases}$
118 $\dfrac{s^3+a_2s^2+a_1s+a_0}{s(s+b)(s+c)(s+d)}$	$\begin{cases} \dfrac{a_0}{bcd} - \dfrac{-b^3+a_2b^2-a_1b+a_0}{b(c-b)(d-b)}e^{-bt} - \dfrac{-c^3+a_2c^2-a_1c+a_0}{c(b-c)(d-c)}e^{-ct} \\[3mm] - \dfrac{-d^3+a_2d^2-a_1d+a_0}{d(b-d)(c-d)}e^{-dt} \end{cases}$
119 $\dfrac{s^3+a_2s^2+a_1s+a_0}{s^2(s+b)(s+c)(s+d)}$	$\begin{cases} \dfrac{a_0}{bcd}t + \left[\dfrac{a_1}{bcd} - \dfrac{a_0(bc+bd+cd)}{b^2c^2d^2}\right] + \dfrac{-b^3+a_2b^2-a_1b+a_0}{b^2(c-b)(d-b)}\varepsilon^{-bt} \\[3mm] + \dfrac{-c^3+a_2c^2-a_1c+a_0}{c^2(b-c)(d-c)}e^{-ct} + \dfrac{d^3+a_2d^2-a_1d+a_0}{d^2(b-d)(c-d)}e^{-dt} \end{cases}$
120 $\dfrac{s^3+a_2s^2+a_1s+a_0}{(s+b)(s+c)(s+d)(s+f)}$	$\begin{cases} \dfrac{-b^3+a_2b^2-a_1b+a_0}{(c-b)(d-b)(f-b)}e^{-bt} + \dfrac{-c^3+a_2c^2-a_1c+a_0}{(b-c)(d-c)(f-c)}e^{-ct} \\[3mm] + \dfrac{-d^3+a_2d^2-a_1d+a_0}{(b-d)(c-d)(f-d)}e^{-dt} + \dfrac{-f^3+a_2f^2-a_1f+a_0}{(b-f)(c-f)(d-f)}e^{-ft} \end{cases}$
121 $\dfrac{s^3+a_2s^2+a_1s+a_0}{s(s+b)(s+c)(s+d)(s+f)}$	$\begin{cases} \dfrac{a_0}{bcdf} - \dfrac{-b^3+a_2b^2-a_1b+a_0}{b(c-b)(d-b)(f-b)}e^{-bt} - \dfrac{-c^3+a_2c^2-a_1c+a_0}{c(b-c)(d-c)(f-c)}e^{-ct} \\[3mm] - \dfrac{-d^3+a_2d^2-a_1d+a_0}{d(b-d)(c-d)(f-d)}e^{-dt} - \dfrac{-f^3+a_2f^2-a_1f+a_0}{f(b-f)(c-f)(d-f)}e^{-ft} \end{cases}$
122 $\dfrac{s^3+a_2s^2+a_1s+a_0}{(s+b)(s+c)(s+d)(s+f)(s+g)}$	$\begin{cases} \dfrac{-b^3+a_2b^2-a_1b+a_0}{(c-b)(d-b)(f-b)(g-b)}e^{-bt} + \dfrac{-c^3+a_2c^2-a_1c+a_0}{(b-c)(d-c)(f-c)(g-c)}e^{-ct} \\[3mm] + \dfrac{-d^3+a_2d^2-a_1d+a_0}{(b-d)(c-d)(f-d)(g-d)}e^{-dt} + \dfrac{-f^3+a_2f^2-a_1f+a_0}{(b-f)(c-f)(d-f)(g-f)}e^{-ft} \\[3mm] + \dfrac{-g^3+a_2g^2-a_1g+a_0}{(b-g)(c-g)(d-g)(f-g)}e^{-gt} \end{cases}$

TABLE 7.3 (continued) LT Pairs

$F(s)$	$f(t)$
123 $\dfrac{s^3 + a_2 s^2 + a_1 s + a_0}{(s+b)(s+c)(s+d)^2}$	$\begin{cases} \dfrac{-b^3 + a_2 b^2 - a_1 b + a_0}{(c-b)(d-b)^2} e^{-bt} + \dfrac{-c^3 + a_2 c^2 - a_1 c + a_0}{(b-c)(d-c)^2} e^{-ct} \\[2mm] + \dfrac{-d^3 + a_2 d^2 - a_1 d + a_0}{(b-d)(c-d)} t e^{-dt} \\[2mm] a_0(2d - b - c) + a_1(bc - d^2) \\[2mm] + \dfrac{+a_2 d(db + dc - 2bc) + d^2(d^2 - 2db - 2dc + 3bc)}{(b-d)^2(c-d)^2} e^{-dt} \end{cases}$
124 $\dfrac{s^3 + a_2 s^2 + a_1 s + a_0}{s(s+b)(s+c)(s+d)^2}$	$\begin{cases} \dfrac{a_0}{bcd^2} - \dfrac{-b^3 + a_2 b^2 - a_1 b + a_0}{b(c-b)(d-b)^2} e^{-bt} - \dfrac{-c^3 + a_2 c^2 - a_1 c + a_0}{c(b-c)(d-c)^2} e^{-ct} \\[2mm] - \dfrac{-d^3 + a_2 d^2 - a_1 d + a_0}{d(b-d)(c-d)} t e^{-dt} - \dfrac{3d^2 - 2a_2 d + a_1}{d(b-d)(c-d)} e^{-dt} \\[2mm] - \dfrac{(-d^3 + a_2 d^2 - a_1 d + a_0)[(b-d)(c-d) - d(b-d) - d(c-d)]}{d^2(b-d)^2(c-d)^2} e^{-dt} \end{cases}$
125 $\dfrac{s^3 + a_2 s^2 + a_1 s + a_0}{(s+b)(s+c)(s+d)(s+f)^2}$	$\begin{cases} \dfrac{-b^3 + a_2 b^2 - a_1 b + a_0}{(c-b)(d-b)(f-b)^2} e^{-bt} + \dfrac{-c^3 + a_2 c^2 - a_1 c + a_0}{(b-c)(d-c)(f-c)}^2 e^{-ct} \\[2mm] + \dfrac{-d^3 + a_2 d^2 - a_1 d + a_0}{(b-d)(c-d)(f-d)^2} e^{-dt} + \dfrac{-f^3 + a_2 f^2 - a_1 f + a_0}{(b-f)(c-f)(d-f)} t e^{-ft} \\[2mm] (-f^3 + a_2 f^2 - a_1 f + a_0)[(b-f)(c-f) \\[2mm] + \dfrac{3f^2 - 2a_2 f + a_1}{(b-f)(c-f)(d-f)} e^{-ft} - \dfrac{+(b-f)(d-f) + (c-f)(d-f)]}{(b-f)^2(c-f)^2(d-f)^2} e^{-ft} \end{cases}$
126 $\dfrac{s}{(s-a)^{3/2}}$	$\dfrac{1}{\sqrt{\pi t}} e^{at}(1 + 2at)$
127 $\sqrt{s-u} - \sqrt{s-b}$	$\dfrac{1}{2\sqrt{\pi t^3}}(e^{bt} - e^{at})$
128 $\dfrac{1}{\sqrt{s}+a}$	$\dfrac{1}{\sqrt{\pi t}} - ae^{a^2 t}\mathrm{erfc}(a\sqrt{t})$
129 $\dfrac{\sqrt{s}}{s - a^2}$	$\dfrac{1}{\sqrt{\pi t}} + ae^{a^2 t}\mathrm{erf}(a\sqrt{t})$
130 $\dfrac{\sqrt{s}}{s + a^2}$	$\dfrac{1}{\sqrt{\pi t}} - \dfrac{2a}{\sqrt{\pi}} e^{-a^2 t} \displaystyle\int_0^{a\sqrt{t}} e^{\lambda^2} \, d\lambda$
131 $\dfrac{1}{\sqrt{s}(s - a^2)}$	$\dfrac{1}{a} e^{a^2 t}\mathrm{erf}(a\sqrt{t})$
132 $\dfrac{1}{\sqrt{s}(s + a^2)}$	$\dfrac{2}{a\sqrt{\pi}} e^{-a^2 t} \displaystyle\int_0^{a\sqrt{t}} e^{\lambda^2} \, d\pi$
133 $\dfrac{b^2 - a^2}{(s - a^2)(b + \sqrt{s})}$	$e^{a^2 t}[b - a\,\mathrm{erf}(a\sqrt{t})] - be^{b^2 t}\mathrm{erfc}(b\sqrt{t})$
134 $\dfrac{1}{\sqrt{s}(\sqrt{s} + a)}$	$e^{a^2 t}\mathrm{erfc}(a\sqrt{t})$
135 $\dfrac{1}{(s+a)\sqrt{s+b}}$	$\dfrac{1}{\sqrt{b-a}} e^{-at}\mathrm{erf}(\sqrt{b-a}\sqrt{t})$

(continued)

TABLE 7.3 (continued)   LT Pairs

$F(s)$	$f(t)$
136 $\dfrac{b^2 - a^2}{\sqrt{s}(s - a^2)(\sqrt{s} + b)}$	$e^{a^2 t}\left[\dfrac{b}{a}\operatorname{erf}(a\sqrt{t}) - 1\right] + e^{b^2 t}\operatorname{erfc}(b\sqrt{t})$
137 $\dfrac{(1-s)^n}{s^{n+(1/2)}}$	$\begin{cases} \dfrac{n!}{(2n)!\sqrt{\pi t}}H_{2n}(\sqrt{t}) \\[2mm] \left[H_n(t) = \text{Hermite polynomial} = e^{x^2}\dfrac{d^n}{dx^n}(e^{-x^2})\right] \end{cases}$
138 $\dfrac{(1-s)^n}{s^{n+(3/2)}}$	$-\dfrac{n!}{\sqrt{\pi}(2n+1)!}H_{2n+1}(\sqrt{t})$
139 $\dfrac{\sqrt{s+2a}}{\sqrt{s}} - 1$	$\begin{cases} ae^{-at}[I_1(at) + I_0(at)] \\ [I_n(t) = j^{-n}J_n(jt) \text{ where } J_n \text{ is the Bessel function of the first kind}] \end{cases}$
140 $\dfrac{1}{\sqrt{s + a}\sqrt{s + b}}$	$e^{-(1/2)(a+b)t}I_0\left(\dfrac{a-b}{2}t\right)$
141 $\dfrac{\Gamma(k)}{(s+a)^k(s+b)^k}(k \geq 0)$	$\sqrt{\pi}\left(\dfrac{t}{a-b}\right)^{k-(1/2)}e^{-(1/2)(a+b)t}I_{k-(1/2)}\left(\dfrac{a-b}{2}t\right)$
142 $\dfrac{1}{(s+a)^{1/2}(s+b)^{3/2}}$	$te^{-(1/2)(a+b)t}\left[I_0\left(\dfrac{a-b}{2}t\right) + I_1\left(\dfrac{a-b}{2}t\right)\right]$
143 $\dfrac{\sqrt{s+2a} - \sqrt{s}}{\sqrt{s+2a} + \sqrt{s}}$	$\dfrac{1}{t}e^{-at}I_1(at)$
144 $\dfrac{(a-b)^k}{(\sqrt{s+a} + \sqrt{s+b})^{2k}}(k > 0)$	$\dfrac{k}{t}e^{-(1/2)(a+b)t}I_k\left(\dfrac{a-b}{2}t\right)$
145 $\dfrac{(\sqrt{s+a} + \sqrt{s})^{-2\nu}}{\sqrt{s}\sqrt{s+a}}$	$\dfrac{1}{a^\nu}e^{-(1/2)(at)}I_\nu\left(\dfrac{1}{2}at\right)$
146 $\dfrac{1}{\sqrt{s^2 + a^2}}$	$J_0(at)$
147 $\dfrac{(\sqrt{s^2 + a^2} - s)^\nu}{\sqrt{s^2 + a^2}}(\nu > -1)$	$a^\nu J_\nu(at)$
148 $\dfrac{1}{(s^2 + a^2)^k}(k > 0)$	$\dfrac{\sqrt{\pi}}{\Gamma(k)}\left(\dfrac{t}{2a}\right)^{k-(1/2)}J_{k-(1/2)}(at)$
149 $(\sqrt{s^2 + a^2} - s)^k (k > 0)$	$\dfrac{ka^k}{t}J_k(at)$
150 $\dfrac{(s - \sqrt{s^2 - a^2})^\nu}{\sqrt{s^2 - a^2}}(\nu > -1)$	$a^\nu I_\nu(at)$
151 $\dfrac{1}{(s^2 - a^2)^k}(k > 0)$	$\dfrac{\sqrt{\pi}}{\Gamma(k)}\left(\dfrac{t}{2a}\right)^{k-(1/2)}I_{k-(1/2)}(at)$
152 $\dfrac{1}{s\sqrt{s+1}}$	$\operatorname{erf}(\sqrt{t}); \operatorname{erf}(y) \triangleq \text{the error function} = \dfrac{2}{\sqrt{\pi}}\displaystyle\int_0^y e^{-u^2}du$
153 $\dfrac{1}{\sqrt{s^2 + a^2}}$	$J_0(at)$; Bessel function of first kind, zero order
154 $\dfrac{1}{\sqrt{s^2 + a^2} + s}$	$\dfrac{J_1(at)}{at}$; J_1 is the Bessel function of first kind, first order
155 $\dfrac{1}{[\sqrt{s^2 + a^2} + s]^N}$	$\dfrac{N}{a^N}\dfrac{J_N(at)}{t}$; $N = 1, 2, 3, \ldots, J_N$ is the Bessel function of first kind, Nth order

TABLE 7.3 (continued) LT Pairs

$F(s)$	$f(t)$
156 $\dfrac{1}{s[\sqrt{s^2 + a^2} + s]^N}$	$\dfrac{N}{a^N} \displaystyle\int_0^t \dfrac{J_N(au)}{u}\, du;\ N = 1, 2, 3, \ldots, J_N$ is the Bessel function of first kind, Nth order
157 $\dfrac{1}{\sqrt{s^2 + a^2}(\sqrt{s^2 + a^2} + s)}$	$\dfrac{1}{a} J_1(at);\ J_1$ is the Bessel function of first kind, first order
158 $\dfrac{1}{\sqrt{s^2 + a^2}[\sqrt{s^2 + a^2} + s]^N}$	$\dfrac{1}{a^N} J_N(at);\quad N = 1, 2, 3, \ldots, J_N$ is the Bessel function of first kind, Nth order
159 $\dfrac{1}{\sqrt{s^2 - a^2}}$	$I_0(at);\quad I_0$ is the modified Bessel function of first kind, zero order
160 $\dfrac{e^{-ks}}{s}$	$S_k(t) = \begin{cases} 0 & \text{when } 0 < t < k \\ 1 & \text{when } t > k \end{cases}$
161 $\dfrac{e^{-ks}}{s^2}$	$\begin{cases} 0 & \text{when } 0 < t < k \\ t - k & \text{when } t > k \end{cases}$
162 $\dfrac{e^{-ks}}{s^\mu}\ (\mu > 0)$	$\begin{cases} 0 & \text{when } 0 < t < k \\ \dfrac{(t - k)^{\mu - 1}}{\Gamma(\mu)} & \text{when } t > k \end{cases}$
163 $\dfrac{1 - e^{-ks}}{s}$	$\begin{cases} 1 & \text{when } 0 < t < k \\ 0 & \text{when } t > k \end{cases}$
164 $\dfrac{1}{s(1 - e^{-ks})} = \dfrac{1 + \coth \frac{1}{2}ks}{2s}$	$S(k, t) = \begin{cases} n & \text{when} \\ & (n - 1)k < t < nk (n = 1, 2, \ldots) \end{cases}$
165 $\dfrac{1}{s(e^{+ks} - a)}$	$S_k(t) = \begin{cases} 0 & \text{when } 0 < t < k \\ 1 + a + a^2 + \cdots + a^{n-1} \\ \quad \text{when } nk < t < (n + 1)k (n = 1, 2, \ldots) \end{cases}$
166 $\dfrac{1}{s} \tanh ks$	$\begin{cases} M(2k, t) = (-1)^{n-1} \\ \quad \text{when } 2k(n-1) < t < 2nk \\ \quad (n - 1, 2, \ldots) \end{cases}$
167 $\dfrac{1}{s(1 + e^{-ks})}$	$\begin{cases} \dfrac{1}{2}M(k, t) + \dfrac{1}{2} = \dfrac{1 - (-1)^n}{2} \\ \quad \text{when } (n - 1)k < t < nk \end{cases}$
168 $\dfrac{1}{s^2} \tanh ks$	$\begin{cases} H(2k, t) & [H(2k, t) = k + (r - k)(-1)^n \text{ where } t = 2kn + r; \\ & \quad 0 \le r \le 2k; n = 0, 1, 2, \ldots] \end{cases}$
169 $\dfrac{1}{s \sinh ks}$	$\begin{cases} 2S(2k, t + k) - 2 = 2(n - 1) \\ \quad \text{when } (2n - 3)k < t < (2n - 1)k\ (t > 0) \end{cases}$
170 $\dfrac{1}{s \cosh ks}$	$\begin{cases} M(2k, t + 3k) + 1 = 1 + (-1)^n \\ \quad \text{when } (2n - 3)k < t < (2n - 1)k\ (t > 0) \end{cases}$
171 $\dfrac{1}{s} \coth ks$	$\begin{cases} 2S(2k, t) - 1 = 2n - 1 \\ \quad \text{when } 2k(n - 1) < t < 2kn \end{cases}$
172 $\dfrac{k}{s^2 + k^2} \coth \dfrac{\pi s}{2k}$	$|\sin kt|$
173 $\dfrac{1}{(s^2 + 1)(1 - e^{-\pi s})}$	$\begin{cases} \sin t & \text{when } (2n - 2)\pi < t < (2n - 1)\pi \\ 0 & \text{when } (2n - 1)\pi < t < 2n\pi \end{cases}$

(continued)

TABLE 7.3 (continued) LT Pairs

$F(s)$	$f(t)$
174 $\quad \dfrac{1}{s}e^{-k/s}$	$J_0(2\sqrt{kt})$
175 $\quad \dfrac{1}{\sqrt{s}}e^{-k/s}$	$\dfrac{1}{\sqrt{\pi t}}\cos 2\sqrt{kt}$
176 $\quad \dfrac{1}{\sqrt{s}}e^{k/s}$	$\dfrac{1}{\sqrt{\pi t}}\cosh 2\sqrt{kt}$
177 $\quad \dfrac{1}{s^{3/2}}e^{-k/s}$	$\dfrac{1}{\sqrt{\pi k}}\sin 2\sqrt{kt}$
178 $\quad \dfrac{1}{s^{3/2}}e^{k/s}$	$\dfrac{1}{\sqrt{\pi k}}\sinh 2\sqrt{kt}$
179 $\quad \dfrac{1}{s^{\mu}}e^{-k/s}\ (\mu > 0)$	$\left(\dfrac{t}{k}\right)^{(\mu-1)/2}J_{\mu-1}(2\sqrt{kt})$
180 $\quad \dfrac{1}{s^{\mu}}e^{k/s}\ (\mu > 0)$	$\left(\dfrac{t}{k}\right)^{(\mu-1)/2}I_{\mu-1}(2\sqrt{kt})$
181 $\quad e^{-k\sqrt{s}}\ (k>0)$	$\dfrac{k}{2\sqrt{\pi t^3}}\exp\left(\dfrac{-k^2}{4t}\right)$
182 $\quad \dfrac{1}{s}e^{-k\sqrt{s}}\ (k \geq 0)$	$\text{erfc}\left(\dfrac{k}{2\sqrt{t}}\right)$
183 $\quad \dfrac{1}{\sqrt{s}}e^{-k\sqrt{s}}\ (k \geq 0)$	$\dfrac{1}{\sqrt{\pi t}}\exp\left(\dfrac{-k^2}{4t}\right)$
184 $\quad s^{-3/2}e^{-k\sqrt{s}}\ (k \geq 0)$	$2\sqrt{\dfrac{t}{\pi}}\exp\left(-\dfrac{k^2}{4t}\right) - k\,\text{erfc}\left(\dfrac{k}{2\sqrt{t}}\right)$
185 $\quad \dfrac{ae^{-k\sqrt{s}}}{s(a+\sqrt{s})}\ (k \geq 0)$	$-e^{ak}e^{a^2 t}\,\text{erfc}\left(a\sqrt{t}+\dfrac{k}{2\sqrt{t}}\right) + \text{erfc}\left(\dfrac{k}{2\sqrt{t}}\right)$
186 $\quad \dfrac{e^{-k\sqrt{s}}}{\sqrt{s}(a+\sqrt{s})}\ (k \geq 0)$	$e^{ak}e^{a^2 t}\,\text{erfc}\left(a\sqrt{t}+\dfrac{k}{2\sqrt{t}}\right)$
187 $\quad \dfrac{e^{-k\sqrt{s(s+a)}}}{\sqrt{s(s+a)}}$	$\begin{cases} 0 & \text{when } 0 < t < k \\ e^{-(1/2)at}I_0(\tfrac{1}{2}a\sqrt{t^2-k^2}) & \text{when } t > k \end{cases}$
188 $\quad \dfrac{e^{-k\sqrt{s^2+a^2}}}{\sqrt{(s^2+a^2)}}$	$\begin{cases} 0 & \text{when } 0 < t < k \\ J_0(a\sqrt{t^2-k^2}) & \text{when } t > k \end{cases}$
189 $\quad \dfrac{e^{-k\sqrt{s^2-a^2}}}{\sqrt{(s^2-a^2)}}$	$\begin{cases} 0 & \text{when } 0 < t < k \\ I_0(a\sqrt{t^2-k^2}) & \text{when } t > k \end{cases}$
190 $\quad \dfrac{e^{-k(\sqrt{s^2+a^2}-s)}}{\sqrt{(s^2+a^2)}}\ (k \geq 0)$	$J_0(a\sqrt{t^2+2kt})$
191 $\quad e^{-ks} - e^{-k\sqrt{s^2+a^2}}$	$\begin{cases} 0 & \text{when } 0 < t < k \\ \dfrac{ak}{\sqrt{t^2-k^2}}J_1(a\sqrt{t^2-k^2}) & \text{when } t > k \end{cases}$
192 $\quad e^{-k\sqrt{s^2+a^2}} - e^{-ks}$	$\begin{cases} 0 & \text{when } 0 < t < k \\ \dfrac{ak}{\sqrt{t^2-k^2}}I_1(a\sqrt{t^2-k^2}) & \text{when } t > k \end{cases}$
193 $\quad \dfrac{a^{v}e^{-k\sqrt{s^2-a^2}}}{\sqrt{(s^2+a^2)}\left(\sqrt{s^2+a^2}+s\right)^{v}}$ $(v > -1)$	$\begin{cases} 0 & \text{when } 0 < t < k \\ \left(\dfrac{t-k}{t+k}\right)^{(1/2)v}J_v(a\sqrt{t^2-k^2}) & \text{when } t > k \end{cases}$

TABLE 7.3 (continued) LT Pairs

$F(s)$	$f(t)$
194 $\dfrac{1}{s}\log s$	$\Gamma'(1) - \log t \quad [\Gamma'(1) = -0.5772]$
195 $\dfrac{1}{s^k}\log s \, (k > 0)$	$t^{k-1}\left\{\dfrac{\Gamma'(k)}{[\Gamma(k)]^2} - \dfrac{\log t}{\Gamma(k)}\right\}$
196 $\dfrac{\log s}{s-a}\,(a>0)$	$e^{at}[\log a - \text{Ei}(-at)]$
197 $\dfrac{\log s}{s^2+1}$	$\cos t\,\text{Si}(t) - \sin t\,\text{Ci}(t)$
198 $\dfrac{s\log s}{s^2+1}$	$-\sin t\,\text{Si}(t) - \cos t\,\text{Ci}(t)$
199 $\dfrac{1}{s}\log(1+ks)(k>0)$	$-\text{Ei}\left(-\dfrac{t}{k}\right)$
200 $\log\dfrac{s-a}{s-b}$	$\dfrac{1}{t}(e^{bt} - e^{at})$
201 $\dfrac{1}{s}\log(1+k^2s^2)$	$-2\text{Ci}\left(\dfrac{t}{k}\right)$
202 $\dfrac{1}{s}\log(s^2+a^2)(a>0)$	$2\log a - 2\text{Ci}(at)$
203 $\dfrac{1}{s^2}\log(s^2+a^2)(a>0)$	$\dfrac{2}{a}[at\log a + \sin at - at\text{Ci}(at)]$
204 $\log\dfrac{s^2+a^2}{s^2}$	$\dfrac{2}{t}(1-\cos at)$
205 $\log\dfrac{s^2-a^2}{s^2}$	$\dfrac{2}{t}(1-\cosh at)$
206 $\arctan\dfrac{k}{s}$	$\dfrac{1}{t}\sin kt$
207 $\dfrac{1}{s}\arctan\dfrac{k}{s}$	$\text{Si}(kt)$
208 $e^{k^2s^2}\text{erfc}(ks)(k>0)$	$\dfrac{1}{k\sqrt{\pi}}\exp\left(-\dfrac{t^2}{4k^2}\right)$
209 $\dfrac{1}{s}e^{k^2s^2}\text{erfc}(ks)(k>0)$	$\text{erf}\left(\dfrac{t}{2k}\right)$
210 $e^{ks}\text{erfc}(\sqrt{ks})(k>0)$	$\dfrac{\sqrt{k}}{\pi\sqrt{t(t+k)}}$
211 $\dfrac{1}{\sqrt{s}}\text{erfc}(\sqrt{ks})$	$\begin{cases} 0 & \text{when } 0 < t < k \\ (\pi t)^{-1/2} & \text{when } t > k \end{cases}$
212 $\dfrac{1}{\sqrt{s}}e^{ks}\text{erfc}(\sqrt{ks})(k>0)$	$\dfrac{1}{\sqrt{\pi(t+k)}}$
213 $\text{erf}\left(\dfrac{k}{\sqrt{s}}\right)$	$\dfrac{1}{\pi t}\sin(2k\sqrt{t})$
214 $\dfrac{1}{\sqrt{s}}e^{k^2/s}\text{erfc}\left(\dfrac{k}{\sqrt{s}}\right)$	$\dfrac{1}{\sqrt{\pi t}}e^{-2k\sqrt{t}}$
215 $-e^{as}\text{Ei}(-as)$	$\dfrac{1}{t+a}; (a>0)$

(continued)

TABLE 7.3 (continued) LT Pairs

$F(s)$	$f(t)$
216 $\dfrac{1}{a} + se^{as}\,\mathrm{Ei}(-as)$	$\dfrac{1}{(t+a)^2};\ (a>0)$
217 $\left[\dfrac{\pi}{2} - \mathrm{Si}(s)\right]\cos s$ $+\,\mathrm{Ci}(s)\sin s$	$\dfrac{1}{t^2+1}$
218 $K_0(ks)$	$\begin{cases} 0 & \text{when } 0 < t < k \\ (t^2-k^2)^{-1/2} & \text{when } t > k \end{cases}$ $\quad$ [$K_n(t)$ is the Bessel function of the second kind of imaginary argument]
219 $K_0(k\sqrt{s})$	$\dfrac{1}{2t}\exp\left(-\dfrac{k^2}{4t}\right)$
220 $\dfrac{1}{s}e^{ks}K_1(ks)$	$\dfrac{1}{k}\sqrt{t(t+2k)}$
221 $\dfrac{1}{\sqrt{s}}K_1(k\sqrt{s})$	$\dfrac{1}{k}\exp\left(-\dfrac{k^2}{4t}\right)$
222 $\dfrac{1}{\sqrt{s}}e^{k/s}K_0\left(\dfrac{k}{s}\right)$	$\dfrac{2}{\sqrt{\pi t}}K_0(2\sqrt{2kt})$
223 $\pi e^{-ks}I_0(ks)$	$\begin{cases} [t(2k-t)]^{-1/2} & \text{when } 0 < t < 2k \\ 0 & \text{when } t > 2k \end{cases}$
224 $e^{-ks}I_1(ks)$	$\begin{cases} \dfrac{k-t}{\pi k\sqrt{t(2k-t)}} & \text{when } 0 < t < 2k \\ 0 & \text{when } t > 2k \end{cases}$
225 $\dfrac{1}{s\sinh(as)}$	$2\displaystyle\sum_{k=0}^{\infty} u[t-(2k+1)a]$

226 $\dfrac{1}{s\cosh s}$	$2\displaystyle\sum_{k=0}^{\infty}(-1)^k u(t-2k-1)$

TABLE 7.3 (continued) LT Pairs

$F(s)$	$f(t)$
227 $\dfrac{1}{s}\tanh\left(\dfrac{as}{2}\right)$	$u(t) + 2\sum_{k=1}^{\infty}(-1)^k u(t - ak)$

Square wave

228 $\dfrac{1}{2s}\left(1 + \coth\dfrac{as}{2}\right)$	$\sum_{k=0}^{\infty} u(t - ak)$

Stepped function

229 $\dfrac{m}{s^2} - \dfrac{ma}{2s}\left(\coth\dfrac{as}{2} - 1\right)$	$mt - ma\sum_{k=1}^{\infty} u(t - ka)$

Saw-tooth function

Slope $= m$

230 $\dfrac{1}{s^2}\tanh\left(\dfrac{as}{2}\right)$	$\dfrac{1}{a}\left[t + 2\sum_{k=1}^{\infty}(-1)^k(t - ka)\cdot u(t - ka)\right]$

Triangular wave

231 $\dfrac{1}{s(1 + e^{-s})}$	$\sum_{k=0}^{\infty}(-1)^k u(t - k)$

(continued)

TABLE 7.3 (continued)   LT Pairs

$F(s)$	$f(t)$

232 $\dfrac{a}{(s^2 + a^2)(1 - e^{-\frac{\pi}{a}s})}$ $\quad\quad\quad \sum_{k=0}^{\infty}\left[\sin a\left(t - k\dfrac{\pi}{a}\right)\right]\cdot u\left(t - k\dfrac{\pi}{a}\right)$

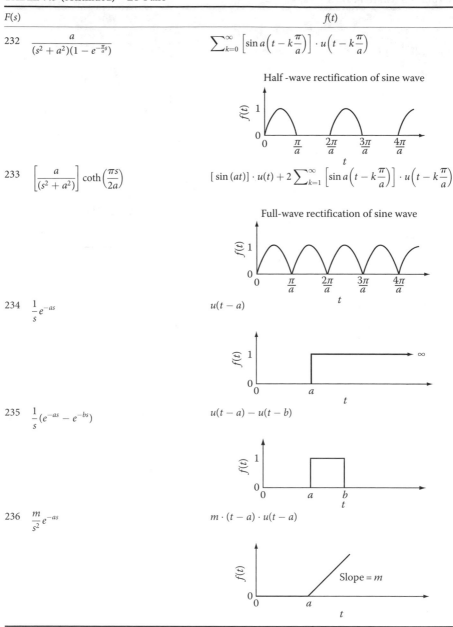

Half-wave rectification of sine wave

233 $\left[\dfrac{a}{(s^2 + a^2)}\right]\coth\left(\dfrac{\pi s}{2a}\right)$ $\quad\quad\quad [\,\sin(at)\,]\cdot u(t) + 2\sum_{k=1}^{\infty}\left[\sin a\left(t - k\dfrac{\pi}{a}\right)\right]\cdot u\left(t - k\dfrac{\pi}{a}\right)$

Full-wave rectification of sine wave

234 $\dfrac{1}{s}e^{-as}$ $\quad\quad\quad u(t - a)$

235 $\dfrac{1}{s}(e^{-as} - e^{-bs})$ $\quad\quad\quad u(t - a) - u(t - b)$

236 $\dfrac{m}{s^2}e^{-as}$ $\quad\quad\quad m\cdot(t - a)\cdot u(t - a)$

TABLE 7.3 (continued) LT Pairs

$F(s)$	$f(t)$

237 $\left[\dfrac{ma}{s}+\dfrac{m}{s^2}\right]e^{-as}$

$mt \cdot u(t-a)$
or
$[ma+m(t-a)] \cdot u(t-a)$

238 $\dfrac{2}{s^3}e^{-as}$

$(t-a)^2 \cdot u(t-a)$

239 $\left[\dfrac{2}{s^3}+\dfrac{2a}{s^2}+\dfrac{a^2}{s}\right]e^{-as}$

$t^2 \cdot u(t-a)$

240 $\dfrac{m}{s^2}-\dfrac{m}{s^2}e^{-as}$

$mt \cdot u(t) - m(t-a) \cdot u(t-a)$

241 $\dfrac{m}{s^2}-\dfrac{2m}{s^2}e^{-as}+\dfrac{m}{s^2}e^{-2as}$

$mt - 2m(t-a) \cdot u(t-a) + m(t-2a) \cdot u(t-2a)$

242 $\dfrac{m}{s^2}-\left(\dfrac{ma}{s}+\dfrac{m}{s^2}\right)e^{-as}$

$mt - [ma+m(t-a)] \cdot u(t-a)$

(continued)

TABLE 7.3 (continued) LT Pairs

$F(s)$	$f(t)$

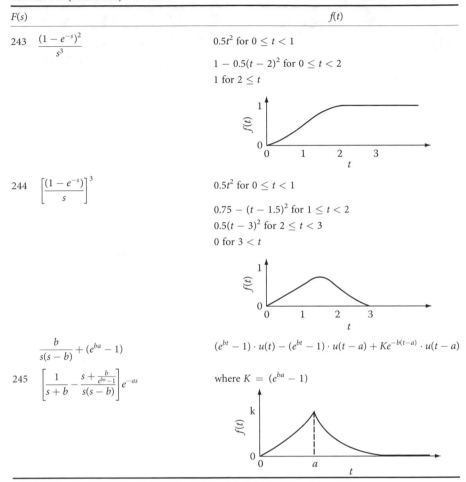

243 $\dfrac{(1 - e^{-s})^2}{s^3}$

$0.5t^2$ for $0 \le t < 1$

$1 - 0.5(t - 2)^2$ for $0 \le t < 2$

1 for $2 \le t$

244 $\left[\dfrac{(1 - e^{-s})}{s}\right]^3$

$0.5t^2$ for $0 \le t < 1$

$0.75 - (t - 1.5)^2$ for $1 \le t < 2$

$0.5(t - 3)^2$ for $2 \le t < 3$

0 for $3 < t$

$\dfrac{b}{s(s - b)} + (e^{ba} - 1)$

$(e^{bt} - 1) \cdot u(t) - (e^{bt} - 1) \cdot u(t - a) + Ke^{-b(t-a)} \cdot u(t - a)$

245 $\left[\dfrac{1}{s + b} - \dfrac{s + \frac{b}{e^{ba}-1}}{s(s - b)}\right] e^{-as}$

where $K = (e^{ba} - 1)$

TABLE 7.4 Properties of LTs

$F(s)$	$f(t)$
1 $\int_0^\infty e^{-st} f(t)dt$	$f(t)$
2 $AF(s) + BG(s)$	$Af(t) + Bg(t)$
3 $sF(s) - f(+0)$	$f'(t)$
4 $s^n F(s) - s^{n-1}f(+0) - s^{n-2}f^{(1)}(+0) - \cdots - f^{(n-1)}(+0)$	$f^{(n)}(t)$
5 $\dfrac{1}{s} F(s)$	$\int_0^t f(\tau)d\tau$
6 $\dfrac{1}{s^2} F(s)$	$\int_0^t \int_0^\tau f(\lambda)d\lambda d\tau$
7 $F_1(s)F_2(s)$	$\int_0^t f_1(t - \tau)f_2(\tau)d\tau = f_1 * f_2$
8 $-F'(s)$	$tf(t)$

TABLE 7.4 (continued) Properties of LTs

$F(s)$	$f(t)$
9 $(-1)^n F^{(n)}(s)$	$t^n f(t)$
10 $\int_s^\infty F(x)dx$	$\dfrac{1}{t}f(t)$
11 $F(s-a)$	$e^{at}f(t)$
12 $e^{-bs}F(s)$	$f(t-b)$, where $f(t) = 0;\ t < 0$
13 $F(cs)$	$\dfrac{1}{c}f\left(\dfrac{t}{c}\right)$
14 $F(cs-b)$	$\dfrac{1}{c}e^{(bt)/c}f\left(\dfrac{t}{c}\right)$
15 $\dfrac{\int_0^a e^{-st}f(t)dt}{1-e^{-as}}$	$f(t+a) = f(t)$ periodic signal
16 $\dfrac{\int_0^a e^{-st}f(t)dt}{1+e^{-as}}$	$f(t+a) = -f(t)$
17 $\dfrac{F(s)}{1-e^{-as}}$	$f_1(t)$, the half-wave rectification of $f(t)$ in No. 16.
18 $F(s)\coth\dfrac{as}{2}$	$f_2(t)$, the full-wave rectification of $f(t)$ in No. 16.
19 $\dfrac{p(s)}{q(s)},\ q(s) = (s-a_1)(s-a_2)\cdots(s-a_m)$	$\sum_1^m \dfrac{p(a_n)}{q'(a_n)}e^{a_n t}$
20 $\dfrac{p(s)}{q(s)} = \dfrac{\phi(s)}{(s-a)^r}$	$e^{at}\sum_{n=1}^r \dfrac{\phi^{(r-n)}(a)}{(r-n)!}\dfrac{t^{n-1}}{(n-1)!}+\cdots$

TABLE 7.5 Proofs of the LT Properties

1. Linearity

   This property is the result of the linear operation of an integral.

2. Time Derivative

$$\mathcal{L}\left\{\frac{df(t)}{dt}\right\} = \int_0^\infty \frac{df(t)}{dt}e^{-st}dt = f(t)e^{-st}|_0^\infty - \int_0^\infty f(t)\frac{de^{-st}}{dt} = -f(0) + s\int_0^\infty f(t)e^{-st}dt$$
$$= sF(s) - f(0);\ \text{only for functions that } \lim_{t\to\infty}f(t)e^{-st} = 0$$

3. Integral $\mathcal{L}\left\{\int_{-\infty}^t f(x)dx\right\} = \int_0^\infty \left[\int_{-\infty}^t f(x)dx\right]e^{-st}dt = \int_0^\infty \left[\int_{-\infty}^t f(x)dx\right]\frac{d}{dt}\left[-\frac{e^{-st}}{s}\right]$

$$= \left[-\frac{e^{-st}}{s}\int_{-\infty}^t f(x)dx\right]\Big|_0^\infty + \int_0^\infty \frac{e^{-st}}{s}\frac{d}{dt}\left[\int_{-\infty}^t f(x)dx\right]dt = -\frac{e^{-s\infty}}{s}\int_{-\infty}^\infty f(x)dx$$

$$+ \frac{e^{-s0}}{s}\int_{-\infty}^0 f(x)dx + \frac{1}{s}\int_0^\infty f(t)e^{-st}dt = \frac{F(s)}{s} + \frac{1}{s}\int_{-\infty}^0 f(t)dt;\ \int_{-\infty}^0 f(t)dt = \text{initial}$$

   value of the integral of $f(t)$ at $t = 0$; the value of $\displaystyle\int_{-\infty}^\infty f(x)dx$ must be finite

   Note that the term $(1/s)\int_{-\infty}^0 f(t)dt$ is the LT of a step function of amplitude $\int_{-\infty}^0 f(t)dt$. This factor is a very important result in network problems since it shows that initial conditions associated with integral functions are automatically as step functions in the LT development.

(continued)

TABLE 7.5 (continued) Proofs of the LT Properties

4. Multiplication by Exponential

$$\mathcal{L}\{e^{-at}f(t)\} = \int_0^\infty e^{-(s+a)t}\,dt = \frac{1}{-(s+a)}\left[e^{-(s+a)t}\big|_0^\infty\right] = \frac{1}{s+a}\ \text{provided that } e^{-(s+a)\infty} \text{ has zero value.}$$

5. Multiplication by t

$$\mathcal{L}\{tf(t)\} = \int_0^\infty tf(t)e^{-st}\,dt = -\int_0^\infty f(t)\frac{de^{-st}}{ds}\,dt = -\frac{d}{ds}\left[\int_0^\infty f(t)e^{-st}\,dt\right] = -\frac{d}{ds}F(s)$$

6. Time Shifting

$$\mathcal{L}\{f(t-a)u(t-a)\} = \int_0^\infty f(t-a)u(t-a)e^{-st}\,dt$$

By setting $x = t - a$, we obtain for $t = 0$ $x = -a$, for $t = \infty$, $x = \infty$, $dx = dt$ and $t = x + a$. Therefore, the above equation becomes

$$\mathcal{L}\{f(x)u(x)\} = \int_{-a}^\infty f(x)u(x)e^{-s(x+a)}\,dx = e^{-sa}\int_0^\infty f(x)e^{-sx}\,dx = e^{-sa}F(s),\ \text{where } a \text{ is a positive constant.}$$

7. Complex Frequency Shift

$$\mathcal{L}\{e^{s_0t}f(t)\} = \int_0^\infty e^{-(s-s_0)t}f(t)\,dt = \int_0^\infty e^{-s't}f(t)\,dt = F(s') = F(s-s_0)$$

8. Scaling

$$\mathcal{L}\{f(at)\} = \int_0^\infty f(at)e^{-st}\,dt = \int_0^\infty f(x)e^{-(s/a)x}\,d(x/a) = \frac{1}{a}F\left(\frac{s}{a}\right)\quad a > 0$$

9. Time Convolution

$$\mathcal{L}\{f(t) * h(t)\} = \mathcal{L}\left\{\int_0^\infty f(x)h(t-x)\,dx\right\} = \int_0^\infty \left[\int_0^\infty f(x)h(t-x)\,dx\right]e^{-st}\,dt$$

$$= \int_0^\infty f(x)\,dx \int_0^\infty h(t-x)e^{-st}\,dt = \int_0^\infty f(x)\,dx \int_0^\infty h(y)e^{-s(y+x)}\,dy$$

$$= \int_0^\infty f(x)e^{-sx}\,dx \int_0^\infty h(y)e^{-sy}\,dy = F(s)H(s)$$

where we set $t - x = y$, and for $t = 0$ $y = -x$ (but $h(y) = 0$ for $y < 0$ and the second integral starts from 0) and for $t = \infty$ $y = \infty$. We also find that $dt = dy$.

10. Initial Value

$$\mathcal{L}\left\{\frac{df(t)}{dt}\right\} = \int_0^\infty \frac{df(t)}{dt}e^{-st}\,dt = sF(s) - f(0)\quad\text{or}\quad sF(s) = \int_0^\infty \frac{df(t)}{dt}e^{-st}\,dt + f(0)$$

Taking the limit of the above equation, we find

$$\lim_{s\to\infty} sF(s) = \int_0^\infty f(t)\left(\lim_{s\to\infty} e^{-st}\right)dt + f(0) = f(0)$$

where the exchange of integral and limit operation were performed due to the linearity of the two operators. The initial value theorem does not apply for functions that have an impulse (delta function) at $t = 0$.

11. Final Value

$$\int_0^\infty \frac{df(t)}{dt}e^{-st}\,dt = sF(s) - x(0)\quad\text{or}\quad \lim_{s\to 0}\int_0^\infty \frac{df(t)}{dt}e^{-st}\,dt = \int_0^\infty \frac{df(t)}{dt}\lim_{s\to 0}(e^{-st})\,dt$$

$$= \int_0^\infty \frac{df(t)}{dt}\,dt = f(\infty) - f(0) = \lim_{s\to 0}[sF(s) - x(0)] = \lim_{s\to 0} sF(s) - x(0)\quad\text{or}$$

$$\lim_{t\to\infty} f(t) = \lim_{s\to 0} sF(s)$$

provided that $f(t)$ exists as the value of t approaches infinity.

8

The z-Transform

The z-transform method provides a powerful tool for solving difference equations of any order and, hence, plays a very important role in digital systems analysis. This chapter includes a study of the z-transform, its properties, and its applications.

The z-transform method provides a technique for transforming a difference equation into an algebraic equation. Specifically, the z-transform converts a sequence of numbers $\{y(n)\}$ into a function of complex variable $Y(z)$, thereby allowing algebraic process and well-defined mathematical procedures to be applied in the solution process. In this sense, the z-transform plays the same general role in the solution of difference equations that the Laplace transform (LT) plays in the solution of differential equations and in a roughly parallel way. Inversion procedures that parallel one another also exist.

8.1 The z-Transform

To understand the essential features of the z-transform, consider a **one-sided sequence** of numbers $\{y(n)\}$ taken at uniform time intervals. This sequence might be the values of a continuous function that has been sampled at uniform time intervals; it could, of course, be a number sequence, for example, the values of the amount that are present in a bank account at the beginning of each month that includes the interest. This number sequence is written as

$$\{y(n)\} = \{y(0), y(1), y(2), \ldots, y(n), \ldots\} \tag{8.1}$$

We now create the series

$$Y(z) = \frac{y(0)}{z^0} + \frac{y(1)}{z} + \frac{y(2)}{z^2} + \cdots = y(0) + y(1)z^{-1} + y(2)z^{-2} + \cdots \tag{8.2}$$

In this expression, z denotes the general complex variable and $Y(z)$ denotes the z-transform of the sequence $\{y(n)\}$. In this more general form, the one-sided z-transform of a sequence $\{y(n)\}$ is written as

$$Y(z) \triangleq \mathfrak{Z}\{y(n)\} = \sum_{n=0}^{\infty} y(n)z^{-n} \tag{8.3}$$

This expression can be taken as the definition of the one-sided z-transform.

Since the exponent of the z's is equal to the distance from the sequence element $y(0)$, we identify the negative exponents as the amount of delay. This interpretation is not explicit in the mathematical form of (8.3), but it is implied when the shifting properties of functions are considered. This same concept will occur when we apply the z-transform to the solution of difference equations. Initially, however, we study the mathematics of the z-transform.

When the sequence of numbers is obtained by sampling a function $y(t)$ every T seconds—for example, by using an analog-to-digital (A/D) converter—the numbers represent sample values $y(nT)$ for $n=0$, 1, 2,.... This suggests that there is a relationship between the LT of a continuous function and the z-transform of a sequence of samples of the function at the time constants $\ldots -nT$, $-(n-1)T, \ldots -T$, $0, T, 2T, \ldots, nT, (n+1)T, \ldots$. To show that there is such a relationship, let $y(t)$ be a function sampled at time constants T seconds apart. The sampled function is (see 6.2)

$$y_s(t) = y(t)comb_T(t) = \sum_{n=-\infty}^{\infty} y(nT)\delta(t - nT) \tag{8.4}$$

The LT of this equation is

$$Y_s(s) \triangleq \mathscr{L}\{y_s(t)\} = \mathscr{L}\left\{ \sum_{n=-\infty}^{\infty} y(nT)\delta(t - nT) \right\} = \sum_{n=-\infty}^{\infty} y(nT)\mathscr{L}\{\delta(t - nT)\} \tag{8.5}$$

$$= \sum_{n=-\infty}^{\infty} y(nT)e^{-nTs}$$

Note that the Laplace operator operates on time t. If we make the substitution $z = e^{sT}$, then

$$Y_s(s)\big|_{z=e^{sT}} = \sum_{n=-\infty}^{\infty} y(nT)z^{-n} = Y(z) \tag{8.6}$$

Hence, $Y(z)$ is the z-transform of the sequence of samples of $y(t)$, namely, $y(nT)$, with $n = 0, 1, 2, \ldots$. From this discussion, we observe that the z-transform may be viewed as the LT of the sampled time function $y_s(t)$, with an appropriate change of variables. This interpretation is in addition to that given in (8.3), which, as already noted, specifies that $Y(z)$ is the z-transform of the number sequence $\{y(nT)\}$ for $n = \cdots -2, -1, 0, 1, 2, \ldots$. The above equation is the two-sided z-transform.

Example 8.1

Deduce the *z*-transform of the discrete function

$$y(n) = \begin{cases} 0 & n \le 0 \\ 1 & n = 1 \\ 2 & n = 2 \\ 0 & n \ge 3 \end{cases}$$

SOLUTION

From the defining equation (8.3), we write

$$Y(z) = \frac{1}{z} + \frac{2}{z^2} = z^{-1} + z^{-2} = \frac{z+2}{z^2}$$

Observe that this function possesses a second-order pole at the origin and a zero at -2. Observe that $y(n)$ could be written as $y(n) = \delta(n - 1) + 2\delta(n - 2)$. The definition of the discrete delta function is (see also Chapter 1)

$$\delta(n) = \begin{cases} 1 & n = 0 \\ 0 & n \ne 0 \end{cases} \tag{8.7}$$

■

Example 8.2

Deduce the *z*-transform of the function

$$f(t) = Ae^{-at} \quad t \ge 0$$

which is sampled every T seconds apart, that is, at time $t = nT$.

SOLUTION

The sampled values are

$$\{f(nT)\} = A, Ae^{-aT}, Ae^{-2aT}, \ldots$$

The *z*-transform of this sequence is written as

$$F(z) = A\left[1 + \frac{e^{-aT}}{z} + \left(\frac{e^{-aT}}{z}\right)^2 + \left(\frac{e^{-aT}}{z}\right)^3 + \cdots\right]$$

The series can be written in closed form by recalling that $1/(1 - x) = 1 + x^1 + x^2 + x^3 + \cdots$

$|x| < 1$. Thus, we have that

$$\mathfrak{Z}\{e^{-anT}u(nT)\} \overset{\Delta}{=} F(z) = \frac{A}{1 - \frac{e^{-aT}}{z}} = \frac{A}{1 - e^{-aT}z^{-1}} = \frac{Az}{z - e^{-aT}} \tag{8.8}$$

The convergence of the series is satisfied if $\left|e^{-aT}z^{-1}\right| = e^{-aT}\left|z^{-1}\right| < 1$ or $|z| > e^{-aT}$. ∎

Example 8.3

Deduce the *z*-transform of the function

$$f(t) = e^{-t} + 2e^{-2t} \quad t \ge 0$$

which is sampled at time intervals $T = 0.1$ s.

SOLUTION

We use the results of Example 8.2 to write

$$F(z) = \frac{z}{z - e^{-0.1}} + \frac{2z}{z - e^{-0.2}} = \frac{z}{z - 0.9048} + \frac{2z}{z - 0.8187} = \frac{3z^2 - 2.6283}{z^2 - 1.7235z + 0.7408} \quad ∎$$

Example 8.4

Find the *z*-transform of the given functions when sampled every *T* seconds:
(a) $f(t) = u(t)$, (b) $f(t) = tu(t)$, (c) $f(t) = e^{-bt}u(t), b > 0$, and (d) $f(t) = \sin \omega t\, u(t)$.

SOLUTION

(a)

$$\mathfrak{Z}\{f(nT)\} = \mathfrak{Z}\{u(nT)\} = \sum_{n=0}^{\infty} u(nT)z^{-n} = (1 + z^{-1} + z^{-2} + \cdots)$$

$$= \frac{1}{1 - z^{-1}} = \frac{z}{z - 1} \quad |z| > 1 \tag{8.9}$$

(b)

$$\mathfrak{Z}\{f(nT)\} = \mathfrak{Z}\{nTu(nT)\} = \sum_{n=0}^{\infty} nTu(nT)z^{-n} = Tz^{-1} + 2Tz^{-2} + 3Tz^{-3} + \cdots$$

$$= -Tz\frac{d}{dz}(z^{-1} + z^{-2} + z^{-3} + \cdots) = -Tz\frac{d}{dz}[z^{-1}(1 + z^{-1} + z^{-2} + z^{-3} + \cdots)]$$

$$= -Tz\frac{d}{dz}\left[z^{-1}\frac{z}{z - 1}\right] = \frac{Tz}{(z - 1)^2} \quad |z| > 1 \tag{8.10}$$

(c)

$$\mathcal{Z}\{f(nT)\} = \mathcal{Z}\{e^{-bnT}u(nT)\} = \mathcal{Z}\{u(nT)c^{-n}\}; \quad c = e^{bT};$$

$$= \sum_{n=0}^{\infty} u(nT)c^{-n}z^{-n} = 1 + c^{-1}z^{-1} + c^{-2}z^{-2} + \cdots = \frac{1}{1 - c^{-1}z^{-1}}$$

$$= \frac{cz}{cz - 1} = \frac{ze^{bT}}{ze^{bT} - 1} = \frac{z}{z - e^{-bT}} \quad |z| > e^{-bT} \quad (8.11)$$

(d)

$$\mathcal{Z}\{f(nT)\} = \mathcal{Z}\{u(nT) \sin \omega nT\} = \mathcal{Z}\left\{ u(nT) \frac{e^{j\omega nT} - e^{-j\omega nT}}{2j} \right\}$$

$$= \sum_{n=0}^{\infty} \frac{u(nT)}{2j} c_1^{-n}z^{-n} - \sum_{n=0}^{\infty} \frac{u(nT)}{2j} c_2^{-n}z^{-n}; \quad c_1 = e^{-j\omega nT}, \quad c_2 = e^{j\omega nT}$$

$$= \frac{1}{2j}\left[\frac{c_1 z}{c_1 z - 1} - \frac{c_2 z}{c_2 z - 1} \right] = \frac{z \sin \omega T}{z^2 - 2z \cos \omega T + 1} \quad |z| > 1 \quad (8.12)$$

The inequalities that appear in the solutions are found from the fact that the summations to converge the factor of the geometric series must have an absolute value less than 1 (see Section 8.2). ∎

8.2 Convergence of the z-Transform

The function $F(z)$ for a specified value of z may be either finite or infinite. The **regions of convergence** (ROC) is the set of values of z in the complex z-plane for which the magnitude of $F(z)$ is finite; whereas the set of values of z for which the magnitude of $F(z)$ is infinite is the **region of divergence**. The region of convergence is determined by considering the defining expression (8.3) and examining the complex values of z for which $\sum_{n=0}^{\infty} f(nT)z^{-n}$ has finite values. If we write z in polar form $z = re^{j\theta}$, then

$$\sum_{n=0}^{\infty} |f(nT)z^{-n}| = \sum_{n=0}^{\infty} |f(nT)(re^{j\theta})^{-n}| = \sum_{n=0}^{\infty} |f(nT)r^{-n}||e^{-j\theta n}|$$

$$= \sum_{n=0}^{\infty} |f(nT)r^{-n}| \quad (8.13)$$

since $|e^{-j\theta n}| = [\cos^2 n\theta + \sin^2 n\theta]^{1/2} = 1$. For this sum to be finite, we find numbers M and R such that $|f(nT)| \leq MR^n$ for $n \geq 0$. Thus,

$$\sum_{n=0}^{\infty} |f(nT)z^{-n}| \leq M \sum_{n=0}^{\infty} R^n r^{-n} = M \sum_{n=0}^{\infty} \left(\frac{R}{r}\right)^n \quad (8.14)$$

For this sum to be finite, it is required that $R/r < 1$ or $r > R$. That is, $F(z)$ is absolutely convergent for all z in the region where $|z| = r > R$. A separate test is required to establish if the boundary belongs or not in the ROC.

Example 8.5

Find the z-transform of the signal specified and discuss its properties.

$$f(n) = \begin{cases} c^n & n = 0, 1, 2, \ldots \\ 0 & n = -1, -2, \ldots \end{cases} \tag{8.15}$$

The real constant c takes the values: (a) $0 < c < 1$ and (b) $c > 1$:

SOLUTION

The time sequences for the two cases are shown in Figure 8.1a and b. The z-transform is given by

$$F(z) = \sum_{n=0}^{\infty} c^n z^{-n} = \sum_{n=0}^{\infty} (c^{-1} z)^{-n} = 1 + c^{-1} z + (c^{-1} z)^2 + \cdots + (c^{-1} z)^n + \cdots \tag{8.16}$$

Initially, we consider the sum of the first n terms (up to $(c^{-1}z)^{n-1}$ term) of this geometric series, which is given by

$$F_n = \frac{1 - (c^{-1}z)^{(n-1)+1}}{1 - c^{-1}z} = \frac{1 - (c^{-1}z)^n}{1 - c^{-1}z} \tag{8.17}$$

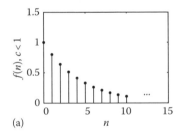

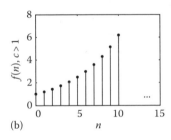

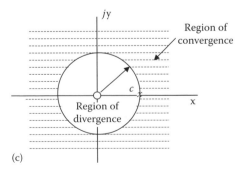

FIGURE 8.1

Next, we set $c^{-1}z = c^{-1}|z|e^{j\theta}$, where θ is the argument of the complex number $c^{-1}z$. Hence, we write $(c^{-1}z)^{-n} = (cz^{-1})^n = (cr^{-1}e^{-j\theta})^n = (c|z|^{-1})^n e^{-j\theta n}$.

Case 1: For values of z for which $|cz^{-1}| < 1$ ($|cz^{-1}| = |c||z^{-1}| = c|z^{-1}|$ for real c), the magnitude of the complex number $(cz^{-1})^n$ approaches zero as $n \to \infty$. As a consequence,

$$F(z) = \lim_{n \to \infty} F_n(z) = \frac{1}{1 - cz^{-1}} = \frac{z}{z - c} \qquad |cz^{-1}| < 1 \qquad \text{or} \qquad |z| > |c| \qquad (8.18)$$

Case 2: For the general case where c is a complex number, the inequality $|cz^{-1}| < 1$ leads to $|c| < |z|$, which implies that the series converges when $|z| > |c|$ and diverges when $|z| < |c|$. Thus, we see that the ROC and divergence in the complex z-plane for $F(z)$ are those shown in Figure 8.1c.

To establish whether the boundary of the circle in Figure 8.1c belongs to the ROC or not, we apply the L'Hospital's rule to (8.17). Thus,

$$\lim_{z \to c} F_n(z) = \lim_{z \to c} \frac{(d/d(cz^{-1}))(1 - (cz^{-1})^n)}{(d/d(cz^{-1}))(1 - (cz^{-1}))} = \lim_{z \to c} \frac{-n(cz^{-1})^{n-1}}{-1} = n$$

and, hence, $\lim_{z \to c} F_n(z) \to \infty$. Clearly, the boundary belongs to the region of divergence. ∎

Example 8.6

Deduce the z-transform and discuss the properties of the impulse functions

$$y(n) = \delta(n) = \begin{cases} 1 & n = 0 \\ 0 & n \neq 0 \end{cases} \qquad (8.19)$$

$$y(n) = \delta(n - N) = \begin{cases} 1 & n = N \\ 0 & n \neq N \end{cases} \qquad (8.20)$$

SOLUTION

From the basic definition of the z-transform, we write that

$$Y(z) = \delta(0)z^{-0} + \delta(1)z^{-1} + \cdots = 1 + 0 + 0 + \cdots = 1$$

Since $Y(z)$ is independent of z, the entire plane is the ROC. On applying the basic definition, (8.20) becomes

$$Y(z) = \sum_{n=0}^{\infty} \delta(n - N)z^{-n} = 0z^{-0} + 0z^{-1} + \cdots + 1z^{-N} + 0z^{-(N+1)} + \cdots = z^{-N}$$

Since $Y(z) \to \infty$, only for $z = 0$, the entire z-plane is the ROC except for an infinitesimal region around the region. ∎

Example 8.7

Deduce the z-transform of the function

$$y(n) = \begin{cases} a^n \sin n\omega & n \geq 0, a > 0 \\ 0 & n < 0 \end{cases} \tag{8.21}$$

Indicate the ROC, the region of divergence, and the poles and zeros in the z-plane.

SOLUTION

The given functions are shown in Figure 8.2a and b for two different values of a. Clearly, for $a = 1$, the function is a sinusoidal discrete signals. The z-transform is given by

$$Y(z) = \sum_{n=0}^{\infty} a^n \sin n\omega\, z^{-n} = \sum_{n=0}^{\infty} a^n \frac{e^{jn\omega} - e^{-jn\omega}}{2j} z^{-n}$$

$$= \frac{1}{2j} \sum_{n=0}^{\infty} (ae^{j\omega}z^{-1})^n - \frac{1}{2j} \sum_{n=0}^{\infty} (ae^{-j\omega}z^{-1})^n = \frac{1}{2j}\left[\frac{1}{1 - ae^{j\omega}z^{-1}} - \frac{1}{1 - ae^{-j\omega}z^{-1}}\right]$$

$$= \frac{z^{-1}a\sin\omega}{1 - 2az^{-1}\cos\omega + a^2 z^{-2}} \quad |z| > a \tag{8.22}$$

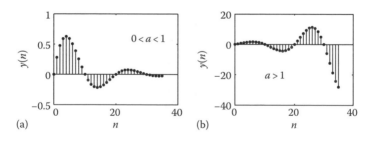

(a)   (b)

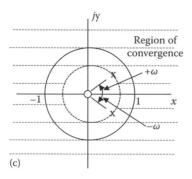

(c)

FIGURE 8.2

Next, multiply the numerator and denominator of this expression by z^2 to find

$$Y(z) = \frac{za \sin \omega}{z^2 - 2a(\cos \omega)z + a^2} = \frac{za \sin \omega}{[z - a \cos \omega - ja \sin \omega][z - a \cos \omega + ja \sin \omega]}$$

$$= \frac{za \sin \omega}{(z - ae^{j\omega})(z - ae^{-j\omega})} \qquad (8.23)$$

The zeros and poles are shown in Figure 8.2c for the case $a < 1$.

To examine the ROC, consider the summations in (8.22). It is seen that each summation converges if $|ae^{j\omega}z^{-1}| = |az^{-1}||e^{j\omega}| = |az^{-1}| < 1$ or $|z| > a$. This region is shown in Figure 8.2c.

Book MATLAB® m-File: ex_8_2_3

```
%m-file:ex8_2_3
n=0:35;
y1=0.9.^n.*sin(0.1*pi*n);
y2=1.1.^n.*sin(0.1*pi*n);
subplot(2,2,1);stem(n,y1,'k.');
xlabel('n');ylabel('y(n)');
hold on;text(25,0.6,'0<a<1');
hold on;text(35,0.8,'a)');
subplot(2,2,2);stem(n,y2,'k.');
xlabel('n');ylabel('y(n)');
hold on;text(10,-20,'a>1')
hold on;text(35,10,'b)');
```
∎

When the sequence $\{y(k)\}$ has values for both positive and negative k, the region of convergence of $Y(z)$ becomes an annular ring around the origin or $Y(z)$ does not have a region of convergence. Let us consider the sequence

$$y(n) = \begin{cases} 3^n & n \geq 0 \\ 4^n & n < 0 \end{cases}$$

This is a bilateral function and the definition of the bilateral z-transform is given by

$$\boxed{\Im\{y(n)\} = \sum_{n=-\infty}^{\infty} y(n)z^{-n}} \qquad (8.24)$$

For our specific function,

$$Y(z) = \sum_{n=0}^{\infty} 3^n z^{-n} + \sum_{n=-\infty}^{-1} 4^n z^{-n} = \sum_{n=0}^{\infty} 3^n z^{-n} + \sum_{n=\infty}^{1} 4^{-n} z^n = \sum_{n=0}^{\infty} 3^n z^{-n} + \sum_{n=1}^{\infty} 4^{-n} z^n$$

The first summation converges as $n \to \infty$ provided that $|3z^{-1}| < 1$ or $|z| > 3$. If we set $R^+ = 3$ for positive n's, we see that the region of convergence for positive n's is $|z| > R^+$.

Similarly, the second summation will converge if $|4^{-1}z| < 1$ or $|z| < 4$, and the region of convergence for negative n's is $|z| < R^-$ with $R^- = 4$. The sequence $y(n)$ and the region of convergence depicted by the double-line region are shown in Figure 8.3a and b, respectively.

The reader can easily show, following parallel steps to the above, that the sequence

$$y(n) = \begin{cases} 4^n & n \geq 0 \\ 3^n & n < 0 \end{cases}$$

has no region of convergence.

From the foregoing discussion, we conclude that

1. The region of convergence of a two-sided sequence is a ring in the z-plane centered at the origin.
2. The region of convergence of a sequence of finite duration is the entire z-plane, except possibly the points $z = 0$ and/or $z = \infty$.
3. If the sequence is right-handed—that is, $n \geq 0$—then the region of convergence is beyond a circle of finite radius.
4. If the sequence is left-handed—that is, $n < 0$—then the region of convergence is within a circle of finite radius.

The following example shows why it is important to specify the region of convergence.

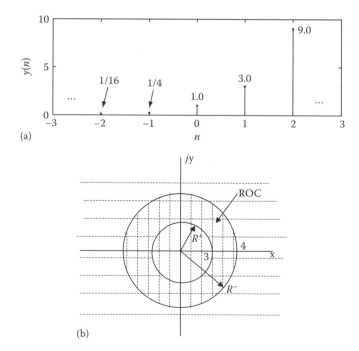

(a)

(b)

FIGURE 8.3

Example 8.8

Specify the ROC for the two sequences

$$
\begin{array}{lll}
x(n) = a^n & a > 1, \quad n = 0, 1, 2, \cdots & \text{(a)} \\
y(n) = -a^n & a > 1, \quad n = -1, -2, -3, \cdots & \text{(b)}
\end{array}
\tag{8.25}
$$

SOLUTION

The z-transform of the first sequence is

$$
X(z) = \sum_{n=0}^{\infty} (az^{-1})^n = \frac{1}{1 - az^{-1}} = \frac{z}{z - a}
\tag{8.26}
$$

For convergence, we must have $|az^{-1}| < 1$, which implies that the region of convergence is $|z| > a$.

The z-transform of the other sequence is

$$
Y(z) = -\sum_{n=-1}^{-\infty} a^n z^{-n} = -\sum_{n=1}^{\infty} a^{-n} z^n = -\sum_{n-0}^{\infty} a^{-n} z^n + 1
$$

$$
= -\frac{1}{1 - a^{-1}z} + 1 = \frac{z}{z - a}
\tag{8.27}
$$

The region of convergence for this function is found as follows: $|a^{-1}z| < 1$ or $|z| < a$. We observe that, although the two sequences are completely different in the time domain (one increases positively with positive time and the other increases negatively for negative time), they have the same function in their z presentation. However, each one has its own ROC, and this distinguishes one sequence from the other. The inverse z-transform is an integral where the integration path is inside the ROC. Therefore, to find the right time function, the integration must be conducted inside the corresponding ROC. ∎

8.3 Properties of the z-Transform

The most important properties of the z-transform for one-sided sequences are given next. The properties are summarized and accompanied by examples to elucidate their use. In the end of this chapter, as Appendix 8.A.1, we provide the proofs. The one-sided sequences are of great importance because all detected signals are of finite extend and their starting point can always be referenced at $t = 0$ $(n = 0)$.

Summary of z-transform properties

1. Linearity

$$
\Im\{ax(n) + by(n)\} = a\Im\{x(n)\} + b\Im\{y(n)\}
$$

2. Right-shifting property
   (a)

$$\Im\{y(n-m)\} = z^{-m}\Im\{y(n)\} = z^{-m}Y(z) \quad \text{zero initial conditions}$$

   (b)

$$\Im\{y(n-m)\} = z^{-m}Y(z) + \sum_{i=0}^{m-1} y(i-m)z^{-i} \quad \text{initial conditions present}$$

3. Left-shifting property

$$\Im\{y(n+m)\} = z^{m}Y(z) - \sum_{i=0}^{m-1} y(i)z^{m-i}$$

4. Time scaling

$$\Im\{a^{n}y(n)\} = Y(a^{-1}z) = \sum_{n=0}^{\infty} (a^{-1}z)^{-n}$$

5. Periodic sequence

$$\Im\{y(n)\} = \frac{z^{N}}{z^{N}-1}\Im\{y_{(1)}(n)\}, \quad N = \text{number of time units in a period}$$
$$y_{(1)}(n) = \text{first period}$$

6. Multiplication by n

$$\Im\{ny(n)\} = -z\frac{dY(z)}{dz}$$

7. Initial value

$$y(n_0) = z^{n_0}Y(z)\big|_{z\to\infty} \quad y(n) = 0 \quad \text{for } n < n_0$$

8. Final value

$$\lim_{n\to\infty} y(n) = \lim_{n\to1} (1-z^{-1})Y(z) \quad \text{provided } y(\infty) \text{ exists}$$

9. Convolution

$$\Im\{y(n)\} = \Im\{h(n) * x(n)\} = \Im\left\{\sum_{m=0}^{\infty} h(n-m)x(m)\right\} = H(z)X(z)$$

10. Bilateral convolution

$$\mathfrak{Z}\{y(n)\} = \mathfrak{Z}\{h(n) * x(n)\} = \mathfrak{Z}\left\{\sum_{m=-\infty}^{\infty} h(n-m)x(m)\right\} = H(z)X(z)$$

Example 8.9 (Shifting property)

Find the z-transform of the functions shown in Figure 8.4a and b using the right-shifting property.

SOLUTION

First, consider the z-transform of the unit step function $u(n)$ shown in Figure 8.4b, which is given by

$$U(z) = \sum_{n=0}^{\infty} u(n)z^{-n} = 1 + z^{-1} + z^{-2} + z^{-3} + \cdots = \frac{1}{1-z^{-1}} = \frac{z}{z-1}$$

The discrete-time function in Figure 8.4a is $y(n) = u(n-2)$. The z-transform of this equation is

$$\mathfrak{Z}\{y(n)\} = \mathfrak{Z}\{u(n-2)\} = z^{-2}\mathfrak{Z}\{u(n)\} = \frac{1}{z^2}\frac{z}{z-1} = \frac{1}{z(z-1)}$$

The pole-zero configuration is shown in Figure 8.4c. We observe from this figure that $U(z)$ does not have poles at zero, whereas the combination of its shifted version $y(n)$ does possess poles (single) at the origin for this particular shifting. ∎

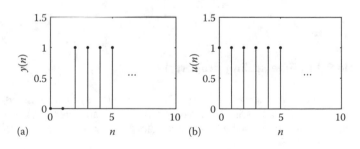

(a)   (b)

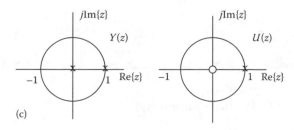

(c)

FIGURE 8.4

Example 8.10 (Shifting property)

Find the z-transform of an *RL* series circuit, with a voltage input and the current as the output. The initial condition is $i(0) = 2$. Discretize the analog system.

Solution

The KVL is

$$L\frac{di(t)}{dt} + Ri(t) = v(t)$$

The discretized form of the above equation is given by

$$L\frac{i(nT) - i(nT - T)}{T} + Ri(nT) = v(nT) \quad \text{or} \quad i(nT) - \frac{1}{1 + T\frac{R}{L}}i(nT - T) = \frac{T}{L}\frac{1}{1 + T\frac{R}{L}}v(nT)$$

Taking into consideration the linearity and shifting property, we obtain

$$I(z) - a[z^{-1}I(z) + i(0 - 0)z^{-0}] = \frac{T}{L}aV(z) \quad a = \frac{1}{1 + T(R/L)}$$

We, finally, obtain the algebraic relation

$$I(z) = \frac{2a}{1 - az^{-1}} + \frac{T}{L}\frac{a}{1 - az^{-1}}V(z) \qquad \blacksquare$$

Example 8.11 (Time scaling property)

The z-transform of $y(n) = \sin n\omega$ for $n = 0, 1, 2, \ldots$ is equal to $z \sin \omega / (z^2 - 2z \cos \omega + 1)$ (see (8.12) with $T = 1$). By applying the scaling property, we can write the z-transform of the function $a^n y(n) = a^n \sin n\omega$ by inserting $a^{-1}z$ for z. This substitution leads to

$$\mathcal{Z}\{a^n \sin n\omega\} = \frac{a^{-1}z \sin \omega}{a^{-2}z^2 - 2a^{-1}z \cos \omega + 1}$$

The result is the same as (8.23), which was deduced by different approach. $\blacksquare$

Example 8.12 (Left-shifting property)

Find the z-transform of the output of the system shown in Figure 8.5.

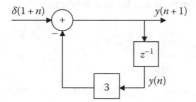

FIGURE 8.5

SOLUTION

From the diagram, the difference equation that describes the system is

$$\delta(n+1) - 3y(n) = y(n+1) \quad \text{or} \quad y(n+1) + 3y(n) = \delta(n+1)$$

Take the z-transform of both sides of the equation, recalling the linearity and the left-shifting property, with the result

$$\Im\{y(n+1)\} + 3\Im\{y(n)\} = \Im\{\delta(n+1)\}$$

which, by applying the left-shifting property, yields

$$zY(z) - zy(0) + 3Y(z) = z \quad \text{or} \quad Y(z) = \frac{z[1 + y(0)]}{z+3} \qquad \blacksquare$$

Example 8.13 (Periodic sequence property)

Find the z-transform of the sequence shown in Figure 8.6.

SOLUTION

Use the periodic sequence Property # 5 with $N=4$ to find

$$Y(z) = \frac{z^4}{z^4 - 1}\Im\{y_{(1)}(n)\} = \frac{z^4}{z^4 - 1}(1 + z^{-1} + z^{-2}) = \frac{z^2(z^2 + z + 1)}{z^4 - 1} \qquad \blacksquare$$

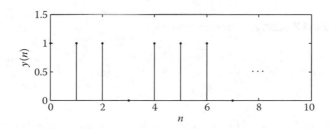

FIGURE 8.6

Example 8.14 (Multiplication by *n* property)

Deduce the *z*-transform of the functions $nu(n), n^2u(n), n(n+1)u(n)$, and $n(n-1)u(n)$.

Solution

Since $\mathfrak{Z}\{u(n)\} = z/(z-1)$, then the multiplication by *n* Property 6, we have

$$\mathfrak{Z}\{nu(n)\} = -z\frac{d}{dz}\left(\frac{z}{z-1}\right) = \frac{z}{(z-1)^2}$$

We can continue this procedure to find

$$\mathfrak{Z}\{n^2u(n)\} = \mathfrak{Z}\{n[nu(n)]\} = -z\frac{d}{dz}\left(\frac{z}{(z-1)^2}\right) = \frac{z(z+1)}{(z-1)^3}$$

Because $n(n+1) = n^2 + n$, we use the linearity property and adding the results above, we find $\mathfrak{Z}\{n(n+1)u(n)\} = \mathfrak{Z}\{n^2u(n)\} + \mathfrak{Z}\{nu(n)\} = 2z^2/(z-1)^3$. Similarly, for the last case, we find $\mathfrak{Z}\{n(n-1)u(n)\} = \mathfrak{Z}\{n^2u(n)\} - \mathfrak{Z}\{nu(n)\} = 2z/(z-1)^3$. ■

Example 8.15 (Initial value property)

To find the initial value of the function given by (8.23), we proceed as follows:

$$y(0) = z^0Y(z)|_{z\to\infty} = \left.\frac{z^{-1}a\sin\omega}{1-2az^{-1}\cos\omega+a^2z^{-2}}\right|_{z\to\infty} = 0$$

which agrees with the value deduced from (8.21). ■

Example 8.16 (Final value property)

We know that the *z*-transform of the function $y(n) = 0.9^nu(n)$ is $Y(z) = z/(z-0.9)$ and, therefore, the final value property is given by

$$\lim_{z\to1}(1-z^{-1})\frac{z}{z-0.9} = \lim_{z\to1}\frac{z-1}{z-0.9} = \frac{1-1}{1-0.9} = 0$$

as it should be. ■

Example 8.17 (Convolution property)

The input signal sequence $x(n)$ and the impulse response $h(n)$ of a system are shown in Figure 8.7a and b. Deduce the output of the system $w(n)$.

Solution

Figure 8.7a and b show the input function $x(n)$ and the impulse response function $h(n)$ of the system. Figure 8.7c shows the reflected function $x(0-m)$ in the m-domain with

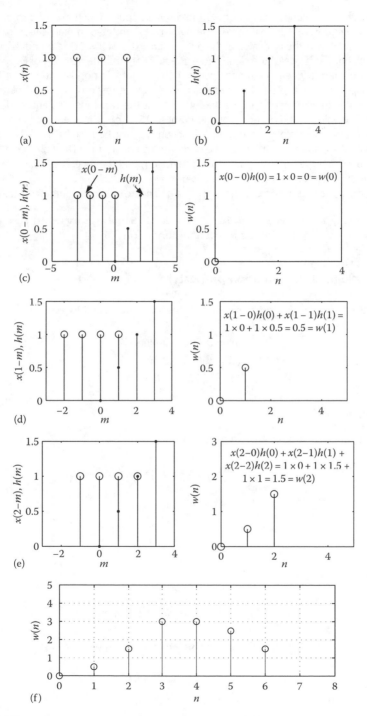

FIGURE 8.7

zero shift. In the same figure, we have plotted the impulse function in the m-domain. From the figure, we see that there is an overlap at 0 and their multiplication gives us zero result. Hence, the first value of the output at $n = 0$ is $w(0) = 0$. The output is plotted on the right-hand side of the same figure as a function of n. The left part of Figure 8.7d shows the input function $x(m)$ shifted by 1 to the right (a positive number +1 was introduced) and was plotted in m-domain. The impulse response was also plotted in the m-domain. Next, we multiplied point by point the two functions and added the result. The value found is shown as the number 0.5 at $n = 1$ in the right-hand figure. However, we have plotted the output function $w(n)$ for both shifts $n = 0$ and $n = 1$. Similarly, the result for shift 2 gives the output to be equal to 1.5. The right-hand figure of Figure 8.7d gives the output $w(n)$ for shifts 0, 1, and 2. Proceed the same way to obtain the final output function, which is given in Figure 8.7e.

The reader can use the following MATLAB algorithm to obtain the output function $w(n)$:

```
x=[1  1  1  1];   h=[0  0.5  1  1.5];w=conv(x,h);n=0:6;
stem(n,w,'k'); % conv is a MATLAB function;          ∎
```

Example 8.18 (Convolution property)

Find the z-transform of the convolution of the following three functions:

$$x(n) = a_0\delta(n) + a_1\delta(n-1)$$
$$y(n) = b_0\delta(n) + b_1\delta(n-1)$$
$$q(n) = c_0\delta(n) + c_1\delta(n-1)$$

SOLUTION

The z-transforms of these three functions are

$$X(z) = a_0 + a_1z^{-1} \quad Y(z) = b_0 + b_1z^{-1} \quad Q(z) = c_0 + c_1z^{-1}$$

We observe that

$$[X(z)Y(z)]Q(z) = X(z)[Y(z)Q(z)]$$

from which it follows that

$$[x(n)*y(n)]*q(n) = x(n)*[y(n)*q(n)]$$

This result shows that convolution is **associative**. Thus, we write

$$Z\{x(n)*y(n)*q(n)\} = X(z)Y(z)Q(z) = \left[a_0b_0 + (a_1b_0 + a_0b_1)z^{-1} + a_1b_1z^{-2}\right]$$
$$\left(c_0 + c_1z^{-1}\right) = a_0b_0c_0 + (a_1b_0c_0 + a_0b_1c_0 + a_1b_0c_1)z^{-1}$$
$$+ (a_1b_0c_1 + a_0b_1c_1 + a_1b_1c_0)z^{-2} + a_1b_1c_1z^{-3}$$

We next apply the definition of the z-transform to see that this result implies

$$x(n) * y(n) * q(n) = \left(\sum_{m=0}^{n} [a_0\delta(n-m) + a_1\delta(n-m-1)][a_0\delta(n-m) + a_1\delta(n-m-1)] \right)$$

$$*[c_0\delta(n) + c_1\delta(n-1)] = \left(\sum_{m=0}^{n} [a_0b_0\delta(n-m)\delta(m) + a_1b_0\delta(n-m-1)\delta(m) \right.$$

$$\left. + a_0b_1\delta(n-m)\delta(m-1) + a_1b_1\delta(n-m-1)\delta(m-1)] \right) * [c_0\delta(n) + c_1\delta(n-1)]$$

$$= [a_0b_0\delta(n) + a_1b_0\delta(n-1) + a_0b_1\delta(n-1) + a_1b_1\delta(n-2)] * [c_0\delta(n) + c_1\delta(n-1)]$$

$$= a_0b_0c_0\delta(n) + (a_1b_0c_0 + a_0b_1c_0 + a_0b_0c_1)\delta(n-1)$$

$$+ (a_1b_0c_1 + a_0b_1c_1 + a_1b_1c_0)\delta(n-2) + a_1b_1c_1\delta(n-3)$$

We have left as an exercise to show that the convolution is also **commutative**.   ∎

Example 8.19 (Convolution property)

Find the output of the relaxed system shown in Figure 8.8a if the input is that shown in Figure 8.8b. Express the system in its discrete form.

SOLUTION

A direct application of the KVL yields the equation

$$\frac{di(t)}{dt} + i(t) = v(t), \qquad \frac{dv_o(t)}{dt} + v_o(t) = v(t)$$

The second equation follows from the first since $v_o(t) = Ri(t)$ with $R = 1\ \Omega$. If we assume the sampling time $T = 1$, then from Section 1.2, we find that $dv_o(t)/dt \cong v_o(n) - v_o(n-1)$ and, thus, the above equation takes the following discretized form:

$$v_o(n) = \frac{1}{2}v(n) + \frac{1}{2}v_o(n-1)$$

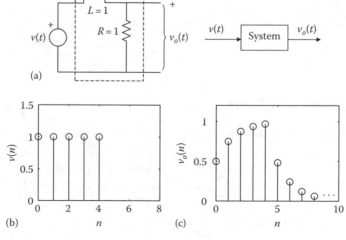

(a)

(b)

(c)

FIGURE 8.8

Next, we proceed to determine the impulse response of the system. We use the fact that the z-transform of the output of the system is equal to the system function H(z) for a delta function input $v(n) = \delta(n)$ (exactly the same approach was used in LT studies). Also, the inverse transform of H(z) is the impulse response h(n). Therefore, H(z) is given by

$$\Im\{v_o(z)\} = \frac{1}{2}\Im\{\delta(n)\} + \frac{1}{2}\Im\{v_o(n-1)\} \quad \text{or} \quad H(z) = \frac{V_o(z)}{V(z)} = \frac{1}{2}\frac{1}{1-\frac{1}{2}z^{-1}}$$

$$= \frac{1}{2}\left[1 + \left(\frac{1}{2}\right)z^{-1} + \left(\frac{1}{2}\right)^2 z^{-2} + \cdots\right], \quad V(z) = 1$$

It follows from (8.2) that

$$h(n) = \frac{1}{2}\left(\frac{1}{2}\right)^n \quad n \geq 0$$

Since the output of the system is equal to the convolution of the input and its impulse response, this result can be used in the expression

$$v_o(n) = \sum_{m=0}^{n} h(n-m)v(m)$$

The outputs at successive time steps are

$$v_o(0) = h(0)v(0) = \frac{1}{2} \times 1 = \frac{1}{2}$$

$$v_o(1) = h(1)v(0) + h(0)v(1) = \frac{1}{4} \times 1 + \frac{1}{2} \times 1 = \frac{3}{4}$$

$$v_o(2) = h(2)v(0) + h(1)v(1) + h(0)v(2) = \frac{1}{8} \times 1 + \frac{1}{4} \times 1 + \frac{1}{2} \times 1 = \frac{7}{8}$$

$$\vdots$$

The output is shown in Figure 8.8c. Observe that the resulting shape of the output is the same as that for the corresponding continuous system. It is expected that if we had selected the sampling time interval to be shorter, the output voltage would be closer to its corresponding analog case and the shorter the sampling time the closer to the analog case would have resulted. ∎

8.4 z-Transform Pairs

Just as with other transforms, the z-transform of a discrete function and its inverse are given as

$$\Im\{f(n)\} \triangleq F(z) = \sum_{n=0}^{\infty} f(n)z^{-n} \quad \text{or} \quad F(z) = \sum_{n=0}^{\infty} f(nT)z^{-n}$$

$$\Im^{-1}\{F(z)\} \triangleq \{f(n)\} = \frac{1}{2\pi j}\oint_C F(z)z^{n-1}dz \tag{8.28}$$

TABLE 8.1 Properties of the z-Transform ($n \geq 0$)

1.	$\mathfrak{Z}\{ax(n) + by(n)\}$	$aX(z) + bY(z)$
2.	$\mathfrak{Z}\{y(n - m)\}$	$z^{-m}Y(z) + \sum_{i=1}^{m} y(-i)z^{-(m-i)}$
3.	$\mathfrak{Z}\{y(n + m)\}$	$z^{m}Y(z) - \sum_{n=0}^{m-1} y(n)z^{m-n}$
4.	$\mathfrak{Z}\{a^{n}y(n)\}$	$Y\left(\dfrac{z}{a}\right)$
5.	$\mathfrak{Z}\{y(n)\}$	$\dfrac{z^{N}}{z^{N} - 1}Y_{1}(z)$; $y(1)$ is the first period of a periodic sequence $y(n) = y(n + N)$
6.	$\mathfrak{Z}\{y(n) * h(n)\}$	$Y(z)H(z)$
7.	$\mathfrak{Z}\{ny(n)\}$	$-z\dfrac{d}{dz}Y(z)$
8.	$\mathfrak{Z}\{n^{m}y(n)\}$	$\left(-z\dfrac{d}{dz}\right)^{m}Y(z)$
9.	$\mathfrak{Z}\{y(n_{0})\}$	$z^{n_{0}}Y(z)\big|_{z \to \infty}$; n_{0} is the initial value of the sequence and $Y(z) = \sum_{n=0}^{\infty} y(n_{0} + n)z^{-n}$
10.	$\lim_{z \to \infty} y(n)$	$\lim_{z \to 1}(1 - z^{-1})Y(z)$

where the integration is done in counterclockwise direction and $\mathfrak{Z}^{-1}$ denotes the inverse z-transformation. Since the inverse z-transform includes integration in the complex z-plane, we will not present it here. However, we would like to point out that the integration path must be inside the ROC for the integral to give us the right answer. In our studies in this chapter, we will rely on tables to obtain the inverse z-transforms. Table 8.1 provides some common z-transform pairs.

8.5 Inverse z-Transform

As already discussed in our studies, we assume that an $F(z)$ corresponds to a sequence $\{f(n)\}$ that is bounded as $n \to \infty$. To find the inverse z-transform, we cast the transform function into a form that is amenable to a simple tabular lookup using Table 8.2. The approach parallels that followed in performing similar operations using the LT. The functions that we will be concerned are rational functions of z- that is are the ratios of two polynomials. Ordinarily, these are **proper fractions** since the degree of the numerator polynomial is less than the degree of the denominator polynomial. If the functions are not proper functions, we will perform long division until the degree of the numerator is less than that of the denominator. This results to a power series and a proper fraction.

8.5.1 Partial Fraction Expansion

If a $F(z)$ is a rational function of z and analytic at infinity, it can be expressed as follows:

$$F(z) = F_{1}(z) + F_{2}(z) + F_{3}(z) + \cdots \tag{8.29}$$

TABLE 8.2 Common z-Transform Pairs

Entry Number	$f(n), f(nT)$ for $n \geq 0$	$F(z) = \sum_{n=0}^{\infty} f(n)z^{-n}$	Radius of Convergence $\lvert z \rvert > R$
1.	$\delta(n)$	1	0
2.	$\delta(n - m)$	z^{-m}	0
3.	1	$\dfrac{z}{z - 1}$	1
4.	n	$\dfrac{z}{(z - 1)^2}$	1
5.	n^2	$\dfrac{z(z + 1)}{(z - 1)^3}$	1
6.	n^3	$\dfrac{z(z^2 + 4z + 1)}{(z - 1)^4}$	1
7.	a^n	$\dfrac{z}{z - a}$	$\lvert a \rvert$
8.	na^n	$\dfrac{az}{(z - a)^2}$	$\lvert a \rvert$
9.	$n^2 a^n$	$\dfrac{az(z + a)}{(z - a)^3}$	$\lvert a \rvert$
10.	$\dfrac{a^n}{n!}$	$e^{a/z}$	0
11.	$(n + 1)a^n$	$\dfrac{z^2}{(z - a)^2}$	$\lvert a \rvert$
12.	$\dfrac{(n + 1)(n + 2)a^n}{2!}$	$\dfrac{z^3}{(z - a)^3}$	$\lvert a \rvert$
13.	$\dfrac{(n + 1)(n + 2) \cdots (n + m)a^n}{m!}$	$\dfrac{z^{m+1}}{(z - a)^{m+1}}$	$\lvert a \rvert$
14.	$\sin n\omega T$	$\dfrac{z \sin \omega T}{z^2 - 2z \cos \omega T + 1}$	1
15.	$\cos n\omega T$	$\dfrac{z(z - \cos \omega T)}{z^2 - 2z \cos \omega T + 1}$	1
16.	$a^n \sin n\omega T$	$\dfrac{az \sin \omega T}{z^2 - 2az \cos \omega T + a^2}$	$\lvert a \rvert^{-1}$
17.	$a^{nT} \sin n\omega T$	$\dfrac{a^T z \sin \omega T}{z^2 - 2a^T z \cos \omega T + a^{2T}}$	$\lvert a \rvert^{-T}$
18.	$a^n \cos n\omega T$	$\dfrac{z(z - a \cos \omega T)}{z^2 - 2az \cos \omega T + a^2}$	$\lvert a \rvert^{-1}$
19.	$e^{-anT} \sin n\omega T$	$\dfrac{ze^{-aT} \sin \omega T}{z^2 - 2e^{-aT}z \cos \omega T + e^{-2aT}}$	$\lvert z \rvert > \lvert e^{-aT} \rvert$
20.	$e^{-anT} \cos n\omega T$	$\dfrac{z(z - e^{-aT} \cos \omega T)}{z^2 - 2e^{-aT}z \cos \omega T + e^{-2aT}}$	$\lvert z \rvert > \lvert e^{-aT} \rvert$
21.	$\dfrac{n(n - 1)}{2!}$	$\dfrac{z}{(z - 1)^3}$	1
22.	$\dfrac{n(n - 1)(n - 2)}{3!}$	$\dfrac{z}{(z - 1)^4}$	1
23.	$\dfrac{n(n - 1)(n - 2) \cdots (n - m + 1)}{m!} a^{n-m}$	$\dfrac{z}{(z - a)^{m+1}}$	1
24.	e^{-anT}	$\dfrac{z}{z - e^{-aT}}$	$\lvert e^{-aT} \rvert$
25.	ne^{-anT}	$\dfrac{ze^{-aT}}{(z - e^{-aT})^2}$	$\lvert e^{-aT} \rvert$

and, therefore,

$$f(nT) = \mathcal{Z}^{-1}\{F_1(z)\} + \mathcal{Z}^{-1}\{F_2(z)\} + \mathcal{Z}^{-1}\{F_3(z)\} + \cdots \tag{8.30}$$

For an expansion of the form

$$F(z) = \frac{F_1(z)}{(z-p)^n} = \frac{A_1}{(z-p)} + \frac{A_2}{(z-p)^2} + \cdots + \frac{A_n}{(z-p)^n} \tag{8.31}$$

the constants A_i are given by

$$A_n = (z-p)^n F(z)|_{z=p}$$

$$A_{n-1} = \frac{d}{dz}[(z-p)^n F(z)]\Big|_{z=p}$$

$$\vdots$$

$$A_{n-k} = \frac{1}{k}\frac{d^k}{dz^k}[(z-p)^n F(z)]\Big|_{z=p} \tag{8.32}$$

$$\vdots$$

$$A_1 = \frac{1}{(n-1)}\frac{d^{n-1}}{dz^{n-1}}[(z-p)^n F(z)]\Big|_{z=p}$$

Example 8.20 (Power series method)

Determine the inverse z-transform of the function

$$F(z) = \frac{1}{1 - 0.1z^{-1}} \tag{8.33}$$

SOLUTION

The function possesses a simple pole at $z = 0.1$ and a zero at $z = 0$. The ROC is a circle in the complex z-plane with radius $|z| > 0.1$. We can proceed by dividing the numerator by the denominator, which results in the following infinite series in powers of z^{-1}:

$$F(z) = 1 + 0.1z^{-1} + (0.1)^2 z^{-2} + (0.1)^3 z^{-3} + \cdots$$

Thus, we have that

$$F(z) = 1 + 0.1z^{-1} + (0.1)^2 z^{-2} + (0.1)^3 z^{-3} + \cdots$$

It follows from (8.2) that the corresponding sequence is

$$f(n) = \begin{cases} 1, 0.1, (0.1)^2, (0.1)^3, \ldots & n \geq 0 \\ 0 & n < 0 \end{cases} \tag{8.34}$$

which is the sequence

$$f(n) = (0.1)^n \quad n \geq 0 \tag{8.35}$$

∎

Example 8.21 (Partial function expansion)

Find the inverse *z*-transform of the function

$$F(z) = \frac{1}{(1 - 0.2z^{-1})(1 + 0.2z^{-1})} = \frac{1}{1 - 0.04z^{-2}} \tag{8.36}$$

SOLUTION

One approach is to proceed as the foregoing example. By long division, the following polynomial results:

$$F(z) = 1 + 0.04z^{-2} + (0.04)^2 z^{-4} + \cdots = (0.2)^{2n}(z^{-1})^{2n}$$

with region of convergence $|z| > 0.2$. This series corresponds to the sequence

$$f(m) = \begin{cases} (0.2)^m & m = 2n, & n \geq 0 \\ 0 & m = 2n + 1, & n > 0 \\ 0 & & n < 0 \end{cases} \tag{8.37}$$

A different approach calls for separating the function $F(z)$ into partial fraction form. Now, two factors must be considered: (1) the resulting function must be a proper function and (2) many entries in Table 8.2 involve z in the numerator of the resulting expression for $F(z)$. This need is achieved by considering $F(z)/z$. Thus, we modify $F(z)$ to $F(z)/z$:

$$\frac{F(z)}{z} = \frac{z}{(z - 0.2)(z + 0.2)} = \frac{A}{z - 0.2} + \frac{B}{z + 0.2}$$

where

$$A = \frac{z}{z + 0.2}\Big|_{z=0.2} = 0.5, \quad B = \frac{z}{z - 0.2}\Big|_{z=-0.2} = \frac{-0.2}{-0.4} = 0.5$$

$$F(z) = \frac{1}{2}\left[\frac{z}{z - 0.2} + \frac{z}{z + 0.2}\right]$$

From appropriate entries of Table 8.2, the inverse transform is

$$f(n) = \begin{cases} \dfrac{1}{2}[(0.2)^n + (0.2)^{-n}] & n \geq 0 \\ 0 & n < 0 \end{cases} \qquad (8.38)$$

The reader can easily verify that (8.37) and (8.38) yield identical results. ∎

Example 8.22

Find the inverse z-transform of the function

$$F(z) = \frac{1}{(1 - 0.2z^{-1})z^{-2}} = \frac{z^3}{z - 0.2} \qquad (8.39)$$

SOLUTION

By long division, we obtain

$$F(z) = z^2 + 0.2z + (0.2)^2 \frac{z}{z - 0.2}$$

From Table 8.2, the inverse transform is

$$f(n) = \delta(n + 2) + 0.2\delta(n + 1) + (0.2)^2(0.2)^n$$

where the last term is applicable for $n \geq 0$. Therefore, this equation is equivalent to

$$f(n) = \begin{cases} 0.2^{n+2} & n \geq -2 \\ 0 & n < -2 \end{cases} \qquad (8.40)$$

It is recalled that (8.39) could be expanded into the form

$$F(z) = z^2 + 0.2z + (0.2)^2 z^0 + (0.2)^3 z^{-1} + (0.2)^4 z^{-2} + \cdots$$

$$= z^2 \left[1 + \frac{0.2}{z} + \frac{0.2^2}{z^2} + \cdots \right]$$

The inverse transform of the bracketed term is 0.2^n, and the factor z^2 indicates a shift to the left of two sample periods. Thus, (8.40) is realized.

From the above, we note that to find the inverse z-transform, we must

- Initially ignore any factor of the form z^n for n an integer
- Expand the remaining part into partial fraction form
- Use z-transform tables or properties to obtain the inverse z-transform of each term in the expansion
- Combine the results and perform the necessary shifting required by z^n omitted in the first step ∎

Example 8.23 (Separate poles)

Find the inverse z-transform of the function

$$F(z) = \frac{z^2 - 3z + 8}{(z-2)(z+2)(z+3)}$$

SOLUTION

Expand $F(z)/z$ in partial fraction form for reasons already discussed:

$$\frac{F(z)}{z} = \frac{z^2 - 3z + 8}{z(z-2)(z+2)(z+3)} = \frac{A}{z} + \frac{B}{z-2} + \frac{C}{z+2} + \frac{D}{z+3}$$

where

$$A = z\frac{F(z)}{z}\bigg|_{z=0} = -\frac{2}{3}$$

$$B = (z-2)\frac{F(z)}{z}\bigg|_{z=2} = \frac{z^2 - 3z + 8}{z(z+2)(z+3)}\bigg|_{z=2} = \frac{3}{20}$$

$$C = (z+2)\frac{F(z)}{z}\bigg|_{z=-2} = \frac{z^2 - 3z + 8}{z(z-2)(z+3)}\bigg|_{z=-2} = \frac{9}{4}$$

$$D = (z+3)\frac{F(z)}{z}\bigg|_{z=-3} = \frac{z^2 - 3z + 8}{z(z-2)(z+2)}\bigg|_{z=-3} = -\frac{26}{15}$$

Therefore,

$$F(z) = -\frac{2}{3} + \frac{3}{20}\frac{z}{z-2} + \frac{9}{4}\frac{z}{z+2} - \frac{26}{15}\frac{z}{z+3}$$

This leads to the following values for $\{f(n)\}$ using Table 8.2:

$$f(n) = -\frac{2}{3}\delta(n) + \frac{3}{20}2^n + \frac{9}{4}(-2)^n - \frac{26}{15}(-3)^n$$

If we set $n=0$, then $f(0)=0$ and MATLAB below ignores this point. However, if we divide numerator by denominator, we obtain the sequence $z^{-1}(1 - 6z^{-1} + 30z^{-2} - 102z^{-3} + \cdots)$ which indicates a shifted sequence to the right and MATLAB ignores the shifting. ∎

Book MATLAB m-File: ex_8_5_4

```
%m-file: ex_8_5_4
num= [0 1 −3 8];den= [conv(conv([1 −2],[1 2]),[1 3])];
[r,p,k] =residue(num,den);
```

For this particular case, the values of r, p, and k are

$$r = [5.2000 \quad -4.5000 \quad 3000]; \quad p = [-3.0000 \quad -2.0000 \quad 2.0000]; \quad k = []$$

Based on the MATLAB results we write

$$F(z) = 5.2\frac{z}{z+3} - 4.5\frac{z}{z+2} + 0.3\frac{z}{z-2}$$

Note the corresponding numbers between r and p vectors. The inverse z-transform of the above equation is

$$f(n) = 5.2(-3)^n - 4.5(-2)^n + 0.3(2)^n$$

Although the two results look different, the sequence $\{f(n)\}$ in the two cases are numerically identical. ∎

Example 8.24 (Multiple poles)

Find the inverse z-transform of the function

$$F(z) = \frac{z^2 - 9}{(z-1)(z-2)^2}$$

Observe that the function has a single- and a second-order pole.

SOLUTION

The function $F(z)/z$ is expanded in partial fraction form as follows:

$$\frac{F(z)}{z} = \frac{z^2 - 9}{z(z-1)(z-2)^2} = \frac{A}{z} + \frac{B}{(z-1)} + \frac{C_1}{(z-2)} + \frac{C_2}{(z-2)^2} \tag{8.41}$$

We can find three of the unknown constants using the relations

$$A = \left.\frac{z^2 - 9}{(z-1)(z-2)^2}\right|_{z=0} = \frac{-9}{(-1)(4)} = \frac{9}{4}$$

$$B = \left.\frac{z^2 - 9}{z(z-2)^2}\right|_{z=1} = \frac{1-9}{1 \times (-1)^2} = -8$$

$$C_2 = \left.\frac{z^2 - 9}{z(z-1)}\right|_{z=2} = \frac{4-9}{2(1)} = -\frac{5}{2}$$

These constants are introduced in (8.41), leaving a relation involving the one remaining constant C_1. One procedure for finding the constant is to select any two appropriate values for z, avoiding the roots of the rational polynomial, thereby creating an equation with the unknown. In particular, if we choose $z = 3$, we obtain the following expression:

$$\frac{3^2 - 9}{3(3-1)(3-2)^2} = \frac{9}{4}\frac{1}{3} - 8\frac{1}{3-1} + C_1\frac{1}{3-2} - \frac{5}{2}\frac{1}{(3-2)^2} \quad \text{or} \quad C_1 = \frac{23}{4}$$

First, we introduce the values of all the constants found in (8.41), next we multiply both sides by z, and finally we take the inverse z-transform using Table 8.2. The result is

$$f(n) = \frac{9}{4}\delta(n) - 8u(n) + \frac{23}{4}2^n - \frac{5}{2}n\,2^{n-1}$$

For $n = 0, 1, 2, 3$, and 4, the value of the sequence is $\{f(n)\} = \{0\ 1\ 5\ 8\ 4\}$.

Book MATLAB m-File: ex_8_5_5

```
%m-file: ex_8_5_5
num = [0 1 0 -9];
den = [1 -5 8 -4];
f = dimpulse(num,den,5);%the number 5 indicates the number
                        %of values desired of f(n);
```

The values found using the above MATLAB m-file are identical to those found above.

∎

*8.5.2 Inverse Transform by Integration

If $F(z)$ is a regular function in the region $|z| > R$, then there exists a single sequence $\{f(nT)\}$ for which $\mathfrak{Z}\{f(nT)\} = F(z)$, namely,

$$f(nT) = \frac{1}{2\pi j_C} \oint F(z)z^{n-1}dz = \sum_{i=1}^{K} \operatorname{res}_{z=z_i}\{F(z)z^{n-1}\} \quad n = 0, 1, 2, \ldots \tag{8.42}$$

The contour of integration C encloses all the singularities of $F(z)$, as shown in Figure 8.9, and it is taken in the counterclockwise direction. The abbreviation res stands for residue of integration.

*8.5.3 Residues for Simple Poles

If $F(z) = H(z)/G(z)$, then the residue at the singularity $z = a$ is given by

$$\lim_{z \to a}\{(z-a)F(z)z^{n-1}\} = \lim_{z \to a}\left\{(z-a)\frac{H(z)}{G(z)}z^{n-1}\right\} \tag{8.43}$$

*8.5.4 Residues for Multiple Poles

The residue at the pole z_i with multiplicity m of the function $F(z)z^{n-1}$ is given by

$$\operatorname{res}_{z=z_i}\{F(z)z^{n-1}\} = \frac{1}{(m-1)!}\lim_{z \to z_i}\frac{d^{m-1}}{dz^{m-1}}\left[(z-z_i)^m F(z)z^{n-1}\right] \tag{8.44}$$

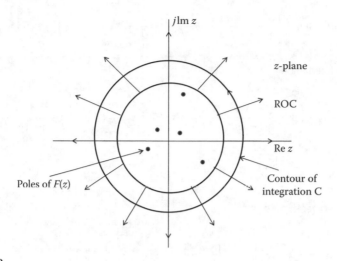

FIGURE 8.9

*8.5.5 Residues for Simple Poles Not Factorable

The residue at the singularity a_m is

$$F(z)z^{n-1}\Big|_{z=a_m} = \frac{H(z)}{(dG(z)/dz)}z^{n-1}\Big|_{z=a_m} \tag{8.45}$$

Example 8.25 (Inverse z-transform by integration)

To find the inverse of the transform

$$F(z) = \frac{(z-1)}{(z+2)(z-(1/2))} \quad |z| > 2$$

we proceed with the following approaches:

1. By fraction expansion

$$\frac{(z-1)}{(z+2)(z-(1/2))} = \frac{A}{z+2} + \frac{B}{z-(1/2)}, \quad A = \frac{(z-1)}{(z-(1/2))}\Big|_{z=-2} = \frac{6}{5},$$

$$B = \frac{(z-1)}{(z+2)}\Big|_{z=1/2} = -\frac{1}{5}$$

$$f(nT) = \mathfrak{Z}^{-1}\left\{\frac{6}{5}\frac{1}{z+2} - \frac{1}{5}\frac{1}{z-(1/2)}\right\} = \frac{6}{5}(-2)^{n-1} - \frac{1}{5}\left(\frac{1}{2}\right)^{n-1} \quad n \geq 1$$

*2. By integration

$$f(nT) = \operatorname{res}_{z=-2}\left\{(z+2)\frac{(z-1)}{(z+2)(z-(1/2))}z^{n-1}\right\} + \operatorname{res}_{z=1/2}\left\{\left(z-\frac{1}{2}\right)\frac{(z-1)}{(z+2)(z-(1/2))}z^{n-1}\right\}$$

$$= \frac{6}{5}(-2)^{n-1} - \frac{1}{5}\left(\frac{1}{2}\right)^{n-1}$$

3. By power expansion

$$f(nT) = \mathfrak{Z}^{-1}\left\{\frac{z-1}{z^2+(3/2)z-1}\right\} = \mathfrak{Z}^{-1}\left\{z^{-1} - \frac{5}{2}z^{-2} + \frac{19}{4}z^{-3} + \cdots\right\} = \mathfrak{Z}^{-1}\left\{z^{-1}\left(1 - \frac{5}{2}z^{-1} + \frac{19}{4}z^{-2} + \cdots\right)\right\}$$

$$= 1, -\frac{5}{2}, \frac{19}{4}, \ldots \quad n = 1, 2, 3, \ldots$$

The multiplier z^{-1} in the above equation indicates one-time delay.   ■

Example 8.26

To find the inverse of the transform

$$F(z) = \frac{5z}{(z-5)^2} \quad |z| > 5$$

1. By expansion

$$\mathfrak{Z}^{-1}\{F(z)\} \triangleq f(nT) = \mathfrak{Z}^{-1}\left\{\frac{5z}{(z-5)^2}\right\} = \mathfrak{Z}^{-1}\left\{\frac{5z}{z^2-10z+25}\right\} = \mathfrak{Z}^{-1}\{5z^{-1} + 50z^{-2} + 375z^{-3} + \cdots\}$$

$$= \mathfrak{Z}^{-1}\{0 \triangleq 5^0 z^{-0} + 1 \triangleq 5z^{-1} + 2 \triangleq 5^2 z^{-2} + 3 \triangleq 5^3 z^{-3} + \cdots\} = n5^n \quad n = 0, 1, 2, \ldots$$

This expansion indicates the difficulty in recognizing the closed form expression.
2. By fraction expansion

$$\mathfrak{Z}^{-1}\{F(z)\} \triangleq f(nT) = \mathfrak{Z}^{-1}\left\{\frac{5z}{(z-5)^2}\right\} = \mathfrak{Z}^{-1}\left\{\frac{Az}{z-5} + \frac{Bz^2}{(z-5)^2}\right\}$$

Therefore,

$$B = \frac{5}{z}\Big|_{z=5} = 1, \quad \frac{5 \times 6}{(6-1)^2} + \frac{6^2}{(6-5)^2} \quad \text{or} \quad A = -1$$

and

$$f(nT) = \mathcal{Z}^{-1}\left\{ -\frac{z}{z-5} + \frac{z^2}{(z-5)^2} \right\} = -(5)^n + (n+1)5^n = n5^n \quad n \geq 0$$

3. By integration

$$\frac{1}{(2-1)!} \frac{d^{2-1}}{dz^{2-1}} \left[(z-5)^2 \frac{5z}{(z-5)^2} z^{n-1} \right]\Big|_{z=5} = 5nz^{n-1}\big|_{z=5} = n5^n \quad n \geq 0 \quad \blacksquare$$

8.6 Transfer Function

The z-transform provides a very important technique in the solution of difference equation. As part of this process, the transfer function plays an important role.

Example 8.27 (Transfer function)

Deduce an expression for the impulse response of the circuit shown in Figure 8.10a using the z-transform technique.

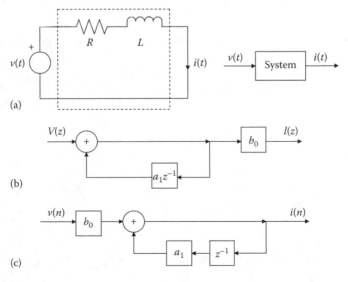

FIGURE 8.10

SOLUTION

The controlling equation of the circuit is

$$L\frac{di(t)}{dt} + Ri(t) = v(t)$$

The appropriate discrete form of this equation for sampling time $T=1$ is

$$L[i(n) - i(n-1)] + Ri(n) = v(n)$$

from which we obtain

$$i(n) = \frac{1}{L+R}v(n) + \frac{L}{L+R}i(n-1) \overset{\Delta}{=} b_0 v(n) + a_1 i(n-1)$$

Taking the z-transform of both sides of the above equation and assuming zero initial conditions, the above equation becomes

$$I(z) = b_0 V(z) + a_1 z^{-1} I(z) \quad \text{or} \quad I(z) = \frac{b_0}{1 - a_1 z^{-1}} V(z) = H(z)V(z) \qquad (8.46)$$

This equation relates the input–output relation explicitly in the transformed domain of a discrete system. The quantity $H(z) = I(z)/V(z)$, or equivalently the ratio of the system output to its input, is the **system transfer function** for the discrete system. Further, the inverse transform of $H(z)$ is the impulse response $h(n)$ of the system. Thus, for our circuit with a delta function excitation, $(\mathfrak{Z}\{\delta(n)\} = 1)$, we have

$$I(z) \overset{\Delta}{=} H(z) = \frac{b_0}{1 - a_1 z^{-1}} \qquad (8.47)$$

The inverse transform is easily found to be

$$h(n) = b_0(a_1)^n \quad n \geq 0 \qquad (8.48)$$

Figure 8.10b and c show the feedback configuration of the discrete system in the transformed and the time domains. ∎

Example 8.28

Determine the response of the first-order system specified by (8.46) to a unit step response by z-transform and convolution methods.

SOLUTION

The unit step sequence, which is written as

$$u(n) = \begin{cases} 1 & n = 0, 1, 2, \dots \\ 0 & n < 0 \end{cases}$$

has the *z*-transformed value

$$U(z) = \frac{z}{z-1} = \frac{1}{1-z^{-1}}$$

The response, by writing *y* for *i*, is given by (see 8.46)

$$Y(z) = b_0 \frac{z}{z-a_1} \frac{z}{z-1} = b_0 \left[\frac{Az}{z-a_1} + \frac{Bz}{z-1} \right]$$

where

$$A = \frac{z}{z-1}\bigg|_{z=a_1} = \frac{a_1}{a_1-1}, \quad B = \frac{z}{z-a_1}\bigg|_{z=1} = \frac{1}{1-a_1}$$

Thus,

$$Y(z) = \frac{b_0}{1-a_1} \left(\frac{-a_1}{1-a_1 z^{-1}} + \frac{1}{1-z^{-1}} \right)$$

The inverse transform is

$$y(n) = \frac{b_0}{1-a_1} [-a_1(a_1)^n + (1)^n] = \frac{b_0}{1-a_1}(1 - a_1^{n+1}) \quad n \geq 0$$

It is recalled that the derivative of the step function response of a system is its impulse response. To show that the result is consistent with the result of Example 8.27, we consider the derivative of *y(n)* in its discrete form representation. Ignoring at first the constant factor $b_0/(1 - a_1)$, we obtain

$$\frac{y(n) - y(n-1)}{1} = 1 - a_1^{n+1} - 1 + a^n = (1 - a_1)a_1^n$$

Therefore, the impulse response is given by

$$h(n) = \frac{b_0}{1-a_1} a_1^n(1 - a_1) = b_0 a_1^n \quad n \geq 0$$

We can proceed to find the output in the foregoing example by using the convolution equation. Here we write, using the results of Example 8.27,

$$y(n) = \sum_{m=0}^{n} h(n-m)u(m) = \sum_{m=0}^{n} b_0 a_1^{n-m} u(m)$$

The output at successive time steps is

$$y(0) = b_0(a_1)^0 = b_0$$
$$y(1) = b_0(a_1 + 1)$$
$$y(2) = b_0(a_1^2 + a_1 + 1)$$

$$\vdots$$

$$y(n) = b_0(a_1^n + a_1^{n-1} + \cdots + 1) = b_0\left(\frac{1 - a_1^{n+1}}{1 - a_1}\right)$$

where the formula for finite geometric series was used in the last summation. This result is identical with that above using z-transform method. ∎

When discrete systems are interconnected, the rules that apply to continuous systems are also applied for discrete systems. For example, if the impulse responses of two systems connected in cascade are known, the impulse response of the total system is

$$h(n) = h_1(n) * h_2(n) \tag{8.49}$$

and in the z-domain

$$\boxed{H(z) = H_1(z)H_2(z) = \mathfrak{Z}\{h_1(n) * h_2(n)\}} \tag{8.50}$$

Example 8.29

Deduce the transfer function for the system shown in Figure 8.11a.

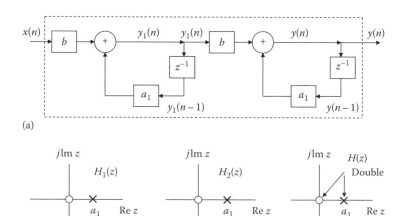

(a)

(b)

FIGURE 8.11

SOLUTION

Consider initially the portion of the system (subsystem) shown between $x(n)$ and $y_1(n)$ described by the difference equation

$$y_1(n) = bx(n) + a_1 y_1(n-1)$$

The z-transform of this expression is

$$Y_1(z) = bX(z) + a_1 z^{-1} Y_1(z) \quad \text{or} \quad Y_1(z) = \frac{b}{1 - a_1 z^{-1}} X(z) = H_1(z)X(z)$$

The portion of the system between $y_1(n)$ and $y(n)$ (a cascaded subsystem with the first subsystem) is described by a similar expression whose z-transform is

$$Y(z) = \frac{b}{1 - a_1 z^{-1}} Y_1(z) = H_2(z)Y_1(z)$$

Substitute the known expression for $Y_1(z)$ into this final expression to obtain

$$Y(z) = \left(\frac{b}{1 - a_1 z^{-1}}\right)\left(\frac{b}{1 - a_1 z^{-1}}\right)X(z) = H_1(z)H_2(z)X(z) = H(z)X(z)$$

where

$$H(z) = H_1(z)H_2(z) = \left(\frac{b}{1 - a_1 z^{-1}}\right)^2$$

The pole-zero configurations for each subsystem and for the combined system are shown in Figure 8.11b. ∎

Example 8.30

Find the transfer function for the first-order system, shown in Figure 8.12a, and sketch the pole-zero configuration.

SOLUTION

The difference equation describing the system is

$$y(n) = b_0 x(n) + b_1 x(n-1)$$

The z-transform of this equation and the transfer function of the system are

$$Y(z) = b_0 X(z) + b_1 z^{-1} X(z) \quad \text{or} \quad H(z) \triangleq \frac{Y(z)}{X(z)} = b_0 + b_1 z^{-1} = b_0 \frac{z + (b_1/b_0)}{z}$$

The pole-zero configurations is shown in Figure 8.12b. ∎

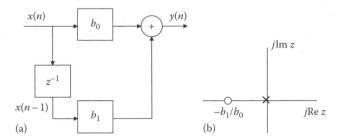

FIGURE 8.12

Note: *In Example 8.27, b_0 as well as a_0 (was one in this case) and a_1 are different than zero. This type of system is known as the first-order **infinite impulse response (IIR)** system. Observe that an analog first-order RL series circuit results is an IIR digital system. In Example 8.30, $a_0(=1)$, b_0, and b_1 are different than zero. This type of first-order digital system is known as the **finite impulse response (FIR)** system.*

Figure 8.13 shows a combined FIR and IIR system. The difference equation describing the total system is found by the following two equations that are found by inspection from Figure 8.13. These are

$$x_1(n) - a_1 y(n-1) = y(n), \quad x_1(n) = b_0 x(n) + b_1 x(n-1)$$

Therefore, the difference equation describing the system, by eliminating $x_1(n)$, is

$$y(n) + a_1 y(n-1) = b_0 x(n) + b_1 x(n-1) \tag{8.51}$$

The a's and b's can take either positive or negative values. Taking the z-transform of both sides of (8.51), we obtain the transfer function of the system, which is

$$\boxed{H(z) \triangleq \frac{Y(z)}{X(z)} = \frac{b_0 + b_1 z^{-1}}{1 + a_1 z^{-1}}} \tag{8.52}$$

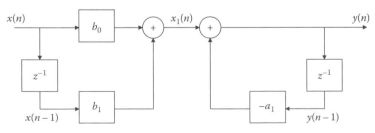

FIGURE 8.13

8.6.1 Higher Order Transfer Functions

The general case of a system is described by the following difference equation:

$$y(n) + a_1 y(n-1) + a_2 y(n-2) + \cdots + a_p y(n-p)$$
$$= b_0 x(n) + b_1 x(n-1) + b_2 x(n-2) + \cdots + b_q x(n-q) \qquad (8.53)$$

Taking the z-transform and solving for the ratio $Y(z)/X(z)$ we obtain the transfer function

$$\boxed{H(z) \triangleq \frac{Y(z)}{X(z)} = \frac{b_0 + b_1 z^{-1} + \cdots + b_q z^{-q}}{1 + a_1 z^{-1} + \cdots + a_p z^{-p}} = \frac{\sum_{n=0}^{q} b_n z^{-n}}{1 + \sum_{n=1}^{p} a_n z^{-n}}} \qquad (8.54)$$

This equation indicates that if we know the transfer function $H(z)$, then the output $Y(z)$ to any input $X(z)$ (or equivalently, $x(n)$) can be determined.

If we set $a_1, a_2, \ldots, a_p$ equal to zero, (8.53) becomes

$$y(n) = b_0 x(n) + b_1 x(n-1) + b_2 x(n-2) + \cdots + b_q x(n-q) \qquad (8.55)$$

This expression defines a **qth-order FIR filter**. The block diagram of a FIR filter is shown in Figure 8.14.

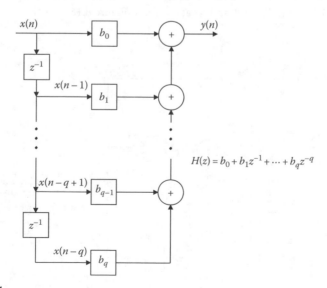

FIGURE 8.14

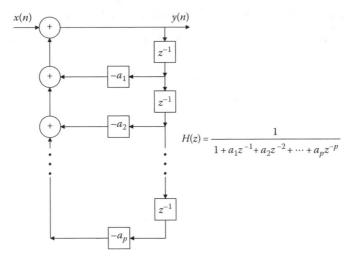

FIGURE 8.15

For the case when $b_0 = 1$ and the rest of the b's are zero, the difference equation (8.53) becomes

$$y(n) + a_1 y(n-1) + a_2 y(n-2) + \cdots + a_p y(n-p) = x(n) \qquad (8.56)$$

This equation defines a **pth-order IIR filter**. A block diagram representation of this equation is shown in Figure 8.15.

Finally, if none of the constants is zero in (8.53), the block diagram representation of the combined FIR and IIR system is that shown in Figure 8.16.

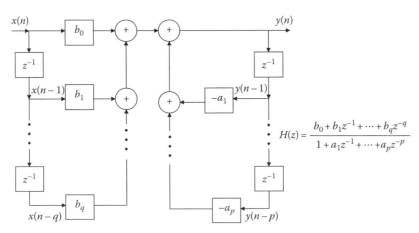

FIGURE 8.16

8.6.1.1 Stability

Using the convolution relation between input and output of a discrete system, we obtain

$$|y(n)| = \left| \sum_{k=0}^{n} h(k)f(n-k) \right| \leq M \sum_{k=0}^{n} |h(k)| < \infty \tag{8.57}$$

where M is the maximum value of $f(n)$. The above inequality specifies that a discrete system is stable if to a finite input the absolute sum of its impulse response is finite. From the properties of the z-transform, the ROC of the impulse response satisfying (8.57) is $|z| > 1$. Hence, the poles of $H(z)$ of a stable system lie inside the unit circle.

8.6.1.2 Causality

A system is causal if $h(n) = 0$, for $n < 0$. From the properties of the z-transform, $H(z)$ is regular in the ROC and at infinity point. For rational functions, the numerator polynomial must be at most of the same degree as the polynomial of the denominator.

The Paley–Wiener theorem provides the necessary and sufficient conditions that a frequency response characteristic $H(\omega)$ must satisfy in order for the resulting filter to be causal.

The conditions are as follows:

1. $H(\omega)$ cannot be zero except at a finite set of points.
2. $|H(\omega)|$ cannot be constant in any finite range of frequencies.
3. The transition from pass band to stop band cannot be infinitely sharp.
4. The real and imaginary parts of $H(\omega)$ are independent and are related by the discrete Hilbert transform.
5. $|H(\omega)|$ and $\varphi(\omega)$ cannot be chosen arbitrarily.

Figure 8.17 illustrates the effect of the position of poles with respect to the unit circle and the multiplicity of the poles.

8.7 Frequency Response of First-Order Discrete Systems

Suppose that the input to the system is the function z^n. Then, using the convolution property of system response, the output is given by

$$y(n) = z^n \star h(n) = \sum_{m=0}^{\infty} h(m)z^{n-m} = z^n \sum_{m=0}^{\infty} h(m)z^{-m} = z^n H(z) \tag{8.58}$$

If we set $z = e^{j\omega}$ in this expression, we have

$$y(n) = e^{j\omega n} H(e^{j\omega}) \tag{8.59}$$

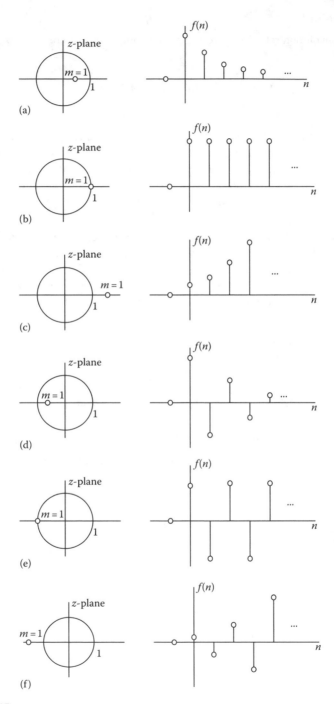

FIGURE 8.17

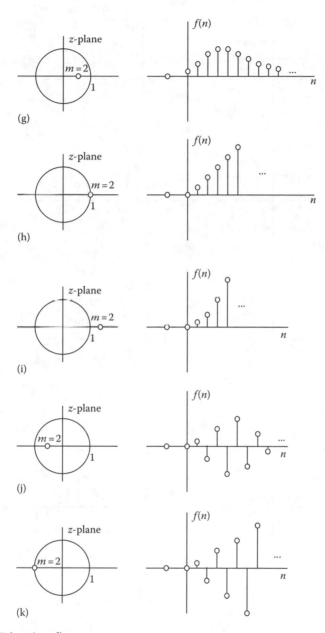

FIGURE 8.17 (continued)

(*continued*)

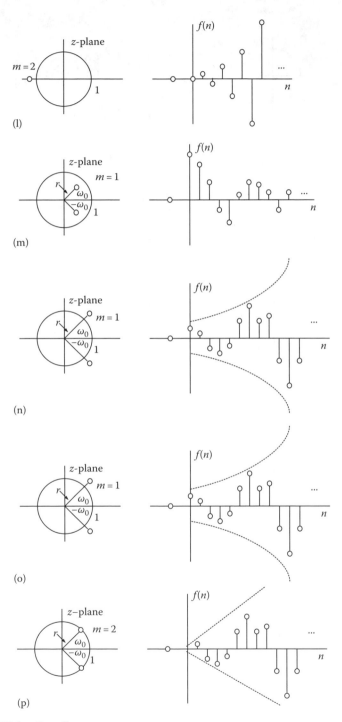

(l)

(m)

(n)

(o)

(p)

FIGURE 8.17 (continued)

Therefore, the transfer function for the first-order systems is

$$H(e^{j\omega}) = \frac{b_0 + b_1 e^{-j\omega}}{1 + a_1 e^{-j\omega}}$$ (8.60)

This function is the **frequency response function**.

If we set $\omega = \omega + 2\pi$ in $H(\cdot)$, we find that $H(e^{j(\omega+2\pi)}) = H(e^{j\omega}e^{j2\pi}) = H(e^{j\omega})$, which indicates that the frequency response function is periodic with period 2π. If, on the other hand, we had introduced $z = e^{j\omega T}$ ($T =$ sampling time), then the frequency response function would be of the form

$$H(e^{j\omega T}) = \frac{b_0 + b_1 e^{-j\omega T}}{1 + a_1 e^{-j\omega T}}$$ (8.61)

If we set $\omega = \omega + (2\pi/T)$ in $H(\cdot)$, we find that $H(e^{j(\omega+2\pi/T)T}) = H(e^{j\omega T}e^{j2\pi}) = H(e^{j\omega T})$, which indicates that the frequency response function is periodic with period $2\pi/T$.

Note: *Discrete systems with unit sampling time ($T = 1$) have periodic frequency response functions with period 2π, and those with time sampling equal to T have periodic frequency response functions with period $2\pi/T$.*

Example 8.31

Find the frequency response of a first-order FIR system and plot its amplitude and phase spectra for $b_0 = 1$ and $b_1 = 0.5$. Plot both cases using $T = 1$ and $T = 2$.

SOLUTION

The frequency response function of a first-order FIR system is found from (8.60) by setting $a_1 = 0$. Hence, we have

$$H(e^{j\omega T}) = \frac{b_0 + b_1 e^{-j\omega T}}{1} = b_0 + b_1 e^{-j\omega T} = b_0 + b_1 \cos \omega T - jb_1 \sin \omega T$$

From the above equation, the amplitude and phase spectra are

$$\sqrt{H(e^{j\omega T})H*(e^{j\omega T})} = \sqrt{(b_0 + b_1 e^{j\omega T})(b_0 + b_1 e^{-j\omega T})} = \sqrt{b_0^2 + 2b_0 b_1 \cos \omega T + b_1^2}$$

$$\mathrm{ph}\{H(e^{j\omega T})\} = -\tan^{-1} \frac{b_1 \sin \omega T}{b_0 + b_1 \cos \omega T}$$

The plots are shown in Figure 8.17. Note that for $T = 1$ the periodicity of the spectra is $2\pi/1 = 2\pi$ and for $T = 2$ the periodicity is $2\pi/2 = \pi$. ∎

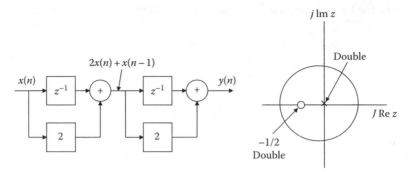

FIGURE 8.18

Example 8.32

Find the frequency characteristics of the system shown in Figure 8.18, which is made up of two FIR systems in cascade.

SOLUTION

The difference equation describing the system is

$$y(n) = 4x(n) + 4x(n-1) + x(n-2)$$

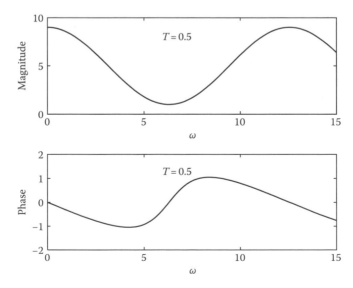

FIGURE 8.19

The system function for this system is

$$H(z) \triangleq \frac{Y(z)}{X(z)} = 4 + 4e^{-j\omega T} + e^{-j2\omega T}$$

The frequency responses are obtained from the relationships

$$\left[[H(z)H*(z)]_{z=e^{j\omega T}} \right]^{1/2} = \left[H(z)H(z^{-1}) \Big|_{z=e^{j\omega T}} \right]^{1/2} = (33 + 40\cos\omega T + 8\cos 2\omega T)^{1/2}$$

Figure 8.19 shows the frequency characteristics of the combined system using the MATLAB functions **abs**(*H*) and **angle**(*H*). Note that the periodicity in this case is equal to $2\pi/0.5 = 4\pi$.

Note: *The amplitude functions are even functions of the frequency and the phase functions are odd functions of the frequency.* ∎

8.7.1 Phase Shift in Discrete Systems

Let assume that a discrete system is described by the difference equation

$$y(n) = x(nT) + 0.6x(nT - T)$$

The system function for this is

$$H(z) \triangleq \frac{Y(z)}{X(z)} = 1 + 0.6z^{-1}$$

The amplitude frequency response function is then given by

$$\left[H(z)H(z^{-1}) \Big|_{z=e^{j\omega T}} \right]^{1/2} = (1.36 + 1.2\cos\omega T)^{1/2}$$

and the phase is

$$\theta(\omega) = \arg H(e^{j\omega T}) = -\tan^{-1}\frac{0.6\sin\omega T}{1 + 0.6\cos\omega T}$$

Since we have shown above that when the input is the complex function $e^{j\omega T}$, the output is $H(e^{j\omega T})e^{j\omega nT} = |H(e^{j\omega T})|e^{j\theta(\omega)}e^{j\omega T}$. This indicates that when the input is $\cos \omega nT$, then the output is the real part of the output, which is

$$y(nT) = |H(e^{j\omega T})| \cos[\omega nT + \theta(\omega)]$$

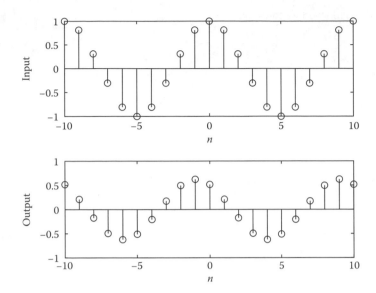

Input
Output
n
1
0.5
0
−0.5
−1
−10
−5
0
5
10

FIGURE 8.20

$e^{j\omega T}$
$x(nT) = \cos \omega nT$
$x(nT) = \sin \omega nT$
$H(z)$
$y(nT) = H(e^{j\omega T})e^{j\omega T} = |H(e^{j\omega T})|e^{j[\omega T + \theta(\omega)]}$
$y(nT) = |H(e^{j\omega T})| \cos[\omega nT + \theta(\omega)]$
$y(nT) = |H(e^{j\omega T})| \sin[\omega nT + \theta(\omega)]$

FIGURE 8.21

If we set $T = 0.5$ and $\omega = 0.4\pi$ in the above equations, then the input and output is shown in Figure 8.20. Observe the phase shift to the left and the decrease of the amplitude of the output signal. The general input–output relations of a discrete system to sinusoidal inputs are shown in Figure 8.21.

8.8 Frequency Response of Higher Order Digital Systems

If we set $z = e^{j\omega T}$ in (8.54), we obtain the general frequency response function:

$$\boxed{H(e^{j\omega T}) = \frac{b_0 + b_1 e^{-j\omega T} + b_2 e^{-j2\omega T} + \cdots + b_q e^{-jq\omega T}}{a_0 + a_1 e^{-j\omega T} + a_2 e^{-j2\omega T} + \cdots + a_p e^{-jp\omega T}} = H(z)|_{z=e^{j\omega T}}} \tag{8.62}$$

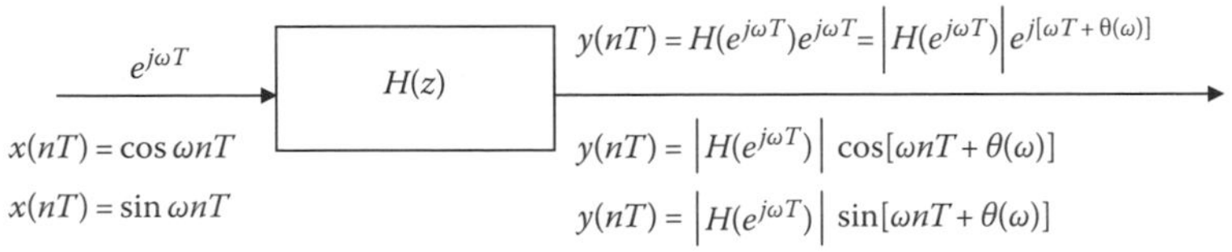

of a combined system of qth order FIR and a pth order IIR systems. If, for example, we want to study a third-order systems, then the transfer function is

$$H(z) = \frac{b_0 + b_1 e^{-j\omega T} + b_2 e^{-j2\omega T}}{a_0 + a_1 e^{-j\omega T} + a_2 e^{-j2\omega T}} \tag{8.63}$$

The amplitude squared is given by

$$|H(z)|^2\Big|_{z=e^{j\omega T}} = H(z)H(z^{-1})\Big|_{z=e^{j\omega T}} = \frac{d_2 z^2 + d_1 z + d_0 + d_1 z^{-1} + d_2 z^{-2}}{c_2 z^2 + c_1 z + c_0 + c_1 z^{-1} + c_2 z^{-2}}\Big|_{z=e^{j\omega T}}$$

$$= \frac{d_0 + \sum_{n=1}^{2} 2d_n \cos n\omega T}{c_0 + \sum_{n=1}^{2} 2c_n \cos n\omega T} \tag{8.64}$$

where

$$c_n = \sum_{k=0}^{k-n} a_k a_{n+k} \qquad d_n = \sum_{k=0}^{k-n} b_k b_{n+k} \tag{8.65}$$

For example,

$$c_0 = \sum_{k=0}^{2-0} a_k a_{0+k} = a_0 a_0 + a_1 a_1 + a_2 a_2, \qquad c_1 = \sum_{k=0}^{2-1} a_k a_{1+k} = a_0 a_1 + a_1 a_2,$$

$$c_2 = \sum_{k=0}^{2-2} a_k a_{2+k} = a_0 a_2$$

Example 8.33

Find and plot the frequency characteristics of a general third-order discrete system with the following constants: $b_0 = 1, b_1 = 0.8, b_2 = 0.1, a_0 = 1, a_1 = -0.6, a_2 = 0.8$, and $T = 2$.

SOLUTION

The frequency characteristics are plotted in Figure 8.22 using the MATLAB functions abs(H) and angle(H). The transfer function with the given data is

$$H(e^{j2\omega}) = \frac{1 + 0.8e^{-j2\omega} + 0.1e^{-j4\omega}}{1 - 0.6e^{-j2\omega} + 0.8e^{-j4\omega}} \qquad \blacksquare$$

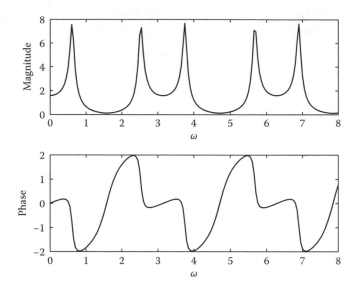

FIGURE 8.22

Example 8.34

Find the frequency response function of the system shown in Figure 8.23.

SOLUTION

We can solve this problem by two different ways. From the figure, the difference equations describing the first and second subsystem are respectively,

$$x_1(n-1) + 2x_1(n) = y(n)$$
$$-[y(n-1) + 2y(n)] + x(n) = x_1(n)$$

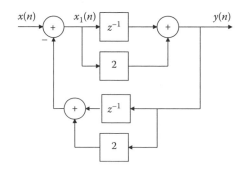

FIGURE 8.23

Substituting $x_1(n)$ from the second equation into the first, we obtain

$$5y(n) + 4y(n-1) + y(n-2) = 2x(n) + x(n-1)$$

By taking the z-transform of the above equation and solving for the ratio output over input, we obtain the transfer function

$$H(z) = \frac{2 + z^{-1}}{5 + 4z^{-1} + z^{-2}}$$

Since the transfer function for each subsystem is the same, $H_s(z) = 2 + z^{-1}$, then the transfer function of the total system is found using the feedback transfer function given in Chapter 8. ∎

8.9 z-Transform Solution of First-Order Difference Equations

We now study the use of z-transform methods in the solution of linear difference equations with constant coefficients. The following examples will explain how to use z-transform method for their solution of difference equations.

Example 8.35

Solve the discrete-time problem defined by the equation

$$y(n) + 2y(n-1) = 3.5u(n) \tag{8.66}$$

with $y(-1) = 0$ and $u(n)$ is the discrete unit step function.

SOLUTION

Begin by taking the z-transform of both sides of (8.66). This is

$$\mathfrak{Z}\{y(n)\} + 2\mathfrak{Z}\{y(n-1)\} = 3.5\mathfrak{Z}\{u(n)\}$$

From Section 8.3 and Table 8.1, we write

$$Y(z) + 2z^{-1}Y(z) = 3.5\frac{z}{z-1} \quad \text{or} \quad Y(z) = 3.5\frac{z}{z-1}\frac{z}{z+2} = \frac{6}{7}\frac{z}{z-1} + \frac{7}{3}\frac{z}{z+2}$$

The inverse z-transform of this equation is

$$y(n) = \frac{7}{6}u(n) + \frac{7}{3}(-2)^n \quad n = 0, 1, 2, \ldots \qquad \blacksquare$$

If you want to use the MATLAB function **residue(.)**, you must solve for the $Y(z)/z$ function and then multiply both sides by z.

Example 8.36

Repeat Example 8.35 with the initial condition $y(-1) = 4$.

SOLUTION

We again begin by taking the z-transform of both sides and use the right-shift property (see Section 8.3):

$$\mathfrak{Z}\{y(n)\} + 2\mathfrak{Z}\{y(n-1)\} = 3.5\mathfrak{Z}\{u(n)\} \quad \text{or} \quad Y(z) + 2z^{-1}Y(z) + 2y(-1) = 3.5U(z)$$

Upon solving for $Y(z)$, we obtain

$$Y(z) = \underbrace{\frac{3.5}{1 + 2z^{-1}} U(z)}_{\substack{\text{zero-state} \\ \text{response}}} - \underbrace{\frac{2y(-1)}{1 + 2z^{-1}}}_{\substack{\text{zero-input} \\ \text{response}}}$$

The inverse transform of the zero-input response is the solution of the homogeneous difference equation $y(n) + 2y(n-1) = 0$, a result that can be readily verified by setting consecutively $n = 0, 1, 2, \ldots$ in the equation. Specifically, the results are

$$\text{zero-input response} = \mathfrak{Z}^{-1}\left\{-\frac{2y(-1)}{1 + 2z^{-1}}\right\} = -2y(-1)(-2)^n u(n)$$

$$\text{zero-state response} = \mathfrak{Z}^{-1}\left\{\frac{3.5}{1 + 2z^{-1}} \frac{1}{1 - z^{-1}}\right\} = Z^{-1}\left\{\frac{3.5z^2}{(z+2)(z+1)}\right\}$$

$$= \mathfrak{Z}^{-1}\left\{7\frac{z}{z+2} - 3.5\frac{z}{z+1}\right\}$$

The complete solution is the sum of these two responses. Hence, we write

$$y(n) = \underbrace{4(-2)^{n+1}}_{\substack{\text{zero input} \\ \text{response}}} + \underbrace{7(-2)^n - 3.5u(n)}_{\substack{\text{zero state} \\ \text{response}}} \quad n = 0, 1, 2, \ldots$$

and

$$y(n) = \underbrace{-(-2)^n}_{\text{transient}} - \underbrace{3.5u(n)}_{\text{steadys state}} \quad n = 0, 1, 2, \ldots \qquad \blacksquare$$

Example 8.37

Determine the output of the discrete approximation of the system shown in Figure 8.24a for a sampling time T. The output for $T = 0.2$ and $T = 1$ are to be plotted, and the results compared with the output of the continuous system. The input is a unit step current source $i(t) = u(t)$, and an initial condition $v_0(0) = 2V$.

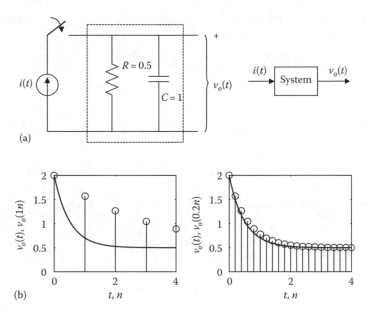

FIGURE 8.24

SOLUTION

The differential equation describing the system is

$$\frac{dv_o(t)}{dt} + \frac{v_o(t)}{0.5} = i(t)$$

The analogous discrete form of this equation is

$$\frac{v_o(nT) - v_o(nT - T)}{T} + 2v_o(nT) = i(nT)$$

From this,

$$v_o(nT) = \frac{T}{1 + 2T} i(nT) + \frac{1}{1 + 2T} v_o(nT - T) = b_0 i(nT) + a_1 v_o(nT - T)$$

The z-transform of this equation gives (see Section 8.3)

$$V_o(z) = \underbrace{\frac{b_0}{1 - a_1 z^{-1}} I(z)}_{\substack{\text{zero state} \\ \text{response}}} + \underbrace{\frac{a_1}{1 - a_1 z^{-1}} v_o(-T)}_{\substack{\text{zero input} \\ \text{response}}}$$

Since the continuous case $v_o(0) = 2$, we must refer back to the difference equation and set the appropriate values to find the value of $v_o(-T)$. Hence, we obtain

$$v_o(0) = b_0 i(0) + a_1 v_0(-T) \quad n = 0, \quad \text{or} \quad v_o(-T) = (2 - b_0)/a_1$$

Thus, we find

$$\text{zero-input response} = \mathfrak{Z}^{-1}\left\{ b_0 \frac{z^2}{(z-1)(z-a_1)} \right\} = \mathfrak{Z}^{-1}\left\{ \frac{b_0}{1-a_1} \frac{z}{z-1} + \frac{b_0 a_1}{a_1 - 1} \frac{z}{z-a_1} \right\}$$

$$= \frac{b_0}{1-a_1} u(n) + \frac{b_0 a_1}{a_1 - 1} a_1^n u(n)$$

$$\text{zero-state response} = \mathfrak{Z}^{-1}\left\{ \frac{(2-b_0)z}{(z-a_1)} \right\} = (2-b_0)a_1^n u(n)$$

The solution is the sum of the zero-input and zero-state solutions.

The solution of the continuous case is easily found to be equal to

$$v_o(t) = 0.5 + 1.5e^{-2t} \quad t \geq 0$$

The results for the continuous and discrete cases $T=1$ and $T=0.2$ are shown in Figure 8.24b. ∎

Example 8.38

Use the z-transform method to find the solution of the following difference equation:

$$y(n) - \frac{3}{4}y(n-1) + \frac{1}{8}y(n-2) = 2\sin\left(\frac{\pi n}{2}\right), \quad y(-1) = 2, \quad y(-2) = 4$$

SOLUTION

Taking the z-transform of both sides, we get

$$Y(z) - \frac{3}{4}z^{-1}[Y(z) + 2z] + \frac{1}{8}z^{-2}[Y(z) + 4z^2 + 2z] = \frac{2z}{z^2 + 1}$$

$$\left(1 - \frac{3}{4}z^{-1} + \frac{1}{8}z^{-2}\right)Y(z) = 1 - \frac{1}{4}z^{-1} + \frac{2z}{z^2 + 1}$$

$$Y(z) = \frac{z^2 - (1/4)z}{z^2 - (3/4)z + (1/8)} + \frac{2z^3}{(z^2 + 1)(z^2 - (3/4)z + (1/8))}$$

Performing partial fraction decomposition, we find

$$Y(z) = \frac{z}{z - (1/2)} + \frac{8}{5}\frac{z}{z - (1/2)} - \frac{8}{17}\frac{z}{z - (1/4)} + \frac{(112/85)z^2 - (96/85)z}{z^2 + 1}$$

Therefore, the inverse z-transform gives the following result:

$$y(n) = \frac{13}{5}\left(\frac{1}{2}\right)^n - \frac{8}{17}\left(\frac{1}{4}\right)^n + \frac{112}{85}\sin\frac{\pi}{2}n - \frac{96}{85}\cos\frac{\pi}{2}n \quad n \geq 0 \qquad ■$$

8.10 Higher Order Difference Equations

The class of linear discrete systems under discussion is described by the general difference equation

$$y(n) + a_1 y(n-1) + a_2 y(n-2) + \cdots + a_p y(n-p) = b_0 x(n) \qquad (8.67)$$

The constant in front of $y(n)$ does not appear since we can always divide by that number on both sides of the equation. We have seen that such an equation can arise when a differential equation is transformed into an equivalent difference equation using approximations of the derivatives. Hence,

$$\frac{dy(t)}{dt} \simeq \frac{y(nT) - y(nT - T)}{T} \qquad (a)$$

$$\frac{d^2 y(t)}{dt^2} \simeq \frac{y(nT) - 2y(nT - T) + y(nT - 2T)}{T^2} \qquad (b) \qquad (8.68)$$

$$\frac{d^3 y(t)}{dt^3} \simeq \frac{y(nT) - 3y(nT - T) + 3Y(nT - 2T) - y(nT - 3T)}{T^3} \qquad (c)$$

We would like to mention again that when the coefficients a_i's are independent of n, the system is time invariant, otherwise is time varying.

For $T = 1$, the second-order difference equation can always be written in the form

$$\boxed{y(n) + a_1 y(n-1) + a_2 y(n-2) = b_0 x(n)} \qquad (8.69)$$

We can assert that a complete and unique solution to the above equation can be found if the initial conditions are known, say,

$$y(-1) = A \quad y(0) = B \qquad (8.70)$$

where A and B are constants. In this connection, we state certain theorems without proof (see Finizio and Ladas[*]).

---

[*] N. Finizio and G. Ladas, *An Introduction to Differential Equations*, Wadsworth Publishing Co., Belmont, CA, 1982.

Definition 8.1 If $\{a(n)\}$ and $\{b(n)\}$ denote two sequences, the determinant

$$C[a(n), b(n)] = \begin{vmatrix} a(n) & b(n) \\ a(n-1) & b(n-1) \end{vmatrix} \tag{8.71}$$

is known as the **Casoratian** or their **Wroskian** determinant. ∎

THEOREM 8.1

Two solutions $y_1(n)$ and $y_2(n)$ of the linear homogeneous difference equation (the input $b_0x(n)$ is set equal to zero) are linearly independent if and only if their Casoratian,

$$C[y_1(n), y_2(n)] = \begin{vmatrix} y_1(n) & y_2(n) \\ y_1(n-1) & y_2(n-1) \end{vmatrix} \tag{8.72}$$

is different from zero for all values of $n = 0, 1, 2, \ldots$ ∎

THEOREM 8.2

If $y_1(n)$ and $y_2(n)$ are two linear independent solutions of the homogeneous equation and $y_p(n)$ is the particular solution to the nonhomogeneous equation (8.69), then the general solution is

$$y(n) = y_h(n) + y_p(n) = C_1 y_1(n) + C_2 y_2(n) + y_p(n) \tag{8.73}$$

where C_1 and C_2 are arbitrary constants and can be determined from the initial conditions. ∎

THEOREM 8.3

The homogeneous difference equation

$$y(n) + a_1 y(n-1) + a_2 y(n-2) = 0 \tag{8.74}$$

with constant and real coefficients and with the roots of the characteristic equation

$$\lambda^2 + a_1 \lambda + a_2 = 0 \tag{8.75}$$

denoted λ_1 and λ_2, has the possible solutions shown in Table 8.3.

TABLE 8.3   Solutions to Homogeneous Difference Equations

Difference Equation $y(n) + a_1 y(n-1) + a_2 y(n-2) = 0$	Characteristic Equation $\lambda^2 + a_1\lambda + a_2 = 0$
$\lambda_1 \neq \lambda_2$	$y(n) = c_1\lambda_1^n + c_2\lambda_2^n$
$\lambda_1 = \lambda_2 = \lambda$	$y(n) = c_1\lambda + c_2 n\lambda^n$
$\lambda_1 = a + jb; \lambda_2 = a - jb$	$y(n) = C_1 e^{an}\cos bn + C_2 e^{an}\sin bn$

Note: All the c_i's and C_i's are constants.

Solution

Assume that $y(n) = c\lambda^n$ is a solution of (8.74), then

$$c\lambda^n + a_1 c\lambda^{n-1} + a_2 c\lambda^{n-2} = 0 \quad \text{or} \quad \lambda^2 + a_1\lambda + a_2 = 0$$

which is (8.75).   ■

Example 8.39

Solve the following difference equation: $y(n) - by(n-1) + ay(n-2) = u(n); a, b > 0$. The constants are to be selected so that the second-order system is (a) **critically damped**, (b) **underdamped**, and (c) **overdamped**. The started conditions are zero: $y(-1) = 0, y(-2) = 0$.

Solution

Taking the z-transform of both sides of the difference equation and solving for $Y(z)$ we obtain

$$Y(z) = \frac{z^2}{z^2 - bz + a}\frac{z}{z - 1} \tag{8.76}$$

The denominator of the first factor, which is the characteristic equation of the difference equation, has two roots that are specified by

$$z_{1,2} = \frac{b \pm \sqrt{b^2 - 4a}}{2} \tag{8.77}$$

Critically damped case ($b^2 = 4a$): We set $b = 0.8$ and, hence, $a = 0.8$. The two roots are 0.4 and 0.4. Thus, (8.76) becomes

$$Y(z) = \frac{z^2}{(z - 0.4)^2}\frac{z}{z - 1} = \frac{Az}{z - 1} + \frac{Bz}{z - 0.4} + \frac{Cz^2}{(z - 0.4)^2}$$

By straightforward methods, we find $A = 25/9$, $B = -1.1111$, and $C = -2/3$. The inverse transform of this equation is

$$y(n) = \frac{25}{9} - 1.1111(0.4)^n - \frac{2}{3}(n + 1)(0.4)^n$$

The response to a step function can also be found using the MATLAB function `[y,t]=stepz(num,den)`. For this particular case and for $n = 0, 1, 2, \ldots 30$, we write

```
[y,t] = stepz([1 0 0],[1 -0.8 0.16],30);
```

Overdamped case $(4a < b^2)$: If we select $b = 0.9$ and $a = 0.1$, the inequality holds. The two roots are 0.1298 and 0.7702. The z-transform of the output is given by

$$Y(z) = 5\frac{z}{z - 1} + 0.0303\frac{z}{z - 0.1298} - 4.0303\frac{z}{z - 0.7702}$$

and its inverse is

$$y(n) = 5 + 0.0303(0.1298)^n - 4.0303(0.7702)^n \quad n = 0, 1, 2, \ldots$$

Underdamped case $(b^2 < 4a)$: In this case two conjugate roots exist, and these roots are poles of $Y(z)$. To proceed we write the denominator of $Y(z)$ in the form

$$z^2 - bz + a = (z - ce^{j\theta})(z - ce^{-j\theta}) \tag{8.78}$$

By expanding the right-hand side and equating like powers of z, we find that

$$a = c^2 \quad \text{and} \quad b = 2c\cos\theta$$

Thus, if a and b are known, then c and θ are readily obtained. By combining (8.78) with (8.76), we write

$$Y(z) = \frac{z^2}{(z - ce^{j\theta})(z - ce^{-j\theta})}\frac{z}{z - 1}$$

This fraction is now expanded into fractional form, which is

$$Y(z) = \frac{1}{1 - 2c\cos\theta + c^2}\frac{z}{z - 1} - \frac{ce^{j2\theta}}{j2\sin\theta(1 - ce^{j\theta})}\frac{z}{z - ce^{j\theta}}$$
$$+ \frac{ce^{-j2\theta}}{j2\sin\theta(1 - ce^{-j\theta})}\frac{z}{z - ce^{-j\theta}}$$

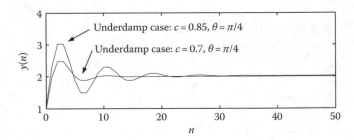

FIGURE 8.25

The inverse z-transform of this equation is

$$y(n) = \frac{1}{1 - 2c\cos\theta + c^2}u(n) - \frac{ce^{j2\theta}}{j2\sin\theta(1 - ce^{j\theta})}c^n e^{jn\theta} + \frac{ce^{-j2\theta}}{j2\sin\theta(1 - ce^{-j\theta})}c^n e^{-jn\theta}$$

This expression can be written in more convenient form by writing $1 - ce^{j\theta} = re^{-j\phi}$; then, $1 - ce^{-j\theta} = re^{j\phi}$. Figure 8.25 shows the underdamp cases for $c = 0.85$ and $\theta = \pi/4$ and $c = 0.7$ and $\theta = \pi/4$. ∎

Example 8.40

Using z-transform, find the solution of the difference equation

$$y(n + 2) - 5y(n + 1) + 6y(n) = 4^n \quad y(0) = 0,\ y(1) = 1$$

SOLUTION

Taking the z-transform of both sides and referring to the z-transform properties, we get

$$z^2 Y(z) - [z^2 y(0) + zy(1)] - 5[zY(z) - zy(0)] + 6Y(z) = \frac{z}{z - 4} \quad \text{or} \quad Y(z) = \frac{z}{(z - 2)(z - 4)}$$

after substituting the initial conditions. Using partial fraction decompositions, we obtain

$$Y(z) = -\frac{1}{2}\left(\frac{z}{z - 2} - \frac{z}{z - 4}\right) \quad \text{or} \quad y(n) = \frac{1}{2}4^n - \frac{1}{2}2^n = \{0, 1, 6, \ldots\} \qquad ∎$$

Example 8.41

Using the z-transform method, find the solution of the difference equation

$$2y(n + 2) - 7y(n + 1) + 3y(n) = 8 \quad y(0) = -1,\ y(1) = 0$$

SOLUTION

Taking the z-transform of both sides, we obtain

$$2z^2Y(z) - 2[z^2y(0) + zy(1)] - 7[zY(z) - zy(0)] + 3Y(z) = \frac{8z}{z-1} \quad \text{or}$$

$$Y(z) = \frac{-z(2z^2 - 9z - 1)}{(z-1)(2z-1)(z-3)} = -\frac{4z}{z-1} + \frac{4z}{2z-1} + \frac{1}{z-3} \quad \text{or}$$

$$y(n) = -1 + 2\left(\frac{1}{2}\right)^n + 3^n$$

Example 8.42

The Fibonacci sequence is obtained by setting the initial conditions $y(0) = y(1) = 1$ and accepting that any number of the sequence is the sum of the two previous numbers. Hence, $y(n+2) = y(n+1) + y(n)$ for $n = 0, 1, 2, 3,\ldots$

SOLUTION

The z-transform of the above equation is

$$z^2Y(z) - [z^2y(0) + zy(1)] = zY(z) - zy(0) + Y(z) \quad \text{or} \quad Y(z) = \frac{z^2}{z^2 - z - 1}$$

Using partial fraction expansion, we find

$$Y(z) = \frac{5 + \sqrt{5}}{10}\left(\frac{z}{z - ((1 + \sqrt{5})/2)}\right) + \frac{5 - \sqrt{5}}{10}\left(\frac{z}{z - ((1 - \sqrt{5})/2)}\right) \quad \text{or}$$

$$y(n) = \frac{5 + \sqrt{5}}{10}\left(\frac{1 + \sqrt{5}}{2}\right)^n + \frac{5 - \sqrt{5}}{10}\left(\frac{1 - \sqrt{5}}{2}\right)^n$$

This represents the law for the general term in the Fibonacci sequence.   ∎

Example 8.43

A ladder network is shown in Figure 8.26. Write the difference equation that characterizes the network. Specify the boundary conditions and solve for the current in any loop. Assume that each resistor is 1 Ω.

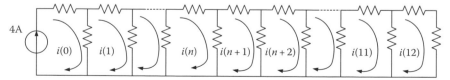

4A $i(0)$ $i(1)$ $i(n)$ $i(n+1)$ $i(n+2)$ $i(11)$ $i(12)$

FIGURE 8.26

SOLUTION

Applying Kirchoff's law for the $(n+1)$th loop, we obtain the equations

$$-Ri(n+1) - Ri(n+1) + Ri(+2) - Ri(n+1) + Ri(n) = 0 \quad \text{or}$$
$$3Ri(n+1) - Ri(n) - Ri(n+2) = 0$$

For the case $R = 1$, we obtain the difference equation

$$3i(n+1) - i(n) - i(n+2) = 0$$

The boundary conditions are

$$i(0) = 4, 3i(12) - i(11) = 0$$

Taking the z-transform, we find

$$[3zI(z) - 3zi(0)] - I(z) - [z^2I(z) - z^2i(0) - zi(1)] = 0$$

Substituting for $i(0)$ and solving for $I(z)$, we obtain

$$I(z) = \frac{4z^2 - 12z + i(1)z}{z^2 - 3z + 1} = \frac{4z(z - 3/2)}{z^2 - 2z(3/2) + 1} - \frac{[6 - i(1)]z}{z^2 - 2z(3/2) + 1}$$

Taking the inverse z-transform and referring to the z-transform table, we get

$$i(n) = 4\cosh \alpha n - \frac{6 - i(1)}{\sinh \alpha}\sinh \alpha n \quad \cosh \alpha = \frac{3}{2} \quad \text{or} \quad \alpha = 0.9624$$

To determine $i(1)$, we use the second boundary condition

$$3\left(4\cosh 12\alpha - \frac{6 - i(1)}{\sinh \alpha}\sinh 12\alpha\right) = \left(4\cosh 11\alpha - \frac{6 - i(1)}{\sinh \alpha}\sinh 11\alpha\right)$$

Solving for $i(1)$, we get $i(1) = 1.52$. Therefore, the current in the $(n+1)$th loop is

$$i(n) = 4\cosh(0.9624n) - \frac{4.48\sinh(0.9624n)}{\sinh(0.9624)} \qquad \blacksquare$$

8.10.1 Method of Undetermined Coefficients

The particular solution of a nonhomogeneous equation is the method of **undetermined coefficients**. The method is particularly efficient for input functions that are linear combinations of the following functions:

1. n^k, where n is a positive integer or zero
2. a^n, where a is a nonzero constant
3. $\cos an$, where a is a nonzero constant

4. sin *an*, where *a* is a nonzero constant
5. A product (finite) of two or more functions of type 1–4

This method works because any derivative of the input function $x(n)$ is also possible as a linear combination of functions of the five types above. For example, the function $2n^2$ or any derivative of $2n^2$ is a linear combination of the sequences n^2, n, and 1, all of which are of type 1. Hence, what is required in any case is the appropriate sequences for which any derivative of the input function $x(n)$ can be constructed by a linear combination of these sequences. Clearly, if $x(n) = \cos 3n$, the appropriate sequences are $\cos 3n$ and $\sin 3n$. The following examples clarify these assertions.

Example 8.44

Consider the system shown in Figure 8.27. Find the general solution if the input is $x(n) = 3^n u(n)$ and initial conditions are $y(-1) = 0$ and $y(-2) = 1$.

SOLUTION

From Figure 8.27, and taking into consideration the time shift of the output function, the controlling difference equation of this system is

$$y(n) - 5y(n-1) + 6y(n-2) = 3^n \quad n = 0, 1, 2, \ldots \tag{8.79}$$

The characteristic equation obtained from the corresponding homogeneous equation is

$$\lambda^2 - 5\lambda + 6 = 0$$

with roots $\lambda_1 = 2$ and $\lambda_2 = 3$. The two solutions are

$$y_1(n) = 2^n \quad y_2(n) = 3^n \tag{8.80}$$

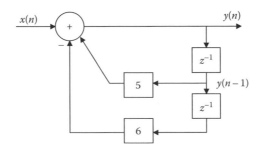

FIGURE 8.27

Since one of the homogeneous solutions is proportional to the input function, the particular solution is of the form $y_p(n) = Bn3^n$, where B is a constant. Introducing the particular solution in (8.79), we obtain

$$Bn3^n - 5b(n-1)3^{n-1} + 6B(n-2)3^{n-2} = 3^n$$

Solving for the unknown, we find that $B = 3$. Hence, the total solution is

$$y(n) = C_1y_1(n) + C_2y_2(n) + y_p(n) = C_1 2^n + C_2 3^n + 3n(3^n) \tag{8.81}$$

Subjecting this solution to the given initial conditions yields

$$\frac{1}{2}C_1 + \frac{1}{3}C_2 = 1$$

$$\frac{1}{4}C_1 + \frac{1}{9}C_2 = \frac{5}{3}$$

From this system, we find $C_1 = 16$ and $C_2 = -21$. The complete solution is

$$y(n) = 16(2^n) - 21(3^n) + 3n(3^n) \tag{8.82}$$

We now solve this problem by the z-transform method. The z-transform of (8.79) is

$$Y(z) - 5[z^{-1}Y(z) + y(-1)] + 6[z^{-2}Y(z) + y(-1)z^{-1} + y(-2)z^{-0}] = \frac{z}{z-3}$$

which can be written as follows:

$$Y(z) = -\frac{6z^2}{(z-2)(z-3)} + \frac{z^3}{(z-2)(z-3)^2}$$

The terms to the right are expanded (two terms and three terms, respectively) as follows:

$$Y(z) = 12\frac{z}{z-2} - 18\frac{z}{z-3} + 4\frac{z}{z-2} - 6\frac{z}{z-3} + 3\frac{z^2}{(z-3)^2}$$

From the tables, the inverse z-transform is

$$y(n) = 12(2^n) - 18(3^n) + 4(2^n) - 6(3^n) + 3(n+1)3^n = 16(2^n) - 21(3^n) + 3n(3^n)$$

which is (8.82) as anticipated.   ∎

Example 8.45

Find the solution for the system shown in Figure 8.28. The initial conditions are $y(-1) = 1$ and $y(-2) = 0$.

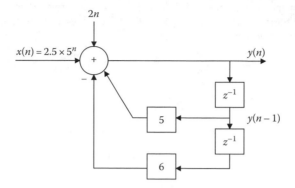

FIGURE 8.28

SOLUTION

The difference equation describing the system of Figure 8.28 is

$$y(n) - 5y(n-1) + 6y(n-2) = 2.5 \times 5^n + 2^n \tag{8.83}$$

As found in the example above, the roots are the same and, hence, the general solution of the homogeneous equation is

$$y_h(n) = C_1 2^n + C_2 3^n \tag{8.84}$$

We observe that the input 2^n is also a solution to the homogeneous equation. This suggests that we try the following particular solution:

$$y_p(n) = A5^n + Bn2^n \tag{8.85}$$

where the constants A and B are undetermined coefficients. These constants can be found by substituting (8.85) into (8.83). When this is done, we obtain

$$A5^n + Bn2^n - 5[A5^{n-1} + B(n-1)2^{n-1}] + 6[A5^{n-2} + B(n-2)2^{n-2}] = 2.5 \times 5^n + 2^n$$

Rearranging terms, we have

$$\left(A - 5A\frac{1}{5} + 6A\frac{1}{25}\right)5^n + \left(Bn + \frac{5B - 5Bn}{2} + \frac{6Bn - 12B}{4}\right)2^n = 2.5 \times 5^n + 2^n$$

By equating coefficients of similar terms, we find that

$$A = \frac{62.5}{6} \quad B = -2$$

and, therefore, the total solution is

$$y(n) = y_h(n) + y_p(n) = C_1 2^n + C_2 3^n + \frac{62.5}{6} 5^n - 2n2^n$$

Applying the initial conditions, the constants are $C_1 = -8.667$ and $C_2 = 6.750$. ∎

Example 8.46

Find the particular solution of the nonhomogeneous equation

$$y(n) - y(n - 2) = 5n^2 \tag{8.86}$$

SOLUTION

The roots of the characteristic equation are readily found to be $\lambda_1 = 1$ and $\lambda_2 = -1$. Since the characteristic equation is $\lambda^2 - 1 = 0$, we can use the MATLAB function **roots([1 0 -1])** to find the roots. Thus, the solution to the homogeneous equation is

$$y_h(n) = C_1(1)^n + C_2(-1)^n \quad n = 0, 1, 2, \ldots$$

We observe that the function $5n^2$ and its derivatives can be found by the linear combination of sequences $n^2, n,$ and 1. However, 1 is a solution of the homogeneous equation; so we choose as a trial, particular solution n times the sequence $n^2, n,$ and 1:

$$y_p(n) = An^3 + Bn^2 + Cn \tag{8.87}$$

Substitute this trial solution in (8.86) and equate coefficients of equal power terms. The coefficients are found to be $A = 5/6$, $B = 5/2$, and $C = 5/3$. ∎
Table 8.4 gives the suggested forms for the particular solutions for a specified $x(n)$.

TABLE 8.4 Method of Undetermined Coefficients

$x(n)$	$y_p(n)$
n^m	$A_1 n^m + A_2 n^{m-1} + \cdots + A_m n + A_{m+1}$
a^n	$A a^n$
$\cos \theta n$ or $\sin \theta n$	$A_1 \cos \theta n + A_2 \sin \theta n$
$n^m a^n$	$a^n(A_1 n^m + A_2 n^{m-1} + \cdots + A_m n + A_{m+1})$
$a^n \cos \theta n$ or $a^n \sin \theta n$	$a^n(A_1 \cos \theta n + A_2 \sin \theta n)$

*8.11 LTI Discrete-Time Dynamical Systems

The mathematical models describing dynamical systems are almost always of finite-order difference equations. If we know the initial conditions at $t = t_0$, their behavior can be uniquely determined for $t \geq t_0$. To see how to develop a dynamic system, let us consider the example below.

Example 8.47

Let a discrete system with input $v(n)$ and output $y(n)$ be described by the difference equation

$$y(n) + 2y(n-1) + y(n-2) = v(n) \tag{8.88}$$

If $y(n_0 - 1)$ and $y(n_0 - 2)$ are initial conditions for $n > n_0$, then $y(n)$ can be found recursively from (8.88). Let us take the pair $y(n-1)$ and $y(n-2)$ as the **state** of the system at time n. Let us call the vector

$$\mathbf{x}(n) = \begin{bmatrix} x_1(n) \\ x_2(n) \end{bmatrix} = \begin{bmatrix} y(n-2) \\ y(n-1) \end{bmatrix} \tag{8.89}$$

the **state vector** for the system. From the definition above, we obtain

$$x_1(n+1) = y(n+1-2) = y(n-1) \tag{8.90}$$

and

$$x_2(n+1) = y(n) = v(n) - y(n-2) - 2y(n-1) \tag{8.91}$$

or

$$x_2(n+1) = v(n) - x_1(n) - 2x_2(n) \tag{8.92}$$

Equations 8.90 and 8.92 can be written in the form

$$\begin{bmatrix} x_1(n+1) \\ x_2(n+1) \end{bmatrix} = \begin{bmatrix} 0 & 1 \\ -1 & -2 \end{bmatrix} \begin{bmatrix} x_1(n) \\ x_2(n) \end{bmatrix} + \begin{bmatrix} 0 \\ 1 \end{bmatrix} v(n) \tag{8.93}$$

or

$$\mathbf{x}(n+1) = \mathbf{A}\mathbf{x}(n) + \mathbf{B}v(n) \tag{8.94}$$

But (8.91) can be written as

$$y(n) = v(n) - 2x_2(n) = \begin{bmatrix} -1 & -2 \end{bmatrix} \begin{bmatrix} x_1(n) \\ x_2(n) \end{bmatrix} + v(n) \tag{8.95}$$

or

$$y(n) = \mathbf{C}x(n) + v(n) \tag{8.96}$$

Hence, the system can be described by vector-matrix difference equation (8.94) and an output equation (8.96) rather than by the second-order difference equation (8.88).

The LTI discrete dynamic system with a number of inputs is, therefore, described by the state equation

$$\mathbf{x}(nT + T) = \mathbf{A}\mathbf{x}(nT) + \mathbf{B}\mathbf{v}(nT) \tag{8.97}$$

and its output is described by the equation

$$\mathbf{y}(nT + T) = \mathbf{C}\mathbf{x}(nT) + \mathbf{D}\mathbf{v}(nT) \tag{8.98}$$

where

$$\mathbf{x}(nT) = N\text{-dimensional column vector}$$
$$\mathbf{v}(nT) = M\text{-dimensional column vector}$$
$$\mathbf{y}(nT) = R\text{-dimensional column vector}$$
$$\mathbf{A} = N \times N \text{ nonsingular matrix}$$
$$\mathbf{B} = N \times M \text{ matrix}$$
$$\mathbf{C} = R \times N \text{ matrix}$$
$$\mathbf{D} = R \times M \text{ matrix}$$

When the input is identically zero, (8.97) reduces to

$$\mathbf{x}(nT + T) = \mathbf{A}\mathbf{x}(nT) \tag{8.99}$$

so that

$$\mathbf{x}(nT + 2T) = \mathbf{A}\mathbf{x}(nT + T) = \mathbf{A}\mathbf{A}\mathbf{x}(nT) = \mathbf{A}^2\mathbf{x}(nT) \tag{8.100}$$

and so on. In general, we have

$$\mathbf{x}(nT + kT) = \mathbf{A}^k\mathbf{x}(nT) \tag{8.101}$$

The state transition matrix from $n_1 T$ to $n_2 T$ $(n_2 > n_1)$ is given by

$$\varphi(n_2 T, n_1 T) = \mathbf{A}^{n_2 - n_1} \tag{8.102}$$

This is a function only of the time difference $n_2 T - n_1 T$. Therefore, it is customary to name the matrix

$$\varphi(nT) = \mathbf{A}^n \tag{8.103}$$

The **state transition matrix** with the understanding that $n = n_2 - n_1$. It follows that the stem states at two times, $n_2 T$ and $n_1 T$, are related by the relation

$$\mathbf{x}(n_2 T) = \boldsymbol{\varphi}(n_2 T, n_1 T)\mathbf{x}(n_1 T) \tag{8.104}$$

when the input is zero. From (8.102), we obtain the following relationships:

(a)

$$\boldsymbol{\varphi}(nT, nT) = \mathbf{I} = \text{identity matrix} \tag{8.105}$$

(b)

$$\boldsymbol{\varphi}(n_2 T, n_1 T) = \boldsymbol{\varphi}^{-1}(n_1 T, n_2 T) \tag{8.106}$$

(c)

$$\boldsymbol{\varphi}(n_3 T, n_2 T)\boldsymbol{\varphi}(n_2 T, n_1 T) = \boldsymbol{\varphi}(n_3 T, n_1 T) \tag{8.107}$$

If the input is not identically zero and $\mathbf{x}(nT)$ is known, then the progress (later states) of the system can be found recursively from (8.97). Proceeding with the recursion, we obtain

$$\mathbf{x}(nT + 2T) = \mathbf{A}\mathbf{x}(nT + T) + \mathbf{B}\mathbf{v}(nT + T)$$
$$= \mathbf{A}\mathbf{A}\mathbf{x}(nT) + \mathbf{A}\mathbf{B}\mathbf{v}(nT) + \mathbf{B}\mathbf{v}(nT + T)$$
$$= \boldsymbol{\varphi}(nT + 2T, nT)\mathbf{x}(nT) + \boldsymbol{\varphi}(nT + 2T, nT + T)\mathbf{B}\mathbf{v}(nT) + \mathbf{B}\mathbf{v}(nT + T)$$

In general, for $k > 0$ we have the solution

$$\mathbf{x}(nT + kT) = \boldsymbol{\varphi}(nT + kT, nT)\mathbf{x}(nT) + \sum_{i=n}^{n+k-1} \boldsymbol{\varphi}(nT + kT, iT + T)\mathbf{B}\mathbf{v}(iT) \tag{8.108}$$

From (8.104), when the input is zero, we obtain the relation

$$\mathbf{x}(n_2 T) = \boldsymbol{\varphi}(n_2 T - n_1 T)\mathbf{x}(n_1 T) = \mathbf{A}^{n_2 - n_1}\mathbf{x}(n_1 T) \tag{8.109}$$

According to (8.108), the solution to the dynamic system when the input is zero is given by

$$\mathbf{x}(nT + kT) = \boldsymbol{\varphi}(nT + kT - nT)\mathbf{x}(nT) + \sum_{i=n}^{n+k-1} \boldsymbol{\varphi}[(n + k - i - 1)T]\mathbf{B}\mathbf{v}(iT) \tag{8.110}$$

or

$$\mathbf{x}(nT + kT) = \boldsymbol{\varphi}(kT)\mathbf{x}(nT) + \sum_{i=n}^{n+k-1} \boldsymbol{\varphi}[(n + k - i - 1)T]\mathbf{B}\mathbf{v}(iT) \quad k > 0 \tag{8.111}$$

To find the solution using the *z*-transform method, we define the one-sided *z*-transform for an $R \times S$ matrix function $\mathbf{f}(nT)$ as the $R \times S$ matrix

$$\mathbf{F}(z) = \sum_{n=0}^{\infty} \mathbf{f}(nT)z^{-n} \tag{8.112}$$

The elements of $\mathbf{F}(z)$ are the transforms of the corresponding elements of $\mathbf{f}(nT)$. Taking the *z*-transform of both sides of (8.97), we obtain

$$z\mathbf{X}(z) - z\mathbf{x}(0) = \mathbf{A}\mathbf{X}(z) + \mathbf{B}\mathbf{V}(z)$$

or

$$\mathbf{X}(z) = (z\mathbf{I} - \mathbf{A})^{-1}z\mathbf{x}(0) + (z\mathbf{I} - \mathbf{A})^{-1}\mathbf{B}\mathbf{V}(z) \tag{8.113}$$

From the output equation (8.98), we see that

$$\mathbf{Y}(z) = \mathbf{C}\mathbf{X}(z) + \mathbf{D}\mathbf{V}(z) \tag{8.114}$$

The state of the system $\mathbf{x}(nT)$ and its output $\mathbf{y}(nT)$ can be found for $n \geq 0$ by taking the inverse transform of (8.113) and (8.114).

For a zero input, (8.113) becomes

$$\mathbf{X}(z) = (z\mathbf{I} - \mathbf{A})^{-1}z\mathbf{x}(0) \tag{8.115}$$

so that

$$\mathbf{x}(nT) = \mathcal{Z}^{-1}\{(z\mathbf{I} - \mathbf{A})^{-1}z\}\mathbf{x}(0) \tag{8.116}$$

If we let $n_1 = 0$ and $n_2 = n$, then (8.109) becomes

$$\mathbf{x}(nT) = \varphi(nT)\mathbf{x}(0) = \mathbf{A}^n\mathbf{x}(0) \tag{8.117}$$

Comparing (8.116) and (8.117) we observe that

$$\varphi(nT) = \mathbf{A}^n = \mathcal{Z}^{-1}\{(z\mathbf{I} - \mathbf{A})^{-1}z\} \quad n \geq 0 \tag{8.118}$$

or equivalently,

$$\boldsymbol{\Phi}(z) = \mathcal{Z}\{\mathbf{A}^n\} = (z\mathbf{I} - \mathbf{A})^{-1}z \tag{8.119}$$

The *z*-transform provides a straightforward method for calculating the state transition matrix.

Next, combine (8.119) and (8.113) to find

$$\mathbf{X}(z) = \boldsymbol{\Phi}(z)\mathbf{x}(0) + \boldsymbol{\Phi}(z)z^{-1}\mathbf{B}\mathbf{V}(z) \tag{8.120}$$

By applying the convolution theorem and the fact that

$$\mathcal{Z}^{-1}\{\boldsymbol{\Phi}(z)z^{-1}\} = \boldsymbol{\varphi}(nT - T)u(nT - T) \tag{8.121}$$

the inverse z-transform of (8.120) is given by

$$\boldsymbol{x}(kT) = \boldsymbol{\varphi}(kT)x(0) + \sum_{i=0}^{k-1} \boldsymbol{\varphi}[(k - i - 1)T]\boldsymbol{Bv}(iT) \tag{8.122}$$

The above equation is identical to (8.111) with $n = 0$.

The behavior of the system with zero input depends on the location of the poles of

$$\boldsymbol{\Phi}(z) = (z\boldsymbol{I} - \boldsymbol{A})^{-1}z \tag{8.123}$$

Because

$$(z\boldsymbol{I} - \boldsymbol{A})^{-1} = \frac{\text{adj}(z\boldsymbol{I} - \boldsymbol{A})}{\det(z\boldsymbol{I} - \boldsymbol{A})} \tag{8.124}$$

where adj( ) denotes the regular adjoint in matrix theory, these poles can only occur at the roots of the polynomial

$$D(z) = \det(z\boldsymbol{I} - \boldsymbol{A}) \tag{8.125}$$

$D(z)$ is known as the **characteristic polynomial** of $\boldsymbol{A}$ (for the system) and its roots are known as the **characteristic values** of $\boldsymbol{A}$. If all the roots are inside the unit circle, the system is stable. If even one root is outside the unit circle, the system is unstable. ∎

Example 8.48

Consider the system

$$\begin{bmatrix} x_1(nT + T) \\ x_2(nT + T) \end{bmatrix} = \begin{bmatrix} 0 & 2 \\ 0.22 & 2 \end{bmatrix} \begin{bmatrix} x_1(nT) \\ x_2(nT) \end{bmatrix} + \begin{bmatrix} 0 \\ 1 \end{bmatrix} v(nT)$$

$$y(nT) = [0.22 \quad 2] \begin{bmatrix} x_1(nT) \\ x_2(nT) \end{bmatrix} + v(nT) \tag{8.126}$$

For this system we have

$$\boldsymbol{A} = \begin{bmatrix} 0 & 2 \\ 0.22 & 2 \end{bmatrix}, \quad \boldsymbol{B} = \begin{bmatrix} 0 \\ 1 \end{bmatrix}, \quad \boldsymbol{C} = [0.22 \quad 2], \quad \boldsymbol{D} = [1]$$

The characteristic polynomial is

$$D(z) = \det(z\mathbf{I} - \mathbf{A}) = \det\left[\begin{bmatrix} z & 0 \\ 0 & z \end{bmatrix} - \begin{bmatrix} 0 & 2 \\ 0.22 & 2 \end{bmatrix}\right] = \det\begin{bmatrix} z & -2 \\ -0.22 & z-2 \end{bmatrix}$$

$$= z(z-2) - 0.44 = z^2 - 2z - 0.44 = (z-2.2)(z+0.2)$$

Hence, we obtain (see (8.123))

$$\boldsymbol{\Phi}(z) = \frac{z}{(z-2.2)(z+0.2)}\begin{bmatrix} z-2 & 2 \\ 0.22 & z \end{bmatrix} = \begin{bmatrix} \dfrac{z(z-2)}{(z-2.2)(z+0.2)} & \dfrac{2z}{(z-2.2)(z+0.2)} \\[2mm] \dfrac{0.22z}{(z-2.2)(z+0.2)} & \dfrac{z^2}{(z-2.2)(z+0.2)} \end{bmatrix}$$

Because $D(z)$ has a root outside the unit circle at 2.2, the system is unstable. Taking the inverse transform, we find that

$$\boldsymbol{\varphi}(nT) = \begin{bmatrix} \dfrac{1}{12}(2.2)^n + \dfrac{11}{12}(-0.2)^2 & \dfrac{5}{6}(2.2)^n - \dfrac{5}{6}(-0.2)^2 \\[2mm] \dfrac{11}{120}(2.2)^n - \dfrac{11}{120}(-0.2)^2 & \dfrac{11}{12}(2.2)^n + \dfrac{1}{12}(-0.2)^2 \end{bmatrix}$$

To check, set $n = 0$ to find $\boldsymbol{\varphi}(0) = \mathbf{I}$ and $\boldsymbol{\varphi}(T) = \mathbf{A}$.

Let $\mathbf{x}(0) = 0$ and the input be the unit impulse $v(nT) = \delta(nT)$ so that $V(z) = 1$. Hence, according to (8.120),

$$\mathbf{X}(z) = \boldsymbol{\Phi}(z)z^{-1}\mathbf{B}V(z) = \frac{1}{(z-2.2)(z+0.2)}\begin{bmatrix} z-2 & 2 \\ 0.22 & z \end{bmatrix}\begin{bmatrix} 0 \\ 1 \end{bmatrix} = \frac{1}{(z-2.2)(z+0.2)}\begin{bmatrix} 2 \\ z \end{bmatrix}$$

The inverse z-transform gives

$$\mathbf{x}(nT) = \frac{5}{6}\begin{bmatrix} (2.2)^{n-1} - (-0.2)^{n-1} \\[2mm] \dfrac{1}{2}(2.2)^n - \dfrac{1}{2}(-0.2)^n \end{bmatrix}$$

And the output is given by

$$y(nT) = \mathbf{C}\mathbf{x}(nT) + \mathbf{D}v(nT) = \begin{cases} 1 & n = 0 \\[2mm] \dfrac{5}{12}(2.2)^{n+1} - \dfrac{5}{12}(-0.2)^{n+1} & n > 0 \end{cases} \qquad\blacksquare$$

*8.12 z-Transform and Random Processes

8.12.1 Power Spectral Densities

The z-transform of the autocorrelation function $R_{xx}(\tau) = E\{x(t+\tau)x(t)\}$ sampled uniformly at nT times is given by

$$S_{xx}(z) = \sum_{n=-\infty}^{\infty} R_{xx}(nT)z^{-n} \tag{8.127}$$

The Fourier transform of the correlation function $R_{xx}(\tau)$ is designated by $S_{xx}(\omega)$. The sampled power spectral density for $x(nT)$ is defined to be

$$S_{xx}(e^{j\omega T}) = S_{xx}(z)\big|_{z=e^{j\omega T}} = \sum_{n=-\infty}^{\infty} R_{xx}(nT)e^{-j\omega nT} \tag{8.128}$$

However, from the sampling theorem we have

$$S_{xx}(e^{j\omega T}) = \frac{1}{T} \sum_{n=-\infty}^{\infty} S_{xx}(\omega - n\omega_s), \quad \omega_s = 2\pi/T \tag{8.129}$$

Because $S_{xx}(\omega)$ is real, nonnegative, and even, it follows from (8.129) that $S_{xx}(e^{j\omega T})$ is also real, nonnegative, and even. If the envelope of $R_{xx}(\tau)$ decays exponentially for $|\tau| > 0$, then the ROC for $S_{xx}(z)$ includes the unit circle. If $R_{xx}(\tau)$ has undamped periodic components, the series in (8.128) converges in the distribution sense that contains impulse functions.

The average power in $x(nT)$ is

$$E\{x^2(nT)\} = R_{xx}(0) = \frac{1}{2\pi j} \oint_C S_{xx}(z) \frac{dz}{z} \tag{8.130}$$

where C is a simple, closed contour lying in the region of convergence and the integration is taken in a counterclockwise sense. If C is the unit circle, then

$$R_{xx}(0) = \frac{1}{\omega_s} \int_{-\omega_s/2}^{\omega_s/2} S_{xx}(e^{j\omega T})d\omega \quad \omega_s = \frac{2\pi}{T} \tag{8.131}$$

$$S_{xx}(e^{j\omega T}) \frac{d\omega}{\omega_s} = \text{average power in } d\omega \tag{8.132}$$

$S_{xy}(z)$ is called the cross power spectral density for two jointly wide-sense stationary processes $x(t)$ and $y(t)$. It is defined by the relation

$$S_{xy}(z) = \sum_{n=-\infty}^{\infty} R_{xy}(nT)z^{-n} \tag{8.133}$$

Because $R_{xy}(nT) = R_{yx}(-nT)$ it follows that

$$S_{xy}(z) = S_{yx}(z^{-1}), \quad S_{xx}(z) = S_{xx}(z^{-1}) \tag{8.134}$$

Equivalently, we have

$$S_{xx}(e^{j\omega T}) = S_{xx}(e^{-j\omega T}) \tag{8.135}$$

If $S_{xx}(z)$ is a rational polynomial, it can be factored in the form

$$S_{xx}(z) = \frac{N(z)}{D(z)} = \gamma^2 G(z)G(z^{-1})$$

$$G(z) = \frac{\prod_{k=1}^{L}(1 - \alpha_k z^{-1})}{\prod_{k=1}^{M}(1 - \beta_k z^{-1})} = \frac{\sum_{k=0}^{L} a_k z^{-k}}{\sum_{k=0}^{M} b_k z^{-k}} \tag{8.136}$$

$$\gamma^2 > 0, |a_k| < 1, |a_k| < 1, a_k \text{ and } b_k \text{ are real}$$

8.12.2 Linear Discrete-Time Filters

Let $R_{xx}(nT)$, $R_{yy}(nT)$, and $R_{xy}(nT)$ be known. Let two systems have transfer functions $H_1(z)$ and $H_2(z)$, respectively. The output of these filters, when the inputs are $x(nT)$ and $y(nT)$ (see Figure 8.29), are

$$v(nT) = \sum_{k=-\infty}^{\infty} h_1(kT)x(nT - kT) \tag{8.137}$$

$$w(nT) = \sum_{k=-\infty}^{\infty} h_2(kT)y(nT - kT) \tag{8.138}$$

Let $n = n + m$ in (8.137), multiply by $y(nT)$, and take the ensemble average to find

$$R_{vy}(mT) = \sum_{k=-\infty}^{\infty} h_1(kT)E\{x(mT + nT - kT)y(nT)\}$$

$$= \sum_{k=-\infty}^{\infty} h_1(kT)R_{xy}(mT - kT) \tag{8.139}$$

Hence, by taking the z-transform, we obtain

$$S_{vy}(z) = H_1(z)S_{xy}(z) \tag{8.140}$$

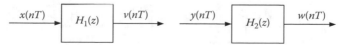

FIGURE 8.29

Similarly, from (8.138), we obtain

$$R_{vw}(mT) = \sum_{k=-\infty}^{\infty} h_2(kT)R_{vy}(mT - kT) \qquad (8.141)$$

and

$$S_{vw}(z) = H_2(z)S_{vy}(z) \qquad (8.142)$$

From (8.140) and (8.142), we obtain

$$S_{vw}(z) = H_1(z)H_2(z^{-1})S_{xy}(z) \qquad (8.143)$$

Also, for $x(nT) = y(nT)$ and $h_1(nT) = h_2(nT) = h(nT)$, (8.143) becomes

$$S_{vv}(z) = H(z)H(z^{-1})S_{xx}(z) \qquad (8.144)$$

and

$$S_{vv}(e^{j\omega T}) = H(e^{j\omega T})H(e^{j\omega T-1})S_{xx}(e^{j\omega T}) = \left|H(e^{j\omega T})\right|^2 S_{xx}(e^{j\omega T}) \qquad (8.145)$$

8.12.3 Optimum Linear Filtering

Let $y(nT)$ be an observed wide-sense stationary process and $x(nT)$ be a desired wide-sense stationary process. The process $y(nT)$ could be the result of the desired signal $x(nT)$ and a noise signal $v(nT)$. It is desired to find a system with transfer function $H(z)$ such that the error $e(nT) = x(nT) - \hat{x}(nT) = x(nT) - \mathcal{Z}^{-1}\{Y(z)H(z)\}$ is minimized. Referring to Figure 8.30 and (8.144), we can write

$$S_{aa}(z) = \frac{1}{H_1(z)H_1(z^{-1})}S_{yy}(z) = \gamma^2 \qquad (8.146)$$

Where $a(nT)$ is taken as white noise (uncorrelated). We, therefore, can write

$$R_{aa}(mT) = \gamma^2\delta(mT) \qquad (8.147)$$

The signal, $a(nT)$, is known as the innovation process associated with $\gamma(nT)$. From Figure 8.30, we obtain

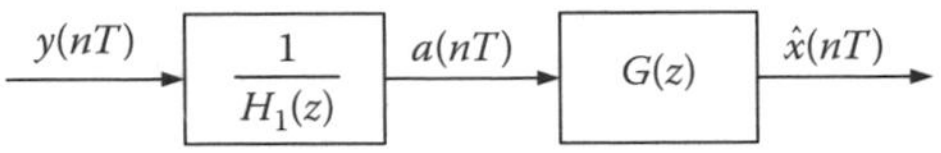

FIGURE 8.30

$$\hat{x}(nT) = \sum_{k=-\infty}^{\infty} g(nT)a(nT - kT) \tag{8.148}$$

The mean square error is given by

$$E\{e^2(nT)\} = E\left\{\left[x(nT) - \sum_{k=-\infty}^{\infty} g(kT)a(nT - kT)\right]^2\right\}$$

$$= E\{x^2(nT)\} - 2E\left\{\sum_{k=-\infty}^{\infty} g(kT)x(nT)a(nT - kT)\right\} + E\left\{\left[\sum_{t=-\infty}^{\infty} g(kT)a(nT - kT)\right]^2\right\}$$

$$= R_{xx}(0) - 2\sum_{k=-\infty}^{\infty} g(kT)R_{xa}(kT) + \gamma^2 \sum_{k=-\infty}^{\infty} g^2(kT)$$

$$= R_{xx}(0) - 2\sum_{k=-\infty}^{\infty} \left[\gamma g(kT) - \frac{R_{xa}(kT)}{\gamma}\right]^2 - \frac{1}{\gamma^2} \sum_{k=-\infty}^{\infty} R_{xa}^2(kT)$$

To minimize the error, we must set the quantity in the brackets equal to zero. Hence,

$$g(nT) = \frac{1}{\gamma^2} R_{xa}(nT) \quad -\infty < n < \infty$$

and its *z*-transform is

$$G(z) = \frac{1}{\gamma^2} S_{xa}(z)$$

But from (8.143) (because $v(nT) = x(nT)$ implies that $H_1(z) = 1$), we have

$$S_{xy}(z) = H_1(z^{-1})S_{xa}(z) \quad \text{or} \quad S_{xa}(z) = \frac{S_{xy}(z)}{H_1(z^{-1})} \tag{8.149}$$

$$G(z) = \frac{1}{\gamma^2} \frac{S_{xy}(z)}{H_1(z^{-1})} \tag{8.150}$$

From Figure 8.30, the optimum filter is given by (see also 8.146)

$$H(z) = \frac{1}{H_1(z)} G(z) = \frac{S_{xy}(z)}{\gamma^2 H_1(z)H_1(z^{-1})} = \frac{S_{xy}(z)}{S_{yy}(z)} \tag{8.151}$$

The mean square error for an optimum filter is

$$E\{e^2(nT)\} = R_{xx}(0) - \frac{1}{\gamma^2} \sum_{k=-\infty}^{\infty} R_{xa}^2(kT) \tag{8.152}$$

Applying Parseval's theorem in the above equation, we obtain

$$E\{e^2(nT)\} = \frac{1}{2\pi j} \oint_C \left[S_{xx}(z) - \frac{1}{\gamma^2} S_{xa}(z)S_{xa}(z^{-1}) \right] \frac{dz}{z} = \oint_C \left[S_{xx}(z) - \frac{S_{xy}(z)S_{xy}(z^{-1})}{S_{yy}(z)} \right] \frac{dz}{z}$$

$$= \frac{1}{2\pi j} \oint_C \left[S_{xx}(z) - H(z)S_{xy}(z^{-1}) \right] \frac{dz}{z} \tag{8.153}$$

where C can be the unit circle.

*8.13 Relationship between the Laplace and z-Transforms

The one-sided LT and its inverse are given by the following equations:

$$F(s) \overset{\Delta}{=} \mathcal{L}\{f(t)\} = \int_0^\infty f(t)e^{-st}dt \quad \text{Re}\{s\} > \sigma_c$$

$$f(t) = \mathcal{L}^{-1}\{F(s)\} = \frac{1}{2\pi j} \int_{c-j\infty}^{c+j\infty} F(s)e^{st}ds \quad c > \sigma_c \tag{8.154}$$

where σ_c is the abscissa of convergence.

The LT of a sampled function

$$f_s(t) \overset{\Delta}{=} f(t)\text{comb}_T(t) = f(t) \sum_{k=-\infty}^{\infty} \delta(t - nT) = \sum_{k=-\infty}^{\infty} f(nT)\delta(t - nT) \tag{8.155}$$

is given by

$$F_s(s) \overset{\Delta}{=} \mathcal{L}\{f_s(t)\} = \sum_{k=-\infty}^{\infty} f(nT)\mathcal{L}\{\delta(t - nT)\}$$

$$= \sum_{k=-\infty}^{\infty} f(nT) \int_{-\infty}^{\infty} \delta(t - nT)e^{-st}dt = \sum_{k=-\infty}^{\infty} f(nT)e^{-nTs} \tag{8.156}$$

From the above equation, we find

$$F(z) = F_s s)|_{s=T^{-1}\ln z} \tag{8.157}$$

and, hence,

$$F(z)|_{z=e^{sT}} = F_s(s) \triangleq \mathcal{L}\{f_s(t)\} = \mathcal{L}\{f(t)\text{comb}_T(t)\} \tag{8.158}$$

If the region of convergence for $F(z)$ includes the unit circle, $|z| = 1$, then

$$F_s(\omega) = F(z)|_{z=e^{j\omega T}} = \sum_{n=-\infty}^{\infty} f(nT)e^{-j\omega T}, \quad F_s(s + j\omega_s) = F_s(s) = \text{periodic,}$$

$$\omega_s = \frac{2\pi}{T} \tag{8.159}$$

The knowledge of $F_s(s)$ in the strip $-\omega_s/2 < \omega \le \omega_s/2$ determines $F_s(s)$ for all s. The transformation $z = e^{sT}$ maps this strip uniquely onto the complex z-plane. Therefore, $F(z)$ contains all the information in $F_s(s)$ without redundancy. Letting $\sigma = s + j\omega$, then

$$z = e^{\sigma T} e^{j\omega T} \tag{8.160}$$

Because $|z| = e^{\sigma T}$, we obtain

$$|z| = \begin{cases} <1 & \sigma < 0 \\ =1 & \sigma = 0 \\ >1 & \sigma > 0 \end{cases} \tag{8.161}$$

Therefore, we have the following correspondence between the s-plane and the z-plane:

1. Points in the left half of the s-plane are mapped inside the unit circle in the z-plane.
2. Points on the $j\omega$-axis are mapped onto the unit circle.
3. Points in the right half of the s plane are mapped outside the unit circle.
4. Lines parallel to the $j\omega$-axis are mapped into circles with radius $|z| = e^{\sigma T}$.
5. Lines parallel to the σ-axis are mapped into rays of the form $\arg z = \omega T$ radius from $z = 0$.
6. The origin of the s-plane corresponds to $z = 1$.
7. The σ-axis corresponds to the positive $u = \text{Re } z$-axis.
8. As ω varies from $-\omega_s/2$ to $\omega_s/2$, $\arg z = \omega T$ varies between $-\pi$ and π radians.

Let $f(t)$ and $g(t)$ be causal functions with LTs $F(s)$ and $G(s)$ that converge absolutely for $\text{Re } s > \sigma_f$ and $\text{Re } s > \sigma_g$, respectively; then

$$\mathcal{L}\{f(t)g(t)\} = \frac{1}{2\pi j} \int_{c-j\infty}^{c+j\infty} F(p)G(s - p)dp \tag{8.162}$$

The contour is parallel to the imaginary axis in the complex p-plane with

$$\sigma = \text{Re } s > \sigma_f + \sigma_g \quad \text{and} \quad \sigma_f < c < \sigma - \sigma_g \tag{8.163}$$

With this choice, the poles of $G(s - p)$ lie at the right of the integration path. For causal $f(t)$, its sampling form is given by (8.155). If

$$g(t) = \text{comb}_T(t) \triangleq \sum_{n=0}^{\infty} \delta(t - nT) \tag{8.164}$$

then its LT is

$$G(s) = \mathcal{L}\{g(t)\} = \sum_{n=0}^{\infty} e^{-nTs} = \frac{1}{1 - e^{-Ts}} \tag{8.165}$$

Because $\sigma_g = 0$, then (8.162) becomes

$$F_s(s) = \frac{1}{2\pi j} \int_{c-j\infty}^{c+j\infty} \frac{F(p)}{1 - e^{-(s-p)T}} \, dp \quad \sigma > \sigma_f, \sigma_f < c < \sigma \tag{8.166}$$

The distance p in Figure 8.31 is given by

$$p = c + \text{Re}^{j\theta} \quad \frac{\pi}{2} \leq \theta \leq \frac{3\pi}{2} \tag{8.167}$$

If the function $F(p)$ is analytic for some $|p|$ greater than a finite number R_0 and has zero at infinity, then in the limit as $R \to \infty$ the integral along the path BDA is identically zero and the integral along the path AEB averages to $F_s(s)$. The contour $C_1 + C_2$ encloses all the poles of $F(p)$. Because of these assumptions, $F(p)$ must have a Laurent series expansion of the form

$$F(p) = \frac{a_{-1}}{p} + \frac{a_{-2}}{p^2} + \cdots = \frac{a_{-1}}{p} + \frac{Q(p)}{p^2} \quad |p| > R_0 \tag{8.168}$$

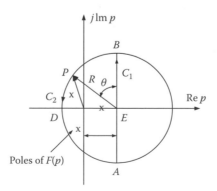

FIGURE 8.31

$Q(p)$ is analytic in this domain and

$$|Q(p)| < M < \infty \quad |p| > R_0 \tag{8.169}$$

Therefore, from (8.168),

$$a_{-1} = \lim_{p \to \infty} pF(p) \tag{8.170}$$

From the initial value theorem,

$$a_{-1} = f(0+) \tag{8.171}$$

Applying Cauchy's residue theorem to (8.166), we obtain

$$F_s(s) = \sum_k \text{Res}\left\{ \frac{F(p)}{1 - e^{pT}e^{-sT}} \right\}\Bigg|_{p=p_k} - \frac{f(0+)}{2} \tag{8.172}$$

By letting $z = e^{sT}$, the above equation becomes

$$F(z) = F_s(s)_{s=\frac{1}{T}\ln z} = \sum_k \text{Res}\left\{ \frac{F(p)}{1 - e^{pT}z^{-1}} \right\}\Bigg|_{p=p_k} - \frac{f(0+)}{2} \quad |z| > e^{\sigma_f T} \tag{8.173}$$

Example 8.49

The LT of $f(t) = tu(t)$ is $1/s^2$. The integrand $|te^{-\sigma t}e^{-j\omega t}| < \infty$ for $\sigma > 0$ implies that the region of convergence is $R\{s\} > 0$. Because $f(t)$ has a double pole at $s = 0$, (8.173) becomes

$$F(z) = \text{Res}\left\{ \frac{1}{p^2(1 - e^{pT}z^{-1})} \right\}\Bigg|_{p=0} - \frac{0}{2} = \frac{d}{dp}\left\{ \frac{p^2}{p^2(1 - e^{pT}z^{-1})} \right\}\Bigg|_{p=0} = \frac{Tz^{-1}}{(1 - z^{-1})^2} \quad \blacksquare$$

Example 8.50

The LT of $f(t) = \exp(-at)u(t)$ is $1/(s+a)$. The ROC is Re $s > -a$ and from (8.173), we obtain

$$F(z) = \text{Res}\left\{ \frac{1}{(p+a)(1 - e^{pT}z^{-1})} \right\}\Bigg|_{p=-a} - \frac{1}{2} = \frac{1}{1 - e^{-aT}z^{-1}} - \frac{1}{2}$$

The inverse transform is

$$f(nT) = -\frac{1}{2}\delta(n) + e^{-anT}u(nT)$$

If we proceed to find the z-transform from $f(nT) = \exp(-anT)u(nT)$, we would have found $F(z) = 1/(1 - e^{-aT} - z^{-1})$. Hence, to make a causal signal $f(t)$ consistent with $F(s)$ and the inverse formula, $f(0)$ should be assigned the value $f(0+)/2$.

It is conventional in calculating with the z-transform of causal functions to assign the value of $f(0+)$ to $f(0)$. With this convention, the formula for calculating $F(z)$ from $F(s)$ reduces to

$$F(z) = \sum_k \text{Res}\left\{\frac{F(p)}{1 - e^{pT}z^{-1}}\right\}\Bigg|_{p=p_k} \qquad |z| > e^{\sigma_r T} \tag{8.174}$$

*8.14 Relationship to the Fourier Transform

The sampled signal can be represented by

$$f_s(t) = \sum_{n=-\infty}^{\infty} f(nT)\delta(t - nT) \tag{8.175}$$

The corresponding Laplace and Fourier transforms are

$$F_s(s) = \sum_{n=-\infty}^{\infty} f(nT)e^{-snT} \tag{8.176}$$

$$F_s(\omega) = \sum_{n=-\infty}^{\infty} f(nT)e^{-j\omega nT} \tag{8.177}$$

If we set $z = e^{sT}$ in the definition of the z-transform, we see that

$$F_s(s) = F(z)|_{z=e^{sT}} \tag{8.178}$$

If the ROC for $F(z)$ includes the unit circle $|z| = 1$, then

$$F_s(\omega) = F(z)|_{z=e^{j\omega T}} \tag{8.179}$$

Because $F_s(s)$ is periodic with period $\omega_s = 2\pi/T$, we need only consider the strip $-\omega_s/2 < \omega \leq \omega_s/2$, which uniquely determines $F_s(s)$ for all s. The transformation $z = \exp(sT)$ maps the strip onto the complex z-plane so that $F(z)$ contains all the information in $F_s(s)$ without the redundancy.

Appendix

TABLE 8.A.1 z-Transform Properties for Positive-Time Sequences

---

1. Linearity

$$Z\{c_i f_i(nT)\} = c_i F_i(z) \; |z| > R_i, \; c_i \text{ are constants}$$

$$\mathscr{Z}\left\{\sum_{i=0}^{\ell} c_i f_i(nT)\right\} = \sum_{i=0}^{\ell} c_i F_i(z) \quad |z| > \max R_i$$

2. Shifting property

$$Z\{f(nT - kT)\} = z^{-k}F(z), \quad f(-nT) = 0, \quad \text{for } n = 1, 2, \ldots$$

$$\mathscr{Z}\{f(nT - kT)\} = z^{-k}F(z) + \sum_{n=1}^{k} f(-nT)z^{-(k-n)}$$

$$\mathscr{Z}\{f(nT + kT)\} = z^{k}F(z) - \sum_{n=0}^{k-1} f(nT)z^{k-n}$$

$$Z\{f(nT + T)\} = z[F(z) - f(0)]$$

3. Time scaling

$$\mathscr{Z}\{a^{nT}f(nT)\} = F(a^{-T}z) = \sum_{n=0}^{\infty} f(nT)(a^{-T}z)^{-n} \quad |z| > a^{T}$$

4. Periodic sequence

$$\mathscr{Z}\{f(nT)\} = \frac{z^{N}}{z^{N}-1}F_{(1)}(z) \quad |z| > R$$

$$N = \text{number of time units in a period}$$

$$R = \text{radius of convergence of } F_{(1)}(z)$$

$$F_{(1)}(z) = z\text{-transform of the first period}$$

5. Multiplication by n and nT

$$\mathscr{Z}\{nf(nT)\} = -z\frac{dF(z)}{dz} \quad |z| > R$$

$$\mathscr{Z}\{nTf(nT)\} = -zT \cdot \frac{dF(z)}{dz} \quad |z| > R$$

$$R = \text{radius of convergence of } F(z)$$

6. Convolution

$$\mathscr{Z}\{f(nT)\} = F(z) \qquad |z| > R_1$$

$$\mathscr{Z}\{h(nT)\} = H(z) \qquad |z| > R_2$$

$$\mathscr{Z}\{f(nT) * h(nT)\} = F(z)H(z) \qquad |z| > \max(R_1, R_2)$$

7. Initial value

$$f(0T) = \lim_{z \to \infty} F(z) \quad |z| > R, \text{ if } F(\infty) \text{ exists}$$

8. Final value

$$\lim_{n \to \infty} f(nT) = \lim_{z \to 1} (z - 1)F(z) \quad \text{if } f(\infty T) \text{ exists}$$

9. Multiplication by $(nT)^{k}$

$$\mathscr{Z}\{n^{k}T^{k}f(nT)\} = -Tz\frac{d}{dz}\mathscr{Z}\{(nT)^{k-1}f(nT)\} \quad k > 0 \text{ and is an integer}$$

10. Complex conjugate signals

$$\mathscr{Z}\{f(nT)\} = F(z) \qquad |z| > R$$

$$\mathscr{Z}\{f^{*}(nT)\} = F^{*}(z^{*}) \qquad |z| > R$$

(continued)

TABLE 8.A.1 (continued) z-Transform Properties for Positive-Time Sequences

11. Transform of product

$$\mathcal{Z}\{f(nT)\} = F(z) \quad |z| > R_f$$
$$\mathcal{Z}\{h(nT)\} = H(z) \quad |z| > R_h$$

$$\mathcal{Z}\{f(nT)h(nT)\} = \frac{1}{2\pi j} \oint_C F(\tau)H\left(\frac{z}{\tau}\right)\frac{d\tau}{\tau}, \quad |\mathcal{Z}| > R_f R_h, R_f < |\tau| < \frac{|\mathcal{Z}|}{R_h}$$

counterclockwise integration

12. Parseval's theorem

$$\mathcal{Z}\{f(nT)\} = F(z) \quad |z| > R_f$$
$$\mathcal{Z}\{h(nT)\} = H(\mathcal{Z}) \quad |z| > R_h$$

$$\sum_{n=0}^{\infty} f(nT)h(nT) = \frac{1}{2\pi j} \oint_C F(z)H(z^{-1})\frac{dz}{z} \quad |z| = 1 > R_f R_h$$

counterclockwise integration

13. Correlation

$$f(nT) \otimes h(nT) = \sum_{m=0}^{\infty} f(mT)h(mT - nT) = \frac{1}{2\pi j} \oint_C F(\tau)H\left(\frac{1}{\tau}\right)\tau^{n-1}d\tau \quad n \geq 1$$

Both $f(nT)$ and $h(nT)$ must exist for $|z| > 1$. The integration is taken in counterclockwise direction.

14. Transform with parameters

$$\mathcal{Z}\left\{\frac{\partial}{\partial a}f(nT, a)\right\} = \frac{\partial}{\partial a}F(z, a)$$

$$\mathcal{Z}\{\lim_{a \to a_0} f(nT, a)\} = \lim_{a \to a_0} F(z, a)$$

$$\mathcal{Z}\left\{\int_{a_0}^{a_1} f(nT, a)da\right\} = \int_{a_0}^{a_1} F(z, a)da \text{ finite interval}$$

TABLE 8.A.2 z-Transform Properties for Positive- and Negative-Time Sequences

1. Linearity

$$\mathcal{Z}_{II}\left\{\sum_{i=0}^{\ell} c_i f_i(nT)\right\} = \sum_{i=0}^{\ell} c_i F_i(z) \quad \max R_{i+} < |z| < \min R_{i-}$$

2. Shifting property

$$\mathcal{Z}_{II}\{f(nT \pm kT)\} = z^{\pm k} F(z) \quad R_+ < |z| < R_-$$

3. Scaling

$$\mathcal{Z}_{II}\{f(nT)\} = F(z) \quad R_+ < |z| < R_-$$
$$\mathcal{Z}_{II}\{a^{nT} f(nT)\} = F(a^{-T}z) \quad |a^T|R_+ < |z| < |a^T|R_-$$

4. Time reversal

$$\mathcal{Z}_{II}\{f(nT)\} = F(z) \quad R_+ < |z| < R_-$$
$$\mathcal{Z}_{II}\{f(-nT)\} = F(z^{-1}) \quad \frac{1}{R_-} < |z| < \frac{1}{R_+}$$

TABLE 8.A.2 (continued) z-Transform Properties for Positive- and Negative-Time Sequences

---

5. Multiplication by nT

$$\mathcal{Z}_{II}\{f(nT)\} = F(z) \qquad R_+ < |z| < R_-$$

$$\mathcal{Z}_{II}\{nTf(nT)\} = -zT\frac{dF(z)}{dz} \qquad R_+ < |z| < R_-$$

6. Convolution

$$\mathcal{Z}_{II}\{f_1(nT) * f_2(nT)\} = F_1(z)F_2(z)$$

$$\text{ROC } F_1(z) \cup \text{ROC } F_2(z) \qquad \max(R_{+f_1}, R_{+f_2}) < |z| < \min(R_{-f_1}, R_{-f_2})$$

7. Correlation

$$R_{f_1 f_2}(z) = \mathcal{Z}_{II}\{f_1(nT) \otimes f_2(nT)\} = F_1(z)F_2(z^{-1})$$

$$\text{ROC } F_1(z) \cup \text{ROC } F_2(z^{-1}) \qquad \max(R_{+f_1}, R_{+f_2}) < |z| < \min(R_{-f_1}, R_{-f_2})$$

8. Multiplication by e^{-anT}

$$\mathcal{Z}_{II}\{f(nT)\} = F(z) \qquad R_+ < |z| < R_-$$

$$\mathcal{Z}_{II}\{e^{-anT}f(nT)\} = F(e^{aT}Z) \qquad |e^{-aT}|R_+ < |z| < |e^{-aT}|R_-$$

9. Frequency translation

$$G(\omega) = \mathcal{Z}_{II}\{e^{j\omega_0 nT}f(nT)\} = G(z)|_{z=e^{j\omega T}} = F(e^{j(\omega-\omega_0)T}) = F(\omega - \omega_0)$$

ROC of $F(z)$ must include the unit circle

10. Product

$$\mathcal{Z}_{II}\{f(nT)\} = F(z) \qquad R_{+f} < |z| < R_{-f}$$

$$\mathcal{Z}_{II}\{h(nT)\} = H(z) \qquad R_{+h} < |z| < R_{-h}$$

$$\mathcal{Z}_{II}\{f(nT)h(nT)\} = G(z) = \frac{1}{2\pi j}\oint_C F(\tau)H\left(\frac{z}{\tau}\right)\frac{d\tau}{\tau}, \qquad R_{+f}R_{+h} < |z| < R_{-f}R_{-h}$$

$$\max\left(R_{+f}, \frac{|z|}{R_{-h}}\right) < |\tau| < \min\left(R_{-f}, \frac{|z|}{R_{+h}}\right)$$

counterclockwise integration

11. Parseval's theorem

$$\mathcal{Z}_{II}\{f(nT)\} = F(z) \qquad R_{+f} < |z| < R_{-f}$$

$$\mathcal{Z}_{II}\{h(nT)\} = H(z) \qquad R_{+h} < |z| < R_{-h}$$

$$\sum_{n=-\infty}^{\infty} f(nT)h(nT) = \frac{1}{2\pi j}\oint_C F(z)H(z^{-1})\frac{dz}{z} \qquad R_{+f}R_{+h} < |z| = 1 < R_{-f}R_{-h}$$

$$\max\left(R_{+f}, \frac{1}{R_{-h}}\right) < |z| < \min\left(R_{-f}, \frac{1}{R_{+h}}\right)$$

counterclockwise integration

12. Complex conjugate signals

$$\mathcal{Z}_{II}\{f(nT)\} = F(z) \qquad R_{+f} < |z| < R_{-f}$$

$$\mathcal{Z}_{II}\{f^*(nT)\} = F^*(z^*) \qquad R_{+f} < |z| < R_{-f}$$

TABLE 8.A.3 Inverse Transforms of the Partial Fractions of $F(z)$

Partial Fraction Term	Inverse Transform Term in $F(z)$ Converges Absolutely for Some $|z| > |a|$
$\dfrac{z}{z-a}$	$a^k, k \geq 0$
$\dfrac{z^2}{(z-a)^2}$	$(k+1)a^k, k \geq 0$
$\dfrac{z^3}{(z-a)^3}$	$\dfrac{1}{2}(k+1)(k+2)a^k, k \geq 0$
$\vdots$	$\vdots$
$\dfrac{z^n}{(z-a)^n}$	$\dfrac{1}{(n-1)!}(k+1)(k+2)\cdots(k+n-1)a^k, k \geq 0$
	Inverse Transform Term in $F(z)$ Converges Absolutely for Some $|z| < |a|$
$\dfrac{z}{z-a}$	$-a^k, k \leq -1$
$\dfrac{z^2}{(z-a)^2}$	$-(k+1)a^k, k \leq -1$
$\dfrac{z^3}{(z-a)^3}$	$-\dfrac{1}{2}(k+1)(k+2)a^k, k \leq -1$
$\vdots$	$\vdots$
$\dfrac{z^n}{(z-a)^n}$	$\dfrac{1}{(n-1)!}(k+1)(k+2)\cdots(k+n-1)a^k, k \leq -1$

TABLE 8.A.4 Inverse Transforms of the Partial Fractions of $F_i(z)$[a]

Elementary Transforms Term $F_i(z)$	Corresponding Time Sequence	
	(I) $F_i(z)$ Converges for $|z| > R_c$	(II) $F_i(z)$ Converges for $|z| < R_c$
1. $\dfrac{1}{z-a}$	$a^{k-1}\big|_{k \geq 1}$	$-a^{k-1}\big|_{k \leq 0}$
2. $\dfrac{z}{(z-a)^2}$	$ka^{k-1}\big|_{k \geq 1}$	$-ka^{k-1}\big|_{k \leq 0}$
3. $\dfrac{z(z+a)}{(z-a)^3}$	$k^2 a^{k-1}\big|_{k \geq 1}$	$-k^2 a^{k-1}\big|_{k \leq 0}$
4. $\dfrac{z(z^2 + 4az + a^2)}{(z-a)^4}$	$k^3 a^{k-1}\big|_{k \geq 1}$	$-k^3 a^{k-1}\big|_{k \leq 0}$

[a] The function must be a proper function.

TABLE 8.A.5 *z*-Transform Pairs

Number	Discrete-Time Function $f(n)$, $n \geq 0$	$\mathcal{F}(z) = \mathcal{Z}[f(n)] = \sum_{n=0}^{\infty} f(n)z^{-n}$ $\|z\| > R$
		z-Transform
1.	$u(n) = \begin{cases} 1, & \text{for } n \geq 0 \\ 0, & \text{otherwise} \end{cases}$	$\dfrac{z}{z-1}$
2.	$e^{-\alpha n}$	$\dfrac{z}{z - e^{-\alpha}}$
3.	N	$\dfrac{z}{(z-1)^2}$
4.	n^2	$\dfrac{z(z+1)}{(z-1)^3}$
5.	n^3	$\dfrac{z(z^2 + 4z + 1)}{(z-1)^4}$
6.	n^4	$\dfrac{z(z^3 + 11z^2 + 11z + 1)}{(z-1)^5}$
7.	n^5	$\dfrac{z(z^4 + 26z^3 + 66z^2 + 26z + 1)}{(z-1)^6}$
8.	n^k	$(-1)^k D^k \left(\dfrac{z}{z-1} \right); D = z\dfrac{d}{dz}$
9.	$u(n-k)$	$\dfrac{z^{-k+1}}{z-1}$
10.	$e^{-\alpha n} f(n)$	$\mathcal{F}(e^\alpha z)$
11.	$n^{(2)} = n(n-1)$	$2\dfrac{z}{(z-1)^3}$
12.	$n^{(3)} = n(n-1)(n-2)$	$3!\dfrac{z}{(z-1)^4}$
13.	$n^{(k)} = n(n-1)(n-2)\cdots(n-k+1)$	$k!\dfrac{z}{(z-1)^{k+1}}$
14.	$n^{[k]} f(n), n^{[k]} = n(n+1)(n+2)\cdots(n+k-1)$	$(-1)^k z^k \dfrac{d^k}{dz^k}[\mathcal{F}(z)]$
15.	$(-1)^k n(n-1)(n-2)\cdots(n-k+1) f_{n-k+1}{}^a$	$zF^{(k)}(z), F^{(k)}(z) = \dfrac{d^k}{dz^k}F(z)$
16.	$-(n-1) f_{n-1}$	$\mathcal{F}^{(1)}(z)$
17.	$(-1)^k (n-1)(n-2)\cdots(n-k) f_{n-k}$	$\mathcal{F}^{(k)}(z)$
18.	$nf(n)$	$-z\mathcal{F}^{(1)}(z)$
19.	$n^2 f(n)$	$z^2 \mathcal{F}^{(2)}(z) + z\mathcal{F}^{(1)}(z)$
20.	$n^3 f(n)$	$-z^3 \mathcal{F}^{(3)}(z) - 3z^2 \mathcal{F}^{(2)}(z) - z\mathcal{F}^{(1)}(z)$
21.	$\dfrac{c^n}{n!}$	$e^{c/z}$
22.	$\dfrac{(\ln c)^n}{n!}$	$c^{1/z}$
23.	$\begin{pmatrix} k \\ n \end{pmatrix} c^n a^{k-n}, \begin{pmatrix} k \\ n \end{pmatrix} = \dfrac{k!}{(k-n)!n!}, n \leq k$	$\dfrac{(az+c)^k}{z^k}$

(continued)

TABLE 8.A.5 (continued)   z-Transform Pairs

Number	Discrete-Time Function $f(n)$, $n \geq 0$	z-Transform $\mathcal{F}(z) = \mathfrak{Z}[f(n)] = \sum_{n=0}^{\infty} f(n)z^{-n}\ |z| > R$
24.	$\binom{n+k}{k} c^n$	$\dfrac{z^{k+1}}{(z-c)^{k+1}}$
25.	$\dfrac{c^n}{n!}, (n = 1, 3, 5, 7, \ldots)$	$\sinh\left(\dfrac{c}{z}\right)$
26.	$\dfrac{c^n}{n!}, (n = 0, 2, 4, 6, \ldots)$	$\cosh\left(\dfrac{c}{z}\right)$
27.	$\sin(\alpha n)$	$\dfrac{z \sin\alpha}{z^2 - 2z\cos\alpha + 1}$
28.	$\cos(\alpha n)$	$\dfrac{z(z - \cos\alpha)}{z^2 - 2z\cos\alpha + 1}$
29.	$\sin(\alpha n + \psi)$	$\dfrac{z^2 \sin\psi + z\sin(\alpha - \psi)}{z^2 - 2z\cos\alpha + 1}$
30.	$\cosh(\alpha n)$	$\dfrac{z(z - \cosh\alpha)}{z^2 - 2z\cosh\alpha + 1}$
31.	$\sinh(\alpha n)$	$\dfrac{z \sinh\alpha}{z^2 - 2z\cosh\alpha + 1}$
32.	$\dfrac{1}{n}, n > 0$	$\ln\dfrac{z}{z-1}$
33.	$\dfrac{1 - e^{-\alpha n}}{n}$	$\alpha + \ln\dfrac{z - e^{-\alpha}}{z-1}, \alpha > 0$
34.	$\dfrac{\sin\alpha n}{n}$	$\alpha + \tan^{-1}\dfrac{\sin\alpha}{z - \cos\alpha}, \alpha > 0$
35.	$\dfrac{\cos\alpha n}{n}, n > 0$	$\ln\dfrac{z}{\sqrt{z^2 - 2z\cos\alpha + 1}}$
36.	$\dfrac{(n+1)(n+2)\ldots(n+k-1)}{(k-1)!}$	$\left(1 - \dfrac{1}{z}\right)^{-k}, k = 2, 3, \ldots$
37.	$\sum_{m=1}^{n} \dfrac{1}{m}$	$\dfrac{z}{z-1}\ln\dfrac{z}{z-1}$
38	$\sum_{m=0}^{n-1} \dfrac{1}{m!}$	$\dfrac{e^{1/z}}{z-1}$
39.	$\dfrac{(-1)^{(n-p)/2}}{2^n(n-p/2)!(n+p/2)!}$, for $n \geq p$ and $n - p = $ even $= 0$, for $n < p$ or $n - p = $ odd	$J_p(z^{-1})$
40.	$\left\{ \begin{array}{l} \binom{\alpha}{n/k} b^{n/k}, \quad n = mk, \quad (m = 0, 1, 2, \ldots) \\ = 0 \qquad\qquad n \neq mk \end{array} \right\}$	$\left(\dfrac{z^k + b}{z^k}\right)^{\alpha}$
41.	$a^n P_n(x) = \dfrac{a^n}{2^n n}\left(\dfrac{d}{dx}\right)^n (x^2 - 1)^n$	$\dfrac{z}{\sqrt{z^2 - 2xaz + a^2}}$
42.	$a^n T_n(x) = a^n \cos(n \cos^{-1} x)$	$\dfrac{z(z - ax)}{z^2 - 2xaz + a^2}$

TABLE 8.A.5 (continued)   z-Transform Pairs

Number	Discrete-Time Function $f(n)$, $n \geq 0$	z-Transform $\mathcal{F}(z) = \mathcal{Z}[f(n)] = \sum_{n=0}^{\infty} f(n)z^{-n} \; |z| > R$
43.	$\dfrac{L_n(x)}{n!} = \sum_{r=0}^{\infty} \binom{n}{r}\dfrac{(-x)^r}{r}$	$\dfrac{z}{z-1}e^{-x/(z-1)}$
44.	$\dfrac{H_n(x)}{n!} = \sum_{k=0}^{[n/2]} \dfrac{(-1)^{n-k}x^{n-2k}}{k(n-2k)!2^k}$	$e^{-x/z-1/2z^2}$
45.	$a^n P_n^m(x) = a^n(1-x^2)^{m/2}\left(\dfrac{d}{dx}\right)^m P_n(x),$ $m = $ integer	$\dfrac{(2m)!}{2^m m!}\dfrac{z^{m+1}(1-x^2)^{m/2}a^m}{(z^2 - 2xaz + a^2)^{m+1/2}}$
46.	$\dfrac{L_n^m(x)}{n!} = \left(\dfrac{d}{dx}\right)^m \dfrac{L_n(x)}{n!},\, m = $ integer	$\dfrac{(-1)^m z}{(z-1)^{m+1}}e^{-x/(z-1)}$
47.	$-\dfrac{1}{n}Z^{-1}\left[z\dfrac{\mathcal{F}'(z))}{\mathcal{F}(z)} - \dfrac{\mathcal{G}'(z))}{\mathcal{G}(z)}\right],$ where $\mathcal{F}(z)$ and $\mathcal{G}(z)$ are rational polynomials in z of the same order	$\ln\dfrac{\mathcal{F}(z)}{\mathcal{G}(z)}$
48.	$\dfrac{1}{m(m+1)(m+2)\cdots(m+n)}$	$(m-1)!z^m\left[e^{1/z} - \sum_{k=0}^{m-1}\dfrac{1}{k!z^k}\right]$
49.	$\dfrac{\sin(\alpha n)}{n!}$	$e^{\cos\alpha/z}\cdot\sin\left(\dfrac{\sin\alpha}{z}\right)$
50.	$\dfrac{\cos(\alpha n)}{n!}$	$e^{\cos\alpha/z}\cdot\cos\left(\dfrac{\sin\alpha}{z}\right)$
51.	$\sum_{k=0}^{n} f_k g_{n-k}$	$\mathcal{F}(z)\mathcal{G}(z)$
52.	$\sum_{k=0}^{n} k f_k g_{n-k}$	$-\mathcal{F}^{(1)}(z)\mathcal{G}(z),\, \mathcal{F}^{(1)}(z) = \dfrac{d\mathcal{F}(z)}{dz}$
53.	$\sum_{k=0}^{n} k^2 f_k g_{n-k}$	$\mathcal{F}^{(2)}(z)\mathcal{G}(z)$
54.	$\dfrac{\alpha^n + (-\alpha)^n}{2\alpha^2}$	$\dfrac{1}{\alpha^2}\dfrac{z^2}{z^2 - \alpha^2}$
55.	$\dfrac{\alpha^n - \beta^n}{\alpha - \beta}$	$\dfrac{z}{(z-\alpha)(z-\beta)}$
56.	$(n+k)^{(k)}$	$kz^k\dfrac{z}{(z-1)^{k+1}}$
57.	$(n-k)^{(k)}$	$kz^{-k}\dfrac{z}{(z-1)^{k+1}}$
58.	$\dfrac{(n \mp k)^{(m)}}{m!}e^{\alpha(n-k)}$	$\dfrac{z^{1\mp k}e^{m\alpha}}{(z-e^\alpha)^{m+1}}$
59.	$\dfrac{1}{n}\sin\dfrac{\pi}{2}n$	$\dfrac{\pi}{2} + \tan^{-1}\dfrac{1}{z}$
60.	$\dfrac{\cos\alpha(2n-1)}{2n-1},\quad n > 0$	$\dfrac{1}{4\sqrt{z}}\ln\dfrac{z + 2\sqrt{z}\cos\alpha + 1}{z - 2\sqrt{z}\cos\alpha + 1}$
61.	$\dfrac{\gamma^n}{(\gamma-1)^2} + \dfrac{n}{1-\gamma} - \dfrac{1}{(1-\gamma)^2}$	$\dfrac{z}{(z-\gamma)(z-1)^2}$

(continued)

TABLE 8.A.5 (continued) z-Transform Pairs

Number	Discrete-Time Function $f(n)$, $n \geq 0$	z-Transform $\mathcal{F}(z) = \mathfrak{Z}[f(n)] = \sum_{n=0}^{\infty} f(n)z^{-n}$ $|z| > R$
62.	$\dfrac{\gamma + a_0}{(\gamma - 1)^2}\gamma^n + \dfrac{1 + a_0}{1 - \gamma}n + \left(\dfrac{1}{1 - \gamma} - \dfrac{a_0 + 1}{(1 - \gamma)^2}\right)$	$\dfrac{z(z + a_0)}{(z - \gamma)(z - 1)^2}$
63.	$a^n \cos \pi n$	$\dfrac{z}{z + a}$
64.	$e^{-\alpha n} \cos an$	$\dfrac{z(z - e^{-\alpha} \cos a)}{z^2 - 2ze^{-\alpha} \cos a + e^{-2\alpha}}$
65.	$e^{-\alpha n} \sinh (an + \psi)$	$\dfrac{z^2 \sinh \psi + ze^{-\alpha} \sinh (a - \psi)}{z^2 - 2ze^{-\alpha} \cosh a + e^{-2\alpha}}$
66.	$\dfrac{\gamma^n}{(\gamma - \alpha)^2 + \beta^2} + \dfrac{(\alpha^2 + \beta^2)^{n/2} \sin (n\theta + \psi)}{\beta[(\alpha - \gamma)^2 + \beta^2]^{1/2}}$ $\theta = \tan^{-1}\dfrac{\beta}{\alpha}$ $\psi = \tan^{-1}\dfrac{\beta}{\alpha - \gamma}$	$\dfrac{z}{(z - \gamma)[(z - \alpha)^2 + \beta^2]}$
67.	$\dfrac{n\gamma^{n-1}}{(\gamma - 1)^3} - \dfrac{3\gamma^n}{(\gamma - 1)^4}$ $+ \dfrac{1}{2}\left[\dfrac{n(n - 1)}{(1 - \gamma)^2} - \dfrac{4n}{(1 - \gamma)^3} + \dfrac{6}{(1 - \gamma)^4}\right]$	$\dfrac{z}{(z - \gamma)^2(z - 1)^3}$
68.	$\sum_{v=0}^{k} (-1)^v \binom{k}{v} \dfrac{(n + k - v)^{(k)}}{k} e^{\alpha(n-v)}$	$\dfrac{z(z - 1)^k}{(z - e^\alpha)^{k+1}}$
69.	$\dfrac{f(n)}{n}$	$\int_z^\infty p^{-1} \mathcal{F}(p)dp + \lim_{n \to 0} \dfrac{f(n)}{n}$
70.	$\dfrac{f_{n+2}}{n + 1}$, $\begin{aligned}f_0 &= 0 \\ f_1 &= 0\end{aligned}$	$z\int_z^\infty \mathcal{F}(p)dp$
71.	$\dfrac{1 + a_0}{(1 - \gamma)[(1 - \alpha)^2 + \beta^2]}$ $+ \dfrac{(\gamma + a_0)\gamma^n}{(\gamma - 1)[(\gamma - \alpha)^2 + \beta^2]}$ $+ \dfrac{[\alpha^2 + \beta^2]^{n/2}[(a_0 + \alpha)^2 + \beta^2]^{1/2}}{\beta[(\alpha - 1)^2 + \beta^2]^{1/2}[(\alpha - \gamma)^2 + \beta^2]^{1/2}}$, $\times \sin(n\theta + \psi + \lambda)$ $\psi = \psi_1 + \psi_2$, $\psi_1 = -\tan^{-1}\dfrac{\beta}{\alpha - 1}$, $\theta = \tan^{-1}\dfrac{\beta}{\alpha}$ $\lambda = \tan^{-1}\dfrac{\beta}{a_0 + \alpha}$, $\psi_2 = -\tan^{-1}\dfrac{\beta}{\alpha - \gamma}$	$\dfrac{z(z + a_0)}{(z - 1)(z - \gamma)[(z - \alpha)^2 + \beta^2]}$
72.	$(n + 1)e^{\alpha n} - 2ne^{\alpha(n + 1)} + e^{\alpha(n-2)}(n - 1)$	$\left(\dfrac{z - 1}{z - e^\alpha}\right)^2$

TABLE 8.A.5 (continued) z-Transform Pairs

Number	Discrete-Time Function $f(n)$, $n \geq 0$	z-Transform $\mathcal{F}(z) = \mathfrak{Z}[f(n)] = \sum_{n=0}^{\infty} f(n)z^{-n} \ |z| > R$
73.	$(-1)^n \dfrac{\cos \alpha n}{n}, \ n > 0$	$\ln \dfrac{z}{\sqrt{z^2 + 2z \cos \alpha + 1}}$
74.	$\dfrac{(n+k)!}{n!} f_{n+k}, \ f_n = 0, \ \text{ for } 0 \leq n < k$	$(-1)^k z^{2k} \dfrac{d^k}{dz^k} [F((z)]$
75.	$\dfrac{f(n)}{n+h}, \ h > 0$	$z^h \int_z^{\infty} p^{-(1+h)} \mathcal{F}(p)\, dp$
76.	$-na^n \cos \dfrac{\pi}{2} n$	$\dfrac{2a^2 z^2}{(z^2 + a^2)^2}$
77.	$na^n \dfrac{1 + \cos \pi n}{2}$	$\dfrac{2a^2 z^2}{(z^2 - a^2)^2}$
78.	$a^n \sin \dfrac{\pi}{4} n \cdot \dfrac{1 + \cos \pi n}{2}$	$\dfrac{a^2 z^2}{z^4 + a^4}$
79.	$a^n \left(\dfrac{1 + \cos \pi n}{2} - \cos \dfrac{\pi}{2} n \right)$	$\dfrac{2a^2 z^2}{z^4 - a^4}$
80.	$\dfrac{P_n(x)}{n!}$	$e^{xz^{-1}} J_0(\sqrt{1 - x^2} z^{-1})$
81.	$\dfrac{P_n^{(m)}(x)}{(n+m)!}, m > 0, P_n^m = 0, \ \text{ for } n < m$	$(-1)^m e^{xz^{-1}} J_m(\sqrt{1 - x^2} z^{-1})$
82.	$\dfrac{1}{(n + \alpha)^{\beta}}, \ \alpha > 0, \ \text{Re } \beta > 0$	$\Phi(z^{-1}, \alpha, \beta), \ \text{where}$ $\Phi(1, \beta, \alpha) = \zeta(\beta, \alpha)$ $= \text{generalized Riemann zeta}$ function
83.	$a^n \left(\dfrac{1 + \cos \pi n}{2} + \cos \dfrac{\pi}{2} n \right)$	$\dfrac{2z^4}{z^4 - a^4}$
84.	$\dfrac{c^n}{n}, \ (n = 1, 2, 3, 4, \dots)$	$\ln z - \ln(z - c)$
85.	$\dfrac{c^n}{n}, \ n = 2, 4, 6, 8, \dots$	$\ln z - \frac{1}{2}\ln(z^2 - c^2)$
86.	$n^2 c^n$	$\dfrac{cz(z + c)}{(z - c)^3}$
87.	$n^3 c^n$	$\dfrac{cz(z^2 + 4cz + c^2)}{(z - c)^4}$
88.	$n^k c^n$	$-\dfrac{d\mathcal{F}(z/c)}{dz}, \ \mathcal{F}(\mathcal{Z}) = \mathcal{Z}[n^{k-1}]$
89.	$-\cos \dfrac{\pi}{2} n \sum_{i=0}^{(n-2)/4} \dbinom{n/2}{2i + 1} a^{n-2-4i}(a^4 - b^4)^i$	$\dfrac{z^2}{z^4 + 2a^2 z^2 + b^4}$
90.	$n^k f(n), k > 0 \text{ and integer}$	$-z \dfrac{d}{dz} \mathcal{F}_1(z), \ \mathcal{F}_1(\mathfrak{Z}) = \mathfrak{Z}[n^{k-1} f(n)]$

(continued)

TABLE 8.A.5 (continued)   z-Transform Pairs

Number	Discrete-Time Function $f(n)$, $n \geq 0$	z-Transform $\mathcal{F}(z) = \mathfrak{Z}[f(n)] = \sum_{n=0}^{\infty} f(n) z^{-n}$ $\lvert z \rvert > R$
91.	$\dfrac{(n-1)(n-2)(n-3)\ldots(n-k+1)}{(k-1)!}\, a^{n-k}$	$\dfrac{1}{(z-a)^k}$
92.	$\dfrac{k(k-1)(k-2)\ldots(k-n+1)}{n!}$	$\left(1+\dfrac{1}{z}\right)^k$
93.	$na^n \cos bn$	$\dfrac{[(z/a)^3 + z/a]\cos b - 2(z/a)^2}{[(z/a)^2 - 2(z/a)\cos b + 1]^2}$
94.	$na^n \sin bn$	$\dfrac{(z/a)^3 \sin b - (z/a)\sin b}{[(z/a)^2 - 2(z/a)\cos b + 1]^2}$
95.	$\dfrac{na^n}{(n+1)(n+2)}$	$\dfrac{z(a-2z)}{a^2}\ln\left(1-\dfrac{a}{z}\right) - \dfrac{2}{a}z$
96.	$\dfrac{(-a)^n}{(n+1)(2n+1)}$	$2\sqrt{z/a}\tan^{-1}\sqrt{a/z} - \dfrac{z}{a}\ln\left(1+\dfrac{a}{z}\right)$
97.	$\dfrac{a^n \sin \alpha n}{n+1}$	$\dfrac{z\cos\alpha}{a}\tan^{-1}\dfrac{a\sin\alpha}{z - a\cos\alpha}$ $+ \dfrac{z\sin\alpha}{2a}\ln\dfrac{z^2 - 2az\cos\alpha + a^2}{z^2}$
98.	$\dfrac{a^n \cos(\pi/2)n \sin \alpha(n+1)}{n+1}$	$\dfrac{z}{4a}\ln\dfrac{z^2 + 2az\sin\alpha + a^2}{z^2 - 2az\sin\alpha + a^2}$
99.	$\dfrac{1}{(2n)!}$	$\cosh(z^{-1/2})$
100.	$\dbinom{-1/2}{n}(-a)^n$	$\sqrt{z/(z-a)}$
101.	$\dbinom{-1/2}{n/2}a^n \cos\frac{\pi}{2}n$	$\dfrac{z}{\sqrt{z^2 - a^2}}$
102.	$\dfrac{B_n(x)}{n!}$ $B_n(x)$ are Bernoulli polynomials	$\dfrac{e^{x/z}}{z(e^{1/z} - 1)}$
103.	$W_n(x) \doteq$ Tchebycheff polynomials of the second kind	$\dfrac{z^2}{z^2 - 2xz + 1}$
104.	$\left\lvert \sin\dfrac{n\pi}{m}\right\rvert$, $m = 1, 2, \ldots$	$\dfrac{z\sin\pi/m}{z^2 - 2z\cos\pi/m + 1}\dfrac{1 + z^{-m}}{1 - z^{-m}}$
105.	$Q_n(x) = \sin(n\cos^{-1} x)$	$\dfrac{z}{z^2 - 2xz + 1}$

Source: Jury, E.I., *Theory and Application of the z-Transform Method*, John Wiley & Sons, Inc., New York, 1964.

ᵃ It may be noted that f_n is the same as $f(n)$.

Hilbert Transforms

9.1 Definition

Hilbert transforms play a useful role in signal and network theory. They are of practical importance in various signal processing systems such as band-pass systems, minimum-phase networks, and spectral analysis. Narrow-band signals contain a small band of frequencies around a center or a carrier frequency. For example, an AM transmission of voice signals usually involves a carrier frequency of several megahertz with bandwidths of 3000 Hz. The function

$$x(t) = A \, \sin \omega_0 t p_T \left(t - \frac{T}{2} \right) \quad T \gg \frac{2\pi}{\omega_0} \tag{9.1}$$

approximately satisfies the narrow-band condition. The function $p_T \left(t - \frac{T}{2} \right)$ indicates a pulse of width T and its center is at $T/2$. Given a real function $x(t)$ in $-\infty < t < \infty$, its Hilbert transform is given by

$$\hat{x}(t) = \mathcal{H}\{x(t)\} = \frac{1}{\pi} P \int_{-\infty}^{\infty} \frac{x(\tau)}{t - \tau} d\tau = x(t) * h(t) \quad h(t) = \frac{1}{\pi t} \tag{9.2}$$

The letter P in front of the integral sign indicates the principal value of the integral. The inverse transform is defined as follows:

$$x(t) = \mathcal{H}^{-1}\{\hat{x}(t)\} = -\frac{1}{\pi} P \int_{-\infty}^{\infty} \frac{\hat{x}(\tau)}{t - \tau} d\tau \tag{9.3}$$

Note that since $\mathcal{F}\{1/\pi t\} = -j \, \text{sgn} \, \omega$, the Fourier transform (FT) of $\hat{x}(t)$ is given by

$$\hat{X}(\omega) = \mathcal{F}\{x(t) * h(t)\} = -jX(\omega)\text{sgn} \, \omega \tag{9.4}$$

This equation shows that we can obtain the Hilbert transform of a function by passing it through an ideal 90° phase shifter. Figure 9.1 shows the amplitude and phase

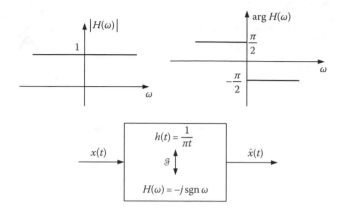

FIGURE 9.1

characteristics of such a system, which is known as a **quadratic filter**. The name comes from its ability to produce a sine signal when its input is a cosine signal.

9.2 Hilbert Transforms, Properties, and the Analytic Signal

Fourier transform

$$\hat{X}(\omega) = \mathcal{F}\{\hat{x}(t)\} = \mathcal{F}\{x(t) * h(t)\} = -jX(\omega)\,\text{sgn}\,\omega \tag{9.5}$$

Transform of cosine signals

$$\mathcal{H}\{\cos(\omega t + \varphi)\} = \sin(\omega t + \varphi) \tag{9.6}$$

Proof

$$\mathcal{H}\{\cos(\omega t + \varphi)\} = \frac{1}{\pi} P \int_{-\infty}^{\infty} \frac{\cos(\omega(t-\tau) + \varphi)}{\tau}\,d\tau = \frac{1}{\pi} P \int_{-\infty}^{\infty} \frac{\cos(\omega t + \varphi)\cos \omega \tau}{\tau}\,d\tau$$

$$+ \frac{1}{\pi} P \int_{-\infty}^{\infty} \frac{\sin(\omega t + \varphi)\sin \omega \tau}{\tau}\,d\tau$$

Since the integrand of the first integral is odd, the integral yields zero. We thus have

$$\mathcal{H}\{\cos(\omega t + \varphi)\} = \frac{\sin(\omega t + \varphi)}{\pi} P \int_{-\infty}^{\infty} \frac{\sin \omega \tau}{\tau}\,d\tau = \sin(\omega t + \varphi)$$

since the integral is equal to π (see Appendix C).

Transform of a transform

$$\mathcal{H}\{\hat{x}(t)\} = \mathcal{H}\{\mathcal{H}\{x(t)\}\} = -x(t) \tag{9.7}$$

Proof

$$\mathcal{H}\{\hat{x}(t)\} = \frac{1}{\pi}P\int_{-\infty}^{\infty}\frac{\hat{x}(\tau)}{\tau}d\tau = -\left\{-\frac{1}{\pi}P\int_{-\infty}^{\infty}\frac{\hat{x}(\tau)}{t-\tau}d\tau\right\} = -x(t)$$

by (9.3). Additionally, owing to (9.7), we also obtain

$$\mathcal{H}\{\sin(\omega t + \varphi)\} = \mathcal{H}\{\mathcal{H}\{\cos(\omega t + \varphi)\}\} = -\cos(\omega t + \varphi) \tag{9.8}$$

Transform of convolution

$$\hat{g}(t) = \mathcal{H}\{g(t)\} = \mathcal{H}\{x(t) * h(t)\} = x(t) * \hat{h}(t) = \hat{x}(t) * h(t) \tag{9.9}$$

Proof

$$\hat{g}(t) = \frac{1}{\pi}P\int_{-\infty}^{\infty}\frac{g(\tau)}{t-\tau}d\tau = \frac{1}{\pi}P\int_{-\infty}^{\infty}P\int_{-\infty}^{\infty}\frac{x(\xi)h(\tau-\xi)}{t-\tau}d\tau d\xi = P\int_{-\infty}^{\infty}x(\xi)d\xi P\int_{-\infty}^{\infty}\frac{1}{\pi}\frac{h(\tau-\xi)}{t-\tau}d\tau$$

$$= P\int_{-\infty}^{\infty}x(\xi)d\xi P\int_{-\infty}^{\infty}\frac{1}{\pi}\frac{h(\eta)}{t-\xi-\eta}d\eta = P\int_{-\infty}^{\infty}x(\xi)\hat{h}(t-\xi)d\xi = x(t) * \hat{h}(t)$$

Similarly, the other equality is proven.

Modulated signals

$$\hat{x}(t) \stackrel{\Delta}{=} \mathcal{H}\{x(t)\} = \mathcal{H}\{a(t)\cos\omega_0 t\} = a(t)\sin\omega_0 t$$

$$\mathcal{F}\{a(t)\} = \begin{cases} A(\omega) & |\omega| < B \\ 0 & \text{otherwise} \end{cases} \tag{9.10}$$

Proof: From (9.4), we find

$$\hat{X}(\omega) \stackrel{\Delta}{=} \mathcal{F}\{\hat{x}(t)\} = -j(\text{sgn }\omega)X(\omega) = -j(\text{sgn }\omega)\mathcal{F}\{a(t)\cos\omega_0 t\}$$

$$= -j\,\text{sgn }\omega_0\left[\frac{A(\omega - \omega_0)}{2} + \frac{A(\omega + \omega_0)}{2}\right]$$

$$= \begin{cases} -\dfrac{j}{2}A(\omega - \omega_0) & \omega > 0 \\ -\dfrac{j}{2}A(\omega + \omega_0) & \omega < 0 \end{cases}$$

Figure 9.2 shows some of the functions.

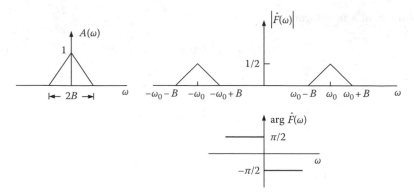

FIGURE 9.2

If we now take the inverse FT of $\hat{X}(\omega)$, we obtain $\hat{x}(t)$. Hence, we obtain the following:

$$\hat{x}(t) = \frac{1}{2\pi} \int_{-\infty}^{0} -j\,\mathrm{sgn}\,\omega X(\omega)e^{j\omega t}\,d\omega = \frac{1}{2\pi} \int_{0}^{\infty} -[jA(\omega - \omega_0)/2]e^{j\omega t}\,d\omega$$

$$+ \frac{1}{2\pi} \int_{-\infty}^{0} [jA(\omega + \omega_0)/2]e^{j\omega t}\,d\omega = -\frac{j}{2}\frac{e^{j\omega_0 t}}{2\pi}\int_{-B}^{B} A(\xi)e^{j\xi t}\,d\xi + \frac{j}{2}\frac{e^{-j\omega_0 t}}{2\pi}\int_{-B}^{B} A(\xi)e^{j\xi t}\,d\xi$$

$$= \frac{ja(t)}{2}(e^{-j\omega_0 t} - e^{j\omega_0 t}) = a(t)\sin\omega_0 t$$

By using (9.7), we find that

$$\mathcal{H}\{a(t)\sin\omega_0 t\} = -a(t)\cos\omega_0 t \qquad (9.11)$$

Let us assume that a modulation signal $a_m(t)$ has the spectrum shown in Figure 9.3a. The single-sideband spectrum with suppress carrier (e.g., $a_m(t)\cos\omega_0 t$) is shown in Figure 9.3b. Based on the figure, we can write the modulated signal as follows:

$$g_{SSB}(t) = \frac{1}{2}e^{j\omega_0 t}\int_{0}^{\infty} A_m(u)e^{jut}\,du + \frac{1}{2}e^{-j\omega_0 t}\int_{-\infty}^{0} A_m(v)e^{jvt}\,dv$$

Observe that we can also write

$$\int_{-\infty}^{0} A_m(v)e^{jvt}\,dv = \frac{1}{2}\left[\int_{-\infty}^{\infty} A_m(v)e^{jvt}\,dv + \int_{-\infty}^{\infty} A_m(v)\mathrm{sgn}\,(-v)e^{jvt}\,dv\right]$$

$$\int_{0}^{\infty} A_m(u)e^{jut}\,du = \frac{1}{2}\left[\int_{-\infty}^{\infty} A_m(u)e^{jut}\,du + \int_{-\infty}^{\infty} A_m(u)\mathrm{sgn}\,(u)e^{jut}\,du\right]$$

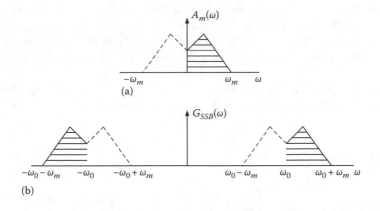

(a)

(b)

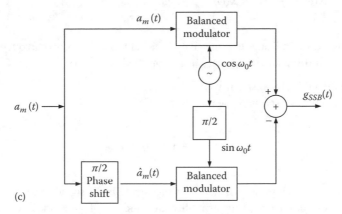

(c)

FIGURE 9.3

Therefore, with the help of (9.4) and the relationship $\text{sgn}(t) = -\text{sign}(-t)$, we find

$$g_{SSB}(t) = \frac{1}{4} e^{-j\omega_0 t} \left[\int_{-\infty}^{\infty} A_m(v) e^{jvt} \, dv + \int_{-\infty}^{\infty} A_m(v) \text{sgn}(-v) e^{jvt} \, dv \right]$$

$$+ \frac{1}{4} e^{j\omega_0 t} \left[\int_{-\infty}^{\infty} A_m(u) e^{jut} \, du + \int_{-\infty}^{\infty} A_m(u) \text{sgn}(u) e^{jut} \, du \right]$$

Since u and v in these equations are dummy variables, we can then write

$$g_{SSB}(t) = \frac{1}{2} \left[\frac{e^{j\omega_0 t} + e^{-j\omega_0 t}}{2} \right] a_m(t) - \frac{1}{2} \left[\frac{e^{j\omega_0 t} - e^{-j\omega_0 t}}{2j} \right] \hat{a}_m(t)$$

$$= \frac{1}{2} [a_m(t) \cos \omega_0 t - \hat{a}_m(t) \sin \omega_0 t]$$

The above shows that the production of a single sideband signal is accomplished by using a quadrature filter that introduces a $\pi/2$ phase shift to every frequency component of the signal as shown in Figure 9.3c. If the two signals at the summation point are subtracted, only the upper sideband is present; if they are added, only the lower sideband is present.

The analytic signal (also pre-envelope)
The **analytic signal** (pre-envelope) is defined by the relation

$$\boxed{\psi(t) = x(t) + j\hat{x}(t)} \tag{9.12}$$

Its FT is

$$\Psi(\omega) = X(\omega) + j\hat{X}(\omega) = X(\omega) + j(-j\,\text{sgn}\,\omega)X(\omega) = \begin{cases} 2X(\omega) & \omega > 0 \\ X(\omega) & \omega = 0 \\ 0 & \omega < 0 \end{cases} \tag{9.13}$$

The above equation indicates that the analytic signal has only positive frequencies. Thus, if we want to find the analytic signal $x(t)$, we proceed as follows:

1. Find the Hilbert transform $\hat{x}(t)$ and use (9.12).
2. Find $X(\omega) \overset{\Delta}{=} \mathscr{F}\{x(t)\}$ and use the inverse FT:

$$\psi(t) = \frac{2}{2\pi} \int\limits_0^\infty X(\omega)e^{j\omega t}\,d\omega \tag{9.14}$$

The analytic signal simplifies the handling of narrow-band signals and systems analogously to the use of phasors, which simplify the handling of sinusoidally varying signals in circuit analysis. To show this, let the analytic signal (complex) be presented by the relation

$$\psi(t) = \tilde{x}(t)e^{j\omega_0 t} \tag{9.15}$$

The complex function $\tilde{x}(t) = \tilde{x}_r(t) + j\tilde{x}_i(t)$ is called the **complex envelope** of the signal. But shifting the spectrum of $x(t)$ at the origin will result in $\Psi(\omega)$, and this implies that the frequency spectrum of $\tilde{x}(t)$ at $\omega = \omega_0$ has a width of 2B centered at $\omega = \omega_0$.

From (9.12), we observe that the signal $x(t)$ is equal to the real part of the analytic signal, hence, by (9.15), we obtain

$$x(t) = \text{Re}\{\psi(t)\} = \text{Re}\{\tilde{x}(t)e^{j\omega_0 t}\} = \tilde{x}_r(t)\cos\omega_0 - \tilde{x}_i(t)\sin\omega_0 \tag{9.16}$$

The function $\tilde{x}_r(t)$ is the **in-phase component** of $x(t)$, and $\tilde{x}_i(t)$ is its **quadrature component**. This equation shows that $x(t)$ is the projection on the real axis of the phasor $\tilde{x}(t)$ lying on the $\tilde{x}_r, \tilde{x}_i$ plane.

If we were to define the complex envelope by

$$\tilde{x}(t) = a(t)e^{j\phi(t)} \tag{9.17}$$

then

$$x(t) = \text{Re}\{\psi(t)\} = a(t)\cos[\omega_0 t + \phi(t)] \tag{9.18}$$

where $a(t)$ is the **envelope** of the signal and $\phi(t)$ is its **phase**.

Suppose we multiply (9.16) by $\cos \omega_0 t$ and then pass the signal through a low-pass filter; the result is $(1/2)\tilde{f}_r(t)$. If we multiply the same equation by $-\sin \omega_0 t$ and use a low-pass filter, we obtain $(1/2)\tilde{f}_i(t)$.

Example 9.1

It is desired to find the complex envelope and the envelope of the narrow-band signal:

$$f(t) = x(t)\cos \omega_0 t - y(t)\sin \omega_0 t \tag{9.19}$$

First, we obtain $\hat{f}(t)$ and then we use (9.12). We write (see (9.10) and (9.11))

$$\hat{f}(t) \stackrel{\Delta}{=} \mathcal{H}\{f(t)\} = x(t)\sin \omega_0 t + y(t)\cos \omega_0 t$$

Then, the pre-envelope is

$$\psi_f(t) = x(t)\cos \omega_0 t - y(t)\sin \omega_0 t + jx(t)\sin \omega_0 t + jy(t)\cos \omega_0 t$$
$$= x(t)(\cos \omega_0 t + j\sin \omega_0 t) + jy(t)(\cos \omega_0 t + j\sin \omega_0 t) = [x(t) + jy(t)]e^{j\omega_0 t}$$

Therefore, the envelope and the complex envelope are, respectively,

$$|\psi_f(t)| = \sqrt{x^2(t) + y^2(t)} \quad \tilde{f}(t) = x(t) + jy(t) \qquad \blacksquare$$

Narrow-band filters

$$\psi_y(t) = h(t) * \psi_x(t) \tag{9.20}$$

Let $h(t)$ be the impulse response of a LTI narrow-band filter. Therefore, if the input is $x(t)$ then the output is $y(t) = x(t) * h(t)$. The analytic output signal is found as follows (see also 9.9):

$$\psi_y(t) = x(t) * h(t) + j\mathcal{H}\{x(t) * h(t)\} = x(t) * h(t) + jh(t) * \hat{x}(t)$$
$$= h(t) * [x(t) + j\hat{x}(t)] = h(t) * \psi_x(t)$$

Example 9.2

To find the spectrum of the analytic function of the function $x(t) = \dfrac{1}{1 + t^2}$ we proceed as follows:

$$\psi(t) = x(t) + j\hat{x}(t) = \frac{1}{1 + t^2} + j\frac{t}{1 + t^2} \Rightarrow \Psi(\omega) = \mathcal{F}\left\{\frac{1}{1 + t^2}\right\} + j\mathcal{F}\left\{\frac{t}{1 + t^2}\right\}$$

But

$$\mathscr{F}\{e^{|t|}\} = \frac{2}{1+\omega^2} \Rightarrow \text{(see Fourier properties) } \mathscr{F}\left\{\frac{2}{1+t^2}\right\} = 2\pi e^{-|\omega|}$$

Also

$$\mathscr{F}\left\{\mathscr{H}\left\{\frac{1}{1+t^2}\right\}\right\} = -j\,\text{sgn}\,(\omega)\mathscr{F}\left\{\frac{1}{1+t^2}\right\} = -j\,\text{sgn}\,(\omega)\pi e^{-|\omega|}$$

Hence,

$$\Psi(\omega) = [1+\text{sgn}\,\omega]\pi e^{-|\omega|} \qquad\qquad (9.21) \quad\blacksquare$$

Hilbert transform and the delta function

From the identity $2u(\omega) = 1 + \text{sgn}\,(\omega)$ and the definition of the relation of the analytic signal's frequency and its time representation, we find

$$\psi(t) = \mathscr{F}^{-1}\{2u(t)\} = \mathscr{F}^{-1}\{1\} + \mathscr{F}^{-1}\{\text{sgn}\,\omega\} = \frac{1}{2\pi}\int_{-\infty}^{\infty} 1 e^{j\omega t}\,d\omega + \frac{1}{2\pi}\int_{-\infty}^{\infty} \text{sgn}\,\omega e^{j\omega t}\,d\omega$$

Using the Fourier properties and the FT table, we write

$$\psi(t) = \delta(t) + j\frac{1}{\pi t} \overset{\Delta}{=} \delta(t) + \hat{\delta}(t) \Rightarrow \mathscr{H}\{\delta(t)\} = \frac{1}{\pi t}$$

The same result is found by taking the inverse Hilbert transform of the delta function:

$$\int_{-\infty}^{\infty} \frac{\delta(\tau)}{\pi(t-\tau)}\,d\tau = \frac{1}{\pi t}$$

Parseval's theorem

From the relation $\mathscr{F}\{\hat{x}(t)\} \overset{\Delta}{=} \hat{X}(t) = -j\,\text{sgn}\,\omega X(\omega)$, we obtain the result $\left|\hat{X}(\omega)\right|^2 = |-\text{sgn}\,\omega X(\omega)|^2 = |X(\omega)|^2$. Therefore, the energy of the signal and the energy of its Hilbert transform are, respectively,

$$E_x = \int_{-\infty}^{\infty} x^2(t)dt = \frac{1}{2\pi}\int_{-\infty}^{\infty} |X(\omega)|^2 d\omega, \quad E_{\hat{x}} = \frac{1}{2\pi}\int_{-\infty}^{\infty} |\hat{X}(\omega)|^2 d\omega = \frac{1}{2\pi}\int_{-\infty}^{\infty} |X(\omega)|^2 d\omega = E_x$$

Orthogonality

$$\int_{-\infty}^{\infty} \hat{x}(t)x(t)dx = 0$$

Example 9.3

From the result $\mathcal{H}\left\{\dfrac{1}{1+t^2}\right\} = \left\{\dfrac{t}{1+t^2}\right\}$, we conclude that the integral $\displaystyle\int_{-\infty}^{\infty} \dfrac{t}{(1+t^2)^2}\, dt$ is zero since the integrand is an odd function. ∎

Autoconvolution

To show the autocorrelation relationship,

$$g_{xx}(t) \overset{\Delta}{=} x(t) * x(t) = -g_{\hat{x}\hat{x}}(t) \overset{\Delta}{=} -\hat{x}(t) * \hat{x}(t) \tag{9.22}$$

we proceed as follows:

From the FT properties, we have $\mathcal{F}\{g_{xx}(t)\} = X(\omega)X(\omega) = X^2(\omega)$. Furthermore,

$$g_{\hat{x}\hat{x}}(t) \overset{\Delta}{=} \hat{x}(t) * \hat{x}(t) = \int_{-\infty}^{\infty} \hat{x}(x)\hat{x}(t-x)dx, \quad \mathcal{F}\{g_{\hat{x}\hat{x}}(t)\} \overset{\Delta}{=} \hat{G}(\omega) = \hat{X}(\omega)\hat{X}(\omega)$$

$$= -j\,\mathrm{sgn}\,(\omega)\hat{X}(\omega)[-j\,\mathrm{sgn}\,(\omega)X(\omega)] = -G_{xx}(\omega) = \hat{G}(\omega) \Rightarrow g_{xx}(t) = -g_{\hat{x}\hat{x}}(t)$$

Example 9.4

Because $\delta(t) * \delta(t) = \delta(t)$ and $\mathcal{F}\{\delta(t)\} = \dfrac{1}{\pi t}\hat{\delta}(t)$ then $\delta(t) = -\hat{\delta}(t) * \hat{\delta}(t) = -\dfrac{1}{\pi t} * \dfrac{1}{\pi t}$ (see Figure 9.4). ∎

Example 9.5

The spectrum of the analytic signal $\psi(t) = e^{j\omega_0 t} = \cos \omega_0 t + j \sin \omega_0 t = \cos \omega_0 t + j\mathcal{H}\{\cos \omega_0 t\}$ is found as follows:

$$\mathcal{F}\{\psi(t)\} = \mathcal{F}\{\cos \omega_0 t\} + j\mathcal{F}\{\sin \omega_0 t\} = \frac{1}{2}\delta(\omega + \omega_0) + \frac{1}{2}\delta(\omega - \omega_0)$$

$$+ j\left[j\frac{1}{2}\delta(\omega + \omega_0) - j\frac{1}{2}\delta(\omega - \omega_0)\right] = \delta(\omega - \omega_0)$$ ∎

Example 9.6

The Hilbert transform of a pulse is

$$\mathcal{H}\{p_a(t)\} \overset{\Delta}{=} \hat{p}_a(t) = \ln\left|\frac{t+a}{t-a}\right| \tag{9.23}$$

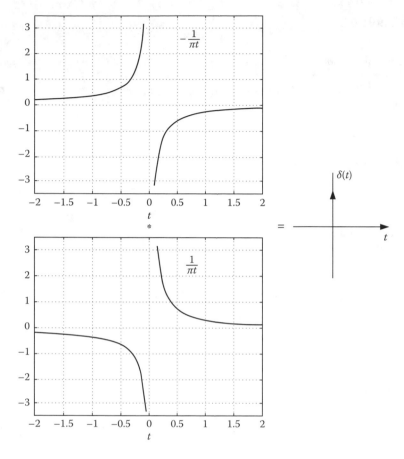

FIGURE 9.4

The above relation is found as follows:

$$H\{p_a(t)\} = -\frac{1}{\pi} P \int_{-\infty}^{\infty} \frac{p_a(x)}{x-t} dx = \frac{1}{\pi} \left[-\int_{-a}^{t-\varepsilon} \frac{1}{x-t} dx - \int_{t+\varepsilon}^{a} \frac{1}{x-t} dx \right]$$

$$= \frac{1}{\pi} \left[-\ln|x-t| \Big|_{-a}^{t-\varepsilon} - \ln|x-t| \Big|_{t+\varepsilon}^{a} \right]$$

$$= \frac{1}{\pi} [-\ln|\varepsilon| + \ln|t+a| - \ln|t-a| + \ln|\varepsilon|] = \underbrace{\ln\left|\frac{\varepsilon}{\varepsilon}\right|}_{\ln 1 = 0} + \ln\left|\frac{t+a}{t-a}\right| \qquad ∎$$

Hilbert transforms of periodic signals

Let $x(t)$ be periodic, that is, $x(t+T) = x(t)$ for any t. The constant T is the period of the periodic signal. The generating function of the signal is defined as follows:

$$x_T(t) = \begin{cases} x(t) & t_0 \le t \le t_0 + T \\ 0 & \text{otherwise} \end{cases} \tag{9.24}$$

Therefore, the periodic signal can be represented by the following relation:

$$x(t) = x_T(t) * \text{comb}_T(t) = x_T(t) \tag{9.25}$$

Cosine expansion

The Hilbert transform of the periodic function is

$$\hat{x}(t) = \mathcal{H}\{A\} + \sum_{n=1}^{\infty} C_n \mathcal{H}\{\cos(n\omega_0 t + \varphi_n)\} = \sum_{n=1}^{\infty} C_n \sin(n\omega_0 t + \varphi_n), \quad \omega_0 = \frac{2\pi}{T} \tag{9.26}$$

The Hilbert transform of a constant is zero since the integrand is an odd function.

Exponential form

The exponential form of a periodic signal is given by

$$x(t) = \sum_{n=-\infty}^{\infty} \alpha_n e^{jn\omega_0 t} \quad \alpha_n = \frac{1}{T} \int_{-T/2}^{T/2} x(t) e^{-jn\omega_0 t} dt \tag{9.27}$$

Its Hilbert transform is given by

$$\hat{x}(t) = \sum_{n=-\infty}^{\infty} \alpha_n \mathcal{H}\{e^{jn\omega_0 t}\} = \sum_{n=-\infty}^{\infty} (-j)\text{sgn}(n)\alpha_n e^{jn\omega_0 t} \tag{9.28}$$

Proof

(a) $n = 0$ $\quad \hat{x}_0(t) = \sum_{n=-\infty}^{\infty} \alpha_n \mathcal{H}\{1\} = 0$

(b) $n > 0$ $\quad \hat{x}_+(t) = \sum_{n=1}^{\infty} \alpha_n \mathcal{H}\{e^{jn\omega_0 t}\} = \sum_{n=1}^{\infty} \alpha_n [\mathcal{H}\{\cos n\omega_0 t\} + j\mathcal{H}\{\sin n\omega_0 t\}]$

$$= \sum_{n=1}^{\infty} \alpha_n [\sin n\omega_0 t - j\cos n\omega_0 t\}] = \sum_{n=1}^{\infty} \alpha_n(-j)[\cos n\omega_0 t + j\sin n\omega_0 t\}]$$

$$= \sum_{n=1}^{\infty} \alpha_n(-j)e^{jn\omega_0 t}$$

$$(c)\, n < 0 \;\; \hat{x}_-(t) = \sum_{n=-\infty}^{-1} \alpha_n \mathcal{H}\{e^{jn\omega_0 t}\} = \sum_{n=1}^{\infty} \alpha_{-n} \mathcal{H}\{e^{-jn\omega_0 t}\} = \sum_{n=1}^{\infty} \alpha_{-n} [\sin n\omega_0 t + j\cos n\omega_0 t]$$

$$= \sum_{n=1}^{\infty} \alpha_{-n}(j) [\cos n\omega_0 t - j\sin n\omega_0 t] = \sum_{n=-\infty}^{-1} \alpha_n(j) [\cos n\omega_0 t + j\sin n\omega_0 t]$$

$$= \sum_{n=-\infty}^{-1} \alpha_n(j) e^{jn\omega_0 t}$$

$$\Rightarrow \hat{x}(t) = \hat{x}_0(t) + \hat{x}_+(t) + \hat{x}_-(t) = \sum_{n=-\infty}^{\infty} (-j)\text{sgn}\,(n)\alpha_n e^{jn\omega_0 t}$$

Differentiation

Starting from a given Hilbert pair, we get a new Hilbert pair using the differentiation of both sides of the relation. Hence,

$$x'(t) \overset{\mathcal{H}}{\Leftrightarrow} \hat{x}'(t) \tag{9.29}$$

Example 9.7

Consider the delta pulse and its Hilbert transform. The double Hilbert transform (the iteration procedure) yields

$$\mathcal{H}\left\{\frac{1}{\pi t}\right\} = \mathcal{H}\{\mathcal{H}\{\delta(t)\} = -\delta(t)$$

or

$$-\delta'(t) = \mathcal{H}\left\{\frac{d}{dt}\frac{1}{\pi t}\right\} = -\mathcal{H}\left\{\frac{1}{\pi t^2}\right\} \, (1)$$

or

$$\mathcal{H}\{\delta'(t)\} = \mathcal{H}\left\{\mathcal{H}\left\{\frac{1}{\pi t^2}\right\}\right\} = -\frac{1}{\pi t^2} \tag{9.30}$$

By differentiating (1) of (9.30), we obtain

$$\mathcal{H}\{\delta''(t)\} = \frac{2}{\pi t^3} \tag{9.31}$$

∎

Example 9.8

Consider the Hilbert pair:

$$\mathcal{H}\left\{\frac{\sin at}{t}\right\} = \frac{1 - \cos at}{t} \tag{9.32}$$

The iteration operation yields

$$\mathcal{H}\left\{\mathcal{H}\left\{\frac{\sin at}{t}\right\}\right\} = -\frac{\sin at}{t} = \mathcal{H}\left\{\frac{1}{t}\right\} - \mathcal{H}\left\{\frac{\cos at}{t}\right\} \tag{9.33}$$

But $\mathcal{H}\{1/t\} = -\pi\delta(t)$ see (9.30) and therefore, the above equation becomes

$$\mathcal{H}\left\{\frac{\cos at}{t}\right\} = -\pi\delta(t) + \frac{\sin at}{t} \tag{9.34}$$

The differentiation of (9.32) yields

$$a\mathcal{H}\left\{\frac{\cos at}{t}\right\} - \mathcal{H}\left\{\frac{\sin at}{t^2}\right\} = -\frac{1}{t^2} + a\frac{\sin at}{t} + \frac{\cos at}{t^2}$$

The insertion of (9.34) produces the equation

$$\mathcal{H}\left\{\frac{\sin at}{t^2}\right\} = -\pi\delta(t) + \frac{1 - \cos at}{t^2} \tag{9.35}$$

∎

Differentiation of the convolution

From Section 9.1, we have the Hilbert pair:

$$x(t) = -\frac{1}{\pi t} * \hat{x}(t) \overset{\mathcal{H}}{\Leftrightarrow} \hat{x}(t) = \frac{1}{\pi t} * x(t) \tag{9.36}$$

We can differentiate the above equations by imposing the derivative only to the first or the second term. Therefore, the first-term differentiation yields

$$x'(t) = -\frac{d}{dt}\left(\frac{1}{\pi t}\right) * x(t) \overset{\mathcal{H}}{\Leftrightarrow} \hat{x}'(t) = \frac{d}{dt}\left(\frac{1}{\pi t}\right) * x(t) \quad \text{or}$$

$$x'(t) = \frac{1}{\pi t^2} * \hat{x}(t) \overset{\mathcal{H}}{\Leftrightarrow} \hat{x}'(t) = -\frac{1}{\pi t^2} * x(t) \tag{9.37}$$

From the second Hilbert pair, we write

$$\hat{x}'(t) = \frac{d}{dt}\left\{\frac{-1}{\pi} P \int_{-\infty}^{\infty} \frac{x(\tau)}{(\tau - t)} d\tau\right\} = -\frac{1}{\pi} P \int_{-\infty}^{\infty} \frac{x(\tau)}{(\tau - t)^2} d\tau = -\frac{1}{\pi t^2} * x(t)$$

which verifies the right-hand expression of (9.37).

Autoconvolution

The **autoconvolution** operation is defined by the formula

$$\rho_{xx}(t) = \int_{-\infty}^{\infty} x(\tau)x(t - \tau)d\tau \tag{9.38}$$

The autoconvolution equality for the Hilbert pair $x(t) \overset{\mathcal{H}}{\Leftrightarrow} \hat{x}(t)$ has the form

$$\rho_{xx}(t) = -\rho_{\hat{x}\hat{x}}(t) \tag{9.39}$$

This means that the autoconvolution of the signal and the autoconvolution of its Hilbert transform differ only with respect to the sign. As proof, consider the Fourier spectra of $x(t)$ and $\hat{x}(t)$. Since the spectra of the two signals are $X(\omega)$ and $-j \operatorname{sgn}(\omega)X(\omega)$, respectively, the spectra of their autoconvolutions are, respectively (the FT of the convolution of two signals is equal to the multiplication of their Fourier transforms (FT property)),

$$x(t) * x(t) \overset{\mathcal{H}}{\Leftrightarrow} X^2(\omega), \quad \hat{x}(t) * \hat{x}(t) \overset{\mathcal{H}}{\Leftrightarrow} [-j \operatorname{sgn}(\omega)X(\omega)]^2 = -X^2(\omega)$$

Both spectra differ only by the sign and, therefore, (9.39) is verified.

Example 9.9

From the Hilbert pair relationship and the convolution of the delta functions, we have the following relationships:

$$\delta(t) \overset{\mathcal{H}}{\Leftrightarrow} \frac{1}{\pi t}, \quad \delta(t) = \delta(t) * \delta(t)$$

From the above equalities and the autoconvolution equality (see 9.39), we obtain the relation

$$\delta(t) = -\frac{1}{\pi t} * \frac{1}{\pi t} \tag{9.40}$$

We observe the compression of the infinite-time support of the function $1/\pi t$ to a single point support of the delta function. ∎

HT of a product of two signals

Using the FT of the product of two signals $f(t)$ and $g(t)$, we obtain

$$x(t) = f(t)g(t) \overset{\mathcal{F}}{\Leftrightarrow} X(\omega) = \frac{1}{2\pi} \int_{-\infty}^{\infty} X(\omega - y)G(y)dy \tag{9.41}$$

The HT of this product is given by the standard formula, and hence,

$$\hat{x}(t) = \mathcal{H}\{f(t)g(t)\} \overset{\mathcal{F}}{\Leftrightarrow} -j \operatorname{sgn}(\omega)\frac{1}{2\pi} \int_{-\infty}^{\infty} F(\omega - y)G(y)dy \tag{9.42}$$

9.3 Hilbert Transform Properties and Hilbert Pairs

TABLE 9.1   Properties of the Hilbert Transformation (HT)

	Original Signal or the Inverse HT $x(t)$	Hilbert Transform $\hat{x}(t) = \mathcal{H}\{x(t)\}$
1. Time-domain definition	$x(t) = \dfrac{1}{\pi}\displaystyle\int_{-\infty}^{\infty} \dfrac{\hat{x}(t)}{\tau - t}d\tau$	$\hat{x}(t) = -\displaystyle\int_{-\infty}^{\infty} \dfrac{x(t)}{\tau - t}d\tau$
2. Convolution notation	$x(t) = -\dfrac{1}{\pi t} * \hat{x}(t)$	$\hat{x}(t) = \dfrac{1}{\pi t} * x(t)$
3. Fourier spectra	$x(t) \overset{\mathscr{F}}{\Leftrightarrow} X(\omega)$ $X(\omega) = j\,\mathrm{sgn}\,(\omega)\hat{X}(\omega)$	$\hat{x}(t) \overset{\mathscr{F}}{\Leftrightarrow} \hat{X}(\omega)$ $\hat{X}(\omega) = -j\,\mathrm{sgn}\,(\omega)X(\omega)$
4. Linearity	$ax_1(t) + bx_2(t)$	$a\hat{x}_1(t) + b\hat{x}_2(t)$
5. Scaling	$x(at)\ a > 0$	$\hat{x}(at)$
6. Time reversal	$x(-at)$	$-\hat{x}(-at)$
7. Time shift	$x(t - a)$	$\hat{x}(t - a)$
8. Evenness and oddness	$x_{1e}(t) + x_{2o}(t)$ $e \equiv \text{even}, o \equiv \text{odd}$	$\hat{x}_{1o}(t) + \hat{x}_{2e}(t)$
9. Time derivative First option	$x'(t) = -\dfrac{1}{\pi t} * \hat{x}'(t)$	$\hat{x}(t) = \dfrac{1}{\pi t} * x'(t)$
Second option	$x'(t) = \dfrac{d}{dt}\left(\dfrac{-1}{\pi t}\right) * \hat{x}(t)$	$\hat{x}'(t) = \dfrac{d}{dt}\left(\dfrac{-1}{\pi t}\right) * x(t)$
10. Convolution	$x(t) * y(t) = -\hat{x}(t) * \hat{y}(t)$	$x(t) * \hat{y}(t) = \hat{x}(t) * y(t)$
11. Autoconvolution	$x(t) * x(t) = -\hat{x}(t) * \hat{x}(t)$	
12. Energy equality	$\int_{-\infty}^{\infty} x^2(t)dt = \int_{-\infty}^{\infty} \hat{x}^2(t)dt$ (the energy of the dc term is excluded)	
13. Multiplication by t	$tx(t)$	$t\hat{x}(t) - \dfrac{1}{\pi}\displaystyle\int_{-\infty}^{\infty} x(\tau)d\tau$

(continued)

TABLE 9.1 (continued) Properties of the Hilbert Transformation (HT)

	Original Signal or the Inverse HT $x(t)$	Hilbert Transform $\hat{x}(t) = \mathcal{H}\{x(t)\}$
14. Analytic signal		
	$\psi(t) = x(t) + j\mathcal{H}\{x(t)\}$	$\mathcal{H}\{\psi(t)\} = -j\psi(t)$
15. Products of analytic signals		
	$\psi(t) = \psi_1(t)\psi_2(t)$	$\mathcal{H}\{\psi(t)\} = \psi_1(t)\mathcal{H}\{\psi_2(t)\}$
		$= \mathcal{H}\{\psi_1(t)\}\psi_2(t)$

16. Iteration	Operation	Fourier image
No	$\mathcal{H}\{x(t)\} = \hat{x}(t)$	$-j\,\mathrm{sgn}(\omega)X(\omega)$
Single	$\mathcal{H}\{\mathcal{H}\{x(t)\}\} = -x(t)$	$[-j\,\mathrm{sgn}(\omega)]^2 X(\omega)$
Double	$\mathcal{H}\{\mathcal{H}\{\mathcal{H}\{x(t)\}\}\} = -\hat{x}(t)$	$[-j\,\mathrm{sgn}(\omega)]^3 X(\omega)$
Triple	$\mathcal{H}\{\mathcal{H}\{\mathcal{H}\{\mathcal{H}\{(x(t)\}\}\}\} = x(t)$	$[-j\,\mathrm{sgn}(\omega)]^4 X(\omega)$

TABLE 9.2 Hilbert Transform Pairs

Function	Hilbert Transform
1. $\sin(\omega t)$	$-\cos(\omega t)$
2. $\cos(\omega t)$	$\sin(\omega t)$
3. $e^{j\omega t}$	$-j\,\mathrm{sgn}(\omega)e^{j\omega t}$
4. $p_a(t)$ (centered pulse of width $2a$)	$\dfrac{1}{\pi}\ln\left\|\dfrac{t+a}{t-a}\right\|$
5. $p_a(t)\mathrm{sgn}(t)$	$-\dfrac{1}{\pi}\ln\|1-(a/t)^2\|$
6. $tp_a(t)\mathrm{sgn}(t)$	$-\dfrac{t}{\pi}\|1-(a/t)^2\|$
7. $\begin{cases} 1-\left\|\frac{t}{a}\right\| & \|t\| \le a \\ 0 & \|t\| > a \end{cases}$	$-\dfrac{1}{\pi}\left\{\ln\left\|\dfrac{t-a}{t+a}\right\|\right\} + \dfrac{t}{a}\ln\left\|\dfrac{t^2}{t^2-a^2}\right\|$
8. $\dfrac{a}{a^2+t^2}$	$\dfrac{t}{a^2+t^2}$
9. $e^{-\pi t^2}$	$\dfrac{1}{\pi}\displaystyle\int_0^\infty e^{-\omega^2/4\pi}\sin\omega t\,d\omega$
10. $1-(t/a)^2 \quad \|t\| \le a$	$\dfrac{-1}{\pi}\left[[1-(t/a)^2]\ln\left\|\dfrac{t-a}{t+a}\right\| - \dfrac{2t}{a}\right]$
11. $e^{-a\|t\|}$	$\dfrac{1}{\pi}\displaystyle\int_0^\infty \dfrac{2a}{a^2-\omega^2}\sin\omega t\,d\omega$
12. $e^{-a\|t\|}u(t)$	$\dfrac{1}{\pi}\displaystyle\int_0^\infty \dfrac{a\sin\omega t - \omega\cos\omega t}{a^2-\omega^2}\,d\omega$
13. $\dfrac{\sin at}{at}$	$\dfrac{\sin^2(at/2)}{(at/2)} = \dfrac{1-\cos at}{at}$

TABLE 9.2 (continued)   Hilbert Transform Pairs

Function	Hilbert Transform
14. $a = $ constant	$\dfrac{1}{\pi t}$
15. $\dfrac{1}{\pi t}$	$-\delta(t)$
16. $\delta'(t)$	$-\dfrac{1}{\pi t^2}$
17. $\dfrac{1}{\pi t^2}$	$\delta'(t)$
18. $\delta(t) * \delta(t) * \delta(t)$	$\dfrac{1}{\pi t} * \dfrac{1}{\pi t} = -\delta(t)$
19. $\delta'(t) * \delta(t) = \delta'(t)$	$\dfrac{1}{\pi t^2} * \dfrac{1}{\pi t} = \delta'(t)$
20. $\dfrac{\sin at}{t}$	$\dfrac{1 - \cos at}{t} = \dfrac{2 \sin^2 (at/2)}{t}$
21. $\dfrac{\cos at}{t}$	$-\pi\delta(t) + \dfrac{\sin at}{t}$
22. $\sum_{n=-\infty}^{\infty} \delta(t - nT)$	$\dfrac{1}{T} \sum_{n=-\infty}^{\infty} \cos\left[\dfrac{\pi}{T}(t - nT)\right]$
23. $\cos^2 (\omega t)$	$0.5 \sin 2\omega t$
24. $\sin^2 \omega t$	$-0.5 \sin 2\omega t$
25. $C_0 + \sum_{n=1}^{\infty} C_n \cos(n\omega t + \varphi_n) = $ Fourier series	$\sum_{n=1}^{\infty} C_n \sin(n\omega t + \varphi_n)$

Appendix A:
Functions of a
Complex Variable*

A.1 Basic Concepts

A complex variable z defined by

$$z = x + jy \tag{A.1}$$

assumes certain values over a region R_z of the complex plane. If a complex quantity $W(z)$ is so connected with z that each z in R_z corresponds with one value of $W(z)$ in R_w, then we say that $W(z)$ is a single-valued function of z

$$W(z) = u(x, y) + jv(x, y) \tag{A.2}$$

which has a **domain** R_z and a **range** R_w (see Figure A.1). The function $W(z)$ can be **single valued** or **multiple valued**. Examples of single-valued functions include

$$W = a_0 + a_1 z + a_2 z^2 + \cdots + a_n z^n \quad n \text{ integer}$$

$$W = e^z$$

Examples of multiple-valued functions are

$$W = z^n \quad n \text{ is not an integer}$$
$$W = \log z$$
$$W = \sin^{-1} z$$

---

* All contour integrals are taken counterclockwise, unless specifically indicated.

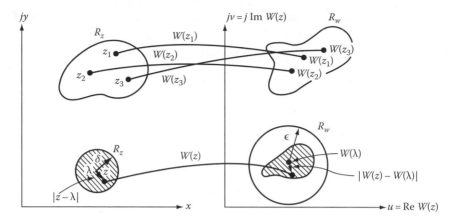

FIGURE A.1   Illustration of the range and domain of complex functions.

Definition A.1   A function $W(z)$ is **continuous** at a point $z = \lambda$ of R_z if, for each number $\varepsilon > 0$, however small, there exists another number $\delta > 0$ such that whenever

$$|z - \lambda| < \delta \quad \text{then} \quad |W(z) - W(\lambda)| < \varepsilon \tag{A.3}$$

The geometric representation of this equation is shown in Figure A.1.

Definition A.2   A function $W(z)$ is **analytic** at a point z if, for each number $\varepsilon > 0$, however small, there exists another number $\delta > 0$ such that whenever

$$|z - \lambda| < \delta \quad \text{then} \quad \left| \frac{W(z) - W(\lambda)}{z - \lambda} - \frac{dW(\lambda)}{dz} \right| < \varepsilon \tag{A.4}$$

Example A.1

Show that the function $W(z) = e^z$ satisfies (A.4).

SOLUTION

From (A.4), we obtain

$$\lim_{z \to \lambda} \frac{W(z) - W(\lambda)}{z - \lambda} = \lim_{z \to \lambda} \frac{e^z - e^\lambda}{z - \lambda}$$

$$= \lim_{z \to \lambda} e^z \left[1 - \frac{(z - \lambda)}{2!} + \frac{(z - \lambda)^2}{3!} - \cdots \right]$$

$$= e^\lambda = \frac{de^z}{dz} \bigg|_{z = \lambda}$$

which proves the assertion.   ∎

In this example, we did not mention the direction from which the z approaches λ. We might surmise from this that the derivative of our analytic function is independent of the path of z as it approaches the limiting point. However, this is not true in general. By setting $\lambda = z$ and $z = z + \Delta z$ in (A.4), we obtain an alternative form of that equation, namely,

$$\frac{dW}{dz} = \lim_{\Delta z \to 0} \left\{ \frac{W(z + \Delta z) - W(z)}{\Delta z} \right\} \tag{A.5}$$

For a function to possess a unique derivative, it is required that

$$\frac{dW}{dz} = \lim_{\Delta z \to 0} \frac{\Delta W}{\Delta z} = \lim_{\substack{\Delta x \to 0 \\ \Delta v \to 0}} \frac{\Delta u + j\Delta v}{\Delta x + j\Delta y}$$

But because

$$\Delta u = \frac{\partial u}{\partial x} \Delta x + \frac{\partial u}{\partial y} \Delta y$$

$$\Delta v = \frac{\partial v}{\partial x} \Delta x + \frac{\partial v}{\partial y} \Delta y$$

the unique derivative becomes

$$\frac{dW}{dz} = \lim_{\substack{\Delta x \to 0 \\ \Delta y \to 0}} \frac{\left(\dfrac{\partial u}{\partial x} + j\dfrac{\partial v}{\partial x} \right) \Delta x + j\left(\dfrac{\partial v}{\partial y} - j\dfrac{\partial u}{\partial y} \right) \Delta y}{\Delta x + j\Delta y}$$

For this to be independent of how Δx and Δy approach zero (that is, for the derivative to be unique), it is necessary and sufficient that $\Delta x + j\Delta y$ cancel in the numerator and the denominator. This requires that

$$\frac{dW}{dz} = \frac{\partial u}{\partial x} + j\frac{\partial v}{\partial y} = \frac{\partial v}{\partial x} - j\frac{\partial u}{\partial y}$$

This condition can be met if

$$\frac{\partial u}{\partial x} = \frac{\partial v}{\partial y} \frac{\partial v}{\partial x} = -\frac{\partial u}{\partial y} \tag{A.6}$$

These are the **Cauchy–Riemann** conditions. If the function satisfies these equations, it possesses a unique derivative and it is analytic at that point. These conditions are necessary and sufficient.

A.1.1 Integration

The integration of a complex function is defined in a manner like that for a real function, except for the important difference that the path of integration as well as the end points must be specified. A number of important theorems relate to integration, as we will discuss later.

Recall that the real integral $\int_a^b f(x)\,dx$ means that the x-axis is broken into tiny elements Δx from a to b, each element is multiplied by the mean value of $f(x)$ in the element, and then the sum of all such products from a to b is taken as $\Delta x \to 0$. The same general procedure is used to define the integral in the complex plane. Instead of being restricted to the x-axis, the path of integration can be anywhere in the z-plane, for example, the arc ABC in Figure A.2. This arc is broken into n elements Δz, and the corresponding mean value of $W(z)$ over each element is written as W_s. Now from the sum $\sum_{s=1}^{n} W_s \Delta z_s$, over all values of s from a to b, take the limit $\Delta z_s \to 0$, $n \to \infty$. This limit, if it exists, is the integral

$$I = \int_a^b W(z)dz \qquad\qquad (A.7)$$

The only innovation introduced here is that the path over which the integral is to be taken must be specified.

Example A.2

Evaluate the integral in (A.7) for the function $W(z) = 1/z$ over the semicircles, as shown in Figure A.3.

Solution

Refer first to Figure A.3a, and introduce the polar coordinates

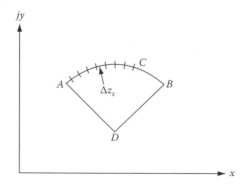

FIGURE A.2   The path of integration in the complex plane.

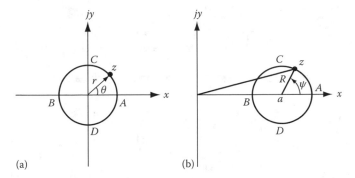

FIGURE A.3 Integral of the function $W = 1/z$ over two paths.

$$z = re^{j\theta} \quad dz = jre^{j\theta}d\theta$$

Then

$$\int W(z)dz = \int \frac{dz}{z} = \int j\,d\theta$$

Over the path ACB, θ varies between 0 and π, and the integral equals $j\pi$. Over the path ADB, θ varies from 0 to $-\pi$, and the integral equals $-j\pi$. Thus, although the end points are the same, the integrals over the two paths are different. (The fact that one integral is numerically the negative of the other has no general significance.)

In evaluating the real integral by starting at A and integrating to B and then back to A, the result will be zero because the integral from A to B is the negative of the integral from B to A. The same result is not necessarily true for complex variables, unless the path from A to B coincides with the path from B to A. In the present complex integral, the integration from A to B via C and then back to A via D yields $j\pi - (-j\pi) = j2\pi$ and no zero.

Now consider the integration over the semicircle displaced from the origin, as shown in Figure A.3b. Introduce the coordinates

$$z = a + Re^{j\psi} \quad dz = jRe^{j\psi}d\psi$$

Then

$$\int\limits_{ACBDA} \frac{dz}{z} = \int\limits_0^{2\pi} \frac{jRe^{j\psi}}{a + Re^{j\psi}} d\psi = \ln(a + Re^{j\psi})|_0^{2\pi} = \ln z|_A^A = 0$$

The results of these calculations emphasize the fact that the two paths possess different features. The difference is that in Figure A.3a, the path encloses a singularity (the function becomes infinite) at the origin, whereas the path in Figure A.3b does not enclose the singularity and $W = 1/z$ is analytic everywhere in the region and on the boundary.

It is easily shown that the integrals of the function

$$W(z) = \frac{1}{z^2}, \quad W(z) = \frac{1}{z^3}, \quad \ldots, \quad W(z) = \frac{1}{z^n}$$

around a contour encircling the origin of the coordinate axis are each equal to zero; that is,

$$\oint \frac{1}{z^2} dz = \oint \frac{1}{z^3} dz = \cdots = \oint \frac{1}{z^n} dz = 0 \tag{A.8}$$

where the contour is taken counterclockwise.   ■

Example A.3

Find the value of the integral $\int_0^{z_0} z\, dz$ from the point $(0, 0)$ to $(2, j4)$.

Solution

Because z is an analytic function along any path, then

$$\int_0^{z_0} z\, dz = \left. \frac{z^2}{2} \right|_0^{2+j4} = -6 + j8$$

Equivalently, we could write

$$\int_0^{z_0} z\, dz = \int_0^2 x\, dz - \int_0^4 y\, dy + j\int_0^4 x\, dy = +\left. \frac{x^2}{2} \right|_0^2 - \left. \frac{y^2}{2} \right|_0^4 + jxy\Big|_0^4$$

$$= 2 - \frac{16}{2} + j2 \times 4 = -6 + j8 \qquad\qquad ■$$

We now state a very important theorem, and this is often referred to as the **principal theorem of complex variable theory**. This is the **Cauchy first integral theorem**.

THEOREM A.1

Given a region of the complex plane within which $W(z)$ is analytic and any closed curve that lies entirely within this region, then

$$\oint_C W(z)dz = 0 \tag{A.9}$$

where the contour C is taken counterclockwise.

The integration over a closed path is called a contour integral. Also, by convention the positive direction of integration is taken so that when traversing the contour, the enclosed region is always to the left. The proof of this theorem depends on the fact that everywhere within C the Cauchy-Riemann equations are satisfied, $W(z)$ possesses a unique derivative at all points of the path.

COROLLARY A.1

If the contour C_2 completely encloses C_1, and if $W(z)$ is analytic in the region between C_1 and C_2 and also on C_1 and C_2, then

$$\oint_{C_1} W(z)dz = \oint_{C_2} W(z)dz \qquad (A.10)$$

Proof: Refer to Figure A.4, which shows the two contours C_1 and C_2 and the two connecting lines $D\,E$ and $G\,A$. In the region closed by the contour $A\,B\,D\,E\,F\,G\,A$, the function $W(z)$ is analytic everywhere, and $\oint Wdz = 0$ over the path. This means that

$$\int_{ABD} + \int_{DE} + \int_{EFG} + \int_{GA} = 0 \qquad (A.11)$$

where $W(z)\,dz$ is to be understood after each integral sign. Now allow A to approach D, and G to approach E, so that DE coincides with AG. Then

$$\int_{DE} = \int_{AG} = -\int_{GA}$$

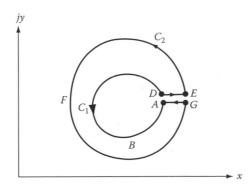

FIGURE A.4 To prove the first corollary.

Also

$$\int_{ABD} = -\int_{C_1} \quad \text{and} \quad \int_{EFG} = \int_{C_2} \tag{A.12}$$

where strict attention has been paid to the convention given in the determination of the positive direction of integration around a contour. Combine (A.11) and (A.12) so that

$$-\int_{C_1} + \int_{C_2} = 0 \quad \text{or} \quad \int_{C_1} W(z)\,dz = \int_{C_2} W(z)\,dz$$

which was to be proved.

This is an important theorem because it allows the evaluation around one contour by replacing that contour with a simpler one, the only restriction being that in the region between the two contours the integral must be regular. It does not require that the function $W(z)$ be analytic within C_1.

COROLLARY A.2

If $W(z)$ has a finite number n of isolated singularities within a region G bounded by a curve C, then

$$\int_C W(z)\,dz = \sum_{s=1}^{N} \oint_{C_s} W(z)\,dz \tag{A.13}$$

where C_s is any contour surrounding the sth singularity. The contours are taken in the counterclockwise direction.

Proof: Refer to Figure A.5. The proof for this case is evident from the manner in which the first corollary was proved.

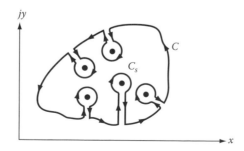

FIGURE A.5 A contour enclosing n isolated singularities.

COROLLARY A.3

The integral $\int_A^B W(z)dz$ depends only on the end points A and B (refer to Figure A.2) and does not depend on the path of integration, provided that this path lies entirely within the region in which W(z) is analytic.

Proof: Consider $A\ C\ B\ D\ A$ of Figure A.2 as a contour that encloses no singularity of $W(z)$. Then

$$\oint_C = 0 = \int_{ADB} + \int_{BCA} \quad \text{or} \quad \int_{ADB} = \int_{ACB} \tag{A.14}$$

Hence, the integral is the same whether taken over path D or C, and thus is independent of the path and depends only on the end points A and B.

THEOREM A.2 The Cauchy Second Integral Theorem

If $W(z)$ is the function $W(z) = f(z)/(z - z_0)$ and the contour encloses the only singularity at z_0, then

$$\oint_C \frac{f(z)}{z - z_0} dz = j2\pi f(z_0) \tag{A.15}$$

or

$$f(z_0) = \frac{1}{2\pi j} \oint_C \frac{f(z)}{z - z_0} dz \tag{A.16}$$

(the contours are taken in the counterclockwise direction.)

Proof: Refer to Figure A.6. Begin with the second corollary and draw a circle C_1 about the point z_0. Then

$$\int_C \frac{f(z)}{z - z_0} dz = \int_{C_1} \frac{f(z)}{z - z_0} dz \tag{A.17}$$

Let $z' = z - z_0 = \rho e^{j\theta}$, which permits writing

$$\int_{C_1} \frac{f(z)}{z - z_0} dz = \int_0^{2\pi} \frac{f(z' + z_0)}{\rho e^{j\theta}} j\rho e^{j\theta} d\theta = j \int_0^{2\pi} f(z' + z_0) d\theta$$

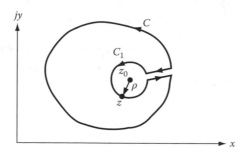

FIGURE A.6 To prove the Cauchy second integral theorem.

In the limit as $\rho \to 0$, $z' \to 0$, and

$$j \int_0^{2\pi} f(z' + z_0) d\theta \big|_{\lim \rho \to 0} = 2\pi j f(z_0)$$

Combine with (A.17) to find

$$\int_C \frac{f(z)}{z - z_0} = 2\pi j f(z_0)$$

which proves the theorem.

A.1.2 Derivative of an Analytic Function $W(z)$

The derivative of an analytic function is also analytic, and consequently itself possesses a derivative. Let C be a contour within and upon which $W(z)$ is analytic. Then, if a is a point inside the contour (the prime indicates first-order derivative)

$$W'(a) = \lim_{|h| \to 0} \frac{W(a + h) - W(a)}{h} \tag{A.18}$$

and can be shown that

$$W'(a) = \frac{1}{2\pi j} \oint_C \frac{W(z) dz}{(z - a)^2} \tag{A.19}$$

where the contour C is taken in a counterclockwise direction. Proceeding, it can be shown that

$$W^{(n)}(a) = \frac{n!}{2\pi j} \oint_C \frac{W(z) dz}{(z - a)^{n+1}} \tag{A.20}$$

The exponent (n) indicates nth derivative and the contour is taken counterclockwise.

A.1.3 Taylor's Theorem

Let $f(z)$ be analytic in the neighborhood of a point $z = a$. Let the contour C be a circle with center point a in the z-plane, and let the function $f(z)$ not have any singularity within and on the contour. Let $z = a + h$ be any point inside the contour; then, by (A.15), we obtain

$$f(a+h) = \frac{1}{2\pi j} \oint_C \frac{f(z)dz}{z - a - h}$$

$$= \frac{1}{2\pi j} \oint_C f(z)dz \left\{ \frac{1}{z-a} + \frac{h}{(z-a)^2} + \cdots + \frac{h^n}{(z-a)^{n+1}} + \frac{h^{n+1}}{(z-a)^{n+1}(z-a-h)} \right\}$$

$$= f(a) + hf^{(1)}(a) + \frac{h^2}{2!}f^{(2)}(a) + \cdots + \frac{h^n}{n!}f^{(n)}(a) + \frac{1}{2\pi j} \oint_C \frac{f(z)h^{n+1}dz}{(z-a)^{n+1}(z-a-h)}$$

But where z is on C, the modulus $f(z)/(z - a - h)$ is continuous and therefore bounded. Its modulus will not exceed some finite number M. Hence, with $|z - a| = R$ for points on the circle, we obtain

$$\left| \frac{1}{2\pi j} \oint_C \frac{f(z)h^{n+1}dz}{(z-a)^{n+1}(z-a-h)} \right| \leq \frac{M 2\pi R}{2\pi} \left(\frac{|h|}{R} \right)^{n+1}$$

where $|h|/R < 1$ and therefore tends to zero as n tends to infinity. Therefore, we have

$$f(a+h) = f(a) + hf^{(1)}(a) + \frac{h^2}{2!}f^{(2)}(a) + \cdots + \frac{h^n}{n!}f^{(n)}(a) + \cdots \tag{A.21}$$

or

$$f(z) = f(a) + (z-a)f^{(1)}(a) + \frac{(z-a)^2}{2!}f^{(2)}(a) + \cdots + \frac{(z-a)^n}{n!}f^{(n)}(a) + \cdots \tag{A.22}$$

where the numbers in the exponents indicate the order of differentiation. The radius of convergence is such that it excludes from the interior of the circle that singularity of the function that is nearest to a.

A.1.4 Laurent's Theorem

Let C_1 and C_2 be two concentric circles, as shown in Figure A.7, with their centers at a. The function $f(z)$ is analytic with the ring and $(a + h)$ is any point in it. From the figure and Cauchy's theorem, we obtain

$$\frac{1}{2\pi j} \oint_{C_2} \frac{f(z)dz}{(z-a-h)} + \frac{1}{2\pi j} \oint_{C_1} \frac{f(z)dz}{(z-a-h)} + \frac{1}{2\pi j} \oint_{C_3} \frac{f(z)dz}{(z-a-h)} = 0$$

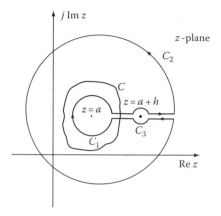

FIGURE A.7   Explaining Laurent's theorem.

where the first contour is counterclockwise and the last two are clockwise. The above equation becomes

$$f(a+h) = \frac{1}{2\pi j} \oint_{C_2} \frac{f(z)dz}{(z-a-h)} - \frac{1}{2\pi j} \oint_{C_1} \frac{f(z)dz}{(z-a-h)} \tag{A.23}$$

where both the contours are taken counterclockwise. For the C_2 contour $|h| < |(z-a)|$ and for the C_1 counter $|h| > |(z-a)|$. Hence, we expand the above integral (contours in the counterclockwise direction) as follows:

$$f(a+h) = \frac{1}{2\pi j} \oint_{C_2} f(z) \left\{ \frac{1}{z-a} + \frac{h}{(z-a)^2} + \cdots + \frac{h^n}{(z-a)^{n+1}} + \frac{h^{n+1}}{(z-a)^{n+1}(z-a-h)} \right\} dz$$

$$+ \frac{1}{2\pi j} \oint_{C_1} f(z) \left\{ \frac{1}{h} + \frac{z-a}{h^2} + \cdots + \frac{(z-a)^n}{h^{n+1}} + \frac{(z-a)^{n+1}}{h^{n+1}(z-a-h)} \right\} dz$$

From Taylor's theorem, it was shown that the integrals of the last term in the two brackets tend to zero as n tends to infinity. Therefore, we have

$$f(a+h) = a_0 + a_1 h + a_2 h^2 + \cdots + \frac{b_1}{h} + \frac{b_2}{h^2} + \cdots \tag{A.24}$$

where (contours in the counterclockwise direction)

$$a_n = \frac{1}{2\pi j} \oint_{C_2} \frac{f(z)dz}{(z-a)^{n+1}} \qquad b_n = \frac{1}{2\pi j} \oint_{C_1} (z-a)^{n+1} f(z)dz$$

The above expansion can be put in more convenient form by substituting $h = z - a$, which gives

$$f(z) = c_0 + c_1(z - a) + c_2(z - a)^2 + \cdots + \frac{d_1}{(z - a)} + \frac{d_2}{(z - a)^2} + \cdots + \frac{d_n}{(z - a)^n} + \cdots$$

(A.25)

Because $z = a + h$, it means that z now is any point within the ring-shaped space between C_1 and C_2 where $f(z)$ is analytic. Equation A.25 is Laurent's expansion of $f(z)$ at a point $z + h$ within the ring. The coefficients c_n and d_n are obtained from (A.24) by replacing a_n, b_n, and z by c_n, d_n, and ζ, respectively. Here ζ is the variable on the contours and z is inside the ring. When $f(z)$ has a simple pole at $z = a$, there is only one term, namely, $d_1/(z - a)$. If there exists an nth-order term, there are n terms of which the last is $d_n/(z - a)^n$; some of the d_n's may be zero.

If m is the highest index of the inverse power of $f(z)$ in (A.25), it is said that $f(z)$ has a pole of order m at $z = a$. Then

$$f(z) = \sum_{n=0}^{\infty} c_n(z - a)^n + \sum_{n=1}^{m} \frac{d_n}{(z - a)^n}$$

(A.26)

The coefficient, d_1, is the **residue** at the pole.

If the series in inverse powers of $(z - a)$ in (A.25) does not terminate, the function $f(z)$ is said to have an **essential singularity** at $z = a$. Thus,

$$f(z) = \sum_{n=0}^{\infty} c_n(z - a)^n + \sum_{n=1}^{\infty} \frac{d_n}{(z - a)^n}$$

(A.27)

The coefficient d_1 is the **residue** of the singularity.

Example A.4

Find the Laurent expansion of $f(z) = 1/[(z - a)(z - b)^n]$ $(n \geq 1, a \neq b \neq 0)$ near each pole.

SOLUTION

First, remove the origin to $z = a$ by the transformation $\zeta = (z - a)$. Hence, we obtain

$$f(z) = \frac{1}{\zeta}\frac{1}{(\zeta + c)^n} = \frac{1}{c^n\zeta}\frac{1}{\left(1 + \frac{\zeta}{c}\right)^n} \qquad c = a - b$$

If $|\zeta/c| < 1$, then we have

$$f(z) = \frac{1}{c^n \zeta}\left[1 - \frac{n\zeta}{c} + \frac{n(n+1)}{2!}\frac{\zeta^2}{c^2} - \cdots\right]$$

$$= \left[-\frac{n}{c^{n+1}} + \frac{n(n+1)\zeta}{2!c^{n+2}} - \cdots\right] + \frac{1}{c^n\zeta}$$

which is the Laurent series expansion near the pole at $z = a$. The residue is $1/c^n = 1/(a-b)^n$.

For the second pole, set $\zeta = (z-b)$ and expand as above to find

$$f(z) = -\left(\frac{1}{c^{n+1}} + \frac{\zeta}{z^{n+2}} + \frac{\zeta^2}{c^{n+3}} + \cdots\right) - \left(\frac{1}{c^n\zeta} + \frac{1}{c^{n-1}\zeta^2} + \cdots + \frac{1}{c\zeta^n}\right)$$

The second part of the expansion is the principal expansion near $z = b$ and the residue is $-1/c^n = -1/(a-b)^n$. ∎

Example A.5

Prove that

$$f(z) = \exp\left[\frac{x}{2}\left(z - \frac{1}{2}\right)\right] = J_0(x) + zJ_1(x) + z^2 J_2(x) + \cdots + z^n J_n(x)$$

$$+ \cdots - \frac{1}{z}J_1(x) + \frac{1}{z^2}J_2(x) - \cdots + \frac{(-1)^n}{z^n}J_n(x) + \cdots$$

where

$$J_n(x) = \frac{1}{2\pi}\int_0^{2\pi} \cos(n\theta - x\sin\theta)d\theta$$

Solution

The function $f(z)$ is analytic except the point $z = a$. Hence, by Laurent's theorem, we obtain

$$f(z) = a_0 + a_1 z + a_2 z^2 + \cdots + \frac{b_1}{z} + \frac{b_2}{z^2} + \cdots$$

where (contours in the counterclockwise direction)

$$a_n = \frac{1}{2\pi j}\oint_{C_2} \exp\left[\frac{x}{2}\left(z - \frac{1}{z}\right)\right]\frac{dz}{z^{n+1}}, \quad b_n = \frac{1}{2\pi j}\oint_{C_1} \exp\left[\frac{x}{2}\left(z - \frac{1}{z}\right)\right]z^{n-1}dz$$

where the contours are circles with center at the origin and are taken counterclockwise. Set C_2 equal to a circle of unit radius and write $z = \exp(j\theta)$. Then we have

$$a_n = \frac{1}{2\pi j} \int_0^{2\pi} e^{jx\sin\theta} e^{-jn\theta} j\,d\theta = \frac{1}{2\pi} \int_0^{2\pi} \cos(n\theta - x\sin\theta)\,d\theta$$

because the last integral vanishes, as can be seen by writing $2\pi - \varphi$ for θ. Thus, $a_n = J_n(x)$, and $b_n = (-1)^n a_n$ because the function is unaltered if $-z^{-1}$ is substituted for z, so that $b_n = (-1)^n J_n(x)$. ∎

A.2 Sequences and Series

Consider a sequence of numbers, such as those that arise in connection with the z-transform. Suppose that the sequence of complex numbers is given as $z_0, z_1, z_2, \ldots$.

The sequence of complex numbers is said to **converge** to the limit L; that is,

$$\lim_{n\to\infty} z_n = L$$

if for every positive δ there exists an integer N such that

$$|Z_n - L| < \delta \quad \text{for all } n > N$$

That is, a convergent sequence is one whose terms approach arbitrarily close to the limit L as n increases. If the series does not converge, it is said to **diverge**.

THEOREM A.3

In order for a sequence $\{z_n\}$ of complex numbers to be convergent, it is necessary and sufficient that for all $\delta > 0$ there exists a number $N(\delta)$ such that for all $n > N$ and all $p = 1, 2, 3, \ldots$ the inequality $|z_{n+p} - z_n| < \delta$ is fulfilled.

The sum of an infinite sequence of complex numbers $z_0, z_1, \ldots$ is given by

$$S = z_0 + z_1 + z_2 + \cdots = \sum_{n=0}^{\infty} z_n \tag{A.28}$$

Consider the partial sum sequence of n terms, which is designated as S_n. The infinite series converges to the sum S if the partial sum sequence S_n converges to S. That is, the series converges if for

$$S_n = \sum_{n=0}^{n} z_n \quad \lim_{n\to\infty} S_n = S \tag{A.29}$$

When the partial sum S_n diverges, the series is said to diverge.

A.2.1 Comparison Test

Let the terms of the numerical series (A.28) for all $n \geq n_0 \geq 1$ satisfy the condition $|z_n| \leq b_n$. Then, the convergence of the series of positive terms, $\sum_{n=1}^{\infty} b_n$, implies an absolute convergence of the above series.

A.2.2 Limit Comparison Test

If the numerical series, $\sum_{n=1}^{\infty} v_n$, converges absolutely and for the terms of the numerical series (A.28) there takes place the relationship

$$\lim_{n \to \infty} \left| \frac{z_n}{v_n} \right| = q = \text{const} < \infty$$

then series (A.28) converges absolutely.

A.2.3 D'Alembert's Test

If for the terms of the numerical series (A.29), the finite limit

$$\lim_{n \to \infty} \left| \frac{z_{n+1}}{z_n} \right| = l$$

then for $0 \leq l < 1$, series (A.28) converges absolutely; for $l > 1$, series (A.28) diverges; and for $l = 1$, an additional test is required.

A.2.4 Root Test

Consider the sequence

$$r_n = \sqrt[n]{|z_n|}$$

If this sequence converges to l as n approaches infinity, then the series (A.28) converges absolutely if $l < 1$ and diverges if $l > 1$.

A.2.5 Uniform Convergence (Weierstrass M Test)

If $|u_n(z)| \leq M_n$, where M_n is independent of z in a region U and $\sum_{n=1}^{\infty} M_n$ converges, then $\sum_{n=1}^{\infty} u_n(z)$ is uniformly convergent in U.

Example A.6

Show that $\sum_{n=1}^{\infty} \dfrac{1}{n^2 + z^2}$ is uniformly convergent in the interval $1 < |z| < 2$.

SOLUTION

$|n^2 + z^2| \geq |n^2| - |z^2| \geq n^2 - 4 \geq \frac{1}{2}n^2$ for $n > 2$ (the convergence is not affected by dropping the first two terms of the series). Therefore, $\dfrac{1}{|n^2 + z^2|} \leq \dfrac{2}{n^2}$ and the series $\sum_{n=3}^{\infty} \dfrac{2}{n^2}$ converges. From the M test with $M_n = \dfrac{2}{n^2}$ implies that the series converges uniformly. ∎

A.2.6 Analyticity of a Sequence of Functions

If the functions of sequence $\{f_k(z)\}$ are analytic in a region U and the sum

$$F(z) = \sum_{k=1}^{\infty} f_k(z)$$

is uniformly convergent, then $F(z)$ is analytic in U.

Proof: Because $F(z)$ is uniformly convergent for any ε, we can find N such that $|F(z) - S_k(z)| < \varepsilon$ for all $n > N$, where $S_k = $ partial sum $= \sum_{k=1}^{\infty} f_k(z)$. Because $F(z)$ is uniformly convergent, it implies that $f_k(z)$ are continuous and, hence, $F(z)$ is continuous. Integrating within the region U, we obtain (integration is performed counterclockwise)

$$\left| \oint_C F(z)dz - \sum_{k=1}^{n} \oint_C f_k(z)dz \right| < \varepsilon \oint_C dz = \varepsilon \ell(C)$$

where $\ell(C)$ is the length of the contour. Since $\varepsilon \to 0$ as $n \to \infty$ implies that $\oint_C F(z)dz = \sum_{k=1}^{\infty} \oint_C f_k(z)dz = 0$, since $f_k(z)$'s are analytic. Hence, $F(z)$ is also analytic.

A.3 Power Series

A series of the form

$$W(z) = a_0 + a_1(z - z_0) + a_2(z - z_0)^2 + \cdots = \sum_{n=0}^{\infty} a_n(z - z_0)^n \qquad (A.30)$$

where the coefficients a_n are given by

$$a_n = \frac{1}{n!} \frac{d^n W(z)}{dz^n} \bigg|_{z=z_0}, \qquad (A.31)$$

which is a **Taylor** series that is expanded about the point $z = z_0$, where z_0 is a complex constant. That is, the Taylor series expands an analytic function as an infinite sum of component functions. More precisely, the Taylor series expands a function $W(z)$, which is analytic in the neighborhood of the point $z = z_0$, into an infinite series whose coefficients are the successive derivatives of the function at the given point. However, we know that the definition of a derivative of any order does not require more than the knowledge of the function in an arbitrarily small neighborhood of the point $z = z_0$. This means, therefore, that the Taylor series indicates that the shape of the function at a finite distance z_0 from z is determined by the behavior of the function in the infinitesimal vicinity of $z = z_0$. Thus, Taylor's series implies that any analytic function has a very strong interconnected structure, and that by studying the function in a small vicinity of the point $z = z_0$, we can precisely predict what happens at the point $z = z_0 + \Delta z_0$, which is a finite distance from the point of study.

If $z_0 = 0$, the expansion is said to be about the origin and is called a **Maclaurin** series. A power series of negative powers of $(z - z_0)$,

$$W(z) = a_0 + a_1(z - z_0)^{-1} + a_2(z - z_0)^{-2} + \cdots \tag{A.32}$$

is called a **negative power** series.

THEOREM A.4

A positive power series converges absolutely in a circle of radius R^+ centered at z_0 where $|z - z_0| < R^+$; it diverges outside of this circle where $|z - z_0| > R^+$. The value of R^+ may be zero, a positive number, or infinity. If $R^+ = $ infinity, the series converges everywhere, and if it is equal to zero, the series converges only at $z = z_0$. The radius R^+ is found from the relation

$$R^+ = \lim_{n \to \infty} \left| \frac{a_n}{a_{n+1}} \right|, \quad \text{if the limit exists} \tag{A.33}$$

or by

$$R^+ = \lim_{n \to \infty} \frac{1}{\sqrt[n]{|a_n|}}, \quad \text{if the limit exists} \tag{A.34}$$

Proof: For a fixed value z, apply the ratio test, where

$$z_n = a_n(z - z_0)^n$$

That is,

$$\left| \frac{z_{n+1}}{z_n} \right| = \left| \frac{a_{n+1}(z - z_0)^{n+1}}{a_n(z - z_0)^n} \right| = \left| \frac{a_{n+1}}{a_n} \right| |z - z_0|$$

For the power series to converge, the ratio test requires that

$$\lim_{n \to \infty} \left| \frac{a_{n+1}}{a_n} \right| |z - z_0| < 1 \quad \text{or} \quad |z - z_0| < \lim_{n \to \infty} \left| \frac{a_n}{a_{n+1}} \right| = R^+$$

That is, the power series converges absolutely for all z that satisfy this inequality. It diverges for all z for which $|z - z_0| > R^+$. The value of R^+ specified by (A.34) is reduced by applying the root test.

Example A.7

Determine the region of convergence for the power series

$$W(z) = \frac{1}{1 + z} = 1 - z + z^2 - z^3 + \cdots$$

SOLUTION

We have $a_n = (-1)^n$, from which

$$R^+ = \lim_{n \to \infty} \left| \frac{(-1)^n}{(-1)^{n+1}} \right| = |1|$$

The series converges for all z for which $|z| < 1$. Hence, this expansion converges for any value of z within a circle of unit radius about the origin. Note that there will be at least one singular point of $W(z)$ on the circle of convergence. In the present case, the point $z = -1$ is a singular point. ∎

Example A.8

Determine the region of convergence for the power series

$$W(z) = e^z = 1 + z + \frac{z^2}{2!} + \frac{z^3}{3!} + \cdots = \sum_{n=1}^{\infty} \frac{1}{n!} z^n$$

SOLUTION

We have $a_n = 1/n!$ from which

$$R^+ = \lim_{n \to \infty} \left| \frac{(n+1)!}{n!} \right| = \lim_{n \to \infty} (n+1) = \infty$$

The circle of convergence is specified by $R^+ = $ infinity; hence, $W(z) = e^z$ converges for all finite values of z. ∎

THEOREM A.5

A negative power series (A.32) converges absolutely outside a circle of radius R^- centered at z_0, where $|z - z_0| > R^-$; it diverges inside of this circle where $|z - z_0| < R^-$. The radius of convergence is determined from

$$R^- = \lim_{n \to \infty} \left| \frac{a_{n+1}}{a_n} \right|, \quad \text{if the limit exists} \tag{A.35}$$

or by

$$R^- = \lim_{n \to \infty} \sqrt[n]{|a_n|}, \quad \text{if the limit exists} \tag{A.36}$$

Proof: The proof of this theorem parallels that of Theorem A.4.

If a function has a singularity at $z = z_0$, it cannot be expanded in a Taylor series about this point. However, if one deletes the neighborhood of z_0, it can be expressed in the form of a Laurent series. The Laurent series is written

$$W(z) = \cdots + \frac{a_{-2}}{(z - z_0)} + \frac{a_{-1}}{(z - z_0)} + a_0 + a_1(z - z_0) + a_2(z - z_0)^2 + \cdots$$

$$= \sum_{n=-\infty}^{\infty} a_n(z - z_0)^n \tag{A.37}$$

If a circle is drawn about the point z_0 such that the nearest singularity of $W(z)$ lies on the circle, then (A.37) defines an analytic function everywhere within this circle except at its center. The portion $\sum_{n=0}^{\infty} a_n(z - z_0)^n$ is regular at $z = z_0$. The portion $\sum_{n=-1}^{-\infty} a_n(z - z_0)^n$ is not regular and is called the principal part of $W(z)$ at $z = z_0$.

The region of convergence for the positive series part of the Laurent series is of the form

$$|z - z_0| < R^+ \tag{A.38}$$

while that for the principal part is given by

$$|z - z_0| > R^- \tag{A.39}$$

The evaluation of R^+ and R^- proceeds according to the methods already discussed. Hence, the region of convergence of the Laurent series is given by those points common to (A.38) and (A.39) or for

$$R^- < |z - z_0| < R^+ \tag{A.40}$$

If $R^- > R^+$, the series converges nowhere. The annular region of convergence for a typical Laurent series is shown in Figure A.8.

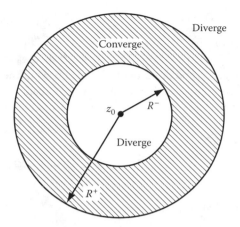

FIGURE A.8

Example A.9

Consider the Laurent series $W(z) = \sum_n a_n z^n$ where

$$a_n \begin{cases} \left(\dfrac{1}{3}\right)^n & \text{for } n = 0, 1, 2, \ldots \\ 2^n & \text{for } n = -1, -2, \ldots \end{cases}$$

Determine the region of convergence.

SOLUTION

By (A.38) and (A.33), we have $R^+ - 3$. By (A.38) and (A.35) we have $R^- = 2$. Hence, the series converges for all z for which $2 < |z| < 3$.

No convenient expression is got for obtaining the coefficients of the Laurent series. However, because there is only one Laurent expansion for a given function, the resulting series, however derived, will be the appropriate one. For example,

$$e^{1/z} = 1 + \frac{1}{z} + \frac{1}{2!z^2} + \frac{1}{3!z^3} + \cdots \tag{A.41}$$

is obtained by replacing z with $1/z$ in the Maclaurin expansion of $\exp(z)$. Note that in this case the coefficients of all positive powers of z in the Laurent expansion are zero. As a second illustration, consider the function $W(z) = (\cos z)/z$. This is found by dividing the Maclaurin series for $\cos z$ by z, with the result

$$\frac{\cos z}{z} = \frac{1}{z}\left(1 - \frac{z^2}{2!} + \frac{z^4}{4!} - \cdots\right) = \frac{1}{z} - \frac{z}{2!} + \frac{z^3}{4!} - \cdots \tag{A.42}$$

In this case, the Laurent expansion includes only one term $1/z$ in descending powers of z, but an infinite number of terms in ascending powers of z. That is, $a_{-1} = 1$ and $a_{-n} = 0$ if $n \neq 1$. ∎

A.4 Analytic Continuation

The Taylor theorem shows that if a function $f(z)$ is given by a power in z, it can also be represented as a power series in $z - z_0 = f[(z - z_0) + z_0]$ where z_0 is any point within the original circle of convergence, and this series will converge within any circle about z_0 that does not pass beyond the original circle of convergence. Actually, it may converge within a circle that does not pass beyond the original circle of convergence. Consider, for example, the function

$$f(z) = 1 + z + z^2 + \cdots = \frac{1}{1 - z} \quad \text{for } |z| < 1$$

Choose $z_0 = j/2$, and the Taylor expansion of

$$f(z) = \frac{1}{1 - \left[\left(z - \frac{1}{2}j \right) + \frac{1}{2}j \right]} = \frac{1}{\left(1 - \frac{1}{2}j \right) - z'} \quad z' = z - \frac{1}{2}j$$

in powers of z' is

$$f(z) = \frac{1}{1 - \frac{1}{2}j} + \frac{z'}{\left(1 - \frac{1}{2}j \right)^2} + \frac{z'^2}{\left(1 - \frac{1}{2}j \right)^3} + \cdots$$

This series must converge and be equal to the original function if $|z'| < 1/2$, because j is the point of circle $|z| = 1$ nearest to $j/2$, a requirement of Taylor's theorem. Actually, this series converges if $|z'| < \left| 1 - \frac{1}{2}j \right| = \frac{1}{2}\sqrt{5}$.

Suppose that the considered series represented no previously known function. In this case, the new Taylor series would define values of an analytic function over a range of z where no function is defined by the original series. Then, we can extend the range of definition by taking a new Taylor series about a point in the new region. This process is called **analytic continuation**. In practice, when continuation is required, the direct use of the Taylor series is laborious and is seldom used. Of more convenience is the following theorem.

THEOREM A.6

If two functions $f_1(z)$ and $f_2(z)$ are analytic in a region D and equal in a region D' within D, they are equal everywhere in D.

A.5 Singularities of Complex Functions

A singularity has already been defined as a point at which a function ceases to be analytic. Thus, a discontinuation function has a singularity at the point of discontinuity, and multivalued functions have a singularity at a branch point. There are two important classes of singularities that a continuous, single-valued function may possess.

Definition A.3   A function has an **essential singularity** at $z = z_0$ if its Laurent expansion about the point z_0 contains an infinite number of terms in inverse powers of $(z - z_0)$.

Definition A.4   A function has a **nonessential singularity** or **pole of order** m if its Laurent expansion can be expressed in the form

$$W(z) = \sum_{n=-m}^{\infty} a_n (z - z_0)^n \tag{A.43}$$

Note that the summation extends from $-m$ to infinity and not from minus infinity to infinity; that is, the highest inverse power of $(z - z_0)$ is m.

An alternative definition that is equivalent to this but somewhat simpler to apply is the following: if $\lim_{z \to z_0} [(z - z_0)^m W(z)] = c$, a nonzero constant (here m is a positive number), then $W(z)$ is said to possess a pole of order m at z_0. The following examples illustrate these definitions:

1. $\exp(1/z)$ (see A.41) has an essential singularity at the origin.
2. $\cos z/z$ (see A.42) has a pole of order 1 at the origin.
3. Consider the following function:

$$W(z) = \frac{e^z}{(z - 4)^2 (z^2 + 1)}$$

Note that functions of this general type exist frequently in the Laplace inversion integral. Because e^z is regular at all finite points of the z-plane, the singularities of $W(z)$ must occur at the points for which the denominator vanishes; that is, for

$$(z - 4)^2 (z^2 + 1) = 0 \quad \text{or} \quad z = 4, +j, -j$$

By the second definition above, it is easily shown that $W(z)$ has a second-order pole at $z = 4$, and first-order poles at the two points $+j$ and $-j$. That is,

$$\lim_{z \to 4} (z - 4)^2 \left[\frac{e^z}{(z - 4)^2 (z^2 + 1)} \right] = \frac{e^4}{17} \neq 0$$

$$\lim_{z \to j} (z - j) \left[\frac{e^z}{(z-4)^2 (z^2+1)} \right] = \frac{e^j}{(j-4)^2 2j} \neq 0$$

4. An example of a function with an infinite number of singularities occurs in heat flow, wave motion, and similar problems. The function involved is

$$W(z) = 1/\sinh az$$

The singularities in this function occur when $\sinh az = 0$ or $az = js\pi$, where $s = 0$, $\pm 1, \pm 2, \dots$. That each of these is a first-order pole follows from

$$\lim_{z \to j(s\pi/a)} \left(z - j \frac{s\pi}{a} \right) \frac{1}{\sinh az} = \frac{0}{0}$$

This can be evaluated in the usual manner by differentiating numerator and denominator (L'Hospital rule) to find

$$\lim_{z \to j(s\pi/a)} \frac{1}{a \cosh az} = \frac{1}{a \cosh js\pi} = \frac{1}{a \cos s\pi} \neq 0$$

Definition A.5 (*Isolated singularities*)   The point $z = z_0$ is an **isolated singularity** of $W(z)$ if we can always find δ such that the circle $|z - z_0| = \delta$ does not contain another singularity. If no such δ exists, the point z_0 is known as a **nonisolated singularity**.

Definition A.6 (*Poles*)   If $\lim_{z \to z_0} (z - z_0)^n W(z) = constant \neq 0$, where n is positive, then the point $z = z_0$ is called a **pole of order** n. If $n = 1$, z_0 is called a **simple pole**.

Example A.10

It is interesting to study the variation of $f(z)$ close to the pole. For example, the function $W(z) = 1/z = (1/r)e^{-j\theta}$ has a simple pole at zero. For any specific angle θ_1, the modulus, $|W(z)|$, increases to infinity as $r \to 0$, and this is true for all the angles from 0 to 2π.

Definition A.7 (*Removable singularities*)   The point $z = z_0$ is a **removable singularity** of $W(z)$, if $\lim_{z \to z_0} W(z)$ exists.

Definition A.8 (*Branch points*)   Multiple-valued functions contain singular points known as the **branch points**.

Example A.11

Investigate the function $W(z) = z^{1/2}$.

SOLUTION

In polar form, we have $W = \sqrt{z} = r^{1/2}\left(\cos\frac{1}{2}\theta + j\sin\frac{1}{2}\theta\right)$ (see Figure A.9) where $z = x + jy$, $r = \sqrt{x^2 + y^2}$, and $\theta = \tan^{-1}(y/x)$. If we increase θ by 2π, we obtain $W = \sqrt{z} = \left[\cos\left(\frac{1}{2}\theta + \pi\right) + j\sin\left(\frac{1}{2}\theta + \pi\right)\right] = -\sqrt{r}\left[\cos\frac{1}{2}\theta + j\sin\frac{1}{2}\theta\right]$, which is evident from Figure A.9b. This implies that $W(z)$ has two values, one value for $0 \leq \theta \leq \pi$ and the other from $\pi \leq \theta \leq 2\pi$. This indicates that $W(z)$ is not analytic on the positive real axis when the angle ranges from $0 \leq \theta \leq 2\pi$. If we create a barrier (or cut) to exist along Ox (see Figure A.9c), then θ cannot take the values $0, 2n\pi, n = 1, 2, \ldots$ Then for the angle $0 < \theta < 2\pi$, W is single valued and continuous and, therefore, analytic. This angle is known as the **principal branch** of the function.

The origin 0 is called the branch point. To make $W = \sqrt{z}$ unique on each branch, the barrier must start from the branch point. The angular position of the barrier is arbitrary.　■

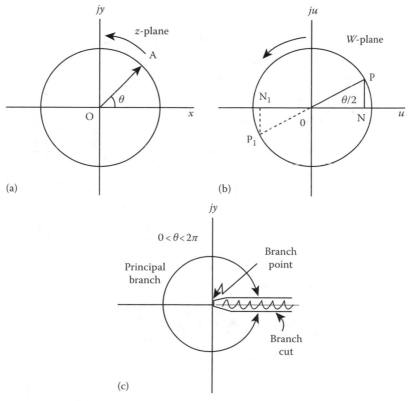

FIGURE A.9　Illustrating $W(z) = z^{1/2}$.

Example A.12

Investigate phase change in relation to branch points.

SOLUTION

If the contour is that shown in Figure A.10a for the function $W(z) = z^{1/2}$, then as z varies on the contour from A to B it sweeps the angle from θ_1 to θ_2. Then, if it varies from B to C continuously in the counterclockwise direction, the angle is swept from θ_2 to θ_1. The angle swept out of OA is zero since it oscillated to B and back. Hence, the value of the function is that given by $W = \sqrt{r}\left[\cos\frac{1}{2}\theta_1 + j\sin\frac{1}{2}\theta_1\right]$.

Let us investigate the function $W(z) = \sqrt{z(z-a)}$, which has two branch points, $z = 0$ and $z = a$. From Figure A.10b we obtain

$$W = \sqrt{r_1 r_2}\, e^{j\frac{1}{2}(\theta_1 + \theta_2)} = \sqrt{r_1 r_2}\left[\cos\frac{1}{2}(\theta_1 + \theta_2) + j\sin\frac{1}{2}(\theta_1 + \theta_2)\right]$$

If we start from P moving counterclockwise, θ_1 varies by 2π and θ_2 returns back to its original value. Hence,

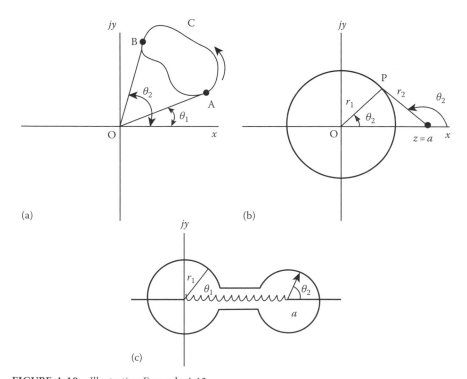

(a)

(b)

(c)

FIGURE A.10   Illustrating Example A.12.

$$W = \sqrt{r_1 r_2}\left[\cos\frac{1}{2}(\theta_1 + 2\pi + \theta_2) + j\sin\frac{1}{2}(\theta_1 + 2\pi + \theta_2)\right]$$

$$= -\sqrt{r_1 r_2}\left[\cos\frac{1}{2}(\theta_1 + \theta_2) + j\sin\frac{1}{2}(\theta_1 + \theta_2)\right]$$

which implies that the function enters a new branch. When P rotates twice, θ_1 becomes $\theta_1 + 4\pi$ and the function has its original value.

From Figure A.10c, we observe that as we trace the contour θ_1 and θ_2 change by 2π each. Therefore,

$$W = \sqrt{r_1 r_2}\left[\cos\frac{1}{2}(\theta_1 + \theta_2 + 4\pi) + j\sin\frac{1}{2}(\theta_1 + \theta_2 + 4\pi)\right]$$

$$= \sqrt{r_1 r_2}\left[\cos\frac{1}{2}(\theta_1 + \theta_2) + j\sin\frac{1}{2}(\theta_1 + \theta_2)\right]$$

and the function regains its original value. Hence, the branch cut can be from 0 to point a as indicated in the figure. ∎

Example A.13 (Essential singularity)

Let us investigate the function $W(z) = e^{1/z}$ as $z \rightarrow 0$. We write

$$W = e^{1/z} = e^1 /_{[re^{j\theta}]} = e^{\frac{1}{r}e^{-j\theta}} = e^{\frac{1}{r}\cos\theta}e^{-j\frac{\sin\theta}{r}} = e^{\frac{\cos\theta}{r}}\left[\cos\frac{\sin\theta}{r} - j\sin\frac{\sin\theta}{r}\right]$$

$$= u + jv = r_1 e^{-j\theta_1} = r_1\cos\theta_1 - jr_1\sin\theta_1$$

where

$$r_1 = e^{\frac{1}{r}\cos\theta}, \quad \theta_1 = \frac{\sin\theta}{r}, \quad u = r_1\cos\theta_1 \quad v - -r_1\sin\theta_1$$

We observe that for a specific value of $0 < \theta < \pi/2$ as $r \rightarrow 0$, $r_1 = e^{\cos\theta/r}$, and $\theta_1 = \sin\theta/r$ increase rapidly and, thus, the point $W = r_1\cos\theta_1 - j\, r_1\sin\theta_1 = r_1[\cos\theta_1 - j\sin\theta_1]$ rotates with an ever-increasing length and angular velocity (assuming $r \rightarrow 0$ linearly with time). Although the function behaves peculiar near the singularity, it is analytic there and it is said that the singularity is isolated. ∎

Definition A.9 (*Singularity at ∞*)   A singularity of $W(z)$ at infinity is the same as that of $W(1/\zeta)$ at $\zeta = 0$, where $z = 1/\zeta$.

Example A.14

The function $W(z) = z^2$ has a double pole at ∞, since $W(1/\zeta) = 1(\zeta)^2$ has a double pole at $\zeta = 0$.

In general, (1) if a function has a branch point at $z = z_0$, it has no other type of singularity there; (2) if a function has an essential singularity at $z = z_0$, it has no pole there. ∎

A.6 Theory of Residues

It has already been shown that the contour integral of any function that encloses no singularities of the integrand will vanish. (In this section, all the contour integrals are taken counterclockwise unless it is indicated otherwise.) Now our purpose is to examine the integral, the path of which encloses one singularity, say at $z = z_0$. The Laurent expansion of such a function is

$$W(z) = \sum_{n=-\infty}^{\infty} a_n (z - z_0)^n$$

and so

$$\oint_C W(z)dz = \sum_{n=-\infty}^{\infty} a_n \oint_{C_n} (z - z_0)^n dz$$

But by (A.11), each term in the sum vanishes except for $n = -1$, with

$$\oint_C (z - z_0)^{-1} dz = 2\pi j$$

In then follows that

$$\oint_C W(z)dz = \sum_{n=-\infty}^{\infty} \oint_{C_n} (z - z_0)^n dz = 2\pi j a_{-1} \tag{A.44}$$

Because the integral $(1/2\pi j) \oint_C W(z)dz$ will appear frequently in subsequent applications, it is given a name; it is called the **residue of $W(z)$** at z_0 and is abbreviated Res(W).

From the second corollary (A.13), it follows that if $W(z)$ has n isolated singularities within C, then

$$\frac{1}{2\pi j} \oint_C W(z)dz = \sum_{s=1}^{n} \frac{1}{2\pi j} \oint_{C_s} W(z)dz = \sum_{s=1}^{n} \text{Res}_s(W) \tag{A.45}$$

or, in other words, the value of the contour integral equals the sum of the residues within C. Observe that to evaluate integrals in the complex plane, it is only necessary to find the residues at the singularities of the integrand within the contour. One obvious way of doing this is (see (A.44)) to find the coefficient a_{-1} in the Laurent expansion about each singularity. However, this is not always an easy task.

Several theorems exist that make evaluating residues relatively easy. We introduce these.

THEOREM A.7

If the $\lim_{z \to z_0}[(z - z_0)W(z)]$ is finite, this limit is the residue of $W(z)$ at $z = z_0$. If the limit is not finite, then $W(z)$ has a pole of at least second order at $z = z_0$ (it may possess an essential singularity here). If the limit is zero, then $W(z)$ is regular at $z = z_0$.

Proof: Suppose that the function is expanded into the Laurent series

$$W(z) = \frac{a_{-1}}{z - z_0} + a_0 + a_1(z - z_0) + a_2(z - z_0)^2 + \cdots$$

Then the expression

$$\lim_{z \to \infty} [(z - z_0)W(z)] = \lim_{z \to \infty} \left[a_{-1} + a_0(z - z_0) + a_1(z - z_0)^2 + \cdots \right] = a_{-1}$$

This proves the theorem.

This process was previously used to ascertain whether or not a function had a first-order pole at $z = z_0$. Thus, referring back to the examples in Section A.5, we have

$$\text{Res}\left(\frac{\cos z}{z}\right)_{z=0} = 1$$

$$\text{Res}\left[\frac{e^z}{(z - 4)^2(z^2 + 1)}\right]_{z=j} = \frac{e^j}{(j - 4)^2 2j}$$

$$\text{Res}\left(\frac{1}{\sinh az}\right)_{z=j(s\pi/a)} = \frac{1}{a \cos s\pi}$$

Many of the singularities that arise in system function studies are first-order poles. The evaluation of the integral is relatively direct.

Example A.15

Evaluate the following integral

$$\frac{1}{2\pi j} \oint_C \frac{e^{zt}}{(z^2 + \omega^2)} dz$$

when the contour C encloses both first-order poles at $z = \pm j\omega$. Note that this is precisely the Laplace inversion integral of the function $1/(z^2 + \omega^2)$.

SOLUTION

This involves finding the following residues

$$\mathrm{Res}\left(\frac{e^{zt}}{z^2 + \omega^2}\right)_{z=j\omega} = \frac{e^{j\omega t}}{2j\omega} \quad \mathrm{Res}\left(\frac{e^{zt}}{z^2 + \omega^2}\right)_{z=-j\omega} = -\frac{e^{-j\omega t}}{2j\omega}$$

Hence,

$$\frac{1}{2\pi j}\oint_C \frac{e^{zt}}{z^2 + \omega^2}dz = \sum \mathrm{Res} = \left(\frac{e^{j\omega t} - e^{-j\omega t}}{2j\omega}\right) = \frac{\sin \omega t}{\omega}$$

A slight modification of the method for finding residues of simple poles

$$\mathrm{Res}W(z_0) = \lim_{z \to z_0} [(z - z_0)W(z)] \qquad (A.46)$$

makes the process even simpler. This is specified by the following theorem.   ■

THEOREM A.8

Suppose that $f(z)$ is analytic at $z = z_0$ and suppose that $g(z)$ is divisible by $z - z_0$ but not by $(z - z_0)^2$. Then

$$\mathrm{Res}\left[\frac{f(z)}{g(z)}\right]_{z=z_0} = \frac{f(z_0)}{g'(z_0)} \quad where \quad g'(z) = \frac{dg(z)}{dz} \qquad (A.47)$$

Proof: We write the relation $(z - z_0)\, h(z) = g(z)$, then $g'(z) = (z - z_0)\, h'(z) + h(z)$ so that for $z = z_0$, $g'(z_0) = h(z_0)$. Then we have

$$\mathrm{Res}\left[\frac{f(z)}{g(z)}\right]_{z=z_0} = \lim_{z \to z_0}\left[(z - z_0)\frac{f(z)}{g(z)}\right] = \lim_{z \to z_0}\left[\frac{f(z)}{h(z)}\right] = \frac{f(z_0)}{h(z_0)} = \frac{f(z_0)}{g'(z_0)},$$

which is the given result.

In reality, this theorem has already been used in the evaluation of $\mathrm{Res}(1/\sinh az)_{z=j(s\pi/a)}$. Here $f(z) = 1$, $g(z) = \sinh az$, and $g'(z) = a \cosh az$.

As a second illustration, consider the previously used function

$$W(z) = \frac{e^z}{(z - 4)^2(z^2 + 1)}$$

Here, we take

$$f(z) = \frac{e^z}{(z-4)^2}, \quad g(z) = z^2 + 1$$

Thus, $g'(z) = 2z$ and the previous result follows immediately with

$$\text{Res}\left[\frac{e^z}{(z-4)^2(z^2+1)}\right] = \frac{e^j}{(j-4)^2 2j}$$

Equation A.46 permits a simple proof of the Cauchy second integral theorem (A.15). This involves choosing $g(z) = (z - z_0)$ in the integral

$$\frac{1}{2\pi j} \oint_C \frac{f(z)}{z - z_0} dz = \frac{f(z_0)}{1} = f(z_0) \tag{A.48}$$

Suppose that (A.48) is differentiated $n - 1$ times with respect to z_0. Then we write

$$\frac{d^{n-1}f(z_0)}{dz_0^{n-1}} \doteq f^{(n-1)}(z_0) = \frac{(n-1)!}{2\pi j} \oint_C \frac{f(z)}{(z-z_0)^n} dz \tag{A.49}$$

This species is any-order derivative of a complex function expressed as a contour integral.

Our discussion so far has concentrated on finding the residue of a first-order pole. However, (A.49) permits finding the residue of a pole of any order. If, for example, $W(z) = [f(z)/(z - z_0)^n]$, then evidently $W(z)$ has a pole of order n at $z = z_0$ because $f(z)$ is analytic at $z = z_0$. Then $f(z) = (z - z_0)^n W(z)$, and (A.49) becomes

$$\text{Res}(W(z))|_{z=z_0} = \frac{1}{2\pi j} \oint_C W(z)dz = \frac{1}{(n-1)!} \frac{d^{n-1}}{dz^{n-1}} [(z - z_0)^n W(z)]_{z=z_0} \tag{A.50}$$

Example A.16

Evaluate the residue at the second-order pole at $z = 4$ of the previously considered function

$$W(z) = \frac{e^z}{(z-4)^2(z^2+1)}$$

Solution

It follows from (A.50) that

$$\text{Res}W(z)|_{z=4} = \frac{1}{1!} \frac{d}{dz} \left[\frac{e^z}{z^2+1}\right]_{z=4} = \frac{9e^4}{289}$$

∎

Example A.17

Evaluate the residue at the third pole of the function

$$W(z) = \frac{e^{zt}}{(z+1)^3}$$

SOLUTION

A direct application of (A.50) yields

$$\text{Res}W(z)\big|_{z=-1} = \frac{1}{2!}\frac{d^2}{dz^2}(e^{zt})\bigg|_{z=-1} = \frac{1}{2}t^2 e^{-t}$$

There is no simple way of finding the residue at an essential singularity. The Laurent expansion must be found and the coefficient a_{-1} is thereby obtained. For example, from (A.41) it is seen that the residue of $\exp(1/z)$ at the origin is unity. Fortunately, an essential singularity seldom arises in practical applications.

Sometimes the function takes the form

$$W(z) = \frac{f(z)}{zg(z)} \tag{A.51}$$

where the numerator and the denominator are prime to each other, $g(z)$ has no zero at $z=0$ and cannot be factored readily. The residue due to the pole at zero is given by

$$\text{Res}W(z) = \frac{f(z)}{g(z)}\bigg|_{z=0} = \frac{f(0)}{g(0)} \tag{A.52}$$

If $z=a$ is the zero of $g(z)$, then the residue at $z=a$ is given by

$$\text{Res}W(z) = \frac{f(a)}{ag'(a)} \tag{A.53}$$

If there are N poles of $g(z)$, then the residues at all simple poles of $W(z)$ are given by

$$\sum \text{Res} = \frac{f(z)}{g(z)}\bigg|_{z=0} + \sum_{m=1}^{N}\left[f(z)\bigg/\left(z\frac{dg(z)}{dz}\right)\right]_{z=a_m} \tag{A.54}$$

If $W(z)$ takes the form $W(z) = f(z)/[h(z)g(z)]$ and the simple poles to the two functions are not common, then the residues at all simple poles are given by

$$\sum \text{Res} = \sum_{m=1}^{N}\frac{f(a_m)}{h(a_m)g'(a_m)} + \sum_{r=1}^{R}\frac{f(b_r)}{h'(b_r)g(b_r)} \tag{A.55}$$

■

Example A.18

Find the sum of the residues $e^{2z}/\sin mz$ at the first $N+1$ poles on the negative axis.

SOLUTION

The simple poles occur at $z = -n\pi/m$, $n = 0, 1, 2, \ldots$. Thus

$$\sum \mathrm{Res} = \sum_{n=0}^{N} \left[\frac{e^{2z}}{m \cos mz} \right]_{z=-n\pi/m} = \frac{1}{m} \sum_{n=0}^{N} (-1)^n e^{-2n\pi/m} \qquad \blacksquare$$

Example A.19

Find the sum of the residues of $e^{2z}/(z \cosh mz)$ at the origin and at the first N poles on each side of it.

SOLUTION

The zeros of $\cosh mz$ are $z = -j(n + 1/2)\pi/m$, n integral. Because $\cosh mz$ has no zero at $z = 0$, then (A.55) gives

$$\sum \mathrm{Res} = 1 + \sum_{n=-N}^{N-1} \left[\frac{e^{2z}}{mz \sinh mz} \right]\Bigg|_{z=-\left(n+\frac{1}{2}\right)\pi j/m} \qquad \blacksquare$$

Example A.20

Find the residue of $ze^z/\sin mz$ at the origin.

SOLUTION

Because near $z = 0$ $\sin mz \approx mz$ there is no pole at the origin and, hence, the integral $(1/2\pi j) \int_C ze^z dz / \sin mz$ is equal to zero for a contour encircling the origin with radius less than π/m. $\qquad \blacksquare$

A.7 Aids to Integration

The following three theorems will substantially simplify the evaluation of certain integrals in the complex plane. Examples will be found in later applications.

THEOREM A.9

*If AB is the arc of a circle of radius $|z| = R$ for which $\theta_1 \leq \theta \leq \theta_2$ and if $\lim_{R\to\infty}(zW(z)) = k$,
a constant that may be zero, then*

$$\lim_{R\to\infty} \int_{AB} W(z)dz = jk(\theta_2 - \theta_1) \tag{A.56}$$

Proof: Let $zW(z) = k + \varepsilon$, where $\varepsilon \to 0$ as R approaches infinity. Then,

$$\int_{AB} W(z)dz = \int_{AB} \frac{k+\varepsilon}{z} dz = (k+\varepsilon)\int_{\theta_1}^{\theta_2} j\, d\theta = (k+\varepsilon)j(\theta_2 - \theta_1)$$

In carrying out this integration, the procedure employed in Example A.2 is used. In the
limit as R approaches infinity, (A.56) follows.

This theorem can be shown to be valid even if there are a finite number of points
on the arc AB for which the $\lim_{R\to\infty}(zW(z)) \neq k$, provided only that the limit remains
finite for finite R at these points. This theorem can also be proved true if we choose
$\lim_{R\to\infty}(z-a)W(z) = k$ when the integral is taken around the arc $\theta_1 \leq \arg(z-a) \leq \theta_2$ of
the circle $|z - av| = r$.

THEOREM A.10

*If A B is the arc of a circle of radius $|z - z_0| = r$ for which $\varphi_1 \leq \varphi \leq \varphi_2$ (as shown in Figure
A.11) and if $\lim_{z\to z_0}[(z - z_0)W(z)] = k$, a constant that may be zero, then*

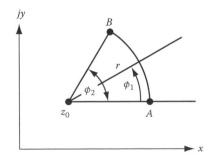

FIGURE A.11

$$\lim_{r \to 0} \int_{AB} W(z)dz = jk(\varphi_2 - \varphi_1) \qquad (A.57)$$

where r and φ are introduced polar coordinates, with the point $z = z_0$ as origin.

Proof: The proof of this theorem follows along similar lines to that of Theorem A.9.

Note specifically that Theorem A.9 will allow the evaluation of integrals over infinitely large arcs, whereas Theorem A.10 will allow the evaluation over infinitely small arcs.

THEOREM A.11

If the maximum value of W(z) along a path C (not necessarily closed) is M, the maximum value of the integral of W(z) along C is Ml, where l is the length of C. When expressed analytically, this specifies that

$$\left| \int_C W(z)dz \right| \le Ml \qquad (A.58)$$

Proof: The proof of this theorem is very simple if recourse is made to the definition of an integral. Thus, from Figure A.12

$$\left| \int_C W(z)dz \right| = \left| \lim_{\substack{\Delta z_s \to 0 \\ n \to \infty}} \sum_{s=1}^{n} W_s \Delta z_s \right| \le M \lim_{\substack{\Delta z_s \to 0 \\ n \to \infty}} \sum_{s=1}^{n} |\Delta z_s| = Ml$$

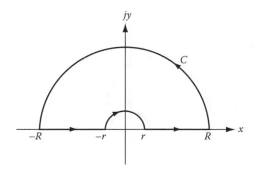

FIGURE A.12

JORDAN'S LEMMA 7.1

If t < 0 and

$$f(z) \to 0 \quad as\ z \to \infty \tag{A.59}$$

then

$$\int_C e^{tz} f(z) dz \to 0 \quad as\ r \to \infty \tag{A.60}$$

where C is the arc shown in Figure A.13a.

Proof: We must assume that the angle of C does not exceed π, $0 \le \arg z \le \pi$. This is not true if $c < 0$. However, the portion of C in the Re $z < 0$ region will have length not exceeding $\pi|c|$. Hence, because of (A.59) the integration over this portion will tend to zero. From (A.59) it follows that, given $\varepsilon > 0$, we can find a constant r_0 such that

$$|f(z)| < \varepsilon \quad \text{for } |z| > r_0$$

Hence, with $z = re^{j\theta}$, $r > r_0$ we obtain $(t < 0)$

$$\left| \int_C e^{tz} f(z) dz \right| = \left| \int_{-\pi/2}^{\pi/2} e^{tr(\cos\theta + j\sin\theta)} f(re^{j\theta}) j re^{j\theta} d\theta \right|$$

$$< \varepsilon r \int_{-\pi/2}^{\pi/2} e^{tr\cos\theta} d\theta \le \varepsilon r 2 \int_{0}^{\pi/2} e^{tr(1 - 2\theta/\pi)} d\theta$$

$$= \frac{\varepsilon r \pi}{|t| r}(1 - e^{rt}) < \frac{\pi \varepsilon}{|t|}$$

Because ε is arbitrarily small, the lemma is verified.

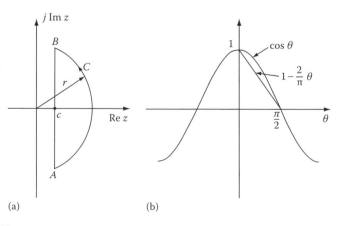

(a)   (b)

FIGURE A.13

From the above lemma we conclude that if $f(z)$ is analytic everywhere in the Re $z \geq c$ region except at a number of poles, then

$$\int_{Br} e^{tz} f(z) dz = -2\pi j \sum_{k=1}^{n} \text{Res}_k \quad t < 0 \tag{A.61}$$

where the Br stands for the Bromwich integration from $c - j\infty$ to $c + j\infty$, which is the line $A\,B$ in Figure A.13a. Res_k are the corresponding residues; the minus sign occurs because of the direction of integration along the Br line from B to A. The lemma can easily be extended for $t > 0$ and C can be an arc lying on the Re $z < c$ plane. The residues are given by

$$\int_{Br} e^{tz} f(z) dz = 2\pi j \sum_{k=1}^{n} \text{Res}_k \quad t > 0 \tag{A.62}$$

THEOREM A.12 MELLIN 1

Let

(a) *$\phi(z)$ be analytic in the strip $\alpha < x < \beta$, both alpha and beta being real*
(b) *$\int_{x-j\infty}^{x+j\infty} |\phi(z)| dz = \int_{-\infty}^{\infty} |\phi(x + jy)| dy$ converges*
(c) *$\phi(z) \to 0$ uniformly as $|y| \to \infty$ in the strip $\alpha < x < \beta$*
(d) *$\theta = $ real and positive: if*

$$f(\theta) = \frac{1}{2\pi j} \int_{c-j\infty}^{c+j\infty} \theta^{-z} \phi(z) dz \tag{A.63}$$

then

$$\phi(z) = \int_{0}^{\infty} \theta^{z-1} f(\theta) d\theta \tag{A.64}$$

THEOREM A.13 MELLIN 2

For θ real and positive, $\alpha < $ Re $z < \beta$, let $f(\theta)$ be continuous or piecewise continuous, and integral (A.64) be absolutely convergent. Then (A.63) follows from (A.64).

THEOREM A.14 MELLIN 3

If in (A.63) and (A.64) we write $\theta = e^{-t}$, t being real, and in (A.64) put p for z and g(t) for $f(e^{-t})$, we get

$$g(t) = \frac{1}{2\pi j} \int_{c-j\infty}^{c+j\infty} e^{zt}\phi(z)dz \qquad (A.65)$$

$$\phi(p) = \int_{-\infty}^{\infty} e^{-pt}g(t)dt \qquad (A.66)$$

A.7.1 Transformation of Contour

To evaluate formally the integral

$$I = \int_{0}^{a} \cos xt\, dt \qquad (A.67)$$

we set $\upsilon = xt$ that gives $dx = d\upsilon/t$ and, thus,

$$I = \frac{1}{t} \int_{0}^{at} \cos\upsilon\, d\upsilon = \frac{\sin at}{t} \qquad (A.68)$$

Regarding this as a contour integral along the real axis for $x = 0$ to a, the change to $\upsilon = xt$ does not change the real axis. However, the contour is unaltered except in length.

Let t be real and positive. If we set $z = \zeta t$ or $\zeta = z/t$, the contour in the ζ-plane is identical in type with that in the z-plane. If it were a circle of radius r in the z-plane, the contour in the ζ-plane would be a circle of radius r/t. When t is complex $z = r_1 e^{j\theta_1}$, $t = r_2 e^{j\theta_2}$, so $\zeta = (r_1/r_2)e^{j(\theta_1 - \theta_2)}$, r_1, θ_1 being variables while r_2 and θ_2 are fixed. If $z = jy = |z|e^{j\theta_1} = |z|e^{j\pi/2}$ and if the phase of t was $\theta_2 = \pi/4$, then the contour in the ζ-plane would be a straight line at $45°$ with respect to the real axis. In effect, any figure in the z-plane transforms into a similar figure in the ζ-plane, whose orientation and dimensions are governed by the factor $1/t = e^{-j\theta_2}/r_2$.

Example A.21

Make the transformation $z = \zeta t$ to the integral $I = \int_C e^{z/t}\frac{dz}{z}$, where C is a circle of radius r_0 around the origin.

SOLUTION

$dz/z = d\zeta/\zeta$ so $I = \int_{C'} e^{\zeta} \frac{d\zeta}{\zeta}$, where C' is a circle around the origin of radius r_0/r $(r = |t|)$. ■

Example A.22

Discuss the transformation $z = (\zeta - a)$, a being complex and finite.

SOLUTION

This is equivalent to a shift of the origin to point $z = -a$. Neither the contour nor the position of the singularities is affected in relation to each other, so the transformation can be made without any alteration in technique. ■

Example A.23

Find the new contour due to transformation $z = \zeta^2$ if the contour was the imaginary axis, $z = jy$.

SOLUTION

Choosing the positive square root we have $\zeta = (jy)^{1/2}$ above and $\zeta = (-jy)^{1/2}$ below the origin. Because

$$\sqrt{j} = \left(e^{j\pi/2}\right)^{1/2} = e^{j\pi/4} \quad \text{and} \quad \sqrt{-j} = \left(e^{-j\pi/2}\right)^{1/2} = e^{-j\pi/4},$$

the imaginary axis of the z-plane transforms to that in Figure A.14. ■

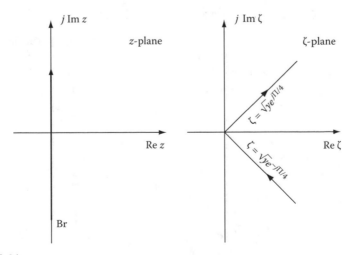

FIGURE A.14

Example A.24

Evaluate the integral $\int_C \frac{dz}{z}$, where C is a circle of radius 4 units around the origin, under the transformation $z = \zeta^2$.

SOLUTION

The integral has a pole at $z = 0$ and its value is $2\pi j$. If we apply the transformation $z = \zeta^2$, then $dz - 2\zeta \, d\zeta$. Also, $\zeta = \sqrt{z} = \sqrt{r}e^{j\theta/2}$ if we choose the positive root. From this relation we observe that as the z traces a circle around the origin, the ζ traces a half-circle from 0 to π. Hence, the integral becomes

$$2 \int_{C'} \frac{d\zeta}{\zeta} = 2 \int_0^\pi \frac{\rho j e^{j\theta}}{\rho e^{j\theta}} d\theta = 2\pi j$$

as was expected.   ∎

A.8 The Bromwich Contour

The Bromwich contour takes the form

$$f(t) = \frac{1}{2\pi j} \int_{c-j\infty}^{c+j\infty} e^{zt} F(z) dz \qquad (A.69)$$

where $F(z)$ is a function of z, all of whose singularities lie on the left of the path, and t is the time, which is always real and positive, $t > 0$.

A.8.1 Finite Number of Poles

Let us assume that $F(z)$ has n poles at $p_1, p_2, \ldots, p_n$ and no other singularities; this case includes the important case of **rational transforms**. To utilize Cauchy's integral theorem, we must express $f(t)$ as an integral along a closed contour. Figure A.15 shows such a situation. We know from Jordan's lemma (see Section A.7) that if $F(z) \to 0$ as $|z| \to \infty$ on the contour C then for $t > 0$

$$\lim_{R \to \infty} \int_C e^{tz} F(z) dz \to 0 \quad t > 0 \qquad (A.70)$$

and because

$$\int_{c-jy}^{c+jy} e^{tz} F(z) dz \to \int_{Br} e^{tz} F(z) dz \quad y \to \infty \qquad (A.71)$$

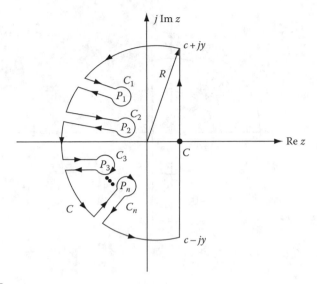

FIGURE A.15

we conclude that $f(t)$ can be written as a limit,

$$f(t) \underset{R \to \infty}{\longrightarrow} \frac{1}{2\pi j} \int_C e^{zt} F(z) dz \tag{A.72}$$

of an integral along the closed path as shown in Figure A.15. If we make R large enough to contain all the poles of $F(z)$, then the integral along C is independent of R. Therefore, we write

$$f(t) = \frac{1}{2\pi j} \int_C e^{zt} F(z) dz \tag{A.73}$$

Using Cauchy's theorem, it follows that

$$\int_C e^{zt} F(z) dz = \sum_{k=1}^{n} \int_{C_k} e^{zt} F(z) dz \tag{A.74}$$

where C_k's are the contours around each pole.

1. For simple poles we obtain

$$f(t) = \sum_{k=1}^{n} F_k(z_k) e^{z_k t} \quad t > 0 \tag{A.75}$$

$$F_k(z_k) = F(z)(z - z_k)|_{z=z_k}$$

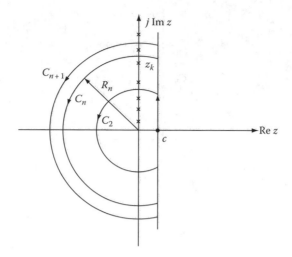

FIGURE A.16

2. For a multiple pole of $m+1$ multiplicity, we obtain

$$\int_{C_k} e^{zt} F(z)dz = \int_{C_k} \frac{e^{zt} F_k(z)}{(z-z_k)^{m+1}} dz = \frac{2\pi j}{m!} \frac{d^m}{dz^m} [e^{zt} F_k(z)]|_{z=z_k} \qquad (A.76)$$

3. Infinitely many poles (see Figure A.16).

If we can find circular arcs with radii tending to infinity such that

$$F(z) \to 0 \quad \text{as } z \to \infty \quad \text{on } C_n \qquad (A.77)$$

Applying Jordan's lemma to the integral along those arcs, we obtain

$$\int_{C_n} e^{zt} F(z)dz \underset{n\to\infty}{\longrightarrow} 0 \quad t > 0 \qquad (A.78)$$

and with C_n' the closed curve, consisting of C_n and the vertical line Re $z = c$, we obtain

$$f(t) = \lim_{n\to\infty} \frac{1}{2\pi j} \int_{C_n'} e^{zt} F(z)dz \quad t > 0 \qquad (A.79)$$

Hence, for simple poles $z_1, z_2, \ldots, z_n$ of $F(z)$, we obtain

$$f(t) = \sum_{k=1}^{\infty} F_k(z_k) e^{z_k t} \tag{A.80}$$

where $F_k(z) = F(z)(z - z_k)$.

Example A.25

Find $f(t)$ from its transformed value $F(z) = 1/(z \cosh az)$, $a > 0$.

SOLUTION

The poles of the above function are

$$z_0 = 0, \ z_k = \pm j \frac{(2k - 1)\pi}{2a} \quad k = 1, 2, 3, \ldots$$

We select the arcs C_n such that their radii are $R_n = jn\pi$. It can be shown that $1/\cosh az$ is bounded on C_n and, therefore, $1/(z \cosh az) \to 0$ as $z \to \infty$ on C_n. Hence,

$$zF(z)|_{z=0} = 1, \quad (z - z_k)F(z)|_{z=z_k} = \frac{(-1)^k 2}{(2k - 1)\pi}$$

and from (A.80) we obtain

$$f(t) = 1 + \frac{2}{\pi} \sum_{k=1}^{\infty} \frac{(-1)^k}{2k - 1} e^{z_k t} + \frac{2}{\pi} \sum_{k=1}^{\infty} \frac{(-1)^k}{2k - 1} e^{-z_k t}$$

$$= 1 + \frac{4}{\pi} \sum_{k=1}^{\infty} \frac{(-1)^k}{2k - 1} \cos \frac{(2k - 1)\pi t}{2a} \qquad \blacksquare$$

A.8.2 Branch Points and Branch Cuts

The singularities that have been considered are those points at which $|W(z)|$ ceases to be finite. At a branch point, the absolute value of $W(z)$ may be finite but $W(z)$ is not single valued, and hence is not regular. One of the simplest functions with these properties is

$$W_1(z) = z^{1/2} = \sqrt{r} e^{j\theta/2} \tag{A.81}$$

which takes on two values for each value of z, one the negative of the other depending on the choice of θ. This follows because we can write an equally valid form for $z^{1/2}$ as

$$W_2(z) = \sqrt{r} e^{j(\theta + 2\pi)/2} = -\sqrt{r} e^{j\theta/2} = -W_1(z) \tag{A.82}$$

Clearly, $W_1(z)$ is not continuous at points on the positive real axis because

$$\lim_{\theta \to 2\pi} \left(\sqrt{r}e^{j\theta/2}\right) = -\sqrt{r} \quad \text{while} \quad \lim_{\theta \to 0} \left(\sqrt{r}e^{j\theta/2}\right) = \sqrt{r}$$

Hence, $W'(z)$ does not exist when z is real and positive. However, the branch $W_1(z)$ is analytic in the region $0 \le \theta \le 2\pi$, $r \to 0$. The part of the real axis where $x \ge 0$ is called a **branch cut** for the branch $W_1(z)$, and the branch is analytic except at points on the cut. Hence, the cut is a boundary introduced so that the corresponding branch is single valued and analytic throughout the open region bounded by the cut.

Suppose that we consider the function $W(z) = z^{1/2}$ and contour C, as shown in Figure A.17a, which encloses the origin. Clearly, after one complete circle in the positive direction enclosing the origin, θ is increased by 2π, given a value of $W(z)$ that changes from $W_1(z)$ to $W_2(z)$; that is, the function has changed from one branch to the second. To avoid this and to make the function analytic, the contour C is replaced by a contour Γ, which consists of a small circle γ surrounding the branch point, a semi-infinite cut connecting the small circle and C, and C itself (as shown in Figure A.17b). Such a contour, which avoids crossing the branch cut, ensures that $W(z)$ is single valued. Because $W(z)$ is single valued and excludes the origin, we would write for this composite contour C

$$\int_C W(z)dz = \int_\Gamma + \int_{l-} + \int_\gamma + \int_{l+} = 2\pi j \sum \text{Res} \tag{A.83}$$

The evaluation of the function along the various segments of C proceeds as before.

Example A.26

If $0 < a < 1$, show that

$$\int_0^\infty \frac{x^{a-1}}{1+x}dx = \frac{\pi}{\sin a\pi}$$

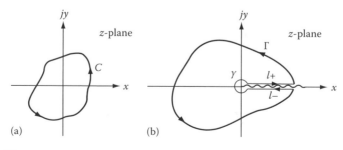

(a)   (b)

FIGURE A.17

Solution

Consider the integral

$$\oint_C \frac{z^{a-1}}{1+z}\,dz = \int_\Gamma + \int_{l-} + \int_\gamma + \int_{l+} = l_1 + l_2 + l_3 + l_4 = 2\pi j \sum \text{Res}$$

which we will evaluate using the contour shown in Figure A.18. Under the conditions

$$\left|\frac{z^a}{1+z}\right| \to 0 \quad \text{as } |z| \to 0 \quad \text{if } a > 0$$

$$\left|\frac{z^a}{z+1}\right| \to 0 \quad \text{as } |z| \to \infty \quad \text{if } a < 1$$

the integral becomes by (A.56)

$$\int_\Gamma \to 0 \qquad \int_{l-} = -e^{2\pi ja} \int_0^\infty$$

and by (A.57)

$$\int_\gamma \to 0 \qquad \int_{l+} = 1 \int_0^\infty$$

Thus,

$$\left(1 - e^{2\pi ja}\right) \int_0^\infty \frac{x^{a-1}}{1+x}\,dx = 2\pi j \sum \text{Res}$$

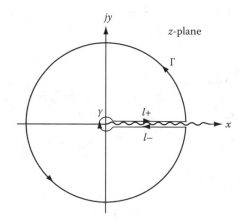

FIGURE A.18

Further, the residue at the pole $z = -1$, which is enclosed, is

$$\lim_{z=e^{j\pi}} (1+z) \frac{z^{a-1}}{1+z} = e^{j\pi(a-1)} = -e^{j\pi a}$$

Therefore,

$$\int_0^\infty \frac{x^{a-1}}{x+1} dx = 2\pi j \frac{e^{j\pi a}}{e^{j\pi a} - 1} = \frac{\pi}{\sin \pi a}$$

If, for example, we have the integral $(1/2\pi j) \int_{Br_1} \frac{e^{zt}dz}{z^{\nu+1}}$ to evaluate with Re $\nu > -1$ and t real and positive, we observe that the integral has a branch point at the origin if ν is a nonintegral constant. Because the integral vanishes along the arcs as $R \to \infty$, the equivalent contour can assume the form depicted in Figure A.19a and marked Br_2. For the contour made up of Br_1, Br_2, the arc is closed and contains no singularities and, hence, the integral around the contour is zero. Because the arcs do not contribute any value, provided Re $\nu > -1$, the integral along Br_1 is equal to that along Br_2, both being described positively. The angle γ between the barrier and the positive real axis may have any value between $\pi/2$ and $3\pi/2$. When the only singularity is a branch point at the origin, the contour of Figure A.19b is an approximate one.   ∎

Example A.27

Evaluate the integral $I = \dfrac{1}{2\pi j} \displaystyle\int_{Br_2} \dfrac{e^z dz}{\sqrt{z}}$, where Br_2 is the contour shown in Figure A.19b.

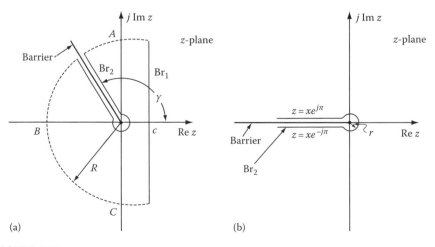

(a)   (b)

FIGURE A.19

SOLUTION

1. Write $z = re^{j\theta}$ on the circle. Hence, we get

$$I_1 = \frac{1}{2\pi j} \int_{-\pi}^{\pi} \frac{e^{re^{j\theta}} d(re^{j\theta})}{\sqrt{re^{j\theta/2}}} = \frac{\sqrt{r}}{2\pi} \int_{-\pi}^{\pi} r^{(\cos\theta + J\sin\theta) + j\theta/2} d\theta$$

2. On the line below, the barrier $z = x\exp(-j\pi)$ where $x = |x|$. Hence, the integral becomes

$$I_2 = \frac{1}{2\pi j} \int_{\infty}^{r} \frac{e^{xe^{-j\pi}} d(xe^{-j\pi})}{\sqrt{xe^{-j\pi/2}}} = \frac{1}{2\pi} \int_{r}^{\infty} e^{-x} x^{-1/2} dx$$

3. On the line above the barrier $z = x\exp(j\pi)$ and, hence,

$$I_3 = \frac{1}{2\pi j} \int_{r}^{\infty} \frac{e^{xe^{j\pi}} d(xe^{j\pi})}{\sqrt{xe^{j\pi/2}}} = \frac{1}{2\pi} \int_{r}^{\infty} e^{-x} x^{-1/2} dx$$

Hence, we have

$$I_2 + I_3 = \frac{1}{\pi} \int_{r}^{\infty} e^{-x} x^{-1/2} dx$$

As $r \to 0$, $I_1 \to 0$ and, hence,

$$I = I_1 + I_2 + I_3 = \frac{1}{\pi} \int_{0}^{\infty} e^{-x} x^{-1/2} dx = \frac{\Gamma\left(\frac{1}{2}\right)}{\pi} = \frac{\sqrt{\pi}}{\pi} = \frac{1}{\sqrt{\pi}} \qquad \blacksquare$$

Example A.28

Evaluate the integral $f(t) = \dfrac{1}{2\pi j} \displaystyle\int_{Br} \dfrac{e^{zt} e^{-a\sqrt{z}}}{\sqrt{z}} dz$, $a > 0$ (see Figure A.20).

SOLUTION

The origin is a point branch and we select the negative axis as the barrier. We select the positive of $\sqrt{z}$ when z takes positive real values so that the integral vanishes as z approaches infinity in the region $\text{Re } z > \gamma$, where γ indicates the region of convergence, $\gamma \le c$. Hence, we obtain

$$z = re^{j\theta} \quad -\pi < \theta \le \pi \quad \sqrt{z} = \sqrt{r} e^{j\theta/2} \qquad (A.84)$$

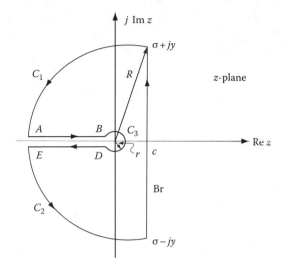

FIGURE A.20

The curve $C = Br + C_1 + C_2 + C_3$ encloses a region with no singularities and, therefore, Cauchy's theorem applies (the intergrand is analytic in the region). Hence,

$$\int_C e^{zt} \frac{e^{-a\sqrt{z}}}{\sqrt{z}} \, dz = 0 \tag{A.85}$$

It is easy to see that the given function converges to zero as R approaches infinity and therefore the integration over $C_1 + C_2$ does not contribute any value. For z on the circle, we obtain

$$\left| \frac{e^{-a\sqrt{z}}}{\sqrt{z}} \right| \to 0$$

Therefore, for fixed $t > 0$, we obtain

$$\left| \int_{C_3} e^{zt} \frac{e^{-a\sqrt{z}}}{\sqrt{z}} \, dz \right| \le 2\pi r \frac{e^{rt}}{\sqrt{r}} = \lim_{r \to 0} 2\pi r \frac{e^{rt}}{\sqrt{r}} = 0$$

because

$$\left| \int_C f(z) \, dz \right| \le ML$$

where L is the length of the contour and $|f(z)| < M$ for z on C.

On AB, $z = -x$, $\sqrt{z} = j\sqrt{x}$, and on DE, $z = -x$, $\sqrt{z} = -j\sqrt{x}$. Therefore, we obtain

$$\int\limits_{AB+DE} e^{zt}\frac{e^{-a\sqrt{z}}}{\sqrt{z}}dz \underset{\substack{r\to 0 \\ R\to\infty}}{\longrightarrow} -\int\limits_{\infty}^{0} e^{-xt}\frac{e^{ja\sqrt{x}}}{j\sqrt{x}}dx - \int\limits_{0}^{\infty} e^{-xt}\frac{e^{-ja\sqrt{x}}}{-j\sqrt{x}}dx \qquad \text{(A.86)}$$

But from (A.69)

$$\int\limits_{Br} e^{zt}\frac{e^{-a\sqrt{z}}}{\sqrt{z}}dz = 2\pi j f(t) \qquad \text{(A.87)}$$

and, hence, (A.85) and (A.87) reduce to

$$f(t) + \frac{1}{2\pi j}\int\limits_{0}^{\infty} e^{-xt}\frac{e^{ja\sqrt{x}} + e^{-ja\sqrt{x}}}{j\sqrt{x}}dx = 0 \qquad \text{(A.88)}$$

If we set $x = y^2$, we have

$$\int\limits_{0}^{\infty} e^{-xt}\frac{\cos a\sqrt{x}}{\sqrt{x}}dx = 2\int\limits_{0}^{\infty} e^{-y^2 t}\cos ay\, dy \qquad \text{(A.89)}$$

But (see Fourier transform of Gaussian function, Chapter 3),

$$2\int\limits_{0}^{\infty} e^{-y^2 t}\cos ay\, dy = \sqrt{\frac{x}{t}}e^{-a^2/4t} \qquad \text{(A.90)}$$

and, hence, (A.89) becomes

$$f(t) = \frac{1}{\sqrt{\pi t}}e^{-a^2/4t} \qquad \text{(A.91)}$$

■

Example A.29

Evaluate the integral $I = \dfrac{1}{2\pi j}\displaystyle\int_{C}\dfrac{e^{zt}dz}{\sqrt{z^2 - 1}}$ where C is the contour shown in Figure A.21.

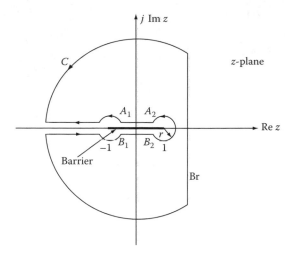

Solution

The Br contour is equivalent to the dumbbell-type contour shown in Figure A.21, $B_1B_2A_2A_1B_1$. Set the phase along the line A_2A_1 equal to zero (it can also be set equal to π). Then on A_2A_1 $z=x$, from $+1$ to -1. Hence, we have

$$I_1 = \frac{1}{2\pi j} \int_1^{-1} \frac{e^{xt}\,dx}{\sqrt{x^2-1}} = \frac{1}{2\pi} \int_{-1}^1 \frac{e^{xt}\,dx}{\sqrt{1-x^2}} \quad |x| < 1 \tag{A.92}$$

By passing around the $z=-1$ point the phase changes by π and, hence, on B_1B_2 $z=x$ $\exp(2\pi j)$. The change by 2π is due to the complete transversal of the contour that contains two branch points. Hence, we obtain

$$I_2 = -\frac{1}{2\pi j} \int_{-1}^1 \frac{e^{xt}\,dx}{\sqrt{x^2-1}} = \frac{1}{2\pi} \int_{-1}^1 \frac{e^{xt}\,dx}{\sqrt{1-x^2}} \tag{A.93}$$

Changing the origin to -1, we set $\zeta = z+1$ or $z = \zeta - 1$, which gives

$$I_3 = \frac{e^{-t}}{2\pi j} \int \frac{e^{\zeta t}\,d\zeta}{\sqrt{[(\zeta - 2)\zeta]}} \tag{A.94}$$

On the small circle with $z=-1$ as center, $\zeta = r\exp(j\theta)$ and we get

$$I_3 = \frac{e^{-t}}{2\pi} \int_\pi^{-\pi} \frac{e^{rt(\cos\theta + j\sin\theta)+(j\theta/2)}\sqrt{r}\,d\theta}{\sqrt{re^{j\theta}-2}} \tag{A.95}$$

When $\theta = 0$ the intergrand has the value $+\sqrt{r}e^{rt}/\sqrt{r-2}$, and for $\theta = 2\pi$ the value is $-\sqrt{r}e^{rt}/\sqrt{r-2}$. Therefore, the intergrand changes sign in rounding the branch point at $z = -1$. Similarly for the branch point at $z = 1$, where the change is from $-$ to $+$. As $r \to 0$, $I_3 \to 0$, and thus I_3 vanishes. The same is true for the branch point at $z = -1$. Therefore, by setting $x = \cos\theta$ we obtain

$$I = I_1 + I_2 = \frac{1}{\pi} \int_{-1}^{1} \frac{e^{xt}}{\sqrt{1-x^2}} dx = \frac{1}{\pi} \int_{0}^{\pi} e^{t\cos\theta} d\theta = \frac{1}{\pi} \int_{0}^{\pi} \sum_{k=0}^{\infty} \frac{(t\cos\theta)^k}{k!} d\theta$$

$$= \frac{1}{\pi}\left[\pi + \pi\frac{1}{2}\frac{t^2}{2!} + \pi\frac{3}{4}\frac{1}{2}\frac{t^4}{4!} + \pi\frac{5}{6}\frac{3}{4}\frac{1}{2}\frac{t^6}{6!} + \cdots \right]$$

$$= 1 + \frac{t^2}{2^2} + \frac{t^4}{2^2 4^2} + \frac{t^6}{2^2 4^2 6^2} + \cdots = \sum_{k=0}^{\infty} \frac{\left(\frac{1}{2}t\right)^{2k}}{(k!)^2} = I_0(t)$$

where $I_0(t)$ is the modified Bessel function of the first kind and zero order. ∎

Example A.30

Evaluate the integral $I = \int_C \frac{e^{zt}}{\sqrt{z^2+1}} dz$ where C is the closed contour shown in Figure A.22.

SOLUTION

The Br contour is equal to the dumbbell-type contour as shown in Figure A.22, $A\,B\,G\,D\,A = C_1$. Hence, we have

$$f(t) = \frac{1}{2\pi j} \int_{C_1} \frac{e^{zt}}{\sqrt{z^2+1}} dz \qquad (A.96)$$

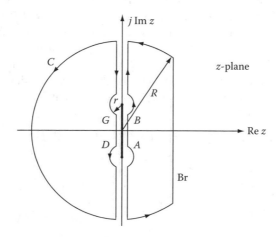

FIGURE A.22

But

$$\left| \frac{e^{zt}}{\sqrt{z^2 + 1}} \right| < \frac{e^{rt}}{\sqrt{r}\sqrt{2 - r}}$$

on the circle on the $+j$ branch point and, therefore, for $t > 0$

$$\left| \int \frac{e^{zt}}{\sqrt{z^2 + 1}} dz \right| < \frac{2\pi\sqrt{r}e^{rt}}{\sqrt{2 - r}} \to 0 \quad \text{as } r \to 0$$

We obtain similar results for the contour around the $-j$ branch point. However, On AB, $z = j\omega$, $\sqrt{1 + z^2} = \sqrt{1 - \omega^2}$; on GD, $z = j\omega$, $\sqrt{1 + z^2} = -\sqrt{1 - \omega^2}$ and, therefore, for $t > 0$ we obtain

$$f(t) = \frac{j}{2\pi j} \int\limits_{-1}^{1} \frac{e^{j\omega t}}{\sqrt{1 - \omega^2}} d\omega + \frac{j}{2\pi j} \int\limits_{1}^{-1} \frac{e^{j\omega t}}{-\sqrt{1 - \omega^2}} d\omega = \frac{1}{\pi} \int\limits_{-1}^{1} \frac{\cos \omega t}{\sqrt{1 - \omega^2}} d\omega$$

If we set $\omega = \sin \theta$ (see also Chapter 1)

$$f(t) = \frac{1}{\pi} \int\limits_{-\pi/2}^{\pi/2} \cos(t \sin \theta) d\theta = J_0(t)$$

where $J_0(t)$ is the Bessel function of the first kind.   ∎

A.9 Evaluation of Definite Integrals

The principles discussed above find considerable applicability in the evaluation of certain definite real integrals. This is a common application of the developed theory, as it is often extremely difficult to evaluate some of these real integrals by other methods. We employ such methods in the evaluation of Fourier integrals. In practice, the given integral is replaced by a complex function that yields the specified integrand in its appropriate limit. The integration is then carried out in the complex plane, with the real integral being extracted for the required result. The following several examples show this procedure.

A.9.1 Evaluation of the Integrals of Certain Periodic Functions (0 to 2π)

An integral of the form

$$I = \int\limits_{0}^{2\pi} F(\cos \theta, \ \sin \theta) d\theta \tag{A.97}$$

where the integral is a **rational function** of $\cos\theta$ and $\sin\theta$ finite on the range of integration, and can be integrated by setting $z = \exp(j\theta)$,

$$\cos\theta = \frac{1}{2}(z + z^{-1}), \quad \sin\theta = \frac{1}{2j}(z - z^{-1}) \tag{A.98}$$

The integral (A.97) takes the form

$$I = \int_C F(z)\,dz \tag{A.99}$$

where $F(z)$ is a rational function of z finite on C, which is a circle of radius unity with center at the origin.

Example A.31

If $0 < a < 1$, find the value of the integral

$$I = \int_0^{2\pi} \frac{d\theta}{1 - 2a\cos\theta + a^2} \tag{A.100}$$

SOLUTION

Introducing (A.98) in (A.100), we obtain

$$I = \int_C \frac{dz}{j(1 - az)(z - a)} \tag{A.101}$$

The only pole inside the unit circle is at a. Therefore, by residue theory we have

$$I = 2\pi j \lim_{z \to a} \frac{z - a}{j(1 - az)(z - a)} = \frac{2\pi}{1 - a^2} \qquad\blacksquare$$

A.9.2 Evaluation of Integrals with Limits $-\infty$ and $+\infty$

We can evaluate the integral $I = \int_{-\infty}^{\infty} F(x)\,dx$ provided that the function $F(z)$ satisfies the following properties:

1. It is analytic when the imaginary part of z is positive or zero (except at a finite number of poles).
2. It has no poles on the real axis.

3. As $|z| \to \infty$, $zF(z) \to 0$ uniformly for all value of arg z such that $0 \leq \text{arg } z \leq \pi$.
4. x is real, $xF(x) \to 0$ as $x \to \pm\infty$, in such a way that $\int_0^\infty F(x)\, dx$ and $\int_{-\infty}^0 F(x)\, dx$ both converge.

The integral is given by

$$I = \int_C F(z)dz = 2\pi j \sum \text{Res} \tag{A.102}$$

where the contour is the real axis and a semicircle having its center in the origin and lying above the real axis.

Example A.32

Evaluate the integral $I = \int_{-\infty}^\infty \dfrac{dx}{(x^2 + 1)^3}$.

SOLUTION

The integral becomes

$$I = \int_C \frac{dz}{(z^2 + 1)^3} = \int_C \frac{dz}{(z + j)^3(z - j)^3}$$

which has one pole at j of order three (see A.50). Hence, we obtain

$$I = \frac{1}{2!} \frac{d^2}{dz^2} \left[\frac{1}{(z + j)^3} \right]\bigg|_{z=j} = -j\frac{3}{16} \qquad\blacksquare$$

Example A.33

Evaluate the integral $I = \int_0^\infty \dfrac{dx}{x^2 + 1}$.

SOLUTION

The integral becomes

$$I = \int_C \frac{dz}{z^2 + 1}$$

where C is the contour of the real axis and the upper semicircle. From $z^2 + 1 = 0$ we obtain $z = \exp(j\pi/2)$ and $z = \exp(-j\pi/2)$. Only the pole $z = \exp(j\pi/2)$ exists inside the contour. Hence, we obtain

$$2\pi j \lim_{z \to e^{j/\pi/2}} \left(\frac{z - e^{j\pi/2}}{(z - e^{j\pi/2})(z - e^{-j\pi/2})} \right) = \pi$$

Therefore, we have

$$\int_{-\infty}^{\infty} \frac{dx}{x^2 + 1} = 2 \int_0^{\infty} \frac{dx}{x^2 + 1} = \pi \quad \text{or} \quad I = \frac{\pi}{2} \qquad \blacksquare$$

A.9.3 Certain Infinite Integrals Involving Sines and Cosines

If $F(z)$ satisfies conditions (1), (2), and (3) above, and if $m > 0$, then $F(z)e^{jmz}$ also satisfies the same conditions. Hence, $\int_0^{\infty} [F(x)e^{jmx} + F(-x)e^{-jmx}]\, dx$ is equal to $2\pi j \sum \text{Res}$, where $\sum \text{Res}$ means the sum of the residues of $F(z)e^{jmz}$ at its poles in the upper half-plane. Therefore,

1. If $F(x)$ is an even function; that is, $F(x) = F(-x)$, then

$$\int_0^{\infty} F(x) \cos mx\, dx = j\pi \sum \text{Res} \qquad (A.103)$$

2. If $F(x)$ is an odd function; that is, $F(x) = -F(-x)$, then

$$\int_0^{\infty} F(x) \sin mx\, dx = \pi \sum \text{Res} \qquad (A.104)$$

Example A.34

Evaluate the integral $I = \int_0^{\infty} \frac{\cos x}{x^2 + a^2}\, dx,\ a > 0$.

Solution

Consider the integral

$$I_1 = \int_C \frac{e^{jz}}{z^2 + a^2}\, dz$$

where the contour is the real axis and the infinite semicircle on the upper side with respect to the real axis. The contour encircles the pole ja. Hence,

$$\int_C \frac{e^{jz}}{z^2 + a^2}\, dz = 2\pi j \frac{e^{jja}}{2ja} = \frac{\pi}{a} e^{-a}$$

However,

$$\int_{-\infty}^{\infty} \frac{e^{jz}}{z^2 + a^2} dz = \int_{-\infty}^{\infty} \frac{\cos x}{x^2 + a^2} dx + j \int_{-\infty}^{\infty} \frac{\sin x}{x^2 + a^2} dx = \int_{-\infty}^{\infty} \frac{\cos x}{(x^2 + a^2)} dx$$

because the integrand of the third integral is odd and therefore is equal to zero. From the last two equations, we find that

$$I = \int_{-\infty}^{\infty} \frac{\cos x}{x^2 + a^2} dx = \frac{\pi}{2a} e^{-a}$$

because the integrand is an even function.   ∎

Example A.35

Evaluate the integral $I = \int_0^\infty \frac{x \sin ax}{x^2 + b^2} dx$, $k > 0$ and $a > 0$.

SOLUTION

Consider the integral

$$I_1 = \int_C \frac{z e^{jaz}}{z^2 + b^2} dz$$

where C is the same type of contour as in Example A.34. Because there is only one pole at $z = jb$ in the upper half of the z-plane, then

$$I_1 = \int_{-\infty}^{\infty} \frac{z e^{jaz}}{z^2 + b^2} dz = 2\pi j \frac{jb e^{jajb}}{2jb} = j\pi e^{-ab}$$

Because the integrand $x \sin ax/(x^2 + b^2)$ is an even function, we obtain

$$I_1 = j \int_{-\infty}^{\infty} \frac{x \sin ax}{x^2 + b^2} dx = j\pi e^{-ab} \quad \text{or} \quad I = \frac{\pi}{2} e^{-ab}$$ ∎

Example A.36

Show that $\int_{-\infty}^{\infty} \frac{x \sin \pi x}{x^2 + 2x + 5} dx = -\pi e^{-2\pi}$.

Integrals of the Form $\int_0^\infty x^{\alpha-1} f(x) dx$, $0 < \alpha < 1$

It can be shown that the above integral has the value

$$I = \int_0^\infty x^{\alpha-1} f(x)dx = \frac{2\pi j}{1 - e^{j2\pi\alpha}} \sum_{k=1}^{N} \text{Res}\left[z^{\alpha-1} f(z)\right]\Big|_{z=z_k} \qquad \text{(A.105)}$$

where $f(z)$ has N singularities and $z^{\alpha-1} f(z)$ has a branch point at the origin. ∎

Example A.37

Evaluate the integral $I = \int_0^\infty \frac{x^{-1/2}}{x+1} dx$

SOLUTION

Because $x^{-1/2} = x^{1/2-1}$, it is simplified that $\alpha = 1/2$. From the integrand, we observe that the origin is a branch point and the $f(x) = 1/(x+1)$ has a pole at -1. Hence, from (A.105) we obtain

$$I = \frac{2\pi j}{1 - e^{j2\pi/2}} \text{Res}\left[\frac{z^{-1/2}}{z+1}\right]\Big|_{z=-1} = \frac{2\pi j}{j(1 - e^{j\pi})} = \pi$$

We can also proceed by considering the integral $I = \int_C \frac{z^{-1/2}}{z+1} dz$. Because $z = 0$ is a branch point, we choose the contour C as shown in Figure A.23. The integrand has a simple pole at $z = -1$ inside the contour C. Hence, the residue at $z = -1 = \exp(j\pi)$ and is

$$\text{Res}\big|_{z=-1} = \lim_{z \to -1} (z+1)\frac{z^{-1/2}}{z+1} = e^{-j\frac{\pi}{2}}$$

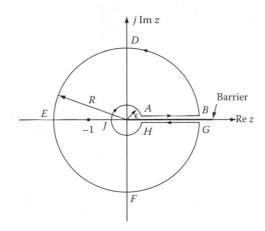

FIGURE A.23

Therefore, we write

$$\oint_C \frac{z^{-1/2}}{z+1} dz = \int_{AB} + \int_{BDEFG} + \int_{GH} + \int_{HJA} = e^{-j\pi/2}$$

The above integrals take the following form:

$$\int_\varepsilon^R \frac{x^{-1/2}}{x+1} dx + \int_0^{2\pi} \frac{(Re^{j\theta})^{-1/2} jRe^{j\theta} d\theta}{1 + Re^{j\theta}}$$

$$+ \int_R^\varepsilon \frac{(xe^{j2\pi})^{-1/2}}{1 + xe^{j2\pi}} dx + \int_{2\pi}^0 \frac{(\varepsilon e^{j\theta})^{-1/2} j\varepsilon e^{j\theta} d\theta}{1 + \varepsilon e^{j\theta}} = j2\pi e^{-j\pi/2}$$

where we have used $z = x \exp(j2\pi)$ for the integral along GH, because the argument of z is increased by 2π in going around the circle $BDEFG$.

Taking the limit as $\varepsilon \to 0$ and $R \to \infty$ and noting that the second and fourth integrals approach zero, we find

$$\int_0^\infty \frac{x^{-1/2}}{x+1} dx + \int_\infty^0 \frac{e^{-j2\pi/2} x^{-1/2}}{x+1} dx = j2\pi e^{-j\pi/2}$$

or

$$(1 - e^{-j\pi}) \int_0^\infty \frac{x^{-1/2}}{x+1} dx = j2\pi e^{-j\pi/2} \quad \text{or} \quad \int_0^\infty \frac{x^{-1/2}}{x+1} dx = \frac{j2\pi(-j)}{2} = \pi \qquad \blacksquare$$

A.9.4 Miscellaneous Definite Integrals

The following examples will elucidate some of the approaches that have been used to find the values of definite integrals.

Example A.38

Evaluate the integral $I = \int_{-\infty}^\infty \frac{1}{x^2 + a^2} dx$, $a > 0$.

SOLUTION

We write (see Figure A.24)

$$\int_C \frac{dz}{z^2 + a^2} = \int_{AB} + \int_{BDA} = 2\pi j \sum \text{Res}$$

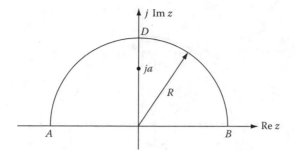

FIGURE A.24

As $R \to \infty$

$$\int_{BDA} \frac{dz}{z^2 + a^2} = \int_0^\pi \frac{R j e^{j\theta}\, d\theta}{R^2 e^{j2\theta} + a^2} \xrightarrow[R \to \infty]{} 0$$

and, therefore, we have

$$\int_{AB} \frac{dx}{x^2 + a^2} = \int_{-\infty}^{\infty} \frac{dx}{x^2 + a^2} = 2\pi j \frac{z - ja}{z^2 + a^2}\bigg|_{z=ja} = 2\pi j \frac{1}{2ja} = \frac{\pi}{a} \qquad ■$$

Example A.39

Evaluate the integral $I = \int_0^\infty \dfrac{\sin ax}{x}\, dx$.

Solution

Because $\sin az / z$ is analytic near $z = 0$, we indent the contour around the origin as shown in Figure A.25. With a positive, we write

$$\int_0^\infty \frac{\sin ax}{x}\, dx = \frac{1}{2} \int_{ABCD} \frac{\sin az}{z}\, dz = \frac{1}{4j} \int_{ABCD} \left[\frac{e^{jaz}}{z} - \frac{e^{-jaz}}{z} \right] dz$$

$$= \frac{1}{4j} \int_{ABCDA} \frac{e^{jaz}}{z}\, dz - \frac{1}{4j} \int_{ABCDFA} \frac{e^{-jaz}}{z}\, dz = \frac{1}{4j} \left[2\pi j \frac{1}{1} - 0 \right] = \frac{\pi}{2}$$

because the lower contour does not include any singularity. Because $\sin ax$ is an odd function of a and $\sin 0 = 0$, we obtain

$$\int_0^\infty \frac{\sin x}{x}\, dx = \begin{cases} \dfrac{\pi}{2} & a > 0 \\[2mm] 0 & a = 0 \\[2mm] -\dfrac{\pi}{2} & a < 0 \end{cases} \qquad ■$$

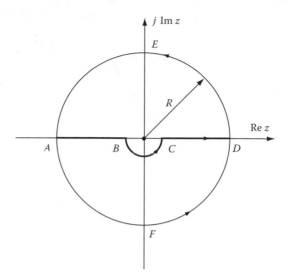

FIGURE A.25

Example A.40

Evaluate the integral $I = \int_0^\infty \dfrac{dx}{1 + x^3}$.

SOLUTION

Because the integrand $f(x)$ is odd, we introduce the $\ln z$. Taking a branch cut along the positive real axis, we obtain

$$\ln z = \ln r + j\theta \quad 0 \leq \theta < 2\pi$$

The discontinuity of $\ln z$ across the cut is (see Figure A.26a)

$$\ln z_1 - \ln z_2 = -2\pi j$$

Therefore, if $f(z)$ is analytic along the real axis and the contribution around an infinitesimal circle at the origin is vanishing, we obtain

$$\int_0^\infty f(x)dx = -\frac{1}{2\pi j} \int_{ABC} f(x)\ln(z)dz$$

If further $f(z) \to 0$ as $|z| \to \infty$, the contour can be completed with CDA (see Figure A.26b). If $f(z)$ has simple poles of order one at points z_k with residues Res(f, z_k), we obtain

$$\int_0^\infty f(x)dx = -\sum_k \text{Res}(f, z_k)\ln z_k$$

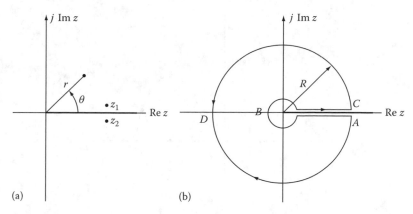

FIGURE A.26

Hence, because $z^3 + 1 = 0$ has poles at $z_1 = e^{j\pi/3}$, $z_2 = e^{j\pi}$, $z_3 = e^{j5\pi/3}$, then the integral is given by

$$I = \int\limits_0^\infty \frac{dx}{x^3 + 1} = -\left[\frac{j\pi/3}{3e^{2j\pi/3}} + \frac{j\pi}{3e^{j2\pi}} + \frac{j5\pi/3}{3e^{j10\pi/3}} \right] = \frac{2\pi\sqrt{3}}{9}$$ ∎

Example A.41

Show that $\int_0^\infty \cos ax^2 dx = \int_0^\infty \sin ax^2 dx = \dfrac{1}{2}\sqrt{\dfrac{\pi}{2a}}$, $a > 0$.

SOLUTION

We first form the integral

$$F = \int\limits_0^\infty \cos ax^2 dx + j \int\limits_0^\infty \sin ax^2 dx = \int\limits_0^\infty e^{jax^2} dx$$

Because $\exp(jaz^2)$ is analytic in the entire z-plane, we can use Cauchy's theorem and write (see Figure A.27)

$$F = \int\limits_{AB} e^{jaz^2} dz = \int\limits_{AC} e^{jaz^2} dz + \int\limits_{CB} e^{jaz^2} dz$$

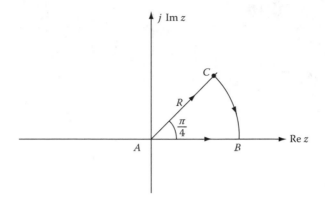

Along the contour $C\,B$, we obtain

$$\left| -\int_0^{\pi/4} e^{jR^2\cos 2\theta - R^2\sin 2\theta} jRe^{j\theta}\,d\theta \right| \leq \int_0^{\pi/4} e^{-R^2\sin 2\theta} R\,d\theta = \frac{R}{2}\int_0^{\pi/2} e^{-R^2\sin\phi}\,d\phi$$

$$\leq \frac{R}{2}\int_0^{\pi/2} e^{-R^2\phi/\pi}\,d\phi = \frac{\pi}{4R}\left(1 - e^{-R^2}\right)$$

where the transformation $2\theta = \phi$ and the inequality $\sin\phi \geq 2\phi/\pi$ were used ($0 \leq \phi \leq \pi/2$). Hence, as R approaches infinity the contribution from $C\,B$ contour vanishes. Hence,

$$F = \int_{AB} e^{jaz^2}\,dz = e^{j\pi/4}\int_0^\infty e^{-ar^2}\,dr = \frac{1+j}{\sqrt{2}}\frac{1}{2}\sqrt{\frac{\pi}{a}}$$

from which we obtain the desired result.   ■

Example A.42

Evaluate the integral $I = \int_{-1}^1 \dfrac{dx}{\sqrt{1-x^2}(1+x^2)}$.

SOLUTION

Consider the integral

$$\oint_C \frac{dz}{\sqrt{1-z^2}(1+z^2)}$$

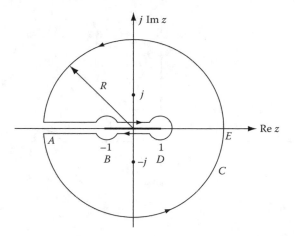

FIGURE A.28

whose contour C is that shown in Figure A.28. On the top side of the branch cut we obtain I and from the bottom we also get I. The contribution of the integral on the outer circle as R approaches infinity vanishes. Hence, due to two poles we obtain

$$2I = 2\pi j \left[\frac{1}{2j\sqrt{2}} + \frac{1}{2j\sqrt{2}} \right] = \pi\sqrt{2} \quad \text{or} \quad I = \frac{\sqrt{2}}{2}\pi \qquad \blacksquare$$

Example A.43

Evaluate the integral $I = \int_{-\infty}^{\infty} \frac{e^{ax}}{e^{bx} + 1} dx,\ a, b > 0.$

SOLUTION

From Figure A.29, we find

$$I = \int_C \frac{e^{az}}{e^{bz} + 1} dz = \int_C \frac{e^{az/b}}{e^z + 1} dz = 2\pi j \sum \text{Res}$$

There is an infinite number of poles: at $z = j\pi/b$, residue is $-\exp(j\pi\, a/b)$; at $z = 3j\pi/b$, residue is $-\exp(3j\pi\, a/b)$, and so on. The sum of residue forms a geometric series and because we assume a small imaginary part of a, $|\exp(j2\pi\, a/b| < 1$. Hence, by considering the common factor $\exp(j\pi\, a/b)$, we obtain

$$I = -\frac{2\pi}{b}j\frac{e^{j\pi a/b}}{1 - e^{j2\pi a/b}} = \frac{1}{b}\frac{\pi}{\sin(\pi a/b)}$$

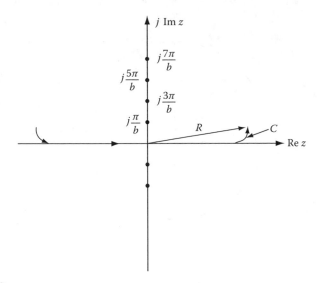

FIGURE A.29

The integral is of the form $\int e^{j\omega x}f(x)dx$ whose evaluation can be simplified by Jordan's lemma (see also A.60)

$$\int_{C} e^{j\omega x} f(x)dx = 0$$

for the contour semicircle C at infinity for which $\mathrm{Im}(\omega x) > 0$, provided $|f(\mathrm{Re}^{j\theta})| < \varepsilon(R) \to 0$ as $R \to \infty$ (note that the bound on $|f(x)|$ must be independent of θ). ∎

Example A.44

A relaxed RL series circuit with an input voltage source $v(t)$ is described by the equation $L di/dt + Ri = v(t)$. Find the current in the circuit using the inverse Fourier transform when the input voltage is a delta function.

Solution

The Fourier transform of the differential equation with delta input voltage function is

$$L j\omega I(\omega) + R I(\omega) = 1 \quad \text{or} \quad I(\omega) = \frac{1}{R + j\omega L}$$

and, hence,

$$i(t) = \frac{1}{2\pi} \int_{-\infty}^{\infty} \frac{e^{j\omega t}}{R + j\omega L} d\omega$$

If $t < 0$, the integral is exponentially small for Im $\omega \to -\infty$. If we complete the contour by a large semicircle in the lower ω-plane, the integral vanishes by Jordan's lemma. Because the contour does not include any singularities, $i(t) = 0$, $t < 0$. For $t > 0$, we complete the contour in the upper ω-plane. Similarly, no contribution exists from the semicircle. Because there is only one pole at $\omega = jR/L$ inside the contour, the value of the integral is

$$i(t) = 2\pi j \frac{1}{2\pi} \frac{1}{jL} e^{j(jR/L)t} = \frac{1}{L} e^{-\frac{R}{L}t},$$

which is known as the **impulse response of the system.** ∎

A.10 Principal Value of an Integral

Refer to the limiting process employed in Example A.39, which can be written in the form

$$\lim_{R \to \infty} \int_{-R}^{R} \frac{e^{jx}}{x} dx = j\pi$$

The limit is called the **Cauchy principal value** of the integral in the equation

$$\int_{-\infty}^{\infty} \frac{e^{jx}}{x} dx = j\pi$$

In general, if $f(x)$ becomes infinite at a point $x = c$ inside the range of integration, and if

$$\lim_{\varepsilon \to 0} \int_{-R}^{R} f(x)dx = \lim_{\varepsilon \to 0} \left[\int_{-R}^{c-\varepsilon} f(x)dx + \int_{c+\varepsilon}^{R} f(x)dx \right]$$

and if the separate limits on the right also exist, then the integral is convergent and the integral is written as $P \int$ where the P indicates the principal value. Whenever each of the integrals

$$\int_{-\infty}^{0} f(x)dx \quad \int_{0}^{\infty} f(x)dx$$

has a value, here $R \to \infty$, the principal value is the same as the integral. For example, if $f(x) = x$, the principal value of the integral is zero, although the value of the integral itself does not exist.

As another example, consider the integral

$$\int_a^b \frac{dx}{x} = \log\frac{b}{a}$$

If a is negative and b is positive, the integral diverges at $x=0$. However, we can still define

$$P\int_a^b \frac{dx}{x} = \lim_{\varepsilon\to 0}\left[\int_a^{-\varepsilon}\frac{dx}{x} + \int_{\varepsilon+}^b\frac{dx}{x}\right] = \lim_{\varepsilon\to 0}\left(\log\frac{\varepsilon}{-a} + \log\frac{b}{a}\right) = \log\frac{b}{|a|}$$

This principal value integral is unambiguous. The condition that the same value of ε must be used in both sides is essential; otherwise, the limit could be almost anything by taking the first integral from a to $-\varepsilon$ and the second from κ to b, and making these two quantities tend to zero in a suitable ratio.

If the complex variables were used, we could complete the path by a semicircle from $-\varepsilon$ to $+\varepsilon$ about the origin, either above or below the real axis. If the upper semicircle were chosen, there would be a contribution $-j\pi$, whereas if the lower semicircle were chosen, the contribution to the integral would be $+j\pi$. Thus, according to the path permitted in the complex plane we should have

$$\int_a^b \frac{dz}{z} = \log\frac{b}{|a|} \pm j\pi$$

The principal value is the mean of these alternatives.

If a path in the complex plane passes through a simple pole at $z=a$, we can define a principal value of the integral along the path by using a hook of small radius ε about the point a and the making ε tend to zero, as already discussed. If we change the variable z to ζ and $dz/d\zeta$ is finite and not equal to zero at the pole, this procedure will define an integral in the ζ-plane, but the values of the integrals will be the same. Suppose that the hook in the z-plane cuts the path at $a-\varepsilon$ and $a+\varepsilon'$, where $|\varepsilon|=|\varepsilon'|$, and in the ζ-plane the hook cuts the path at $\alpha-\kappa$ and $\alpha+\kappa'$. Then, if κ and κ' tend to zero so that $\varepsilon/\varepsilon' \to 1$, κ and κ' will tend to zero so that $\kappa/\kappa' \to 1$.

To illustrate this discussion, suppose we want to evaluate the integral

$$I = \int_0^\pi \frac{d\theta}{a - b\cos\theta}$$

where a and b are real and $a>b>0$. A change of variable by writing $z=\exp(j\theta)$ transforms this integral to (where a new constant α is introduced)

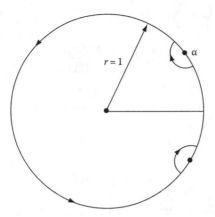

FIGURE A.30

$$I = \int_0^\pi \frac{2e^{j\theta}\,d\theta}{2ae^{j\theta} - b(e^{j2\theta}+1)} = -\frac{1}{j}\int_C \frac{2\,dz}{bz^2 - 2az + b} = -\frac{1}{j}\int_C \frac{2\,dz}{b(z-\alpha)\left(z-\frac{1}{\alpha}\right)}$$

where the path of integration is around the unit circle. Because the contour would pass through the poles, hooks are used to isolate the poles as shown in Figure A.30. Because no singularities are closed by the path, the integral is zero. The contributions of the hooks are $-j\pi$ times the residues, where the residues are

$$-\frac{1}{j}\frac{\frac{2}{b}}{\alpha - \frac{1}{\alpha}} - \frac{1}{j}\frac{\frac{2}{b}}{\frac{1}{\alpha} - \alpha}$$

These are equal and opposite, and they cancel each other. Therefore, the principal value of the integral around the unit circle is zero. This approach for finding principal values succeeds only at simple poles.

A.11 Integral of the Logarithmic Derivative

Of importance in the study of mapping from z-plane to $W(z)$-plane is the integral of the logarithmic derivative. Consider, therefore, the function

$$F(z) = \log W(z) \qquad (\text{A.106})$$

Then

$$\frac{dF(z)}{dz} = \frac{1}{W(z)}\frac{dW(z)}{dz} = \frac{W'(z)}{W(z)}$$

The function to be examined is the following:

$$\int_C \frac{dF(z)}{dz}\,dz = \int_C \frac{W'(z)}{W(z)}\,dz \qquad (A.107)$$

The integrand of this expression will be analytic within the contour C except for the points at which $W(z)$ is either zero or infinity.

Suppose that $W(z)$ has a pole of order n at z_0. This means that $W(z)$ can be written as

$$W(z) = (z - z_0)^n g(z) \qquad (A.108)$$

with n positive for a zero and n negative for a pole. We differentiate this expression to get

$$W'(z) = n(z - z_0)^{n-1} g(z) + (z - z_0)^n g'(z)$$

and so

$$\frac{W'(z)}{W(z)} = \frac{n}{z - z_0} + \frac{g'(z)}{g(z)} \qquad (A.109)$$

For n positive, $W'(z)/W(z)$ will possess a pole of order one. Similarly, for n negative $W'(z)/W(z)$ will possess a pole of order one, but with a negative sign. Thus, for the case of n positive or negative, the contour integral in the positive sense yields

$$\int_C \frac{W'(z)}{W(z)}\,dz = \pm \int_C \frac{n}{z - z_0}\,dz + \int_z \frac{g'(z)}{g(z)}\,dz \qquad (A.110)$$

But because $g(z)$ is analytic at the point z_0, then $\int_C [g'(z)/g(z)]dz = 0$, and by (6.1)

$$\int_C \frac{W'(z)}{W(z)}\,dz = \pm 2\pi jn \qquad (A.111)$$

Thus, the existence of a zero of $W(z)$ introduces a contribution $2\pi jn_z$ to the contour integral, where n_z is the multiplicity of the zero of $W(z)$ at z_0. Clearly, if a number of zeros of $W(z)$ exist, the total contribution to the contour integral is $2\pi jN$, where N is the weighted value of the zeros of $W(z)$ (weight 1 to a first-order zero, weight 2 to a second-order zero, and so on).

For the case where n is negative, which specifies that $W(z)$ has a pole of order n at z_0, then in (A.111) n is negative and the contribution to the contour integral is now $-2\pi n_p$ for each pole of $W(z)$; the total contribution is $-2\pi jP$, where P is the weighted number of poles. Clearly, because both zeros and poles of $F(z)$ cause poles of $W'(z)/W(z)$ with opposite signs, then the total value of the integral is

$$\int_C \frac{W'(z)}{W(z)} dz = \pm 2\pi j(N - P) \qquad (A.112)$$

Note further that

$$\int_C W'(z) dz = \int_C \frac{dW(z)}{dz} dz = \int d[\log W(z)] = \int d[\log |W(z)| + j \arg W(z)]$$

$$= \log |W(z)| \big|_0^{2\pi} + j[\arg W(2\pi) - \arg W(0)]$$

$$= 0 + j[\arg W(2\pi) - \arg W(0)]$$

so that

$$[\arg W(0) - \arg W(2\pi)] = 2\pi(N - P) \qquad (A.113)$$

This relation can be given a simple graphical interpretation. Suppose that the function $W(z)$ is represented by its pole and zero configuration on the z-plane. As z traverses the prescribed contour on the z-plane, $W(z)$ will move on the $W(z)$-plane according to its functional dependence on z. But the left-hand side of this equation denotes the total change in the phase angle of $W(z)$ as z transverses around the complete contour. Therefore, the number of times that the moving point representing $W(z)$ revolves around the origin in the $W(z)$-plane as z moves around the specified contour is given by $N - P$.

The foregoing is conveniently illustrated graphically. Figure A.31a shows the prescribed contour in the z-plane, and Figure A.31b shows a possible form for the variation of $W(z)$. For this particular case, the contour in the z-plane encloses one zero and no poles; hence, $W(z)$ encloses the origin once in the clockwise direction in the $W(z)$-plane.

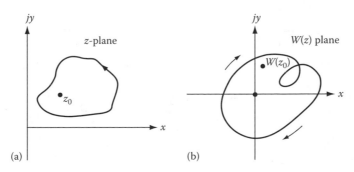

FIGURE A.31

Note that corresponding to a point z_0 within the contour in the z-plane, the point $W(z_0)$ is mapped inside the $W(z)$-plane. In fact, every point on the inside of the contour in the z-plane maps onto the inside of the $W(z)$-contour in the $W(z)$-plane (for single-valued functions). Clearly, there is one point in the z-plane that maps into $W(z) = 0$, the origin.

On the other hand, if the contour includes a pole but no zeros, it can be shown by a similar argument that any point in the interior of the z-contour must correspond to a corresponding point outside of the $W(z)$-contour in the $W(z)$-plane. This is manifested by the fact that the $W(z)$-contour is traversed in a counterclockwise direction. With both zeros and poles present, the situation depends on the values of N and P.

Of special interest is the locus of the network function that contains no poles in the right-hand plane or on the $j\omega$-axis. In this case, the frequency locus is completely traced as z varies along the ω-axis from $-j\infty$ to $+j\infty$. To show this, because $W(z)$ is analytic along the this path, $W(z)$ can be written for the neighborhood of a point z_0 in a Taylor series

$$W(z) = \alpha_0 + \alpha_1(z - z_0) + \alpha_2(z - z_0)^2 + \cdots$$

For the neighborhood $z \to \infty$, we examine $W(z')$, where $z' = 1/z$. Because $W(z)$ does not have a pole at infinity, then $W(z')$ does not have a pole at zero. Therefore, we can expand $W(z')$ in a Maclaurin series

$$W(z') = \alpha_0 + \alpha_1 z' + \alpha_2(z')^2 + \cdots$$

which means that

$$W(z) = \alpha_0 + \frac{\alpha_1}{z} + \frac{\alpha_2}{z^2} + \cdots$$

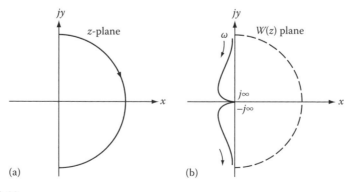

(a)   (b)

FIGURE A.32

But as z approaches infinity, $W(\infty)$ approaches infinity. In a real network function when z^* is written for z, then $W(z^*) = W^*(z)$. This condition requires that $\alpha_0 = \alpha_0 + j0$ be a real number irrespective of how z approaches infinity; that is, as z approaches infinity, $W(z)$ approaches a fixed point in the $W(z)$-plane. This shows that as z varies around the specific contour in the z-plane, $W(z)$ varies from $W(-j\infty)$ to $W(+j\infty)$ as z varies along the imaginary axis. However, $W(-j\infty) = W(+j\infty)$, from the above, which thereby shows that the locus is completely determined. This is illustrated in Figure A.32.

Appendix B: Series and Summations

Series

The expression in parentheses following certain of the series indicates the region of convergence. If not otherwise indicated, it is to be understood that the series converges for all finite values of x.

Binomial

$$(x+y)^n = x^n + nx^{n-1}y + \frac{n(n-1)}{2!}x^{n-2}y^2$$
$$+ \frac{n(n-1)(n-2)}{3!}x^{n-3}y^3 + \cdots \qquad (y^2 < x^2)$$

$$(1 \pm x)^n = 1 \pm nx + \frac{n(n-1)x^2}{2!} \pm \frac{n(n-1)(n-2)x^3}{3!} + \cdots \quad (x^2 < 1)$$

$$(1 \pm x)^{-n} = 1 \mp nx + \frac{n(n+1)x^2}{2!} \mp \frac{n(n+1)(n+2)x^3}{3!} + \cdots \quad (x^2 < 1)$$

$$(1 \pm x)^{-1} = 1 \mp x + x^2 \mp x^3 + x^4 \mp x^5 + \cdots \qquad (x^2 < 1)$$

$$(1 \pm x)^{-2} = 1 \mp 2x + 3x^2 \mp 4x^3 + 5x^4 \mp 6x^5 + \cdots \qquad (x^2 < 1)$$

Reversion of Series

Let a series be represented by

$$y = a_1x + a_2x^2 + a_3x^3 + a_4x^4 + a_5x^5 + a_6x^6 + \cdots \quad (a_1 \neq 0)$$

to find the coefficients of the series

$$x = A_1y + A_2y^2 + A_3y^3 + A_4y^4 + \cdots$$

$$A_1 = \frac{1}{a_1} \quad A_2 = -\frac{a_2}{a_1^3} \quad A_3 = \frac{1}{a_1^5}\left(2a_2^2 - a_1 a_3\right)$$

$$A_4 = \frac{1}{a_1^7}\left(5a_1 a_2 a_3 - a_1^2 a_4 - 5a_2^3\right)$$

$$A_5 = \frac{1}{a_1^9}\left(6a_1^2 a_2 a_4 + 3a_1^2 a_3^2 + 14a_2^4 - a_1^3 a_5 - 21a_1 a_2^2 a_3\right)$$

$$A_6 = \frac{1}{a_1^{11}}\left(7a_1^3 a_2 a_5 + 7a_1^3 a_3 a_4 + 84a_1 a_2^3 a_3 - a_1^4 a_6 - 28a_1^2 a_2^2 a_4 - 28a_1^2 a_2 a_3^2 - 42a_2^5\right)$$

$$A_7 = \frac{1}{a_1^{13}}\left(8a_1^4 a_2 a_6 + 8a_1^4 a_3 a_5 + 4a_1^4 a_4^2 + 120a_1^2 a_2^3 a_4 + 180a_1^2 a_2^2 a_3^2 + 132a_2^6 - a_1^5 a_7\right.$$
$$\left. -36a_1^3 a_2^2 a_5 - 72a_1^3 a_2 a_3 a_4 - 12a_1^3 a_3^3 - 330a_1 a_2^4 a_3\right)$$

Taylor

1.

$$f(x) = f(a) + (x - a)f'(a) + \frac{(x - a)^2}{2!}f''(a) + \frac{(x - a)^3}{3!}f'''(a)$$
$$+ \cdots + \frac{(x - a)^n}{n!}f^{(n)}(a) + \cdots \quad \text{(Taylor's series)}$$

(Increment form)

2.

$$f(x + h) = f(x) + hf'(x) + \frac{h^2}{2!}f''(x) + \frac{h^3}{3!}f'''(x) + \cdots$$
$$= f(h) + xf'(h) + \frac{x^2}{2!}f''(h) + \frac{x^3}{3!}f'''(h) + \cdots$$

3. If $f(x)$ is a function possessing derivatives of all orders throughout the interval $a \le x \le b$, then there is a value X, with $a < X < b$, such that

$$f(b) = f(a) + (b - a)f'(a) + \frac{(b - a)^2}{2!}f''(a)$$
$$+ \cdots + \frac{(b - a)^{n-1}}{(n - 1)!}f^{(n-1)}(a) + \frac{(b - a)^n}{n!}f^{(n)}(X)$$

$$f(a + h) = f(a) + hf'(a) + \frac{h^2}{2!}f''(a) + \cdots + \frac{h^{n-1}}{(n - 1)!}f^{(n-1)}(a)$$
$$+ \frac{h^n}{n!}f^{(n)}(a + \theta h), \quad b = a + h, \ 0 < \theta < 1$$

or

$$f(x) = f(a) + (x - a)f'(a) + \frac{(x - a)^2}{2!}f''(a) + \cdots + (x - a)^{n-1}\frac{f^{(n-1)}(a)}{(n - 1)!} + R_n$$

where

$$R_n = \frac{f^{(n)}[a + \theta \cdot (x - a)]}{n!}(x - a)^n, \quad 0 < \theta < 1$$

The above forms are known as Taylor's series with the remainder term.

4. Taylor's series for a function of two variables

If $\left(h\dfrac{\partial}{\partial x} + k\dfrac{\partial}{\partial y}\right)f(x,\ y) = h\dfrac{\partial f(x,\ y)}{\partial x} + k\dfrac{\partial f(x,\ y)}{\partial y}$;

$$\left(h\frac{\partial}{\partial x} + k\frac{\partial}{\partial y}\right)^2 f(x,\ y) = h^2\frac{\partial^2 f(x,\ y)}{\partial x^2} + 2hk\frac{\partial^2 f(x,\ y)}{\partial x \partial y} + k^2\frac{\partial^2 f(x,\ y)}{\partial y^2},$$

etc., and if $h((\partial/\partial x) + k(\partial/\partial y))^n f(x,\ y)|_{y=b}^{x=a}$ with the bar and the subscripts means that after differentiation we are to replace x by a and y by b,

$$f(a + h,\ b + k) = f(a,\ b) + \left.\left(h\frac{\partial}{\partial x} + k\frac{\partial}{\partial y}\right)f(x,\ y)\right|_{y=b}^{x=a}$$

$$+ \cdots + \left.\frac{1}{n!}\left(h\frac{\partial}{\partial x} + k\frac{\partial}{\partial y}\right)^n f(x,\ y)\right|_{y=b}^{x=a} + \cdots$$

Maclaurin

$$f(x) = f(0) + xf'(0) + \frac{x^2}{2!}f''(0) + \frac{x^3}{3!}f'''(0) + \cdots + x^{n-1}\frac{f^{(n-1)}(0)}{(n-1)!} + R_n$$

where

$$R_n = \frac{x^n f^{(n)}(\theta x)}{n!}, \quad 0 < \theta < 1$$

Exponential

$$e = 1 + \frac{1}{1!} + \frac{1}{2!} + \frac{1}{3!} + \frac{1}{4!} + \cdots$$

$$e^x = 1 + x + \frac{x^2}{2!} + \frac{x^3}{3!} + \frac{x^4}{4!} + \cdots \quad \text{(all real values of } x\text{)}$$

$$a^x = 1 + x\log_e a + \frac{(x\log_e a)^2}{2!} + \frac{(x\log_e a)^3}{3!} + \cdots$$

$$e^x = e^a\left[1 + (x - a) + \frac{(x - a)^2}{2!} + \frac{(x - a)^3}{3!} + \cdots\right]$$

Logarithmic

$$\log_e x = \frac{x-1}{x} + \frac{1}{2}\left(\frac{x-1}{x}\right)^2 + \frac{1}{3}\left(\frac{x-1}{x}\right)^3 + \cdots \qquad \left(x > \frac{1}{2}\right)$$

$$\log_e x = (x-1) - \frac{1}{2}(x-1)^2 + \frac{1}{3}(x-1)^3 - \cdots \qquad (2 \geq x > 0)$$

$$\log_e x = 2\left[\frac{x-1}{x+1} + \frac{1}{3}\left(\frac{x-1}{x+1}\right)^3 + \frac{1}{5}\left(\frac{x-1}{x+1}\right)^5 + \cdots\right] \qquad (x > 0)$$

$$\log_e (1+x) = x - \frac{1}{2}x^2 + \frac{1}{3}x^3 - \frac{1}{4}x^4 + \cdots \qquad (-1 < x < 1)$$

$$\log_e (n+1) - \log_e (n-1) = 2\left[\frac{1}{n} + \frac{1}{3n^3} + \frac{1}{5n^5} + \cdots\right]$$

$$\log_e (a+x) = \log_e a + 2\left[\frac{x}{2a+x} + \frac{1}{3}\left(\frac{x}{2a+x}\right)^3 + \frac{1}{5}\left(\frac{x}{2a+x}\right)^5 + \cdots\right]$$
$$a > 0, \ -a < x < +\infty)$$

$$\log_e \frac{1+x}{1-x} = 2\left[x + \frac{x^3}{3} + \frac{x^5}{5} + \cdots + \frac{x^{2n-1}}{2n-1} + \cdots\right], \qquad -1 < x < 1$$

$$\log_e x = \log_e a + \frac{(x-a)}{a} - \frac{(x-a)^2}{2a^2} + \frac{(x-a)^3}{3a^3} - \cdots, \qquad 0 < x \leq -\alpha$$

Trigonometric

$$\sin x = x - \frac{x^3}{3!} + \frac{x^5}{5!} - \frac{x^7}{7!} + \cdots \qquad \text{(all real values of } x)$$

$$\cos x = 1 - \frac{x^2}{2!} + \frac{x^4}{4!} - \frac{x^6}{6!} + \cdots \qquad \text{(all real values of } x)$$

$$\tan x = x + \frac{x^3}{3} + \frac{2x^5}{15} + \frac{17x^7}{315} + \frac{62x^9}{2835} + \cdots + \frac{2^{2n}(2^{2n}-1)B_n}{(2n)!}x^{2n-1} + \cdots$$
$$\left[x^2 < \frac{\pi^2}{4}, \text{ and } B_n \text{ represents the } n\text{th Bernoulli number.}\right]$$

$$\cot x = \frac{1}{x} - \frac{x}{3} - \frac{x^2}{45} - \frac{2x^5}{945} - \frac{x^7}{4725} - \cdots - \frac{2^{2n}B_n}{(2n)!}x^{2n-1} - \cdots$$
$$\left[x^2 < \pi^2, \text{ and } B_n \text{ represents the } n\text{th Bernoulli number.}\right]$$

$$\sec x = 1 + \frac{x^2}{2} + \frac{5}{24}x^4 + \frac{61}{720}x^6 + \frac{277}{8064}x^8 + \cdots + \frac{E_n x^{2n}}{(2n)!} + \cdots$$

$$\left[x^2 < \frac{\pi^2}{4}, \text{ and } E_n \text{ represents the } n\text{th Euler number.} \right]$$

$$\csc x = \frac{1}{x} + \frac{x}{6} + \frac{7}{360}x^3 + \frac{31}{15,\,120}x^5 + \frac{127}{604,\,800}x^7 + \cdots + \frac{2(2^{2n-1}-1)}{(2n)!}B_n x^{2n-1} + \cdots$$

$$[x^2 < \pi^2, \text{ and } B_n \text{ represents the } n\text{th Bernoulli number.}]$$

$$\sin x = x \left(1 - \frac{x^2}{\pi^2} \right) \left(1 - \frac{x^2}{2^2 \pi^2} \right) \left(1 - \frac{x^2}{3^2 \pi^2} \right) \cdots \qquad (x^2 < \infty)$$

$$\cos x = \left(1 - \frac{4x^2}{\pi^2} \right) \left(1 - \frac{4x^2}{3^2 \pi^2} \right) \left(1 - \frac{4x^2}{5^2 \pi^2} \right) \cdots \qquad (x^2 < \infty)$$

$$\sin^{-1} x = x + \frac{x^3}{2 \cdot 3} + \frac{1 \cdot 3}{2 \cdot 4 \cdot 5}x^5 + \frac{1 \cdot 3 \cdot 5}{2 \cdot 4 \cdot 6 \cdot 7}x^7 + \cdots \quad \left(x^2 < 1, \ -\frac{\pi}{2} < \sin^{-1} x < \frac{\pi}{2} \right)$$

$$\cos^{-1} x = \frac{\pi}{2} - \left(x + \frac{x^3}{2 \cdot 3} + \frac{1 \cdot 3}{2 \cdot 4 \cdot 5}x^5 + \frac{1 \cdot 3 \cdot 5x^7}{2 \cdot 4 \cdot 6 \cdot 7} + \cdots \right) \quad (x^2 < 1, \ 0 < \cos^{-1} x < \pi)$$

$$\tan^{-1} x = x - \frac{x^3}{3} + \frac{x^5}{5} - \frac{x^7}{7} + \cdots \qquad (x^2 < 1)$$

$$\tan^{-1} x = \frac{\pi}{2} - \frac{1}{x} + \frac{1}{3x^2} - \frac{1}{5x^5} + \frac{1}{7x^7} - \cdots \qquad (x > 1)$$

$$\tan^{-1} x = -\frac{\pi}{2} - \frac{1}{x} + \frac{1}{3x^2} - \frac{1}{5x^5} + \frac{1}{7x^7} - \cdots \qquad (x < -1)$$

$$\cot^{-1} x = \frac{\pi}{2} - x + \frac{x^3}{3} - \frac{x^5}{5} + \frac{x^7}{7} - \cdots \qquad (x^2 < 1)$$

$$\log_e \sin x = \log_e x - \frac{x^2}{6} - \frac{x^4}{180} - \frac{x^6}{2835} - \cdots \qquad (x^2 < \pi^2)$$

$$\log_e \cos x = -\frac{x^2}{2} - \frac{x^4}{12} - \frac{x^6}{45} - \frac{17x^8}{2520} - \cdots \qquad \left(x^2 < \frac{\pi^2}{4} \right)$$

$$\log_e \tan x = \log_e x + \frac{x^2}{3} + \frac{7x^4}{90} + \frac{62x^6}{2835} + \cdots \qquad \left(x^2 < \frac{\pi^2}{4} \right)$$

$$e^{\sin x} = 1 + x + \frac{x^2}{2!} - \frac{3x^4}{4!} - \frac{8x^5}{5!} - \frac{3x^6}{6!} + \frac{56x^7}{7!} + \cdots$$

$$e^{\cos x} = e \left(1 - \frac{x^2}{2!} + \frac{4x^4}{4!} - \frac{31x^6}{6!} + \cdots \right)$$

$$e^{\tan x} = 1 + x + \frac{x^2}{2!} + \frac{3x^3}{3!} + \frac{9x^4}{4!} + \frac{37x^5}{5!} + \cdots \qquad \left(x^2 < \frac{\pi^2}{4} \right)$$

$$\sin x = \sin a + (x - a) \cos a - \frac{(x-a)^2}{2!} \sin a - \frac{(x-a)^3}{3!} \cos a + \frac{(x-a)^4}{4!} \sin a + \cdots$$

Hyperbolic and Inverse Hyperbolic

Table of expansion of certain functions into power series

$$\sinh x = x + \frac{x^3}{3!} + \frac{x^5}{5!} + \frac{x^7}{7!} + \cdots + \frac{x^{2n+1}}{(2n+1)!} + \cdots \qquad |x| < \infty$$

$$\cosh x = 1 + \frac{x^2}{2!} + \frac{x^4}{4!} + \frac{x^6}{6!} + \cdots + \frac{x^{2n}}{(2n)!} + \cdots \qquad |x| < \infty$$

$$\tanh x = x - \frac{1}{3}x^3 + \frac{2}{15}x^5 - \frac{17}{315}x^7 + \frac{62}{2835}x^9 - \cdots$$
$$+ \frac{(-1)^{n+1}2^{2n}(2^{2n}-1)}{(2n)!}B_n x^{2n-1} \pm \cdots \qquad |x| < \frac{\pi}{2}^{*}$$

$$\coth x = \frac{1}{x} + \frac{x}{3} - \frac{x^3}{45} + \frac{2x^5}{945} - \frac{x^7}{4725} + \cdots$$
$$+ \frac{(-1)^{n+1}2^{2n}}{(2n)!}B_n x^{2n-1} \pm \cdots \qquad 0 < |x| < \pi^{*}$$

$$\operatorname{sech} x = 1 - \frac{1}{2!}x^2 + \frac{5}{4!}x^4 - \frac{61}{6!}x^6 + \frac{1385}{8!}x^8 - \cdots + \frac{(-1)^n}{(2n)!}E_n x^{2n} \pm \cdots \qquad |x| < \frac{\pi}{2}^{\dagger}$$

$$\operatorname{cosech} x = \frac{1}{x} - \frac{x}{6} + \frac{7x^3}{360} - \frac{31x^5}{15,120} + \cdots$$
$$+ \frac{2(-1)^n(2^{2n-1}-1)}{(2n)!}B_n x^{2n-1} + \cdots \qquad 0 < |x| < \pi^{*}$$

$$\operatorname{arg\,sinh} x = x - \frac{1}{2 \cdot 3}x^3 + \frac{1 \cdot 3}{2 \cdot 4 \cdot 5}x^5 - \frac{1 \cdot 3 \cdot 5}{2 \cdot 4 \cdot 6 \cdot 7}x^7 + \cdots$$
$$+ (-1)^n \cdot \frac{1 \cdot 3 \cdot 5(2n-1)}{2 \cdot 4 \cdot 6 \ldots 2n(2n+1)}x^{2n+1} \pm \cdots \qquad |x| < 1$$

$$\operatorname{arg\,cosh} x = \pm\left[\ln(2x) - \frac{1}{2 \cdot 2x^2} - \frac{1 \cdot 3}{2 \cdot 4 \cdot 4x^4} - \frac{1 \cdot 3 \cdot 5}{2 \cdot 4 \cdot 6 \cdot 6x^6} - \cdots\right] \qquad x > 1$$

$$\operatorname{arg\,tanh} x = x + \frac{x^3}{3} + \frac{x^5}{5} + \frac{x^7}{7} + \cdots + \frac{x^{2n+1}}{2n+1} + \cdots \qquad |x| < 1$$

$$\operatorname{arg\,coth} x = \frac{1}{x} + \frac{1}{3x^3} + \frac{1}{5x^5} + \frac{1}{7x^7} + \cdots + \frac{1}{(2n+1)x^{2n+1}} + \cdots \qquad |x| > 1$$

---

[*] B_n denotes Bernoulli's numbers.
[†] E_n denotes Euler's numbers.

Arithmetic Progression of the first order (first differences constant) to n terms,

$$a + (a+d) + (a+2d) + (a+3d) + \cdots + \{a + (n-1)d\} \equiv na + \frac{1}{2}n(n-1)d$$
$$\equiv \frac{n}{2}(\text{1st term} + n\text{th term}).$$

Geometric Progression to n terms,

$$a + ar + ar^2 + ar^3 + \cdots + ar^{n-1} \equiv a(1 - r^n)/(1 - r)$$
$$\equiv a(r^n - 1)/(r - 1)$$

If $r^2 < 1$, the limit of the sum of an infinite number of terms is $a/(1 - r)$.

The reciprocals of the terms of a series in arithmetic progression of the first order are in **harmonic progression**. Thus,

$$\frac{1}{a}, \; \frac{1}{a+d}, \; \frac{1}{a+2d}, \; \cdots \frac{1}{a + (n-1)d}$$

are in Harmonic Progression.

The arithmetic mean of n quantities is

$$\frac{1}{n}(a_1 + a_2 + a_3 + \cdots + a_n).$$

The **geometric mean** of n quantities is

$$(a_1 a_2 a_3 \ldots a_n)^{1/n}$$

Let the **harmonic mean** of n quantities be H. Then

$$\frac{1}{H} = \frac{1}{n}\left(\frac{1}{a_1} + \frac{1}{a_2} + \frac{1}{a_3} + \cdots + \frac{1}{a_n}\right)$$

The arithmetic mean of a number of positive quantities is $\geqq$ their geometric mean, which in turn is $\geqq$ their harmonic mean.

$$1 + 2 + 3 + \cdots + n = \frac{n}{2}(n + 1) = \sum_{k=0}^{n} k$$

$$1^2 + 2^2 + 3^2 + \cdots + n^2 = \frac{n}{6}(n + 1)(2n + 1)$$
$$= \frac{n}{6}(2n^2 + 3n + 1) = \sum_{k=0}^{n} k^2$$

$$1^3 + 2^3 + 3^3 + \cdots + n^3 = \frac{n^2}{4}(n+1)^2$$

$$= \frac{n^2}{4}(n^2 + 2n + 1) = \sum_{k=0}^{n} k^3$$

$$1 + 3 + 5 + 7 + 9 + \cdots + (2n - 1) = n^2 = \sum_{k=0}^{2n-1}(2k + 1)$$

$$1 + 8 + 16 + 24 + 32 + \cdots + 8(n - 1) = (2n - 1)^2$$

$$1 + 3x + 5x^2 + 7x^3 + \cdots = \frac{1 + x}{(1 - x)^2}$$

$$1 + ax + (a + b)x^2 + (a + 2b)x^3 + \cdots = 1 + \frac{ax + (b - a)x^2}{(1 - x)^2}$$

$$1 + 2^2 x + 3^2 x^2 + 4^2 x^3 + \cdots = \frac{1 + x}{(1 - x)^3}$$

$$1 + 3^2 x + 5^2 x^2 + 7^2 x^3 + \cdots = \frac{1 + 6x + x^2}{(1 - x)^3}$$

$$\frac{a[1 - (n + 1)a^n + na^{n+1}]}{(1 - a)^2} = \sum_{k=0}^{n} k a^k$$

$$\frac{a\left[(1 + a) - (n + 1)^2 a^n + (2n^2 + 2n - 1)a^{n+1} - n^2 a^{n+2}\right]}{(1 - a)^3} = \sum_{k=0}^{n} k^2 a^k$$

$$\frac{a}{(1 - a)^2} = \sum_{k=0}^{\infty} k a^k \quad |a| < 1$$

$$\frac{a^2 + a}{(1 - a)^3} = \sum_{k=0}^{\infty} k^2 a^k \quad |a| < 1$$

Appendix C:
Definite Integrals

$$\int_0^\infty x^{n-1}e^{-x}dx = \int_0^1 \left(\log\frac{1}{x}\right)^{n-1} dx - \frac{1}{n}\prod_{m=1}^\infty \frac{\left(1+\frac{1}{m}\right)^n}{1+\frac{n}{m}}$$

$$= \Gamma(n), n \neq 0, -1, -2, -3, \ldots \qquad \text{(Gamma function)}$$

$$\int_0^\infty t^n p^{-t}dt = \frac{n!}{(\log p)^{n+1}} \qquad (n = 0, 1, 2, 3, \ldots \text{ and } p > 0)$$

$$\int_0^\infty t^{n-1}e^{-(a+1)t}dt = \frac{\Gamma(n)}{(a+1)^n} \qquad (n > 0, a > -1)$$

$$\int_0^1 x^m \left(\log\frac{1}{x}\right)^n dx = \frac{\Gamma(n+1)}{(m+1)^{n+1}} \qquad (m > -1, n > -1)$$

$\Gamma(n)$ is finite if $n > 0$, $\Gamma(n+1) = n\Gamma(n)$

$$\Gamma(n) \cdot \Gamma(1-n) = \frac{\pi}{\sin n\pi}$$

$\Gamma(n) = (n-1)!$ if $n = \text{integer} > 0$

$$\Gamma\left(\frac{1}{2}\right) = 2\int_0^\infty e^{-t^2}dt = \sqrt{\pi} = 1.7724538509\ldots$$

$$\Gamma\left(n+\frac{1}{2}\right) = \frac{1 \cdot 3 \cdot 5 \cdot 7 \dots (2n-1)}{2^n} \sqrt{\pi},$$

where n is an integer > 0

$$\int_0^1 x^{m-1}(1-x)^{n-1} dx = B(m,n) \qquad\qquad\qquad\qquad \text{(Beta function)}$$

$$B(m,n) = B(n,m) = \frac{\Gamma(m)\Gamma(n)}{\Gamma(m+n)}$$

where m and n are positive real numbers

$$\int_0^1 x^{m-1}(1-x)^{n-1} dx = \int_0^\infty \frac{x^{m-1} dx}{(1+x)^{m+n}} = \frac{\Gamma(m)\Gamma(n)}{\Gamma(m+n)}$$

$$\int_a^b (x-a)^m (b-x)^n dx = (b-a)^{m+n+1} \frac{\Gamma(m+1) \cdot \Gamma(n+1)}{\Gamma(m+n+2)} \quad (m > -1, n > -1, b > a)$$

$$\int_0^\infty \frac{dx}{x^m} = \frac{1}{m-1} \qquad\qquad\qquad\qquad\qquad\qquad\qquad (m > 1)$$

$$\int_0^\infty \frac{dx}{(1+x)x^p} = \pi \csc p\pi \qquad\qquad\qquad\qquad\qquad\qquad (p < 1)$$

$$\int_0^\infty \frac{dx}{(1-x)x^p} = -\pi \cot p\pi \qquad\qquad\qquad\qquad\qquad\quad (p < 1)$$

$$\int_0^\infty \frac{x^{p-1} dx}{1+x} = \frac{\pi}{\sin p\pi}$$

$$\qquad\qquad = B(p, 1-p) = \Gamma(p)\Gamma(1-p) \qquad\qquad\qquad\quad (0 < p < 1)$$

$$\int_0^\infty \frac{x^{m-1} dx}{1+x^n} = \frac{\pi}{n \sin \frac{m\pi}{n}} \qquad\qquad\qquad\qquad\qquad\quad (0 < m < n)$$

$$\int_0^\infty \frac{x^a dx}{(m+x^b)^c} = \frac{a+1}{m^{\frac{b-c}{b}}} \left[\frac{\Gamma\left(\frac{a+1}{b}\right)\Gamma\left(c - \frac{a+1}{b}\right)}{\Gamma(c)}\right] \qquad \left(a > -1, b > 0, m > 0, c > \frac{a+1}{b}\right)$$

$$\int_0^\infty \frac{dx}{(1+x)\sqrt{x}} = \pi$$

$$\int_0^\infty \frac{a\,dx}{a^2+x^2} = \frac{\pi}{2}, \quad \text{if } a>0; 0, \quad \text{if } a=0; \ -\frac{\pi}{2}, \quad \text{if } a<0$$

$$\int_0^a x^m(a^2-x^2)^{\frac{n}{2}}dx = \begin{cases} \frac{1}{2}a^{m+n+1}B\left(\frac{m+1}{2},\frac{n+2}{2}\right) \\[2ex] \frac{1}{2}a^{m+n+1}\dfrac{\Gamma\left(\frac{m+1}{2}\right)\Gamma\left(\frac{n+2}{2}\right)}{\Gamma\left(\frac{m+n+3}{2}\right)} \end{cases}$$

$$\int_0^1 \frac{dx}{\sqrt{(1-x^n)}} = \frac{\sqrt{\pi}}{n}\frac{\Gamma\left(\frac{1}{n}\right)}{\Gamma\left(\frac{1}{n}+\frac{1}{2}\right)} \qquad\qquad (n>0)$$

$$\int_0^1 \frac{x^m\,dx}{\sqrt{(1-x^n)}} = \frac{\sqrt{\pi}}{n}\frac{\Gamma\left(\frac{m+1}{n}\right)}{\Gamma\left(\frac{m+1}{n}+\frac{1}{2}\right)} \qquad\qquad (m+1, n>0)$$

$$\int_0^1 x^m(1-x^2)^p\,dx = \frac{\Gamma(p+1)\Gamma\left(\frac{m+1}{2}\right)}{2\Gamma\left(p+\frac{m+3}{2}\right)} \qquad\qquad (p+1, m+1>0)$$

$$\int_0^1 x^m(1-x^n)^p\,dx = \frac{\Gamma(p+1)\Gamma\left(\frac{m+1}{n}\right)}{n\Gamma\left(p+1+\frac{m+1}{n}\right)} \qquad\qquad (p+1, m+1, n>0)$$

$$\int_0^1 \frac{x^m\,dx}{\sqrt{(1-x^2)}} = \frac{2\cdot4\cdot6\cdots(m-1)}{3\cdot5\cdot7\cdots m} \qquad\qquad (m\text{ an odd integer}>1)$$

$$= \frac{1\cdot3\cdot5\cdots(m-1)}{2\cdot4\cdot6\cdots m}\frac{\pi}{2} \qquad\qquad (m\text{ an even, positive integer})$$

$$= \frac{\sqrt{\pi}}{2}\frac{\Gamma\left(\frac{m+1}{2}\right)}{\Gamma\left(\frac{m}{2}+1\right)} \qquad\qquad (m\text{ any value}>-1)$$

$$\int_0^\infty \frac{x^{p-1}\,dx}{1+x} = \frac{\pi}{\sin(\pi-p\pi)} = \frac{\pi}{\sin p\pi} \qquad\qquad (0<p<1)$$

$$\int_0^\infty \frac{dx}{(1+x)\sqrt{x}} = \pi$$

$$\int_0^\infty \frac{x^{p-1}\,dx}{a+x} = \frac{\pi a^{p-1}}{\sin p\pi} \qquad\qquad (0 < p < 1)$$

$$\int_0^\infty \frac{dx}{1+x^p} = \frac{\pi}{p \sin \frac{\pi}{p}} \qquad\qquad (p > 1)$$

$$\int_0^\infty \frac{x^p\,dx}{(1+ax)^2} = \frac{p\pi}{a^{p+1} \sin p\pi} \qquad\qquad (0 < p < 1)$$

$$\int_0^\infty \frac{x^p\,dx}{1+x^2} = \frac{\pi}{2 \cos \frac{p\pi}{2}} \qquad\qquad (-1 < p < 1)$$

$$\int_0^\infty \frac{x^{p-1}\,dx}{1+x^q} = \frac{\pi}{q \sin \frac{p\pi}{q}} \qquad\qquad (0 < p < q)$$

$$\int_0^\infty \frac{x^{m-1}\,dx}{(1+x)^{m+n}} = \frac{\Gamma(m)\Gamma(n)}{\Gamma(m+n)} \qquad\qquad (m, n > 0)$$

$$\int_0^\infty \frac{x^{m-1}\,dx}{(a+bx)^{m+n}} = \frac{\Gamma(m)\Gamma(n)}{a^n b^m \Gamma(m+n)} \qquad\qquad (a, b, m, n > 0)$$

$$\int_0^\infty \frac{dx}{(a^2+x^2)^n} = \frac{1\cdot3\cdot5\cdots(2n-3)}{2\cdot4\cdot6\cdots(2n-2)}\frac{\pi}{2a^{2n-1}} \qquad\qquad (a > 0;\, n = 2, 3, \ldots)$$

$$\int_0^\infty \frac{dx}{(a^2+x^2)(b^2+x^2)} = \frac{\pi}{2ab(a+b)} \qquad\qquad (a, b > 0)$$

$$\int_0^{\pi/2} (\sin^n x)\,dx = \begin{cases} \int_0^{\pi/2} (\cos^n x)\,dx & \\[2mm] \frac{1\cdot3\cdot5\cdot7\cdots(n-1)}{2\cdot4\cdot6\cdot8\cdots(n)}\frac{\pi}{2} & (n \text{ is an even integer,} \quad n \neq 0) \\[2mm] \frac{2\cdot4\cdot6\cdot8\cdots(n-1)}{1\cdot3\cdot5\cdot7\cdots(n)} & (n \text{ is an odd integer,} \quad n \neq 1) \\[2mm] \frac{\sqrt{\pi}\,\Gamma\left(\frac{n+1}{2}\right)}{2\,\Gamma\left(\frac{n}{2}+1\right)} & (n > -1) \end{cases}$$

$$\int_0^\infty \frac{\sin mx\,dx}{x} = \frac{\pi}{2}, \quad \text{if } m > 0; \quad \text{if } m = 0; \quad -\frac{\pi}{2}, \quad \text{if } m < 0$$

$$\int_0^\infty \frac{\cos x \, dx}{x} = \infty$$

$$\int_0^\infty \frac{\tan x \, dx}{x} = \frac{\pi}{2}$$

$$\int_0^\pi \sin ax \cdot \sin bx \, dx = \int_0^\pi \cos ax \cdot \cos bx \, dx = 0 \qquad\qquad (a \neq b; \, a, b \text{ integers})$$

$$\int_0^{\pi/a} [\sin(ax)][\cos(ax)] \, dx = \int_0^\pi [\sin(ax)][\cos(ax)] \, dx = 0$$

$$\int_0^\pi [\sin(ax)][\cos(bx)] \, dx = \frac{2a}{a^2 - b^2}, \quad \text{if } a - b \text{ is odd, or zero if } a - b \text{ is even}$$

$$\int_0^\infty \frac{\sin x \cos mx \, dx}{x} = 0, \quad \text{if } m < -1 \text{ or } m > 1; = \frac{\pi}{4}, \quad \text{if } m = \pm 1; = \frac{\pi}{2}, \quad \text{if } m^2 < 1$$

$$\int_0^\infty \frac{\sin ax \sin bx}{x^2} \, dx = \frac{\pi a}{2} \quad (a \leq b)$$

$$\int_0^\pi \sin^2 mx \, dx = \int_0^\pi \cos^2 mx \, dx = \frac{\pi}{2}$$

$$\int_0^\infty \frac{\sin^2 x \, dx}{x^2} = \frac{\pi}{2}$$

$$\int \frac{\cos mx}{1 + x^2} \, dx = \frac{\pi}{2} e^{-|m|}$$

$$\int_0^\infty \cos(x^2) \, dx = \int_0^\infty \sin(x^2) \, dx = \frac{1}{2} \sqrt{\frac{\pi}{2}}$$

$$\int_0^\infty \frac{\sin x \, dx}{\sqrt{x}} = \int_0^\infty \frac{\cos x \, dx}{\sqrt{x}} = \sqrt{\frac{\pi}{2}}$$

$$\int_0^{\pi/2} \frac{dx}{1 + a\cos x} = \frac{\cos^{-1} a}{\sqrt{1-a^2}} \qquad\qquad (a < 1)$$

$$\int_0^{\infty} \frac{dx}{a + b\cos x} = \frac{\pi}{\sqrt{a^2 - b^2}} \qquad\qquad (a > b \geq 0)$$

$$\int_0^{2\pi} \frac{dx}{1 + a\cos x} = \frac{2\pi}{\sqrt{1-a^2}} \qquad\qquad (a^2 < 1)$$

$$\int_0^{\infty} \frac{\cos ax - \cos bx}{x}\, dx = \log\frac{b}{a}$$

$$\int_0^{\pi/2} \frac{dx}{a^2 \sin^2 x + b^2 \cos^2 x} = \frac{\pi}{2ab}$$

$$\int_0^{\pi/2} \frac{dx}{(a^2 \sin^2 x + b^2 \cos^2 x)^2} = \frac{\pi(a^2 + b^2)}{4a^3 b^3} \qquad\qquad (a, b > 0)$$

$$\int_0^{\pi/2} \sin^{n-1} x \, \cos^{m-1} x \, dx = \frac{1}{2} B\left(\frac{n}{2}, \frac{m}{2}\right) \qquad\qquad (m \text{ and } n \text{ are positive integers})$$

$$\int_0^{\pi/2} (\sin^{2n+1} \theta)d\theta = \frac{2 \cdot 4 \cdot 6 \cdots (2n)}{1 \cdot 3 \cdot 5 \cdots (2n+1)} \qquad\qquad (n = 1, 2, 3, \dots)$$

$$\int_0^{\pi/2} (\sin^{2n} \theta)d\theta = \frac{1 \cdot 3 \cdot 5 \cdots (2n-1)}{2 \cdot 4 \cdots (2n)} \left(\frac{\pi}{2}\right) \qquad\qquad (n = 1, 2, 3, \dots)$$

$$\int_0^{\pi/2} \sqrt{\cos\theta}\,d\theta = \frac{(2\pi)^{\frac{3}{2}}}{\left[\Gamma\left(\frac{1}{4}\right)\right]^2}$$

$$\int_0^{\pi/2} (\tan^h \theta)d\theta = \frac{\pi}{2\cos\left(\frac{h\pi}{2}\right)} \qquad\qquad (0 < h < 1)$$

$$\int_0^{\infty} \frac{\tan^{-1}(ax) - \tan^{-1}(bx)}{x}\, dx = \frac{\pi}{2}\log\frac{a}{b} \qquad\qquad (a, b > 0)$$

The area enclosed by a curve is defined through the equation $x^{b/c} + y^{b/c} = a^{b/c}$

where

$a > 0$

c is a positive odd integer

b is a positive even integer given by

$$\frac{\left[\Gamma\left(\frac{c}{b}\right)\right]^2}{\Gamma\left(\frac{2c}{b}\right)}\left(\frac{2ca^2}{b}\right)$$

$I = \iiint_R x^{h-1} y^{m-1} z^{n-1} \, dv$, where R denotes the region of space bounded by the coordinate planes and that portion of the surface $(x/a)^p + (y/b)^q + (z/c)^k = 1$, which lies in the first octant and where $h, m, n, p, q, k, a, b,$ and c denote positive real numbers given by

$$\int_0^a x^{h-1} \, dx \int_0^{b\left[1-\left(\frac{x}{a}\right)^p\right]^{\frac{1}{q}}} y^m \, dy \int_0^{c\left[1-\left(\frac{x}{a}\right)^p-\left(\frac{y}{b}\right)^q\right]^{\frac{1}{k}}} z^{n-1} \, dz = \frac{a^h b^m c^n}{pqk} \frac{\Gamma\left(\frac{h}{p}\right)\Gamma\left(\frac{m}{q}\right)\Gamma\left(\frac{n}{k}\right)}{\Gamma\left(\frac{h}{p}+\frac{m}{q}+\frac{n}{k}+1\right)}$$

$$\int_0^{\pi/2} \frac{dx}{a^2 \sin^2 x + b^2 \cos^2 x} = \frac{\pi}{2ab} \qquad\qquad (ab > 0)$$

$$\int_0^{\pi} \frac{dx}{a^2 \sin^2 x + b^2 \cos^2 x} = \frac{\pi}{ab} \qquad\qquad (ab > 0)$$

$$\int_0^{\pi/2} \frac{\sin^2 x \, dx}{a^2 \sin^2 x + b^2 \cos^2 x} = \int_0^{\pi/2} \frac{dx}{a^2 + b^2 \, \text{ctn}^2 x} = \frac{\pi}{2a(a+b)} \qquad (a, b > 0)$$

$$\int_0^{\pi/2} \frac{\cos^2 x \, dx}{a^2 \sin^2 x + b^2 \cos^2 x} = \int_0^{\pi/2} \frac{dx}{b^2 + a^2 \tan^2 x} = \frac{\pi}{2b(a+b)} \qquad (a, b > 0)$$

$$\int_0^{\pi/2} \frac{dx}{(a^2 \sin^2 x + b^2 \cos^2 x)^2} = \frac{\pi}{4} \frac{(a^2 + b^2)}{a^3 b^3} \qquad\qquad (ab > 0)$$

$$\int_0^{\pi/2} \frac{\sin^2 x \, dx}{(a^2 \sin^2 x + b^2 \cos^2 x)^2} = \frac{\pi}{4a^3 b} \qquad\qquad (ab > 0)$$

$$\int_0^{\pi/2} \frac{\cos^2 x \, dx}{(a^2 \sin^2 x + b^2 \cos^2 x)^2} = \frac{\pi}{4ab^3} \qquad\qquad (ab > 0)$$

$$\int_0^\infty \sin(a^2 x^2)\,dx = \int_0^\infty \cos(a^2 x^2)\,dx = \frac{\sqrt{\pi}}{2a\sqrt{2}} \qquad (a > 0)$$

$$\int_0^\infty \sin\frac{\pi x^2}{2}\,dx = \int_0^\infty \cos\frac{\pi x^2}{2}\,dx = \frac{1}{2} \qquad \text{(Fresnel's integrals)}$$

$$\int_0^\infty \sin(x^p)\,dx = \Gamma\left(1 + \frac{1}{p}\right)\sin\frac{\pi}{2p} \qquad (p > 1)$$

$$\int_0^\infty \cos(x^p)\,dx = \Gamma\left(1 + \frac{1}{p}\right)\cos\frac{\pi}{2p} \qquad (p > 1)$$

$$\int_0^\infty \sin a^2 x^2 \cos mx\,dx = \frac{\sqrt{\pi}}{2a}\sin\left(\frac{\pi}{4} - \frac{m^2}{4a^2}\right) \qquad (a > 0)$$

$$\int_0^\infty \cos a^2 x^2 \cos mx\,dx = \frac{\sqrt{\pi}}{2a}\cos\left(\frac{\pi}{4} - \frac{m^2}{4a^2}\right) \qquad (a > 0)$$

$$\int_0^\infty \frac{\sin^{2p} mx}{x^2}\,dx = \frac{1 \cdot 3 \cdot 5 \cdots (2p-3)}{2 \cdot 4 \cdot 6 \cdots (2p-2)}\frac{|m|\pi}{2} \qquad (p = 2, 3, 4, \ldots)$$

$$\int_0^\infty \frac{\sin^3 mx}{x^3}\,dx = \frac{3}{8}m^2\pi \qquad (m > 0)$$

$$\int_0^\infty \frac{\sin mx \cos nx}{x}\,dx = \pi/2 \qquad (m > n > 0)$$

$$= \pi/4 \qquad (m = n > 0)$$

$$= 0 \qquad (n > m > 0)$$

$$\int_0^\infty \frac{\sin mx \sin nx}{x}\,dx = \frac{1}{2}\log\frac{m+n}{m-n} \qquad (m > n > 0)$$

$$\int_0^\infty \frac{\cos mx \cos nx}{x}\,dx = \infty$$

$$\int_0^\infty \frac{\sin^2 ax \, \sin mx}{x} \, dx = \frac{\pi}{4} \qquad\qquad (2a > m > 0)$$

$$= \frac{\pi}{4} \qquad\qquad (2a = m > 0)$$

$$= 0 \qquad\qquad (m > 2a > 0)$$

$$\int_0^\infty \frac{\sin mx \, \sin nx}{x^2} \, dx = \frac{\pi m}{2} \qquad\qquad (n \geqq m > 0)$$

$$= \frac{\pi n}{2} \qquad\qquad (m \geqq n > 0)$$

$$\int_0^\infty \frac{\sin^2 ax \, \sin mx}{x^2} \, dx = \frac{m + 2a}{4} \log |m + 2a| + \frac{m - 2a}{4} \log |m - 2a| - \frac{m}{2} \log m \quad (m > 0)$$

$$\int_0^\infty \frac{\cos mx}{a^2 + x^2} \, dx = \frac{\pi}{2a} e^{-ma} \qquad\qquad (a > 0; m \geqq 0)$$

$$\int_0^\infty \frac{\sin^2 mx}{a^2 + x^2} \, dx = \frac{\pi}{4a} (1 - e^{-2ma}) \qquad\qquad (a > 0; m \geqq 0)$$

$$\int_0^\infty \frac{\cos^2 mx}{a^2 + x^2} \, dx = \frac{\pi}{4a} (1 + e^{-2ma}) \qquad\qquad (a > 0; m \geqq 0)$$

$$\int_0^\infty \frac{x \sin mx}{a^2 + x^2} \, dx = \frac{\pi}{2} e^{-ma} \qquad\qquad (a \geqq 0; m > 0)$$

$$\int_0^\infty \frac{\sin mx}{x(a^2 + x^2)} \, dx = \frac{\pi}{2a^2} (1 - e^{-ma}) \qquad\qquad (a > 0; m \geqq 0)$$

$$\int_0^\infty \frac{\sin mx \, \sin nx}{a^2 + x^2} \, dx = \frac{\pi}{2a} e^{-ma} \sinh na \qquad\qquad (a > 0; m \geqq n \geqq 0)$$

$$= \frac{\pi}{2a} e^{-na} \sinh ma \qquad\qquad (a > 0; n \geqq m \geqq 0)$$

$$\int_0^\infty \frac{\cos mx \, \cos nx}{a^2 + x^2} \, dx = \frac{\pi}{2a} e^{-ma} \cosh na \qquad\qquad (a > 0; m \geqq n \geqq 0)$$

$$= \frac{\pi}{2a} e^{-na} \cosh ma \qquad\qquad (a > 0; n \geqq m \geqq 0)$$

$$\int_0^\infty \frac{x \sin mx \cos nx}{a^2 + x^2} \, dx = \frac{\pi}{2} e^{-ma} \cosh na \qquad\qquad (a > 0; \, m > n > 0)$$

$$= -\frac{\pi}{2} e^{-na} \sinh ma \qquad\qquad (a > 0; \, n > m > 0)$$

$$\int_0^\infty \frac{\cos mx}{(a^2 + x^2)^2} \, dx = \frac{\pi}{4a^3} (1 + ma) e^{-ma} \qquad\qquad (a, m > 0)$$

$$\int_0^\infty \frac{x \sin mx}{(a^2 + x^2)^2} \, dx = \frac{\pi m}{4a} e^{-ma} \qquad\qquad (a, m > 0)$$

$$\int_0^\infty \frac{x^2 \cos mx}{(a^2 + x^2)^2} \, dx = \frac{\pi}{4a} (1 - ma) e^{-ma} \qquad\qquad (a, m > 0)$$

$$\int_0^\infty \frac{\sin^2 ax \cos mx}{x^2} \, dx = \frac{\pi}{2} \left(a - \frac{m}{2} \right) \qquad\qquad \left(a > \frac{m}{2} > 0 \right)$$

$$= 0 \qquad\qquad \left(\frac{m}{2} \geqq a \geqq 0 \right)$$

$$\int_0^\infty \frac{1 - \cos mx}{x^2} \, dx = \frac{\pi |m|}{2}$$

$$\int_0^\infty \frac{\sin^2 ax \sin mx}{x^3} \, dx = \frac{\pi \, am}{2} - \frac{\pi \, m^2}{8} \qquad\qquad \left(a \geqq \frac{m}{2} > 0 \right)$$

$$= \frac{\pi \, a^2}{2} \qquad\qquad \left(\frac{m}{2} \geqq a > 0 \right)$$

$$\int_0^\infty \frac{\sin mx}{\sqrt{x}} \, dx = \int_0^\infty \frac{\cos mx}{\sqrt{x}} \, dx = \frac{\sqrt{\pi}}{\sqrt{(2m)}} \qquad\qquad (m > 0)$$

$$\int_0^\infty \frac{\sin mx}{x\sqrt{x}} \, dx = \sqrt{(2\pi m)} \qquad\qquad (m > 0)$$

$$\int_0^\infty \frac{\sin mx}{x^p} \, dx = \frac{\pi m^{p-1}}{2 \sin \left(\frac{p\pi}{2} \right) \Gamma(p)} \qquad\qquad (0 < p < 2; \, m > 0)$$

$$\int_0^\infty e^{-ax} \, dx = \frac{1}{a} \qquad\qquad (a > 0)$$

$$\int_0^\infty \frac{e^{-ax} - e^{-bx}}{x} dx = \log\frac{b}{a} \qquad\qquad (a, b > 0)$$

$$\int_0^\infty x^n e^{-ax} dx = \frac{\Gamma(n+1)}{a^{n+1}} \qquad\qquad (n > -1, a > 0)$$

$$= \frac{n!}{a^{n+1}} \qquad\qquad (n \text{ pos. integ.}, a > 0)$$

$$\int_0^\infty e^{-a^2 x^2} dx = \frac{1}{2a}\sqrt{\pi} = \frac{1}{2a}\Gamma\left(\frac{1}{2}\right) \qquad\qquad (a > 0)$$

$$\int_0^\infty x e^{-x^2} dx = \frac{1}{2}$$

$$\int_0^\infty x^2 e^{-x^2} dx = \frac{\sqrt{\pi}}{4}$$

$$\int_0^\infty x^{2n} e^{-ax^2} dx = \frac{1 \cdot 3 \cdot 5 \cdots (2n-1)}{2^{n+1} a^n}\sqrt{\frac{\pi}{a}}$$

$$\int_0^1 x^m e^{-ax} dx = \frac{m!}{a^{m+1}}\left[1 - e^{-a}\sum_{r=0}^m \frac{a^r}{r!}\right]$$

$$\int_0^\infty e^{\left(-x^2 - \frac{a^2}{x^2}\right)} dx = \frac{e^{-2a}\sqrt{\pi}}{2} \qquad\qquad (a \geq 0)$$

$$\int_0^\infty e^{-nx}\sqrt{x}\, dx = \frac{1}{2n}\sqrt{\frac{\pi}{n}}$$

$$\int_0^\infty \frac{e^{-nx}}{\sqrt{x}} dx = \sqrt{\frac{\pi}{n}}$$

$$\int_0^\infty e^{-ax}\cos mx\, dx = \frac{a}{a^2 + m^2} \qquad\qquad (a > 0)$$

$$\int_0^\infty e^{-ax}\sin mx\, dx = \frac{m}{a^2 + m^2} \qquad\qquad (a > 0)$$

$$\int_0^\infty xe^{-ax}[\sin(bx)]dx = \frac{2ab}{(a^2 + b^2)^2} \qquad (a > 0)$$

$$\int_0^\infty xe^{-ax}[\cos(bx)]dx = \frac{a^2 - b^2}{(a^2 + b^2)^2} \qquad (a > 0)$$

$$\int_0^\infty x^n e^{-ax}[\sin(bx)]dx = \frac{n![(a - ib)^{n+1} - (a + ib)^{n+1}]}{2(a^2 + b^2)^{n+1}} \qquad (i^2 = -1, a > 0)$$

$$\int_0^\infty x^n e^{-ax}[\cos(bx)]dx = \frac{n![(a - ib)^{n+1} + (a + ib)^{n+1}]}{2(a^2 + b^2)^{n+1}} \qquad (i^2 = -1, a > 0)$$

$$\int_0^\infty \frac{e^{-ax}\sin x}{x} dx = \cot^{-1} a \qquad (a > 0)$$

$$\int_0^\infty e^{-a^2 x^2} \cos bx \, dx = \frac{\sqrt{\pi}}{2a} e^{\frac{-b^2}{4a^2}} \qquad (ab \neq 0)$$

$$\int_0^\infty e^{-t\cos\phi} t^{b-1}[\sin(t\sin\phi)]dt = [\Gamma(b)]\sin(b\phi) \qquad \left(b > 0, -\frac{\pi}{2} < \phi < \frac{\pi}{2}\right)$$

$$\int_0^\infty e^{-t\cos\phi} t^{b-1}[\cos(t\sin\phi)]dt = [\Gamma(b)]\cos(b\phi) \qquad \left(b > 0, -\frac{\pi}{2} < \phi < \frac{\pi}{2}\right)$$

$$\int_0^\infty \frac{e^{-ax^c} - e^{-bx^c}}{x} dx = \frac{1}{c}\log\frac{b}{a} \qquad (a, b, c > 0)$$

$$\int_0^\infty \frac{1 - e^{-ax^2}}{x^2} dx = \sqrt{(a\pi)} \qquad (a > 0)$$

$$\int_0^\infty \exp\left[-a^2 x^2 - \frac{b^2}{x^2}\right] dx = \frac{\sqrt{\pi}}{2a} e^{-2ab} \qquad (a, b > 0)$$

$$\int_0^\infty \frac{dx}{e^{ax} - 1} = \infty \qquad (a > 0)$$

$$\int_0^\infty \frac{x\,dx}{e^{ax}-1} = \frac{\pi^2}{6a^2} \qquad (a>0)$$

$$\int_0^\infty \frac{e^{-ax}-e^{-bx}}{x}\,dx = \log\frac{b}{a} \qquad (a,b>0)$$

$$\int_0^\infty \frac{dx}{e^{ax}+1} = \frac{\log 2}{a} \qquad (a>0)$$

$$\int_0^\infty \frac{x\,dx}{e^{ax}+1} = \frac{\pi^2}{12a^2} \qquad (a>0)$$

$$\int_0^\infty \frac{e^{-ax}}{x}\sin mx\,dx = \tan^{-1}\frac{m}{a} \qquad (a>0)$$

$$\int_0^\infty \frac{e^{-ax}}{x}\cos mx\,dx = \infty$$

$$\int_0^\infty \frac{e^{-ax}}{x}(1-\cos mx)dx = \frac{1}{2}\ln\left(1+\frac{m^2}{a^2}\right) \qquad (a>0)$$

$$\int_0^\infty \frac{e^{-ax}}{x}(\cos mx-\cos nx)dx = \frac{1}{2}\ln\frac{a^2+n^2}{a^2+m^2} \qquad (a>0)$$

$$\int_0^\infty \frac{e^{-ax}-e^{-bx}}{x}\cos mx\,dx = \frac{1}{2}\ln\frac{b^2+m^2}{a^2+m^2} \qquad (a,b>0)$$

$$\int_0^\infty e^{-ax}\cos^2 mx\,dx = \frac{a^2+2m^2}{a(a^2+4m^2)} \qquad (a>0)$$

$$\int_0^\infty e^{-ax}\sin^2 mx\,dx = \frac{2m^2}{a(a^2+4m^2)} \qquad (a>0)$$

$$\int_0^\infty \frac{e^{-ax}}{x}\sin^2 mx\,dx = \frac{1}{4}\ln\left(1+\frac{4m^2}{a^2}\right) \qquad (a>0)$$

$$\int_0^\infty \frac{e^{-ax}}{x^2} \sin^2 mx \, dx = m \tan^{-1} \frac{2m}{a} - \frac{a}{4} \ln\left(1 + \frac{4m^2}{a^2}\right) \qquad (a > 0)$$

$$\int_0^\infty e^{-ax} \sin mx \sin nx \, dx = \frac{2amn}{\{a^2 + (m-n)^2\}\{a^2 + (m+n)^2\}} \qquad (a > 0)$$

$$\int_0^\infty e^{-ax} \sin mx \cos nx \, dx = \frac{m(a^2 + m^2 - n^2)}{\{a^2 + (m-n)^2\}\{a^2 + (m+n)^2\}} \qquad (a > 0)$$

$$\int_0^\infty e^{-ax} \cos mx \cos nx \, dx = \frac{a(a^2 + m^2 + n^2)}{\{a^2 + (m-n)^2\}\{a^2 + (m+n)^2\}} \qquad (a > 0)$$

$$\int_0^\infty \frac{e^{-ax}}{x} \sin mx \sin nx \, dx = \frac{1}{4} \log \frac{a^2 + (m+n)^2}{a^2 + (m-n)^2} \qquad (a > 0)$$

$$\int_0^\infty e^{-a^2 x^2} \cos mx \, dx = \frac{\sqrt{\pi}}{2a} e^{-m^2/(4a^2)} \qquad (a > 0)$$

$$\int_0^\infty x e^{-a^2 x^2} \sin mx \, dx = \frac{m\sqrt{\pi}}{4a^3} e^{-m^2/(4a^2)} \qquad (a > 0)$$

$$\int_0^\infty \frac{e^{-a^2 x^2}}{x} \sin mx \, dx = \frac{\pi}{2} \operatorname{erf}\left(\frac{m}{2a}\right) \qquad (a > 0)$$

$$\int_0^\infty \frac{e^{-ax}}{\sqrt{x}} \cos mx \, dx = \frac{\{a + \sqrt{(a^2 + m^2)}\}^{1/2} \sqrt{\pi}}{(a^2 + m^2)^{1/2} \sqrt{2}} \qquad (a > 0)$$

$$\int_0^\infty e^{-ax} \sin \sqrt{(mx)} \, dx = \frac{\sqrt{(\pi m)}}{2a\sqrt{a}} e^{-m/(4a)} \qquad (a, m > 0)$$

$$\int_0^\infty \frac{e^{-ax}}{\sqrt{x}} \cos \sqrt{(mx)} \, dx = \frac{\sqrt{\pi}}{\sqrt{a}} e^{-m/(4a)} \qquad (a, m > 0)$$

$$\int_0^\infty e^{-ax} \sin (px + q) \, dx = \frac{a \sin q + p \cos q}{a^2 + p^2} \qquad (a > 0)$$

$$\int_0^\infty e^{-ax} \cos(px + q)dx = \frac{a \cos q - p \sin q}{a^2 + p^2} \qquad (a > 0)$$

$$\int_0^\infty t^{b-1} \cos t \, dt = [\Gamma(b)] \cos\left(\frac{b\pi}{2}\right) \qquad (0 < b < 1)$$

$$\int_0^\infty t^{b-1}(\sin t)dt = [\Gamma(b)] \sin\left(\frac{b\pi}{2}\right) \qquad (0 < b < 1)$$

$$\int_0^1 (\ln x)^n dx = (-1)^n \cdot n!$$

$$\int_0^1 \left(\ln\frac{1}{x}\right)^{\frac{1}{2}} dx = \frac{\sqrt{\pi}}{2}$$

$$\int_0^1 \left(\ln\frac{1}{x}\right)^{-\frac{1}{2}} dx = \sqrt{\pi}$$

$$\int_0^1 \left(\ln\frac{1}{x}\right)^n dx = n!$$

$$\int_0^1 x \ln(1 - x)dx = -\frac{3}{4}$$

$$\int_0^1 x \ln(1 + x)dx = \frac{1}{4}$$

$$\int_0^1 \frac{\ln x}{1 + x}dx = -\frac{\pi^2}{12}$$

$$\int_0^1 \frac{\ln x}{1 - x}dx = -\frac{\pi^2}{6}$$

$$\int_0^1 \frac{\ln x}{1 - x^2}dx = -\frac{\pi^2}{8}$$

$$\int_0^1 \ln\left(\frac{1+x}{1-x}\right) \cdot \frac{dx}{x} = \frac{\pi^2}{4}$$

$$\int_0^1 \frac{\ln x\, dx}{\sqrt{1-x^2}} = -\frac{\pi}{2}\ln 2$$

$$\int_0^1 x^m \left[\ln\left(\frac{1}{x}\right)\right]^n dx = \frac{\Gamma(n+1)}{(m+1)^{n+1}}, \quad \text{if } m+1>0, n+1>0$$

$$\int_0^1 \frac{(x^p - x^q)dx}{\ln x} = \ln\left(\frac{p+1}{q+1}\right) \qquad\qquad (p+1>0, q+1>0)$$

$$\int_0^1 \frac{dx}{\sqrt{\ln\left(\frac{1}{x}\right)}} = \sqrt{\pi}$$

$$\int_0^\infty \ln\left(\frac{e^x + 1}{e^x - 1}\right) dx = \frac{\pi^2}{4}$$

$$\int_0^{\pi/2} \ln \sin x\, dx = \int_0^{\pi/2} \ln \cos x\, dx = -\frac{\pi}{2}\ln 2$$

$$\int_0^{\pi/2} \ln \sec x\, dx = \int_0^{\pi/2} \ln \csc x\, dx = \frac{\pi}{2}\ln 2$$

$$\int_0^\pi x \ln \sin x\, dx = -\frac{\pi^2}{2}\ln 2$$

$$\int_0^{\pi/2} \sin x \ln \sin x\, dx = \ln 2 - 1$$

$$\int_0^{\pi/2} \ln \tan x\, dx = 0$$

$$\int_0^\pi \ln(a \pm b \cos x)dx = \pi \log\left(\frac{a + \sqrt{a^2 - b^2}}{2}\right) \qquad\qquad (a \geqq b)$$

$$\int_0^\infty \frac{dx}{\cosh ax} = \frac{\pi}{2a}$$

$$\int_0^\infty \frac{x\,dx}{\sinh ax} = \frac{\pi^2}{4a^2}$$

$$\int_0^\infty e^{-ax} \cosh bx\,dx = \frac{a}{a^2 - b^2} \qquad (0 \le |b| < a)$$

$$\int_0^\infty e^{-ax} \sinh bx\,dx = \frac{b}{a^2 - b^2} \qquad (0 \le |b| < a)$$

$$\int_{+\infty}^1 \frac{e^{-xu}}{u}\,du = \gamma + \ln x - x + \frac{x^2}{2 \cdot 2!} - \frac{x^3}{3 \cdot 3!} + \frac{x^4}{4 \cdot 4!} - \cdots,$$

$$\text{where } \gamma = \lim_{z \to \infty} \left(1 + \frac{1}{2} + \frac{1}{3} + \cdots + \frac{1}{z} - \ln z\right)$$

$$= 0.5772157\ldots \qquad (0 < x < \infty)$$

$$\int_0^{\pi/2} \frac{dx}{\sqrt{1 - k^2 \sin^2 x}} = \frac{\pi}{2}\left[1 + \left(\frac{1}{2}\right)^2 k^2 + \left(\frac{1 \cdot 3}{2 \cdot 4}\right)^2 k^4\right.$$

$$\left. + \left(\frac{1 \cdot 3 \cdot 5}{2 \cdot 4 \cdot 6}\right)^2 k^6 + \cdots\right], \quad \text{if } k^2 < 1$$

$$\int_0^{\pi/2} \sqrt{1 - k^2 \sin^2 x}\,dx = \frac{\pi}{2}\left[1 - \left(\frac{1}{2}\right)^2 k^2 - \left(\frac{1 \cdot 3}{2 \cdot 4}\right)^2 \frac{k^4}{3}\right.$$

$$\left. - \left(\frac{1 \cdot 3 \cdot 5}{2 \cdot 4 \cdot 6}\right)^2 \frac{k^6}{5} - \cdots\right], \quad \text{if } k^2 < 1$$

$$\int_0^\infty e^{-x} \ln x\,dx = -\gamma = -0.5772157\ldots$$

$$\int_0^\infty \left(\frac{1}{1 - e^{-x}} - \frac{1}{x}\right) e^{-1}\,dx = \gamma = 0.5772157\ldots \qquad \text{(Euler's constant)}$$

$$\int_0^\infty \frac{1}{x}\left(\frac{1}{1 + x} - e^{-x}\right) dx = \gamma = 0.5772157\ldots$$

Appendix D:
Suggestions and
Explanations for
MATLAB® Use

It is suggested that before the reader, who does not have much experience with MATLAB, start using the text, he or she goes over this appendix and tries to execute the presented material in MATLAB.

Creating a directory

It was found by the author that it is less confusing if, for a particular project, we create our own directory where our own developed MATLAB m-files are stored. However, any time we need any one of these files, we must include the directory in the MATLAB path. Let us assume that we have the following directory path: c:\ap\sig-syt\ssmatlab. The following two approaches can be used:

```
        ≫cd 'c: \ap\sig-syst\ssmatlab'
or
        ≫path (path, 'c: \ap\sig-syst\ssmatlab') %remember
%to introduce the path any time you start new
%MATLAB operations; the symbol is necessary
        % for the MATLAB to ignore the explanations;
```

Help

In case we know the name of a MATLAB function and we would like to know how to use it, we write the following command in the command window:

```
≫help sin
```

or

```
≫help exp
```

etc.

For the case we want to look for a keyword, we write

```
≫look for filter
```

Save and load

When we are in the command window and we have created many variables and, for example, we would like to save two of them in a particular directory and in a particular file, we proceed as follows:

```
≫cd 'c: \ap\matlabdata'
≫save data1 x dt %it saves in the matlabdata
%directory the
% file data1 having the two variables x and dt;
```

Let us assume now that we want to bring these two variables in the working space to use them. We first change directory, as we did earlier, and then we write in the command window

```
≫ load data1
```

Then, the two variables will appear in the working space ready to be used.

MATLAB as calculator

```
≫pi^pi−10;
≫cos (pi/4) ;
≫ans*ans %the result will be (√2̄/2) × (√2̄/2) = 1/2
  because the
        %first output is eliminated, only the
        %last output is
        %kept in the form of ans;
```

Variable names

```
≫x= [1 2 3 4 5] ;
≫dt =0.1;
≫cos (pi*dt) ; %since no assignment takes place there
%is no variable;
```

Complex numbers

```
≫z=3+j*4;%note the multiplication sign;
≫zs=z*z;%or z^2 will give you the same results;
≫rz=real(z);iz=imag(z):%will give rz=3, and iz=4;
≫az=angle(z); abz=abs(z);%will give az=0.9273
%rad, and abz=5;
≫x=exp(−z)+4;%x=3.9675+j0.0377;
```

Array indexing

```
≫x=2:1:6;%x is an array of the numbers {2, 3, 4, %5, 6};
≫y=2:−1:−2:%y is an array of the numbers {2, 1, 0, %−1,
                                                   %−2};
≫z=[1 3 y];%z is an array of the numbers {1, 3, %2, 1, 0,
                                               %−1, −%2};
%note the required space between array
%numbers;
≫xt2=2*x;%xt2 is an array of numbers of x each
%one multiplied by 2;
≫xty=x.*y;%xty is an array of numbers which are
%the result of
          %multiplication of corresponding
          %elements, that is
          %{4, 3, 0, −5, −12};
```

Extracting and inserting numbers in arrays

```
≫x=2:1:6;
≫y=[x zeros(1,3)]; %y is an array of the numbers
                   %{2, 3, 4, 5, 6, 0, 0, 0};
≫z=y(1,3:7); %1 stands for row 1 which y is and
             %3:7 instructs to
             %keep columns
             %3 through 7 the result is the array
             %{4, 5, 6, 0, 0};
lx=length(x); %lx is the number equal to the
              %number of columns of
              %the row vector x, that is lx=5;
x(1,2:4)=4.5*(1:3);%this assignment substitutes
      %the elements of x
      %at column
      %positions 2,3 and 4 with the numbers 4.5*
      %[1 2 3]=4.5, 9,
      %and 13.5, note the columns of 2:4 and 1:3
      %are the same;
```

```
x(1,2:2:length(x))=pi;% substitutes the columns 2 and
%4 of x with
%the value of pi, hence the array is {2, 3.1416,
%4,3.1416 6}
```

Vectorization

```
≫n=0:0.2:1;
≫s=sin(0.2*pi*n);% the result of these two
      %commands gives the
      %signal s (sine function) at times (values
      %of n)0, 0.2,
      %0.4, 0.6, 0.4, 1;
```

This approach is preferable since MATLAB executes faster than the vectorization approach rather than the loop approach, which is

```
≫s=[];% initializes all values of vector s to
%zero;
≫for n=0:5% note that the index must be integer;
≫s(n+1)=sin(0.2*pi*n*0.2); % since we want values
            %of s every 0.2
            %seconds we must multiply n by 0.2; note
            %also that
            %for n=0 the variable becomes s(1) and
            %this
            %because the array in MATLAB always
            %starts
            %counting columns from 1;
≫end;
```

The results are identical with the previous one.

Matrices

If **a** and **b** are matrices such that **a** is a 2×3 matrix and **b** is a 3×3 matrix, then $c = a*b$ is a 2×3 matrix.

```
≫a=[1 2; 4 6]; %a is a 2×2 matrix ⎡1 2⎤;
                                    ⎣4 6⎦
≫b=a';%b is a transposed 2×2 matrix of
%a and is ⎡1 2⎤;
         ⎣2 6⎦
≫da=det(a);%da is a number equal to the
  %determinant of a, da=−2;
```

```
≫c = a (:) ;%c is a vector which is made up of the
%columns of a,
        %c = [1 4 2 6];
≫ia = inv (a) ; %ia is a matrix which is the inverse
%of a;
≫sa1 = sum (a,1) ;%sa1 is a row vector made up of
%the sum of the
            %rows,sa1= [5 8];
≫sa2 = sum (a,2) ;%sa2 is a column vector made up
    %by the sum of the
    %columns, sa2= [3 10]';
```

Produce a periodic function

```
≫x = [1 2 3 4];
≫xm = x'*ones (1,5) ;%xm is 4x5 matrix and each of
%its column is x';
≫xp = xm (:)';% xp is a row vector, xp= [x x x x x];
```

Script files

Script files are m-files; when their names are introduced in the command window, we receive the results. We must, however, have the directory that includes the file in the MATLAB search directories. You can modify the file in any desired way and get new results. Suppose that at any time we ask for the file pexp.m, the magnitude and angle of the exponential function $e^{j\omega}$ are plotted. To accomplish this, we first go to the command window and open a new m-file. At the window, we type the file as shown below. As soon as we finish typing, we click on "Save as" and save the file in, say, c:\a\pssmatlab. If we want to see the results, at the command window we just write pexp and hit the enter key.

Script file pexp.m

```
≫w = 0:pi/500:pi-pi/500;%they are 500 at pi/500
%appart;
≫x = exp (j*w) ;ax = abs (x) ;anx = angle (x) ;
≫subplot (2,1,1) ;plot (w,ax,'k') %'k' means plot
%line in black;
≫xlabel ('omega rad/s') ;ylabel
('Magnitude') ;
≫subplot (2,1,2) ;plot (w,anx,'k') ;
≫xlabel ('omega rad/s') ;ylabel ('Angle') ;
```

If we have the function $(2e^{j\omega}/(e^{j\omega} - 0.5))$ and want to plot the results as above, we substitute in the script file the function x with the function: $x = 2*\exp(j*w)/(\exp(j*w) - 0.5)$.

In the above MATLAB expression, note the dot before the slash. This instructs MATLAB to operate at each value of *w* separately and, thus, give results at each frequency point.

Functions

We will present here an example of how to write functions. The reader should also study the functions, which are presented through out the book. In Fourier series, for example, we have to plot functions of the following form:

$$s(t) = \sum_{n=0}^{N} A_n \cos n\omega_0 t$$

and we want to plot this sum of cosines each one having different amplitudes and frequencies. Let $A = [1\ 0.6\ 0.4\ 0.1]$, $\omega_0 = 2$, and $0 \le t \le 4$. We approach this solution by vectorizing the summation. The MATLAB function is of the following form:

```
≫function[s] = sumofcos(A,N,w0,rangeoft)
≫n = 0:N−1;
≫s = A*cos(w0*n'*rangeoft)
%when we want to use this function at the command window
%to find s we write for example:
≫A = [1 0.6 0.4 0.1];N = 4;w0 = 2;rangeoft = 0:0.05:6;
≫ [s] = sumofcos(A,N,w0,rangeoft);
At the enter key, click the vector s that is one of the variables in
the command window and it can be plotted at the wishes of the reader;
we must secure that the directory in which sumofcos function exists
is in the MATLAB path; after you type the function in the editing
window you "save as" in the directory, for example, c:\ap\ssmatlab
and filename: sumofcos.m
```

It is recommended that the reader set small numbers for N ($N = 4$) and range of t (0:0.2:1) and produce first the matrix $\cos(w_0^* n'^* t)$ and, then, see the result $A^*\cos(w_0^* n'^* t)$.

Complex expressions

We can produce results by writing, for example,

```
≫x = [1 3 1 5 3 4 5 8];
≫plot(abs(fft(x,256)),'r');%will plot in red color the
                          %spectrum of the
                          %vector x of 256 points;
```

Axes

```
≫axis([xmin xmax ymin ymax]);%sets the max and min values
%of the axes;
≫grid on;%turns on grid lines in the graph;
```

Two-dimensional (2D) graphics

To plot a sine and a cosine signal
```
≫x = linspace(0,2*pi,40);%produces 40 equal spaced points
                        %between 0 and
                        %2π;
≫y = sin(x);plot(x,y,'r');%will plot the sine signal with
                        %color red;
≫y1 = cos(x);plot(x,y1,'g');%will plot the cosine signal
                        %with color
                        %green;
```

For other color lines: 'y'=yellow,'c'=cyan,'b'=blue,'w'=white, 'k'=black
Type of lines:'g:'=green dotted line,'r-'=red dashed line, 'k-x'=black
        dashed line with x's,'k−.'=black dash-dot line, '+'=plus sign,'ko'=black circles

Add Greek letters: \omega=will produce Greek lower case omega, \Omega = will produce capital case Greek omega. The same is true for the rest of the Greek letters. For example, if we want to write the frequency in a figure under the x-axis, in the command window we write:≫xlabel('\omega rad/s');. For an omega with a subscript 01 we write:≫xlabel('\omega_{01} rad/s');

Add grid lines: ≫grid on;%this is done after the command plot;

Adjusting axes: ≫axis square;%sets the current plot to be %square rather than the default rectangle; ≫axis off;%turn off all axis labeling, %grid,and tick marks;leave the title and %any labels placed by the 'text' and %'*gtext*' commands; ≫axis on;%turn on axis labaling,tick marks %and grid; ≫axis([xmin xmax ymin ymax]);set the %maximum and minimum values of the axes %using values given in the row vector;

Subplots (Example):
```
≫n = 0:100;x = sin(n*pi*n);y = cos(n*pi*n);
    z = x.*y;...
≫w = x+y;
≫subplot(2,2,1);plot(n,x);
    subplot(2,2,2);...
```

```
≫plot(n,y);
≫subplot(2,2,3);plot(n,z);
  subplot(2,2,4); . . .
≫plot(n,w);
```

Log plotting: ≫`semilogx(x);%will plot the vector x in log`
`%scale in x-axis`
`%and linear scale in y-axis;`
≫`semilogy(x);%will plot the vector x in log`
`%scale in y-direction and linear scale in`
`%the x-axis;`
≫`loglog(x);%will plot the vector x in log`
`%scale both axes;`

Histogram: ≫`x=randn(1,1000);hist(x,40);colormap([0 0 0]);`
`%will`
`%plot a Gausian histogram of 40 bars white;if`
`%instead we entered the vector [1 1 1] the`
`%bares would be black;the vector [1 0 0] will`
`%give red and the vector [0.5 0.5 0.5] will`
`%give gray;`
≫`x=−3.0:0.05:3;y=exp(−x.*x);bar(x,y); . . .`
≫`colormap([.5 .5 .5]); %will produce bar-`
`%figure of the bell curve with gray color;`
≫**`sairs`**`(x,y,'k');%will produce a stair-like`
`%black curve;`

Add words: ≫`gtext('the word');`

After the return, the figure will appear and a crosshair. Move the center at the point in the figure where the word must start and click.

Add legend: ≫`plot(x1,y1,'+',x2,y2,'*');%there will be two`
`%curves in the graph;`
≫`legend('Function 1','Function 2');`

The following rectangle will appear in the figure:

+ Function1
* Function2

Three-dimensional (3D) plots

Mesh-type figures

If, for example, we desire to plot the function $f(x) = e^{-(x2+y2)}$ in the ranges $2 \le x \le 2$, $2 \le y \le 2$, we proceed as follows:

```
≫x = −2:0.1:2;y = −2:0.1:2; [X,Y] = meshgrid(x,y);
≫f = exp(−(X.*X+Y.*Y));mesh(X,Y,f);colomap([0 0 0]);
```

The above commands will produce a mesh-type 3D figure with black lines.

D.1 General Purpose Commands

Managing commands and function

help   Online help for MATLAB functions and m-files, e.g., ≫ help plot
path   Shows the path to MATLAB directories which are available at the command window

Managing variables and the workplace

clear   Removes all the variables and items in the memory. Let us assume that the memory contains the variables x, y, z then ≫clear x z; only y will remain in the memory
length A number that gives the length of a vector. ≫x = [1 3 2 5]; then ≫length(x); will give the number 4. If we write ≫y = length(x); then the variable y is equal to 4.
size   Array dimensions. ≫x = [1 3 2 5]; then size(x) will give the numbers 1 4, which means 1 row and 4 columns. Let us write ≫x = [1 2; 3 5; 6 4]; then size(x) will give the numbers 3 2, which means that x is a matrix of 3 × 4 dimensions
who   Produces a list of the variables in the memory
format This command is used as follows for display: ≫format short,pi; will produce the number 1.1416, ≫format long,pi; will produce the number 3.14159265358979, ≫format long,single(pi); will produce the number 3.1415927

D.2 Operators and Special Characters

Operators and special characters

+   Plus
−   Minus
*   Number and matrix multiplications
.*   Array multiplication. ≫x = [1 2 3]; y = [2 3 4]; z = x.*y; hence z = [2 6 12]
.^   Array power. ≫x = [2 3 4]; y = x.^3; hence y = [8 27 64]. ≫x = [2 4; 1 5]; y = x.^2; hence y = [4 16; 1 25]
/   Right division
./   Array division. ≫x = [2 4 6]; y = [4 4 12]; z = x./y; hence z = [0.5 1 0.5]
:   Colon. ≫x = [1 3 6 2 7 8]; y = x(1,3:6); hence y = [6 2 7 8]
.   Decimal point
...   Continuation. ≫x = [1 4 6 7 8 9...≫2 5 8 1]; the vector x is interpreted by MATLAB as a row vector having 10 elements
%   Comments. ≫x = [1 4 2 6]; %this is a vector. MATLAB ignores "this is a vector"

'   Transpose of a matrix or vector. $\gg x = [2\ 6\ 3]; y = x';$ will have $y = \begin{bmatrix} 2 \\ 6 \\ 3 \end{bmatrix}$

&   Logical AND
| Logical OR
~   Logical NOT
xor   Logical exclusive (XOR)

Control flow

for   Repeat statements a specific number of times.
          $\gg$for $n = 0:3;$
          $\gg x(n+1) = \sin(n*pi*0.1);$ %observe the $n+1$, if the $+1$ was not there $x(0)$
          $\gg$end; %was not defined by MATLAB

          Then $x = [0\ 0.3090\ 0.5878\ 0.8090]$

          $\gg$for $n = 0:2$
          $\gg$ for $m = 0:1$
          $\gg x(n+1,m+1) = n\wedge2 + m\wedge2;$
          $\gg$ end;
          $\gg$end;

          Then $x = \begin{bmatrix} 0 & 1 \\ 1 & 2 \\ 4 & 5 \end{bmatrix}$

while   Repeat statements an indefinite times of times.
          $\gg a = 1;$ num $= 0;$
          $\gg$while $(1+a) < = 2$ & $(1+a) > = 1.0001$
          $\gg a = s/2;$
          $\gg$num $=$ num $+ 1;$
          $\gg$end;
          We obtain $a = 0.0001$, and num $= 14$
if   Conditionally execute statements.
          if expression
                    commands evaluated if true
          else
                    commands evaluated if false
          end

If there are more than one alternative, the if-else-end statement takes the form

elseif
          if expression 1
                    commands evaluated if expression 1 is true

        elseif expression 2
                commands evaluated if expression 2 is true
        elseif...
        .

        .

        .

        else
                commands evaluated if no other expression is true
        end

D.3 Elementary Matrices and Matrix Manipulation

Elementary matrices and arrays

eye(n,n)   Identity matrix (its diagonal elements are 1 and all the others are 0)

linspace   Linspace($x1,x2$) generates 100 equally spaced points between $x1$ and $x2$
            Linspace($x1,x1,N$) generates N equally spaced points between $x1$ and $x2$

ones   ones(1,5) generates a row vector with its elements only ones
            ones(2,4) generates a 2×4 matrix with all its elements ones

rand   Uniformly distributed random numbers. $\gg x - \text{rand}(1,5)$; x is a row vector of
            five elements of random numbers. $\gg x = \text{rand}(2,3)$; x is a 2×3 matrix whose
            elements are random numbers

randn   Normally distributed random numbers. Applications are similar to rand
            above

zeros   Creates arrays and matrices of all zeros. $\gg x = \text{zeros}(1,4)$; x is a row vector of
            four elements all with zero value. $\gg x = \text{zeros}(3,4)$; x is a 3×4 matrix with all
            of its elements zero

: (colon)   Regularly spaced vector. $\gg x = [1\ 4\ 2\ 5\ 8\ 3]$; $y = x(1,3:6)$; hence $y = [2\ 5\ 8\ 3]$

eps   Floating-point relative accuracy. To avoid NA response in case there exists a
            zero over zero expression at a point, as in the sine function, we for example
            write $\gg n = -4:4$; $x = \sin(n*\text{pi}*.1)./\ldots((n*\text{pi} + \text{eps})$;

i,j   Imaginary unit

pi   Ratio of a circle's circumference to its diameter

Matrix Manipulation

diag   Diagonal matrices and diagonals of a matrix. $\gg x = [1\ 3\ 5;\ldots 2\ 6\ 9;\ 4\ 7\ 0]$;
            $y = \text{diag}(x)$; will give a column vector $y = [1\ 6\ 0]^{\text{T}}$. $\gg y1 = \ldots \text{diag}(x,1)$; will
            give a column vector $y1 = [3\ 9]^{\text{T}}$, which is the diagonal above the main
            diagonal
            $\gg y2 = \text{diag}(x, -1)$; will give the column vector $y2 = [2\ 7]$, which is the
            diagonal just below the main diagonal
            $y3 = \text{diag}(\text{diag}(x))$; will give a 3×3 matrix with the diagonal 1, 6, 0 and the rest
            of the elements zero

fliplr   Flips vectors and matrices left–right

flipud   Flip matrices and vectors up–down

tril   Lower triangular part of a matrix including the main diagonal and the rest are zero. If $x = [1\ 3\ 5;\ 2\ 6\ 9;\ 4\ 7\ 0]$, then $y = \text{tril}(x)$ is the matrix $[1\ 0\ 0;\ 3\ 6\ 0;\ 4\ 7\ 0]$

triu   Upper triangular part of a matrix

toeplitz Produces a Toeplitz matrix given a vector. $\gg x = [1\ 5\ 2];\ y = \text{Toeplitz}(x)$ produces the matrix $y = [1\ 5\ 2;\ 5\ 1\ 5;\ 2\ 5\ 1]$

D.4 Elementary Mathematics Function

Elementary functions

abs   Absolute value of a number and the magnitude of a complex number

acos,acosh   Inverse cosine and inverse hyperbolic cosine

acot,acoth   Inverse cotangent and inverse hyperbolic cotangent

acsc,acsch   Inverse cosecant and inverse hyperbolic cosecant

angle   Phase angle of a complex number. $\text{angle}(1 + j) = 0.7854$

asec,asech   Inverse secant and inverse hyperbolic secant

asin,asinh   Inverse sine and inverse hyperbolic sine

atan,atanh   Inverse tangent and inverse hyperbolic tangent

ceil   Round toward infinity. For example, $\text{ceil}(4.22) = 5$

conj   Complex conjugate. $\text{conj}(2 + j*3) = 2 - j*3$

cos,cosh   Cosine and hyperbolic cosine

cot,coth   Cotangent and hyperbolic cotagent

csc,csch   Cosecant and hyperbolic cosecant

exp   Exponential. For example, $\exp(-1) = 1/e = 0.3679$

fix   Rounds toward zero. For example, $\text{fix}(-3.22) = -3$

floor   Round toward minus infinity. For example, $\text{floor}(-3.34) = -4$, and floor $(3.65) = 3$

imag   Imaginary part of a complex number. For example, $\text{imag}(2 + j*5) = 5$

log   Natural logarithm. For example, $\log (10) = 2.3026$

log 2   Based 2 logarithm. For example, $\log 2(10) = 3.3219$

log 10   Common (base 10) logarithm. For example, $\log 10(10) = 1$

mod   Modulus (signed remainder after division). For example, $\text{mod}(10,3) = 1$, $\text{mod}(10,4) = 2$. In general $\text{mod}(x,y) = x - n*y$

real   Real part of complex number

rem   Remainder after division. For example, $\text{rem}(10,3) = 1, \text{rem}(10,5) = 0$, $\text{rem}(10,4) = 2$

round   Round to the nearest integer. For example, $\text{round}(3.22) = 3, \text{round}(3.66) = 4$

sec,sech   Secant and hyperbolic secant

sign   Signum function. $\text{sign}(x) = 0$ for $x = 0$, $\text{sign}(x) = 1$ for $x > 0$ and sign $(x) = -1$ for $x < 1$

sin,sinh   Sine and hyperbolic sine

sqrt   Square root, e.g., $\text{sqrt}(4) = 2$

tan,tanh   Tangent and hyperbolic tangent

erf,erfc   Error and coerror function

gamma   Gamma function, e.g., $\text{gamma}(6) = 120$ or $1*2*3*4*(6 - 1) = 120$

D.5 Numerical Linear Algebra

Matrix analysis

det   Matrix determinant $\gg a = [1\ 2;\ 3\ 4]$; $\det(a) = 1 \times 4 - 2 \times 3 = -2$

norm   The norm of a vector, e.g., $\text{norm}(v) = \text{sum}(\text{abs}(v).^2)^{\wedge}(1/2)$

rank   Rank of a matrix. Rank(A) provides the number of independent columns or rows of matrix A

trace   Sum of the diagonal elements, e.g., $\text{trace}([1\ 3;\ 4\ 12]) = 13$

eig   Eigenvalues and eigenvectors. $\gg[v,d] = \text{eig}([1\ 3;\ 5\ 8])$; therefore,

$$v = \begin{bmatrix} 0.8675 & -0.3253 \\ 0.4974 & -0.9456 \end{bmatrix}, \quad d = \begin{bmatrix} -0.7202 & 0 \\ 0 & 9.7202 \end{bmatrix}$$

inv   Matrix inversion, e.g., $\gg A = [1\ 3;\ 5\ 8]$; $B = \text{inv}(A)$; therefore,

$$B = \begin{bmatrix} -1.1429 & 0.4286 \\ 0.7143 & -0.1429 \end{bmatrix}, \quad A*B = \begin{bmatrix} 1.0000 & 0 \\ 0 & 1.0000 \end{bmatrix}$$

D.6 Data Analysis

Basic operations

max   Maximum element of an array. $\gg v = [1\ 3\ 5\ 2\ 1\ 7]$; $x = \max(v)$; therefore, $x = 7$

mean   Average or mean value of an array, e.g., $\text{mean}([1\ 3\ 5\ 2\ 8]) = 19/5 = 3.8$

median   Median value of an array, e.g., $\text{median}([1\ 3\ 5\ 2\ 8]) = 3$

min   Minimum element of an array

sort   Sorts elements in ascending order, e.g., $\text{sort}([1\ 3\ 5\ 2\ 8]) = [1\ 2\ 3\ 5\ 8]$

std   Standard deviation

sum   Sum of an array elements, e.g., $\text{sum}([1\ 3\ 5\ 2\ 8]) = 19$

Filtering-convolution

conv   Convolution and polynomial multiplication, e.g., $\text{conv}([1\ 1\ 1]) = [1\ 2\ 3\ 2\ 1]$, if we have to multiply these two polynomials $(x^2 + 2x + 1)*(x + 2)$ we convolve their coefficients $\text{conv}([1\ 2\ 1], [1\ 2]) = [1\ 4\ 5\ 2]$, therefore, we write the polynomial $x^3 + 4x^2 + 5x + 2$

conv2   2D convolution

filter   Filter data with infinite impulse response (IIR) or finite impulse response (FIR) filter. Let the FIR filter be given by $y(n) = 0.8x(n) + 0.2x(n - 1) - 0.05x(n - 2)$. Let the input data be $x = [0.5\ -0.2\ 0.6\ 0.1]$. Hence, $a = [1]$, $b = [0.5\ 0.2\ -0.05]$ and the output is given by $y = \textbf{filter}(a,b,x)$. The result is $y = [0.6250\ -0.4063\ 0.8906\ -0.1230]$.

Fourier transforms

abs   Absolute value and complex magnitude, e.g., $\text{abs}(4 + j*3) = 5$, $\text{abs}([-0.2\ 3.2]) = [0.2\ 3.2]$

angle   Phase angle, e.g., angle($4 + j*3$) $= 0.6435$ rad

fft   One-dimensional (1D) fast Fourier transform. $\gg x = [1\ 1\ 1\ 0]$; $y = $ fft(x);. Hence, $y = [3\ 0 - 1.0000i\ 1.0000\ 0 + 1.0000i]$. If we had written $z = $ fft$(x,8)$ we would have obtained $z = [3\ 1.7071\ -1.7071i\ 0\ -1.0000i\ 0.2929\ +0.2929i\ 1\ 0.2929 - 0.2929i\ 0\ +1.0000i\ 1.7071\ +1.7071i]$

fft2   1D fast Fourier transform

fftshift   Shift DC component of fast Fourier transform to the center of spectrum. For example, we write in the command window: $\gg x = [1\ 1\ 1\ 1\ 1 \ldots 0]$; $y = $ fft $(x,256)$; then the command plot(abs(fftshift(y))) will center the spectrum. We can also write plot(abs(fftshift(fft$(x,256)$))))

ifft   Inverse 1D fast Fourier transform

ifft2   Inverse 2D fast Fourier transform

D.7 Two- and Three-Dimensional Plotting

Two-dimensional plots

plot   Linear plot. If we have three vectors of equal length such as x with numbers of equal distance, y and z, we can create the following simple plots: plot(y) will plot the values of y at numbers 1, 2,… in the x-direction, plot(x,y) will plot the y values versus the equal-distance values of the vector x in the x-direction, plot(x,y,x,z) will plot both vectors y and z on the same graph, we can plot the two vectors by writing $\gg$plot(x,y); hold on; plot(x,z); if we would like the second graph to have different colors we write plot$(x,z,$ "g") for green color

loglog   Log–log scale plot. For example, loglog(y) will produce the plot

semilogx   Semilog scale plot. The log scale will be on the x-axis and the linear scale on the y-axis. The plot is accomplished by writing semilogx(y)

semilogy   Semilog scale plot. The log scale will be on the y-axis and the linear scale on the x-axis. The plot is accomplished by writing semilogy(y)

axis   Controls axis scaling. For example, if we want the axes to have specific ranges, we write after we created a plot using the MATLAB default axis([minx maxx miny max])

grid   Grid lines. After we created the plot, we write grid on

subplot   Create axes in tiled positions. For example, when we write subplot(3,1,1) we expect 3×1 plots in one page starting plot one. Next, we write subplot(3,1,2) and then proceed to plot the second plot, etc. If we write subplot(3,2,1), we expect $3 \times 2 = 6$ plots on the page. After we write subplot(3,2,1), we proceed to plot the first of the 3×2 matrix format plots. For example, if we write subplot(3,2,2) and proceed to plot the figure, we create a plot at line two and the second plot.

legend   Graph legend. For example, if we have two lines on the plot, one red and one green, and write legend("one," "two"), then a rectangle frame will appear on the graph with a red line and the letters *one* and under a green line with the letters *two*

title   Graph title. For example, if we write title("This is a graph"), then the script in parenthesis will appear on the top of the graph

xlabel   *X*-axis label. For example, if we write xlabel("n time") the *n time* will appear under the *x*-axis

gtext   Place text with mouse. After we have created a plot, if we write in the command window gtext("this is the 1st graph") at the return, a crosshair will appear on the graph and at the click, the phrase in parenthesis will appear on the graph

Index